THE PHYSIOLOGY OF INSECTA

Second Edition

VOLUME VI

CONTRIBUTORS

Moises Agosin

John Buck

Walter Ebeling

R. H. Hackman

Margaret Keister

Michael Locke

P. J. Mill

P. L. Miller

Albert S. Perry

THE PHYSIOLOGY OF INSECTA

Second Edition

Edited by MORRIS ROCKSTEIN

Department of Physiology and Biophysics
University of Miami School of Medicine
Miami, Florida

Volume VI

1974

ACADEMIC PRESS   New York and London

A Subsidiary of Harcourt Brace Jovanovich, Publishers

ACADEMIC PRESS, INC.
111 Fifth Avenue, New York, New York 10003

United Kingdom Edition published by
ACADEMIC PRESS, INC. (LONDON) LTD.
24/28 Oval Road, London NW1

Library of Congress Cataloging in Publication Data

Rockstein, Morris, ed.
The physiology of Insecta.

Includes bibliographies.
1. Insects—Physiology.   I.   Title.
QL495.R58   1973   595.7'01   72-9986
ISBN 0-12-591606-X

PRINTED IN THE UNITED STATES OF AMERICA

CONTENTS

The Insect and the Internal Environment: Homeostasis III

Chapter 1. The Physiology of Insecticide Resistance by Insects

Albert S. Perry and Moises Agosin

Chapter 2. The Structure and Formation of the Integument in Insects

Michael Locke

v

Chapter 7. RESPIRATION: SOME EXOGENOUS AND ENDOGENOUS
EFFECTS ON RATE OF RESPIRATION

Margaret Keister and John Buck

LIST OF CONTRIBUTORS

Numbers in parentheses indicate the pages on which the authors' contributions begin.

MOISES AGOSIN (3), Department of Zoology, University of Georgia, Athens, Georgia

JOHN BUCK (469), Laboratory of Physical Biology, National Institutes of Health, Bethesda, Maryland

WALTER EBELING (271), Department of Entomology (Los Angeles Division), University of California, Riverside, California

R. H. HACKMAN (215), C.S.I.R.O., Division of Entomology, Canberra City, Australia

MARGARET KEISTER* (469), Laboratory of Physical Biology, National Institutes of Health, Bethesda, Maryland

MICHAEL LOCKE (123), Department of Zoology, University of Western Ontario, London, Ontario

P. J. MILL (403), Department of Pure and Applied Zoology, The University of Leeds, Leeds, England

P. L. MILLER (345), Department of Zoology, Oxford University, Oxford, England

ALBERT S. PERRY (3), Vector Biology and Control Branch, Tropical Disease, U.S. Department of Health, Education and Welfare, Public Healt. Service, Center for Disease Control, Savannah, Georgia

* Present address: 3400 Nimitz Road, Kensington, Maryland.

PREFACE

Since the first edition of this multivolume treatise appeared well over eight years ago there has been a notable expansion of scientific endeavor in each of the now numerous aspects of insect physiology. Accordingly, revising this major reference work has been a challenging undertaking both to the original authors as well as to the several new contributors in areas in which the growth of research has led to such an increase in the relevant body of knowledge as to warrant this additional coverage. Consequently, the original three-volume work has now grown "like Topsy" to a thoroughly revised six-volume work.

Thus, just as previous volumes of this edition have included entirely new and distinct chapters on Environmental Aspects—Radiation, Circadian Rhythms and Photoperiodism in Insects, Insect Pheromones, The Pharmacology of the Insect Nervous System, and Protein Synthesis in Insects, this volume includes one new chapter on Microsomal Mixed-Function Oxidases by Drs. M. Agosin and A. L. Perry. The Physiology of Insect Behavior, originally a single chapter by Professors Markl and Lindauer, has now been expanded and appears as two chapters: one on the actual mechanisms of insect behavior by Professor Markl and the second on social behavior and mutual communication by Professor Lindauer.

Just as the original chapter on The Circulatory System of Insects by Dr. Jack Jones has been subdivided into three chapters in Volume V, a single chapter in the first edition on Respiration: Aerial Gas Transport by Dr. P. L. Miller has now been expanded and appears as two chapters in this volume: Respiration: Aerial Gas Transport by Dr. Miller and Respiration in Aquatic Insects by Dr. P. J. Mill.

Once again, I am deeply indebted to my former teacher and mentor, Professor A. Glenn Richards, for his frank criticism and suggestions concerning the first edition which have assisted me immeasurably in the improvement of both the content scope of this edition. Again I acknowledge the technical and editorial assistance of Mrs. Estella Cooney, Mrs. Ricki Davidson, and my daughter Susan Sorkin, and the cooperation of and the concern for quality of content and accuracy by each of the contributors to this second edition of "The Physiology of Insecta."

MORRIS ROCKSTEIN

PREFACE TO FIRST EDITION

This multivolume treatise brings together the known facts, the controversial material, and the many still unsolved and unsettled problems of insect physiology in chapters written by the outstanding workers in each of a wide range of areas of insect function.

It is designed to meet a manifest need which has arisen from the phenomenal increase in research activity on insects (during the past two decades, especially) for an authoritative, comprehensive reference work in insect physiology.

Although the insect physiologist usually considers himself either a comparative physiologist or a general physiologist studying a particular process in insects, the fact is that each is a biologist whose primary interest is in the *total organism* in relation to a specific function. This viewpoint is reflected in the organization and arrangement of the chapters by sections and volumes, Thus, instead of that classic arrangement of chapters which emphasizes organ or systemic physiology, this treatise has been organized into three main sections, each covering a major division of insect biology: the ontogeny of insects from reproduction to senescence of the individual; the insect's perception of and reaction to its external environment; and the mechanisms by which the internal homeostatic state is maintained. The last-mentioned division, especially, includes many classic functions—from the role of the nervous system to nutrition, metabolism, respiration, circulation, maintenance of salt and water balance, and cuticular functions. In addition, under this major division the heretofore unemphasized areas of immunological responses and mechanisms of insect resistance to insecticides have been included, since the contributions of research investigators to these fields in recent times are widely recognized.

I hope that this diversified subject matter will serve an equally varied group of students of biology. To the student of comparative physiology as well as to the entomologist, the organization of the new extensive literature on insect physiology into one large work should be especially useful. To the applied entomologist, the chapters concerned with insect functions in relation to the external environment should prove especially interesting; they provide a basis for understanding the distribution, epidemiology, and bionomics of insects in general, but especially of those insects of medical and economic importance. Those chapters concerned with the maintenance of the constancy of the internal environment should be equally helpful, forming a rational basis for control of insect pests. Finally, the details of structure, both gross and histological, necessarily included in those chapters covering neurophysiology, circulation, respiration, digestion, and cuticular functions, should be of special interest to the anatomist or taxonomist concerned with the physiological implications of his own research interests in insects.

The responsibilities of editing an opus of this size include securing the complete cooperation and sustained efforts of one's co-authors. To this I can attest without qualification. I must also acknowledge the critical, but always helpful suggestions—especially in the early planning and in the reading of some of the manuscripts—of my many colleagues, namely, Dr. A. Glenn Richards, Dr. V. B. Wigglesworth, Dr. Carroll M. Williams, Dr. Leigh E. Chadwick, Dr. Vincent G. Dethier, Dr. Herbert H. Ross, Dr. Curtis W. Sabrosky, and the late Dr. R. N. Snodgrass.

To Miss Norma Moskovitz, special expression of appreciation is due for her untiring efforts and sustained dedication to achieving a final product of exacting technical standards.

On a more personal level, the early encouragement of the late Elaine S. Rockstein and the patience and forbearance of my oftimes neglected daughters Susan and Madelaine, especially during the past year, must be gratefully acknowledged as well.

MORRIS ROCKSTEIN

CONTENTS OF OTHER VOLUMES

Volume IV

The Insect and the Internal Environment—Homeostasis—I

Volume V

The Insect and the Internal Environment—Homeostasis—II

THE PHYSIOLOGY OF INSECTA

Second Edition

VOLUME VI

The Insect and the Internal
Environment—Homeostasis—III

Chapter 1

THE PHYSIOLOGY OF INSECTICIDE RESISTANCE BY INSECTS

Albert S. Perry and Moises Agosin

I. Introductory and Historical

Interest in the problem of insect resistance to insecticides gained considerable momentum between the years 1950 and the early 1960's, but has subsided materially in later years. That is because, other than screening for new insecticides, no economical solution has been found to date to remedy the resistance situation and no new, challenging, and thought-provoking ideas have been forthcoming to give impetus to renewed interest in this scientific endeavor. Recently, however, such interest has experienced spontaneous rejuvenation due to knowledge gained from mammalian pharmacology, specifically, from experimental evidence on the role of microsomal enzymes in drug metabolism, the induction of this important enzyme system by drugs and insecticides, and its inhibition by so-called synergists.

Perhaps the most significant commentary on the resistance problem was made in 1915 by Stephen A. Forbes when he said: "The struggle between man and insects began long before the dawn of civilization, has continued without cessation to the present time, and will continue, no doubt, as long as the human race endures . . . We commonly think of ourselves as the lords and conquerors of nature, but insects had thoroughly mastered the world and taken full possession of it long before man began the attempt . . . We cannot even protect our very persons from their annoying and pestiferous attacks, and since the world began, we have never yet exterminated—we probably shall never exterminate—so much as a single species" (quoted by Decker, 1958).

It is true that through the centuries man has fought indefatigably to restrain, control, or eradicate his insect enemies, but there is little doubt that, through intensification of this gallant fight with formidable new chemical weapons, man has been instrumental in the past three decades in speeding up the course of certain evolutionary processes which otherwise might have gone unnoticed. For many insect species, the environmental changes of recent years have been most sudden and overwhelming, as one insecticide after another, with complete disregard for selectivity, has been poured in their paths—so the insects have responded valiantly with a weapon of their own—a phenomenon well known to all of us as "resistance."

Actually, insect resistance to insecticides is not entirely of recent origin. Melander (1914) is generally credited with the first publication on insect resistance and the now famous quotation, "Can insects become resistant to sprays?" although observations on resistance had been made as early as 1887 (Babers and Pratt, 1951). The earlier publications on insect resistance to kerosene, lime-sulfur, hydrocyanic acid, tartar emetic, lead arsenate, cryolite, barium fluorosilicate, phenothiazine, and other compounds, as well as the early work on resistance to chlorohydrocarbon insecticides have been reviewed and well documented by Babers (1949) and Babers and Pratt (1951). Comprehensive monographs on the resistance problem have been published by Brown (1958) and Brown and Pal (1971). Several other reviews covering specific items and discussing the resistance phenomenon from various points of view have been published by Agosin (1963), Brooks (1966), Brown (1960, 1964), Casida (1963), Chadwick (1957), Garms (1961), Hewlett (1960), Hoskins (1964), Hoskins and Gordon (1956), Kearns (1955), March (1959), Metcalf (1955b, 1967), Micks (1960), O'Brien (1966), Oppenoorth (1965), Pal and Kalra (1965), Perry (1958, 1960c, 1964, 1966), Plapp (1970a), Quarterman and Schoof (1958), Sakai (1960), Winteringham and Barnes (1955), and others. In addition, chapters on resistance may be found in books by Metcalf (1955b) and O'Brien (1960, 1967).

II. The Origin and Development of Resistance

It is widely accepted that the development of insecticide resistance by insects is due to the selection of variants in the population carrying preadaptive genes (Crow, 1957). In fact, insecticide resistance provides some of the best examples of Darwinian evolution, but at an accelerated pace, since the intensive selection which precedes the development of resistance by insects results, in numerous instances, in the accentuation of many characters which otherwise might have gone unnoticed.

The rate at which resistance develops in a population may depend on (a) "the frequency of resistant genes present in the normal population, (b) the nature of these genes (either single or multiple, dominant or recessive), (c) the intensity of selection pressure, and (d) the rate at which the species breeds" (Kerr, 1963). In some instances of geographical isolation, confined populations may not contain the prerequisite resistant genes, and therefore do not develop resistance (Elliott, 1959), and highly inbred laboratory strains may fail to develop resistance if the gene pool contains no resistance factors (Crow, 1966; D'Allessandro et al., 1949; Harrison, 1952a; Merrell and Underhill, 1956).

It is quite apparent from toxicological observations that there exists a dynamic variation in tolerance to a poison among individuals of a given population. However, truly sublethal doses of insecticides do not induce resistance in susceptible populations of the house fly (Beard, 1952, 1965b; Brown, 1964), or increase the level of DDT dehydrochlorinase (Moorefield, 1958). Rather, the cumulative effect of daily sublethal amounts of DDT, γ-BHC, dieldrin, or diazinon renders the house fly more susceptible (Hadaway, 1956).

It is also well established that insecticides are not mutagenic. Genetic variability for resistance already exists in natural populations not previously exposed to insecticides. This is also evident from the fact that resistance develops slowly in highly inbred lines. If new mutations were to arise, inbred lines under selection pressure would develop resistance as rapidly as heterogeneous populations (Crow, 1966). Genetic variability in DDT dehydro-chlorinase content is also evident from analysis of individuals in a random population of house flies selected with DDT, showing a clear genetic relationship between DDT dehydrochlorinase and resistance (Lovell and Kearns, 1959).

The acceleration of resistance development in a population may have a nongenetic explanation. It may be the property of the chemistry or the physiology of the insect that when a certain level of resistance is reached a large change in resistance can follow from a small genetic change (Crow, 1966).

In this context, we propose that resistance is biphasic. Phase I is due to selection of variants in the population according to genetic principles, attaining a resistance level commensurate with the gene pool for resistance initially present in the population. Assuming that resistance is due, in most instances, to enzymatic detoxication of the chemical, genetic selection will then result in an increase in enzyme activity. But how does enzyme activity increase beyond the preadaptive level? In phase II, acceleration of resistance takes place by induction of preexisting detoxifying enzymes toward enhanced activity, resulting in faster breakdown of the chemical. The inducer is the insecticide itself which is being used for the control of that particular insect. Evidence for phase II comes from extensive studies on the induction of microsomal mixed-function oxidases in mammals and in some insects (see Chapter 10, Volume V). This part of the total resistance depends on the presence of the insecticide in the insect's environment. Removal of the insecticide from that ecosystem results in quick reversion of resistance to the preexisting genetic level. Prolonged breeding in complete absence of the chemical will cause further reversion toward susceptibility—the rate of return being dependent on the resistance genotypes, including the incorporation of modifier genes during the selection process (Crow, 1966; Keiding, 1967; Plapp, 1970a).

The induction phenomenon corresponds to the expression of a regulatory mechanism affecting nucleic acid and protein synthesis. For instance, DDT has been shown to increase not only the synthesis of messenger RNA but also to produce new RNA species with higher template activity (Balazs and Agosin, 1968; Ishaaya and Chefurka, 1968; Litvak and Agosin, 1968). tRNA may also be involved since translation processes dictate the characteristics of protein synthesis.

III. Neurophysiological Aspects of Insecticide Resistance

The mode of action of the chlorohydrocarbon (CH) insecticides in insects is not fully understood. However, toxicological manifestations and physiological effects, either directly or indirectly, implicate the insect's nervous system as the site of primary attack.

The unstabilizing effect of DDT on peripheral nerves of insects is well established (Roeder and Weiant, 1948, 1951; Hodgson and Smyth, 1955). Even if this effect were not the primary cause of DDT poisoning, differences in response among insect strains might shed some light on the resistance mechanism. Thus, Pratt and Babers (1953a) demonstrated that the thoracic ganglia of DDT-resistant (R) house flies are less affected by, and recover more rapidly from the direct action of DDT (fewer flies showing tremors) than ganglia of the susceptible (S) strain (Table I). This difference cannot be due to DDT breakdown during the initial 5 minutes after treatment and must, therefore, involve an intrinsic factor characteristic of the ganglion itself. A similar conclusion was reached by Wiesmann (1955a) and Weiant (1955) who demonstrated stronger tremors in isolated legs or sensory cells of S flies than in those of DDT-R flies after exposure to DDT, benzene, or toluene vapors. Unlike DDT, however, neither aldrin nor dieldrin produced symptoms in isolated legs of the American cockroach *Periplaneta americana* (Giannotti et al., 1956).

DDT lowers the tarsal sucrose threshold of the R strain. The same result occurs either by topical application, injection, or tarsal contact. Hence, cuticular penetration of DDT is not the cause of this difference (Smyth and Roys, 1955). Resistance, therefore, must be due to an intrinsic mechanism by which the receptors are protected from the unstabilizing effect of DDT. Similarly, Barton-Browne and Kerr (1967) concluded that resistance to DDT in their strains of flies could be fully expressed at the level of chemoreceptor hairs on the labella and was due to the R fly's ability to recover from the poisoning effect.

Nerve sensitivity to lindane and dieldrin is lower in R flies than in S flies, as determined by the longer latent period required for initiation of

TABLE I

DURATION OF LEG TREMORS INDUCED BY DDT APPLIED TO THORACIC GANGLIA[a]

Fly strain	DDT applied (μg)	Flies tremoring (%)		
		5 minutes	60 minutes	120 minutes
Susceptible	0.57	88	76	—
Resistant	0.57	64	16	—
Susceptible	1.14	100	—	93
Resistant	1.14	60	—	27

[a] From Pratt and Babers (1953a).

bursts of impulses (Yamasaki and Narahashi, 1958). This difference is not as marked with DDT, and this relationship does not hold true for diazinon resistance (Narahashi, 1964).

An explanation of this latent period would be plausible if the site of the resistance mechanism were at the site of action of the toxicant. An interesting experiment by Earle (1963) indicates that such might be the case. Earle found that a 1-minute exposure to aldrin vapors was sufficient to kill 90% of an S strain of house flies, whereas exposure of 120 minutes resulted in little or no mortality of the R strain. It has been fairly well established that house flies do not metabolize dieldrin; hence, a protective effect such as storage away from the site of action is unlikely to act with such rapidity. It is probable then, as Earle (1963) concluded, that the resistance mechanism resides at the site of action, but that other possibilities might also exist. The possibility of a dieldrin-impermeable neural lamella in R cockroaches *Blattella germanica* was considered by Ray (1963). His results showed no difference in permeability of nerve tissue to dieldrin between S and R cockroaches.

The use of crossing experiments, with visible mutants as genetic markers, enabled Tsukamoto et al. (1965) to ascertain the lower nerve sensitivity to DDT of R strains of house flies as an incompletely recessive genetic characteristic carried on the second chromosome. This resistance mechanism is not the same as the one caused by detoxication since the latter is carried on the fifth chromosome (Tsukamoto and Suzuki, 1964; Oppenoorth, 1954). The genetic factor for low nerve sensitivity to γ-BHC is associated with neither the second nor the fifth linkage group (Tsukamoto et al., 1965).

Electron microscopic and autoradiographic studies on the distribution of dieldrin in the intact nerve tissue of dieldrin-R and dieldrin-S *B. germanica* revealed that the interstrain difference in susceptibility is due to the accumulation of less dieldrin in the R strain (Telford and Matsumura, 1971). Otherwise, the nervous system of the R strain shows no visually recognizable morphological characteristics distinct from those of the S strain. On the other hand, Eaton and Sternberg (1967) could find no relationship between the DDT content of the central nervous system of susceptible *P. americana* and the pattern of nerve activity observed; also, no [^{3}H]dieldrin deposition (determined by electron microscopy) was observed in the nerve cell body of dieldrin-R house flies, and no radioactivity was detected in mitochondria or tracheae enmeshed in ganglionic tissue (Sellers and Guthrie, 1971).

Another aspect of a physicochemical nature which might throw some light on nerve sensitivity to toxicants is the binding of insecticides to subcellular components of nerve tissue. Studies by Matsumura and O'Brien (1966) and Matsumura and Hayashi (1966a,b) suggest that DDT and dieldrin form complexes with components of insect nerve. This "charge-transfer" complex with a component of the axon (O'Brien and Matsumura, 1964) resulting in greatly enhanced efflux of potassium ions from the nerve cord might be related to the mode of action of these compounds. An extension of this work by Hatanaka et al. (1967) throws some doubt on the validity of this assumption, especially in the use of Sephadex for demonstrating the formation of stable complexes. Holan (1969) presents evidence that neither DDT nor the active cyclopropane substituents of DDT give UV spectra showing complex formation with pure lipid nerve components. Rather, the author suggests complex formation with protein and not with lipid components. On the other hand, Wilson et al. (1971) confirmed O'Brien and Matsumura's (1964) charge-transfer complex by analogy with the UV spectral phenomena of an alcohol–water suspension of DDT and the interaction of the p-chlorophenyl groups of DDT with tetracyanoethylene.

Many of the above examples would seem to indicate the presence, in resistant strains of insects, of a lipid barrier which prevents the rapid penetration of the toxicant into the delicate nerve structures.

Another barrier which may hinder the action of a toxicant is the ionic nerve barrier. Insect nerves are surrounded by a homogenous noncellular elastic sheath, the neural lamella, which is structurally different from the protective coating of vertebrate nerves (Richards, 1943, 1944). This sheath has been shown by Hoyle (1953) to be impermeable to ionized substances, especially potassium ions, and Winton et al. (1958) have provided histochemical evidence for this lipid sheath barrier in P. americana. It remained for Twarog and Roeder (1956, 1957) to demonstrate the active role of the Hoyle sheath surrounding nerve fibers and ganglia in protecting the nerve against penetration of various ions and drugs. By desheathing the nerve cord of P. americana, acetylcholine, which is inactive when injected into the insect, produces a rapid asynchronous burst of action potentials followed by synaptic depression and block. It is postulated that this sheath, too, may be responsible for altered nerve responses to foreign chemicals as a mechanism of resistance to insecticides. However, this assumption is not without criticism. Treherne (1962) has indicated that stripping off of the nerve sheath involves much more than the mere removal of a superficial membrane. Removal of this sheath will result in dramatic changes in the chemical composition of the fluid surrounding nerve fibers and other nerve cells due to the disruption of the equilibrium between this fluid and the hemolymph. In particular, the decrease in sodium and calcium ions will affect nerve transmission.

Indeed, a most attractive hypothesis advanced in explanation of the unstabilizing action of DDT is that entry of the toxicant into the lipoprotein structure of the nerve membrane destroys the capacity of the membrane to maintain the normal ionic balance, especially that of calcium, so essential for normal function between nerve and environment. This hypothesis, first stated and supported experimentally by Welsh and Gordon (1947) and Gordon and Welsh (1948), has been elaborated by Mullins (1954, 1955) in a theoretical model membrane composed of a lattice of cylindrical lipoprotein molecules. These molecules are oriented in such a manner as to provide free interspaces for the passage of ions and small molecules from one side of the membrane to the other. Foreign molecules may penetrate freely through the interspaces and interfere with membrane permeability, thus causing narcosis but no membrane damage, or they may cause distortion of the surrounding lipoprotein molecules, thus increasing the size of adjacent interspaces and leading to sodium ion leaks and excitation. A concise critique of this theory has been given by Kearns (1956). Recently, Mullins' theory has been put into practice in the rational synthesis of hydrocarbon insecticides specifically designed against DDT-resistant strains of insects (Holan, 1969). The prediction of activity was made on the premise that the substitution of the more chemically stable dichlorocyclopropane for the easily degradable trichloroethane group of DDT would prevent the enzymatic detoxication of the insecticide. This proved to be correct. Based on biological and structural findings, Holan proposes a modification of Mullins' theory as follows: some of the insecticide distributes itself at the lipid-protein nerve membrane interface. There, because of its three-dimensional conformation, the base containing the phenyl ring forms a molecular complex with the overlaying protein layer. The cyclopropane ring fits into the channel of a pore in the lipid portion of the membrane, but keeps the pore open to sodium ions with the consequent delay in the falling phase of the sodium ion potential. In this hypothesis, the complexing of DDT analogues with proteins is controlled chiefly by substituents in the rings and is largely independent of insecticidal activity. This explains the tenacity of DDT and its inactive metabolites in the environment.

NEUROHORMONAL EFFECTS

The hypothesis of a lipid barrier which prevents the entry of a toxicant into the nerve structures of resistant insects implies, of course, that entry of the poison into nervous tissue is a prerequisite for insecticidal action. But are there alternative explanations of the action of insecticides on nerve function, perhaps hormonal effects?

The blood of DDT-prostrate cockroaches *P. americana* contains a pharmacologically active substance which is toxic to both S and R strains of house flies (Sternburg and Kearns, 1952b). When injected into house flies it elicits symptoms resembling those of DDT. Prostration ordinarily is not permanent in either strain, indicating that the toxin is not DDT or a DDT derivative, as is also shown by chemical analysis (Sternburg et al., 1959; Shankland and Kearns, 1959; Sternburg, 1960). The same or a similar neuroactive substance can be produced by electrical stimulation and by tetraethylpyrophosphate (TEPP). Similarly, blood from TEPP- or DDT-prostrate P. americana which contains high titers of corpora allata and corpora cardiaca hormones stimulates the nerve cord activity of normal cockroaches (Colhoun, 1958, 1959b). The neurohormone is derived from the corpora cardiaca but not from the corpora allata (Colhoun, 1959a; Milburn et al., 1960).

DDT appears to induce a hormonal-type effect on the chordotonal receptor cells of P. americana, but this effect is not produced by lindane (Becht, 1958). Also, mechanical or electrical stimulation produces paralysis in cockroach nymphs (Beament, 1958). The paralyzing agent is carried in the blood. It can be shown that this neuroactive agent is depleted in the corpora cardiaca after subjecting the insect to stress (Hodgson and Geldiay, 1959). These neurohormones, whether responding to stress (Ozbas and Hodgson, 1958) or inducing spontaneous excitation or depression (Milburn et al., 1960) are not necessarily the same in all situations, a fact which makes the stress theory much more complex.

Extensive studies have not led to the elucidation of structure of the DDT-induced stress factor. It is dialyzable, unstable, but has been collected in sizable quantities from DDT-poisoned crayfish (Hawkins and Sternburg, 1964). Spot color tests suggest that the factor is an aromatic amine containing an ester group. Similarly, the nerve cord and the head of P. americana contain the highest concentrations of factor S, a neuroactive substance which excites motor neurone activity in the cockroach at low concentrations (Cook, 1967; Cook et al., 1969). The chemical and biological properties of this substance are indicative of a biogenic amine which appears to be distinct from other known neuropharmacologically active agents.

Stress induced by DDT or immobilization on P. americana differs in its effects on oxygen consumption, heartbeat frequency, and loss of weight, but both conditions stimulate the production of a characteristic fluorescent compound which is not produced by organophosphorus insecticides (Patel and Cutkomp, 1967, 1968).

Neurohormonal effects may have no bearing on insecticide resistance, although Heslop and Ray (1959) have demonstrated that the ability of P. americana to withstand stress by prolonged mechanical agitation can be directly correlated with their degree of DDT-resistance. The evidence and role of pharmacologically active substances released during various forms of stress has been reviewed by Sternburg (1963).

The "neurotoxin" theory could, in principle, be compatible with nerve membrane interference but it does not offer an explanation for the primary lesion of DDT action, which, presumably, is the unstabilizing effect on sensory neurones and nerve cells.

IV. Physiological Protective Mechanisms

A. Detoxication Mechanisms in Insects

Insects, like most other organisms, must chemically alter and dispose of a large variety of compounds to maintain their normal body functions. In addition, they must neutralize, inactivate, or eliminate many foreign organic substances, including poisons, in avoiding toxic damage. According to Williams (1959), all foreign organic compounds are susceptible to metabolic attack *in vivo*. These "biotransformations" may involve oxidation, reduction, hydrolytic processes, and synthetic or conjugation reactions.

Comprehensive reviews on conjugation mechanisms in insects have been published by Smith (1955, 1962). These reactions include the following.

1. Conjugation of benzoic acid with glycine to form hippuric acid as shown in mosquito larvae, silkworms, and locusts.

2. Formation of ethereal sulfates such as the conversion of aminophenols and naphthols by certain beetles and by house flies.

3. β-Glucoside formation, e.g., conjugation of phenols to form phenyl-β-glucosides in a similar manner that β-glucuronides are formed in vertebrate detoxication processes (Smith, 1964). Insects do not synthesize glucuronic acid, although Terriere *et al.* (1961) have demonstrated glucuronide-like metabolites resulting from naphthalene metabolism in house flies and blow flies, and Smith (1955) found an active β-glucuronidase enzyme in the crop fluid of locusts. Glucuronic acid conjugation has also been reported to occur in *Prodenia* larvae (Hassan *et al.,* 1965; Zayed *et al.,* 1965) and in silkworms (Inagami, 1955).

In most other insects so far examined, including representatives of the orders Thysanura, Orthoptera, Coleoptera, Lepidoptera, and Diptera, phenols, as well as other compounds, combine with glucose to form the corresponding glucoside. The source of glucose for this type of conjugation is uridine diphosphate glucose (Smith and Turbert, 1961). Similar conjugation reactions have now been shown to occur with the newer type insecticides, especially the carbamates.

4. Phosphate conjugation: Smith and Turbert (1964) detected an acidic conjugate in the excreta of flies treated with 1-naphthol that had properties similar to 1-naphthyl dihydrogen phosphate. This rather unusual type of conjugation has been reported to occur freely in the house fly, blow fly, and the New Zealand grass grub *Costelytra zealandica* treated with various

phenols (Binning *et al.,* 1967), the reactions yielding substantial amounts of monoaryl phosphates. These authors question whether previous reports of glucuronide conjugation in insects might have been based on the non-recognition of phosphates.

5. Cysteine and sulfhydryl conjugation: Detoxication mechanisms involving cysteine conjugation have been detected in locusts by Smith (1962). Chlorobenzene is metabolized by locusts and excreted as *o-, m-,* and *p*-chlorophenylcysteine (Gessner and Smith, 1960).

6. Acetylation: Insects, like mammals, acetylate certain amino compounds to their corresponding acetamido derivatives. For example, 4-6-dinitro-*o*-cresol is first reduced by locusts to 6-amino-4-nitro-*o*-cresol followed by acetylation to 6-acetamido-4-nitro-*o*-cresol. Primary amines such as chlorophenyl cysteine conjugates are excreted as the corresponding acetyl derivatives by locusts, and a *Drosophila* mutant excretes hydroxytryptamine as *N*-acetylhydroxytryptamine glucoside.

7. Methylation: Certain heterocyclic nitrogen compounds are methylated by coleopterous insects.

8. Detoxication involving sulfur compounds: A number of heavy metal poisons such as those containing Pb, As, Ni, Co, etc., are detoxified by clothes moths larvae *Tineola biselliella* through conversion to insoluble sulfides. The sulfur used in these conjugations is derived from the keratin in the wool eaten by the larvae. In other insects and in the blue tick *Boophilus decoloratus,* high levels of SH compounds afford specific protection to the organism against arsenicals.

B. Metabolic Fate of the Chlorohydrocarbon Insecticides

1. Conversion of the Insecticide to Nontoxic Metabolites

a. Metabolism of DDT in Vivo. In the preceding sections we have considered some physicochemical characteristics of the resistance problem through elucidation of the mode of action of the insecticide. By far, the most prevalent type of resistance is that of removing the insecticide from the site of action or preventing its access to a sensitive site by detoxication, although with a few insecticides such as dieldrin and other cyclodienes, detoxication is not a major factor in resistance.

The first study of DDT metabolism in an insect was that of Ferguson and Kearns (1949) who injected DDT into the large milkweed bug *Oncopeltus fasciatus* and found that 80–100% of the dose was metabolized within 90 minutes. The metabolite did not respond to the colorimetric method of Schechter *et al.* (1945) and was, therefore, neither 2,2-bis(*p*-chlorophenyl)-1,1-dichloroethylene (DDE) nor bis(*p*-chlorophenyl) acetic

acid (DDA). The latter compounds were also metabolized to unknown derivatives. On the other hand, immersion of the insects in boiling water prior
to administration of DDT resulted in complete recovery of the unchanged
DDT, indicating the involvement of an enzyme mechanism.

Perhaps the most challenging and fruitful investigations in elucidating
the mechanism of DDT resistance have been those dealing with DDT metabolism in the house fly, *M. domestica*. The house fly dealt the first blow
to our hopes that certain insect control problems should have never gotten
out of hand if proper suppressive measures were taken. But the development
of resistance also gave impetus to a wide variety of fundamental research
in insect morphology, physiology, biochemistry, genetics, and behavior. It
also stimulated the chemists to synthesize new compounds and correlate their
chemical structure with biological activity.

The early assumptions that control failure with DDT was due to such
factors as improper application of the chemical, adulterated insecticide,
favorable environmental conditions for fly breeding, etc., were soon dis-
carded with the simultaneous discovery by Sternburg *et al.* (1950) and Perry
and Hoskins (1950) that resistance resulted from the ability of the house
fly to metabolize DDT to the nontoxic derivative DDE. This finding was
soon corroborated by many other investigators who variously showed that
only resistant flies metabolized DDT to DDE (Sternburg and Kearns, 1950;
Winteringham *et al.*, 1951; Fletcher, 1952) or that both susceptible and
resistant strain could accomplish this detoxication but at different rates
(Perry and Hoskins, 1951; Lindquist *et al.*, 1951a; Winteringham, 1952a;
March, 1952a; Tahori and Hoskins, 1953). The same type of conversion
occurs also with the bromine analogue of DDT (Winteringham *et al.*, 1951).

In one instance (Terriere and Schonbrod, 1955) large quantities of water-
soluble conjugates were found in the excreta of DDT-treated R and S flies,
and continuing formation of unidentifiable DDT-derivatives were reported
by Tahori and Hoskins (1953).

Several insect species exhibit natural tolerance to DDT. The Mexican
bean beetle, *Epilachna varivestis,* the red-banded leaf roller, *Argyrotaenia
velutinana,* the red-legged grasshopper, *Melanoplus femur-rubrum,* and the
differential grasshopper, *M. differentialis,* are able to convert DDT to DDE
and other metabolites before selection pressure has been applied (Sternburg
and Kearns, 1952a). The high degree of tolerance in the grasshopper species
appears to be due, in great part, to the slow absorption of DDT through
the cuticle and the intestinal wall.

Similarly, slower penetration of DDT plays a major role in the resistance
of the boll weevil, *Anthonomus grandis* (Blum *et al.,* 1959a), and the
tobacco budworm, *Heliothis virescens* (Pate and Vinson, 1968), but more

extensive dehydrochlorination of DDT to DDE is the predominant mechanism in the pink bollworm, *Pectinophora gossypiella* (Bull and Adkisson, 1963). In other strains of *H. virescens,* rate of absorption and metabolism of DDT, as well as excretion of DDT, DDE, and DDA, determine the extent of resistance (Vinson and Brazzel, 1966).

Both susceptible and resistant strains of the spotted root maggot *Euxesta notata* metabolize DDT to DDE at approximately equal rates, but absorption of DDT is much slower in the resistant strain (Hooper, 1965). Here, too, absorption of DDT plays a major role in resistance.

Most of the available information on insecticide metabolism has been obtained by investigating interstrain differences within species, but intriguing biochemical differences occur also among closely related species. For example, DDT-resistant cabbage worm *Pieris rapae* larvae metabolize DDT to DDE quite freely, whereas a closely related species, *Barathra brassicae,* is unable to accomplish this conversion (Kojima *et al.,* 1958). In this instance, penetration is practically the same in both species.

In other insects, the resistance mechanism is less understood. Thus, fifth-instar larvae of the European corn borer, *Pyrausta nubilalis,* convert significant amounts of DDT to DDE but show only slight tolerance to the insecticide (Lindquist and Dahm, 1956), and two strains of the codling moth, *Carpocapsa pomonella,* metabolize DDT to DDE but the rate of detoxication is considered insufficient to account for the resistance (Rose and Hooper, 1969).

Aside from the resistance potential present in many species, DDT has also had the effect of reducing predator and parasite populations of insects. Interestingly, however, the lady beetle, *Coleomegilla maculata,* an important predator of several insect pests, is highly resistant to DDT. Its resistance is due to conversion of DDT to DDE and to excretion of both DDT and DDE in the feces and eggs (Atallah and Nettles, 1966). Enhanced DDT tolerance has also been noted in *Macrocentrus ancylivorus,* a parasite of the oriental fruit moth, (Pilou and Glasser, 1951; Robertson, 1957), and in *Bracon mellitor,* an ectoparasite of the boll weevil (Adams and Cross, 1967).

Investigations on the metabolism of DDT in some 30 insect species (Hoskins and Witt, 1958) revealed that, in general, DDT metabolism falls into three classes: (1) absorbed DDT remains largely unchanged, (2) much of the absorbed insecticide is metabolized to DDE, and (3) the chief metabolites do not respond to the colorimetric method of Schechter *et al.* (1945). With the advent of newer analytical techniques such as thin-layer and gas-liquid chromatography, it is now possible to add (4) metabolism to 2,2-bis(*p*-chlorophenyl)-1,1-dichloroethane (DDD or TDE), 2,2-bis(*p*-chlorophenyl)-1,1,1-trichloroethanol (dicofol), 4,4-dichlorobenzophenone (DBP), bis(*p*-chlorophenyl) acetic acid (DDA), bis(*p*-chlorophenyl) methanol (DBH), 4, 4′-dichlorodiphenylmethane (DCPM), and *p*-chlorobenzoic acid.

The various pathways of DDT metabolism and the structural formulas of metabolites are depicted in Fig. 1.

DDT metabolism in susceptible and DDT-resistant body lice, *Pediculus humanus humanus,* pose an even greater dilemma. The resistant (Korean) strain tolerates oral doses of 200 ppm in the blood whereas the susceptible (Orlando) strain succumbs to 10–20 ppm (Perry and Buckner, 1958). Nevertheless, homogenates of both strains metabolize DDT *in vitro* at approximately equal rates and yield the same metabolites (Perry *et al.,* 1963; Miller and Perry, 1964). What then is the physiological basis of this resistance? Unfortunately, the answer is not known, but one can speculate that an effective nerve barrier in the resistant strain prevents the penetration of DDT into sensitive neurons since rate of enzymatic attack is no different between the two strains.

Another type of DDT metabolism is exemplified by the vinegar fly, *Drosophila melanogaster.* DDT-resistant strains of this species metabolize DDT *in vivo* to the corresponding ethanol derivative 2,2-bis(*p*-chlorophenyl)-1,1,1-trichloroethanol (dicofol) (Tsukamoto, 1959, 1960). This resistance mechanism is controlled by a dominant gene *Rst* (2) DDT, located on the second chromosome (Tsukamoto and Ogaki, 1953). Similar results have been obtained with [^{14}C]DDT in two other strains of *D. melanogaster,* but in addition to dicofol, another metabolite 2,2-bis(*p*-chlorophenyl)-1,1-dichloroethanol (FW-152) was detected (Menzel *et al.,* 1961).

Dicofol also is one of the metabolites resulting from the *in vivo* degradation of DDT by *T. infestans* (Dinamarca *et al.,* 1962), in addition to DDE and three other polar metabolites. Slower penetration and enhanced detoxication of DDT appear to account for the resistance of fifth-instar nymphs of this species.

Although the weight of evidence supports a causal relationship between dehydrochlorination of DDT and resistance in the house fly, there are several areas of disagreement. Among the latter are that (a) large quantities of unchanged DDT sufficient to kill several S flies remain in the tissues of surviving R flies; (b) DDT metabolism is too slow to account for the resistance of several strains; (c) certain R strains show cross resistance to compounds such as 1,1-bis(*p*-chlorophenyl)-2-nitropropane (Prolan) and 1,1-dianisyl neopentane which cannot be dehydrochlorinated.

The presence of unchanged DDT need not necessarily imply that the toxicant is in a biochemically active state, for it may be bound or stored in nonsensitive tissues away from the site of action, or it may be distributed in various organs throughout the body (Table II), so that the concentration at any one locus may be below the threshold of poisoning.

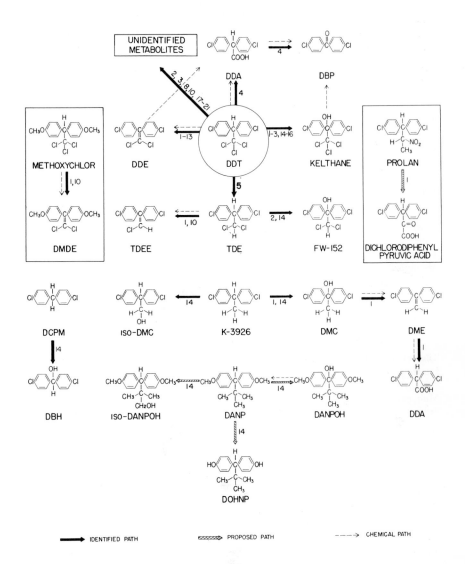

Fig. 1. Metabolism of DDT and related compounds by various insect species. Adapted and modified from Tsukamoto (1961). Key: 1, *Musca domestica;* 2, *Blattella germanica;* 3, *Periplaneta americana;* 4, *Pediculus humanus;* 5, Anopheline and Culicine spp.; 6, *Malacosoma americana;* 7, *Pyrausta nubilalis;* 8, *Plodia interpunctella;* 9, *Pieris, rapae;* 10, *Epilachna varivestis;* 11, *Argyrotaenia velutinana;* 12, *Melanoplus differentialis;* 13, *Melanoplus femur-rubrum;* 14, *Drosophila melanogaster;* 15, *Drosophila virilis;* 16, *Triatoma infestans;* 17, *Nymphalis antiopa;* 18, *Leucophea maderae;* 19, *Oncopeltus fasciatus;* 20, *Ornithodorus coriaceous;* 21, *Anthonomus grandis.*

TABLE II

DISTRIBUTION OF DDT, INCLUDING METABOLITES, IN VARIOUS TISSUES
FOLLOWING TOPICAL APPLICATION OF DDT

	DDT and metabolites recovered					
				Applied dose (%)		
	Absorbed dose (%)				Treated leg[d]	
Tissue	a	b	c		S[e]	R[e]
Head	5.8	38.0	58.6	Head	5.5	0.5
Wings and legs	11.5			Thorax	18.0	1.5
Thoracic cuticle	18.5	26.0		Abdomen	11.0	1.5
Abdominal cuticle	33.5					
					Treated labella	
					S	R
Thoracic muscle	16.2	5.0				
Intestinal tract	14.4	5.0	26.5	Head	11.0	3.0
Hemolymph	—	13.0		Thorax	33.0	3.0
Malpighian tubules	0.0	—		Abdomen	30.0	2.5
Sex organs	0.0	10.0				
Nerve tissue	0.0	3.0	14.8			
	DDE = 100%	Radioactive DDT	DDE = 73%	Radioactive DDT		
	DDT = 0		DDT = 23%			

[a] From Sternburg and Kearns (1950).
[b] From Lindquist et al. (1951b).
[c] From Tahori and Hoskins (1953).
[d] From Le Roux and Morrison (1954).
[e] S, susceptible; R, resistant.

Rate of DDT-metabolism varies from strain to strain, but even when exposed to sublethal doses it can be shown that R flies metabolize DDT at a faster rate than S flies (Menn et al., 1957).

Cross-resistance is a unique phenomenon in insect toxicology both among closely related compounds and among those relatively distant in structure. For example, Prolan-R flies metabolize Prolan to neutral and acidic derivatives (Perry and Buckner, 1959), but they also dehydrochlorinate DDT via a different mechanism (Brown and Perry, 1956). *Drosophila melanogaster* adults metabolize 1,1-dianisyl neopentane to the trichloroethanol derivative (cf. Fig. 1) in addition to detoxifying DDT (Tsukamoto, 1961).

i. Metabolism of DDT by Cockroaches. The metabolites of DDT responding to the colorimetric method of Schechter *et al.* (1945) form only a small portion of the total metabolic pool in *P. americana* after topical application of DDT. Approximately 55% of the applied dose is in the form of unidentified metabolites (Vinson and Kearns, 1952). Injection of [^{14}C]DDT in *P. americana* also yields a water-soluble metabolite which appears to be a DDT-derivative conjugated with a carbohydrate fragment (Butts *et al.*, 1953). Conversion of DDT to DDE occurs mainly in the fat-body, alimentary canal, and cuticular components of the American roach (Cochran, 1956), but fat-body appears to contribute much to the differential susceptibility to DDT exhibited by the sexes and the developmental stages of this insect.

Application of radiochromatographic techniques to the study of DDT metabolism in *P. americana* revealed the presence in the excreta of six metabolites of differing R_f values (Robbins and Dahm, 1955). One of these metabolites, produced by both susceptible and DDT-resistant American roaches, corresponds exactly to the R_f value of 4,4-dichlorobenzophenone (Hoskins and Witt, 1958). Small amounts of DDE but larger quantities of DDT and polar metabolites predominate in the excreta of DDT-treated *Leucophaea maderae* (Lindquist and Dahm, 1956).

Metabolism of DDT in the German roach *B. germanica* is somewhat similar, with both susceptible and resistant strains converting small amounts of DDT to DDE and to predominantly unidentified metabolites (Babers and Roan, 1953; Hooper, 1969).

In addition to DDE, dicofol has been recorded as a DDT metabolite in the German roach (Tsukamoto, 1961; Agosin *et al.*, 1961b), and both S and R strains metabolize TDE to the ethylene derivative, TDEE (Hooper, 1969).

The paucity of data regarding metabolism studies with cockroaches stems, perhaps, from the fact that DDT has never been extensively used for roach control. Although resistant populations of *B. germanica* have been obtained in the laboratory by selection pressure (Grayson, 1951, 1953) and have been recorded worldwide (Brown, 1963), little information is available regarding the biochemical changes brought about by this selection. Nevertheless, it is evident that conversion of DDT to DDE or to dicofol plays only a minor role in metabolism. Detoxication of DDT to polar derivatives which are excreted in the feces seems to be more important, but in neither of the cases cited above can the mechanism of resistance be directly correlated with absorption or detoxication of DDT.

ii. DDT Metabolism by Mosquitoes. Among the numerous species which have developed resistance to insecticides, mosquitoes assume a prominent position. Eighty or more anopheline and culicine species are known to be resistant to one or more of the commonly used insecticides, and there is reason to believe that this list will keep growing as more areas around the world come under present-day mosquito control operations.

The first appearance of DDT resistance in culicine mosquitoes was recorded in 1946 and in anopheline species in 1951 (Brown, 1958). Although the resistance problem became acute in certain parts of the world, especially in connection with malaria eradication programs, little progress has been made in ascertaining the nature of this resistance.

DDT resistance has been more extensively investigated in the yellow fever mosquito *Aedes aegypti*. The ease of mass rearing, the availability of many resistant strains from different genetic background and geographic distribution, and the introduction of marker genes in larvae and adults make this insect an ideal species for resistance studies.

DDT-resistant larvae of *A. aegypti* and *A. taeniorhynchus* detoxify more DDT to DDE than their respective susceptible counterparts (Brown and Perry, 1956). On the other hand, resistant *A. nigromaculis* larvae absorb more DDT than susceptible larvae but convert about the same amount of DDT to nontoxic metabolites (Gjullin *et al.*, 1952). Several DDT-resistant *A. aegypti* strains produce much larger amounts of DDE than their susceptible counterparts, while others, notably those of Asiatic origin, show little increase in DDE production (Chattoraj and Brown, 1960; Abedi *et al.*, 1963). In one strain, at least, relaxation of DDT pressure caused an increase in susceptibility to the toxicant but no corresponding decrease in detoxication, as would be expected if the resistance mechanism were primarily detoxication of DDT.

DDT-resistant *A. aegypti* larvae originating from a strain selected with malathion respond to DDT by excreting long streamers of peritrophic membrane containing large quantities of DDT and DDE (Abedi and Brown, 1961). This hypersecretion might constitute a defense mechanism by removing unabsorbed DDT from the alimentary tract, thus reducing the internal concentration of the toxicant. This, however, is an isolated case involving two strains of Caribbean origin and it does not extend to resistant strains of different geographical distribution.

Among culicine species, DDT-resistant *Culex pipiens fatigans* larvae are very efficient in converting DDT to DDE (Bami *et al.*, 1957; Kimura *et al.*, 1965), and a similar R species of *Culex pipiens quinquefasciatus* from

California contained practically all DDE and only minute amounts of DDT within larval tissues (Hoskins *et al.,* 1958). Two polar metabolites similar to those found in the German roach were also detected in the water. In contrast with the above results, a highly DDT-resistant strain of *Culex pipiens molestus* produces only little more DDE than its susceptible counterpart (Perry, 1960a).

The results of Plapp *et al.* (1965) and Plapp and Hennessy (1966) suggest that the mechanism of resistance to DDT in *Culex tarsalis* is unknown and does not involve dehydrochlorination of the insecticide. Kalra *et al.* (1967) also concluded that mechanisms other than dehydrochlorination contribute to DDT resistance in *Culex fatigans.* Alternately, Kimura *et al.* (1965) attribute the DDT resistance of their *Culex tarsalis* strain to enhanced DDE production, whereas enhanced DDT metabolism was characteristic of both R and S larvae of *Culex fatigans.*

In addition to DDE, Hooper (1967, 1968) detected the metabolite DDD (TDE) as a by-product of DDT metabolism by *Culex fatigans,* a finding confirmed by Kalra *et al.* (1967). Earlier, DDD was demonstrated in the excreta of the stable fly *Stomoxys calcitrans* (Stenersen, 1965), but was presumed to be due entirely to the anaerobic action of bacteria.

The degradation of DDT to DDD is an interesting phenomenon since it has been reported in a number of organisms, including mammals, birds, fish, microorganisms (bacteria and fungi), and by the action of ferrous deuteroporphyrins, reduced porphyrins and aqueous media (see Singh and Malaiyandi, 1969, for references). In addition, DDD has been reported as a product of DDT metabolism by HeLa cells (Huang *et al.,* 1970), by stored wheat grains (Rowlands, 1968), and by ionizing radiation (Sherman *et al.,* 1971).

DDT resistance in anopheline mosquitoes is less understood. Field-collected adults of *Anopheles atroparvus, A. maculipenis typicus, A. labranchiae, A. superpictus,* and *A. claviger* from Italy exhibited normal susceptibility to DDT and little conversion of DDT to DDE (Perry, 1960a). Laboratory colonies of DDT-resistant *A. atroparvus* and *A. stephensi* produced little DDE in proportion to their resistance and showed no significant difference in DDT metabolism from their susceptible counterparts (Frontali and Carta, 1959; Perry, 1960a). Hence, resistance cannot be correlated with DDT dehydrochlorination in these species. Only in certain strains of *A. sacharovi* (Perry, 1960a) and in *A. sundaicus* (Kearns, 1957) have substantial quantities of DDE been recovered so that the resistance mechanism might possibly correlate with DDT detoxication.

So far, the available data indicate little correlation between DDT resistance and DDT metabolism in mosquitoes. However, some exceptions, i.e., in some strains of *A. aegypti,* do exist. Lack of agreement in results from various laboratories likewise precludes forming a generalized hypothesis regarding the resistance mechanism in *Culex* species, and the dearth of information on anophelines makes the problem even less amenable to interpretation in that group.

iii. Metabolism of DDT in Vitro. The conversion of DDT to DDE by resistant house flies had, for many years, been regarded by some investigators as the consequence of survival of the insect rather than the cause of its resistance. Adequate proof to the contrary could not be provided until the isolation by Sternburg *et al.* (1953) of the enzyme DDT dehydrochlorinase (DDTase) which, in the presence of glutathione, catalyzes the dehydrochlorination of DDT *in vitro* (Fig. 2). Under identical experimental conditions only DDT-R strains were shown to contain this enzyme (Sternburg *et al.*, 1954). House fly DDT dehydrochlorinase has been purified by various procedures (Moorefield, 1956; Lipke and Kearns, 1959a; Dinamarca *et al.*, 1969) and its kinetics have been studied spectrophotometrically in some detail with purified preparations in which DDT was dissolved in egg yolk lipoprotein (Lipke and Kearns, 1959b; Lipke, 1960). With this technique it was possible to demonstrate DDT dehydrochlorinase in adult susceptible flies also, particularly in the larval stage (Moorefield, 1958).

Lipke and Kearns (1959b) found a molecular weight of 36,000 for house fly DDTase. Reinvestigation of the problem by Dinamarca *et al.* (1969) established the fact that DDT dehydrochlorinase has a tetrameric structure, with each monomer having a molecular weight of 30,000. The tetramer is formed in the presence of DDT, possibly because the enzyme is a lipoprotein (Dinamarca *et al.*, 1971). Glutathione, which is not used during the reaction, is necessary to maintain the tetrameric structure in the presence of DDT (Dinamarca *et al.*, 1969). Recently, DDT isozymes have been described by Goodchild and Smith (1970) but it is not clear if these isozymes are related to the tetrameric form of DDTase.

With present-day gas chromatographic methods the detection of enzyme activity from a single susceptible fly is quite feasible (Oppenoorth and Voerman, 1965) and differences between strains can be made more manifest (Oppenoorth, 1965; Grigolo and Oppenoorth, 1966; Khan and Terriere, 1968).

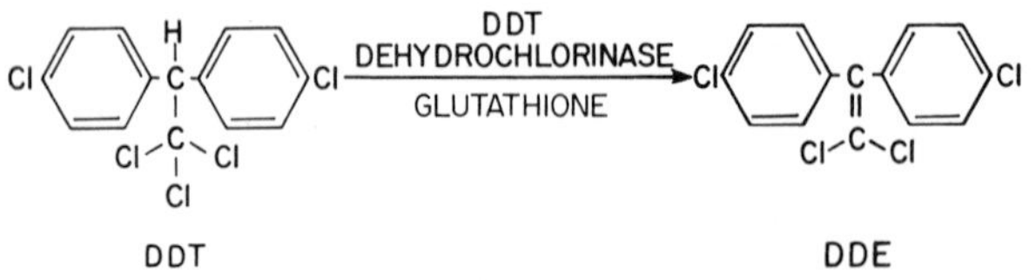

Fig. 2. Enzymatic degradation of DDT by the house fly.

The amount of DDT dehydrochlorinase in house fly eggs is negligible. During metamorphosis of the resistant strain there is a progressive increase in enzyme concentration throughout larval life, and an abrupt drop of as much as 50% activity in the pupal stage. This lower level of activity is maintained throughout the pupal and adult life (Moorefield and Kearns, 1957) (Fig. 3). In the adult fly, high DDTase titers are found in fat-body and brain tissue, intermediate amounts in cuticle, muscle, and hemolymph, and little or none in the intestinal tract and ovaries (Miyake *et al.*, 1957).

The occurrence of high levels of DDTase in fat-body is highly significant since, in general, insect fat-body is rich in detoxifying enzymes, and rather than being a storage depot, fat-body is an active tissue similar in function to mammalian liver. The high concentration of DDTase in nerve tissue is also important, especially from the resistance standpoint, since it might provide a site for local detoxication. Consistent with this hypothesis is the demonstration by Wigglesworth (1956) that the underlying cytoplasmic sheath of insect nerve is exceptionally rich in oxidases, dehydrogenases, and various esterases and, presumably, could protect the sensitive axons by local detoxication.

DDT dehydrochlorinase is also present in tissue homogenates of the Mexican bean beetle, and catalyzes the dehydrochlorination of DDT, DDD, and methoxychlor (Chattoraj and Kearns, 1958; Swift and Forgash, 1959; Tombes and Forgash, 1961). Unlike the distribution of DDTase in the house fly, the highest concentration of the enzyme in the Mexican bean beetle

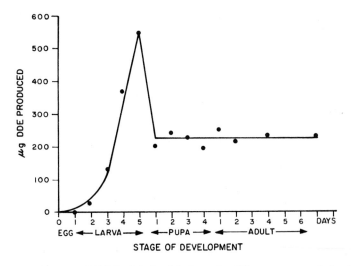

Fig. 3. Distribution of DDT dehydrochlorinase in different stages of development of the DDT-resistant house fly. (From Moorefield and Kearns, 1957.)

is found in the reproductive organs, followed by the alimentary canal, exoskeleton, flight muscle, central nervous system, and fat-body.

Enzymic degradation of DDT in the body louse yields three metabolites: DDE, DDA, and 4,4-dichlorobenzophenone (DBP). Accordingly, three or more enzymes are involved in these biotransformations (Perry et al., 1963), two of which have been isolated and partially purified (Miller and Perry, 1964). The pathway of DDT metabolism in this insect is shown in Fig. 1. The conversion of DDT to DDE is analogous to that of the house fly, but DDE is not an intermediate in the breakdown of DDT to DDA, and the degradation of DDA to DBP most likely involves a microsomal mixed-function oxidase.

The role of DDT dehydrochlorination in mosquito resistance became more apparent with the isolation of the enzyme DDT dehydrochlorinase from *A. aegypti* larvae by Kimura and Brown (1964). This was made possible by protecting the enzyme with glutathione, nitrogen, and low temperature during the process of preparation. The mosquito enzyme has an optimum pH of 7.4 and, in addition to attacking DDT, dehydrochlorinates DDD and methoxychlor. In a large number of *A. aegypti* strains of American origin, the DDT dehydrochlorinase activity was found to be proportional to the resistance level.

Such correlation does not hold true for anopheline species. Relatively large amounts of DDE are formed *in vivo* by both S and R strains of *A. stephensi, A. quadrimaculatus, A. gambiae,* and *A. albimanus,* but no correlation is evident between resistance and metabolism (Lipke and Chalkley, 1964). Furthermore, little or no relation exists between the presence of glutathione-dependent DDT dehydrochlorinase in cell-free extracts of these species and the degree of their resistance to DDT. In fact, the highest enzyme titers were found in DDT-susceptible *A. albimanus* and *A. stephensi.*

Another type of enzymatic degradation of DDT involves hydroxylation at the tertiary carbon with the production of dicofol. The enzyme responsible for this metabolic pathway was first isolated and characterized by Agosin *et al.* (1961b). The enzyme resides in the microsomal fraction of tissue homogenates and requires NADPH, O_2, nicotinamide, and magnesium ions for activation. Its presence has been shown in *B. germanica, P. americana, M. domestica* (both DDT-S and DDT-R strains), and *Culex quinquefasciatus.* Dicofol is also a product of *in vivo* DDT metabolism in *D. melanogaster* (Tsukamoto, 1959), *T. infestans* (Dinamarca et al., 1962; Agosin et al., 1964), and the grain weevil *Sitophilus granarius* (Rowlands and Lloyd, 1969).

The properties of the microsomal enzyme catalyzing the hydroxylation of DDT by *T. infestans* were described by Agosin *et al.* (1969). Evidence from experiments with phenobarbital indicates that more than one enzyme is involved in the metabolism of DDT by microsomal mixed-function oxidases.

The *in vivo* availability of NADPH might be a limiting factor in the production of dicofol, since NADPH is also necessary for the synthesis of reduced glutathione, an essential cofactor for DDT dehydrochlorinase (Dinamarca *et al.*, 1962). Hence, the microsomal hydroxylating system might be in competition with DDT dehydrochlorinase for NADPH (Fig. 4).

The role of DDT hydroxylation in resistance was first demonstrated by Morello (1964) who showed that hydroxylation of DDT is blocked by SKF 525-A (β-diethylaminoethyl diphenylpropyl acetate) and iproniazid (2-isopropyl-1-isonicotinoyl hydrazine) with a concomitant increase in mortality of resistant fifth-instar *T. infestans* nymphs, while 3-methylcholanthrene increases the DDT-hydroxylating capacity of the insect resulting in somewhat greater tolerance for the insecticide. Earlier, Arias and Terriere (1962) had shown a higher rate of naphthalene hydroxylation in microsomes of DDT-R house flies than in those of DDT-S counterparts, an observation substantiated by Schonbrod *et al.* (1965).

Microsomal hydroxylation of DDT was soon corroborated in a number of other DDT-resistant strains of house flies (Tsukamoto and Casida, 1967; Gil *et al.*, 1968; Oppenoorth and Houx, 1968; Khan and Terriere, 1968; Plapp and Casida, 1969) and it now appears to be a major pathway of DDT detoxication in certain insects.

Recently, Kapoor *et al.* (1970) compared the metabolism of DDT, methoxychlor, and methiochlor in various organisms in a model ecosystem for evaluating pesticide biodegradability. R and S house flies and the salt marsh caterpillar *Estigmene acrea* were included in this study. Piperonyl butoxide (a methylenedioxyphenyl synergist) was applied to the flies' abdomens 1 hour prior to topical application of the insecticides to indicate the extent of metabolism by the mixed-function oxidase system. The data in Table III show that methoxychlor is substantially more effective than DDT against the R strain and this can perhaps be related to its lower rate of dehydro-

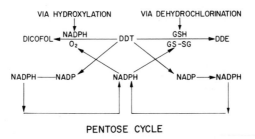

Fig. 4. Role of NADPH in hydroxylation and dehydrochlorination of DDT. (From Dinamarca *et al.*, 1962.)

TABLE III

TOXICITY AND METABOLISM OF DDT, METHOXYCHLOR, AND METHIOCHLOR IN
RESISTANT HOUSE FLIES[a]

Compound recovered	LD_{50} γ/female fly		% Radioactivity recovered			
			DDT treatment			
	Alone	+PB[b]	External	Internal	Excreta	Total
DDT	3.4	0.8	9.1	12.4	1.0	22.5
DDE	—	—	—	12.6	8.7	21.3
Dicofol	—	—	—	11.9	7.6	19.5
Conjugates	—	—	—	23.1	7.5	30.6
			Methoxychlor treatment			
Methoxychlor	0.96	0.09	9.7	8.5	12.0	30.2
O-Demethylmethoxychlor	—		3.5	—	3.5	
Conjugates	—		5.9	30.9	36.8	
			Methiochlor treatment			
Methiochlor	56	10	50.3	3.7	4.9	58.9
$CH_3SC_6H_4C{=}CCl_2C_6H_4SCH_3$	—		1.4	0.2	1.6	
$CH_3SOC_6H_4HCCCl_3C_6H_4SCH_3$	—		2.7	0.8	3.5	
$CH_3SOC_6H_4HCCCl_3C_6H_4SOCH_3$	—		0.6	0.2	0.8	
$CH_3SOC_6H_4HCCCl_3C_6H_4SO_2CH_3$	—		trace	0.3	0.3	
$CH_3SO_2C_6H_4HCCCl_3C_6H_4SO_2CH_3$	—		1.0	2.4	3.4	

[a] From Kapoor et al. (1970).
[b] PB, piperonyl butoxide.

chlorination by DDTase (Metcalf and Fukuto, 1968). The effective
inhibition of methoxychlor and methiochlor degradation by piperonyl
butoxide clearly indicates that detoxication of these compounds proceeds
via oxidative processes rather than by dehydrochlorination. Methiochlor
which is first metabolized by oxidation to 2-(p-methylsulfinylphenyl)-2-
(p-methylthiophenyl)-1,1,1-trichloroethane and by further oxidation to the
bissulfoxide and bissulfone derivatives, becomes than more susceptible to
attack by DDTase. Methoxychlor is metabolized by O-demethylation to form
2-(p-hydroxyphenyl)-2-(p-methoxyphenyl)-1,1,1-trichloroethane and 2,2-
bis(p-hydroxyphenyl)-1,1,1-trichloroethane which are largely eliminated in
conjugated form. Hence, both pathways play a role in the detoxication of
these compounds.

b. Metabolism of Benzene Hexachloride (BHC). The metabolic fate of BHC (actually, the compound is hexachlorocyclohexane) in insects also has been a subject of considerable investigation, and in conformity with DDT, the house fly has emerged as the organism of choice, for no other insect to date has shown such an amazing potential for chemical biotransformation of xenobiotics.

i. Rate of BHC Metabolism. Normally, γ-BHC (lindane) does not accumulate in tissues of warm-blooded animals to the same extent as DDT, but house fly pupae and adults originating from larvae reared on a lindane-treated medium contain significant amounts of the toxicant (Bradbury *et al.*, 1953). Adult flies metabolize the insecticide fairly rapidly, and the rate of metabolism is faster in R than in S flies (Oppenoorth, 1954). These observations suggest an increased detoxication capacity in the adult stage, in contrast with the pattern observed with DDT metabolism (Moorefield and Kearns, 1957). In addition to γ-BHC, the α-, β-, and δ-isomers which are less toxic are also metabolized at a faster rate by R than by S strains (Oppenoorth, 1955; Bradbury and Standen, 1956a). This fact clearly demonstrates a greater metabolic capacity in the R strain and a possible causal relationship between metabolism and resistance. The results of Bradbury and Standen (1960) with ^{14}C-labeled γ-BHC administered to four strains of flies add support to this hypothesis by showing a fitness in the relationship

$$\ln Y = \ln Y_0 - rt$$

where Y is the amount of γ-BHC present at time t, Y_0 the amount picked up in the original 15-minute exposure, and r represents rate of metabolism. In the four strains studied, r increased with an increase in the level of resistance.

Oppenoorth (1956) summarized his extensive work on the metabolism of BHC isomers in eight strains of flies as follows: (a) injected γ-BHC is metabolized by both S and R strains, but at a faster rate by the latter; (b) rate of absorption of the different isomers decreases in the order $γ > δ > α$; (c) rate of metabolism decreases in the order $γ \sim α > δ$; and (d) good correlation exists between degree of resistance, absorption rate, breakdown capacity, and amount of unchanged chemical found in the flies.

Bradbury (1957) extended his work on the comparative metabolism of γ-BHC in several other species and showed (Table IV) that house flies, both S and R strains, are in a class by themselves in their capacity to metabolize the toxicant. Even γ-BHC-resistant *A. gambiae* adults converted no more of the absorbed dose (10% in 24 hours) than their susceptible counterparts (Bradbury and Standen, 1956b), and a similar situation occurred with *Cimex lectularius* (Bradbury and Standen, 1960). According to Busvine and

TABLE IV

COMPARATIVE METABOLISM OF ^{14}C-LABELED γ-BHC IN VARIOUS INSECTS[a]

Species	Dosage (μg/gm)	BHC recovered (μg/gm)			
		External	Internal	Water-soluble metabolites	% metabolized
Cockroach	42	10	19	9	22.0
Bean weevil	111	7	94	8	7.7
Grain weevil	34	14	13	1	4.2
Locust	100	16	83	4	4.7
Mosquito	213	17	141	10	5.1
Khapra beetle	166	81	45	11	12.9
House fly (S)[b]	193	22	54	99	57.9
House fly (R)[b]	230	13	21	174	80.2

[a] From Bradbury (1957).
[b] S, susceptible, R, resistant.

Townsend (1963), the significance of BHC degradation in resistant house flies can be explained if rate of detoxication is considered in relation to the resistance spectrum which Busvine (1954) showed to be linked with cyclodiene insecticides. Resistance, then, has two components, one associated with cross-resistance to cyclodienes which are not metabolized, and another, due to enhanced degradation of γ-BHC which is characteristic of the house fly but not of other insects.

ii. Nature of BHC Metabolites. In the presence of alcoholic alkali, BHC isomers readily undergo dehydrochlorination liberating 3 moles of HCl per mole of BHC to yield principally 1,2,4-trichlorobenzene and smaller amounts of 1,2,3- and 1,3,5-trichlorobenzene (Cristol, 1947; Gunther and Blinn, 1947).

The early work with house flies (Bradbury and Standen, 1955, 1958) indicated that trichlorobenzene was, at best, only a minor metabolite of [^{14}C]lindane and that the expired air contained no radioactivity, but the internal tissues contained equal amounts of polar and nonpolar metabolites. Both S and R strains metabolized the α-, β-, and δ-isomers with equal facility, but the γ-isomer was metabolized twice as fast by the R strain.

γ-Pentachlorocyclohexene (γ-PCCH), the monodehydrochlorinated product of γ-BHC, was reported to be a major product of lindane metabolism in house flies (Sternburg and Kearns, 1956). Both S and R strains produce this metabolite, which is further converted to other unidentified polar products. However, Bradbury and Standen (1958) consider γ-PCCH to be a minor metabolite, and isotope dilution techniques indicate that this metabolite is formed in relatively small amounts (Bridges, 1959). A second isomer of PCCH, also produced in small amounts, was recently detected by Reed and Forgash (1968). While the production of γ-PCCH is perhaps toxicologically insignificant, there appears to be a correlation between the formation of the second PCCH-isomer and lindane resistance. House flies treated with either lindane, γ-PCCH, or γ-PCCH-isomer yield a common metabolite which was identified by mass spectroscopy and gas-liquid chromatography as 1,2,4,5-tetrachlorobenzene. In addition, 1,2,3,4-tetrachlorobenzene, pentachlorobenzene, 1,2,4- and 1,2,3-trichlorobenzene were also detected. While these metabolites appeared only in small quantities following lindane application, they appeared in much larger quantities ($\geq 50\%$) following treatment with the two PCCH-isomers (Reed and Forgash, 1969).

The relationship of lindane metabolism to organic soluble products and resistance is, at best, only of minor significance since, by far, water-soluble metabolites predominate in all fly strains studied (Reed and Forgash, 1970; Bradbury, 1957; Bradbury and Standen, 1955, 1958, 1960). Water-soluble compounds also account for most of the lindane metabolic pool in mammals (Davidow and Frawley, 1951; van Asperen, 1954; van Asperen and Oppenoorth, 1954; Koransky et al., 1964; Sims and Grover, 1965; Grover and Sims, 1965).

iii. Enzymic Metabolism of BHC in Vitro. As discussed earlier, the rate of detoxication of BHC-isomers by house flies decreases in the order $\gamma > \alpha > \delta$. Alkaline dechlorination to trichlorobenzene decreases in the order $\alpha > \delta > \gamma > \beta$ (Kauer et al., 1947). Hence, *in vivo* metabolism of BHC does not follow the pattern of alkaline dechlorination, a situation analogous to that found with some DDT derivatives.

Faster metabolism of the α- and δ-isomers by R flies (note exception for α-BHC, Bridges and Cox, 1959) implies a causal relationship to resistance, but the same interpretation for the γ-isomer might be open to objection on grounds that the greater metabolizing capacity of the more toxic isomer is due to the longer survival of the resistant fly. As in other instances, isolation of the detoxifying enzyme *in vitro* is a prerequisite in characterizing the resistance mechanism.

An important step in this direction was the finding that alkaline hydrolysis of BHC metabolites produces dichlorothiophenols (Bradbury and Standen, 1959). From this it is inferred that metabolism of BHC to water-soluble metabolites involves the formation of a C—S bond. The nature of the initial metabolic product suggests a conjugation with aryl mercapturic acid (Bradbury and Standen, 1960), similar to the linking of sulfur with nuclear carbon in mammalian detoxication mechanisms (Williams, 1959).

Homogenates, acetone powders, and crude soluble enzyme preparations of BHC-resistant flies yield thiophenols as end-products in *in vitro* BHC metabolism. Reduced glutathione is essential for activation of the enzyme. The reaction might proceed by substitution of an aromatic chlorine followed by hydrolysis of the conjugate to yield dichlorothiophenols (Fig. 5); γ- and δ-pentachlorocyclohexene (PCCH), the monodehydrochlorination products of BHC-isomers, are also converted to water-soluble metabolites by the same enzyme system (Ishida and Dahm, 1965a,b; Sims and Grover, 1965) and at a much faster rate than γ-BHC. This is, perhaps, the reason why some investigators find only traces of PCCH in their studies of BHC metabolism. The water-soluble metabolites arising from PCCH metabolism in grass grubs and blow flies are chromatographically indistinguishable from *S*-2,4-dichlorophenylglutathione (Clark *et al.*, 1969). These observations support the contention of Bradbury and Standen (1959) that a pentachlorocyclohexylglutathione is the initial metabolite of γ-BHC degradation, and suggests that the detoxication enzyme belongs to the group of glutathione *S*-aryltransferases (Clark *et al.*, 1967; Fukami and Shishido, 1966; Ishida and Dahm, 1965a,b; Ishida, 1968) and might also include DDT dehydrochlorinase (Lipke and Kearns, 1960; Ishida and Dahm, 1965b; Ishida, 1968). The scheme outlined in Fig. 5 will also permit the production of PCCH by metabolism of the thioether conjugate. PCCH is then further metabolized to water-soluble compounds which are excreted. This process releases the HSR moiety which again can conjugate with additional BHC to repeat the cycle.

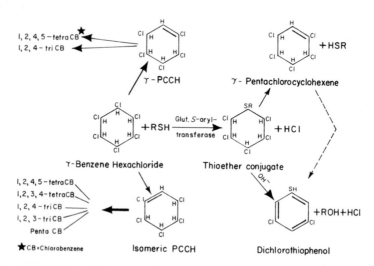

Fig. 5. Metabolism of benzene hexachloride by the house fly. Major metabolic routes are designated by heavy arrows. (From Bradbury and Standen, 1959, 1960; Reed and Frogash, 1968, 1969).

If detoxication by these enzymes is causally related to resistance, then their inhibition becomes of considerable importance. Insect glutathione S-aryltransferase (Clark et al., 1967) and DDT dehydrochlorinase (Balabaskaran et al., 1968) are inhibited by phthaleins and sulfonphthaleins, and these dyes also inhibit the metabolism or γ- and δ-BHC (Clark et al., 1969). Furthermore, bromophenol blue inhibits house fly enzymes that detoxify γ-BHC and DDT (Ishida and Dahm, 1965a). Inhibitors such as these might be used to distinguish between isozymes among different species. Indeed, Clark et al. (1969), using the inhibitor principle, found distinguishable characteristics between house fly and blow fly enzymes that detoxify γ-BHC and DDT.

The properties and distribution of the BHC-metabolizing enzyme (Ishida and Dahm, 1965a,b) show that it is found in the soluble fraction of house fly homogenates, has a molecular weight of 54,000 (later corrected to 36,000 by Ishida, 1968), and is specific in its requirement for reduced glutathione. Enzyme activity is particularly high in house fly homogenates as compared with similar preparations from other insect species and from mammals (Table V), and the level of activity is higher in the adult fly than in the egg, larval, and pupal stages (Fig. 6). This is in contrast with the distribution of DDT dehydrochlorinase showing highest activity in the larval stage (cf. Fig. 3).

A significant feature of these studies is that no correlation is evident between enzymic metabolism of γ-BHC to water-soluble metabolites and resistance. Neither is the metabolism of γ-BHC to organic-soluble products a significant factor in house fly resistance (Reed and Forgash, 1970). As more species are studied the detoxication hypothesis is weakened, except for the fact that the house fly stands alone with its tremendous potential for insecticide degradation. Alternatively, a supplementary defense mechanism might be essential for the insect's survival, and together with detoxication, the combination might provide the organism with a formidable weapon to withstand the toxicant.

c. *Metabolism of Chlordane and Toxaphene.* Very little is known about the metabolic fate of chlordane and toxaphene in insects. The paucity of data might be due, perhaps, to their limited use in pest control operations and, until recently, to the unavailability of suitable analytical methods for the detection of metabolites.

Technical chlordane consists of at least five compounds, three of which are relatively toxic to insects (Davidow, 1950; March, 1952b). For an understanding of the biochemistry of resistance to this insecticide it is essential to study the metabolism of each component separately. This has not yet been accomplished.

TABLE V

COMPARATIVE RATES OF BHC- AND PCCH-METABOLIZING
ENZYMES IN VARIOUS SPECIES[a]

Enzyme source	Sex	Enzyme activity (units/mg N)			
		α-BHC	γ-BHC	γ-PCCH	δ-PCCH
Insects					
Musca domestica (S)[b]	M	92	22	290	3,650
	F	158	38	560	6,390
Musca domestica (R)[b]	M	332	59	920	10,600
	F	228	24	590	7,420
Drosophila melanogaster	M and F	0	0	23	990
Stomoxys calcitrans	M and F	13	0	63	0
Periplaneta americana	M	40	2.8	50	—
	F	14	0	44	—
Blattella germanica	M and F	0	0	23	—
Leucophaea maderae	F	0	0	2.4	—
Ostrinia nubilalis (larvae)		0	0	23	—
Ostrinia nubilalis (adults)	M and F	0	0	18	—
Apis mellifera	F	0	0	0	—
Diabrotica virgifera		0	0	0	—
Lachesilla pedicularia (adults)	M and F	0	0	0	—
Mammals					
Rattus norvegicus (rat liver)	M	0	0	19	—
Lepus cuniculus (rabbit liver)	M	2.4	0	16	274
	F	3.2	0	13	227
Lepus cuniculus (rabbit kidney)	M	1.5	0	3.7	105
	F	1.1	0	2.7	79

[a] From Ishida and Dahm (1965a).
[b] S, susceptible; R, resistant.

Bioassay procedures (Hoffman and Linquist, 1952) have established that chlordane-resistant house flies are capable of metabolizing a large percentage of topically applied chlordane to innocuous derivatives. Chlordane-resistant German roaches, *B. germanica,* metabolize 70% of the absorbed dose to unidentified products. However, the remaining unchanged insecticide within the tissues is sufficient to kill more than 20 susceptible roaches. Here, too, it appears that two or more defense mechanisms operate simultaneously— one mechanism neutralizes the insecticide by detoxication, while the other removes the toxicant from circulation by storage in nonsensitive sites. Analogous results obtained with chlordane-resistant house flies (Perry, 1953) are amenable to the same interpretation, but Earle (1963) contends that chlordane detoxication does not contribute materially to the resistance mechanism in house flies.

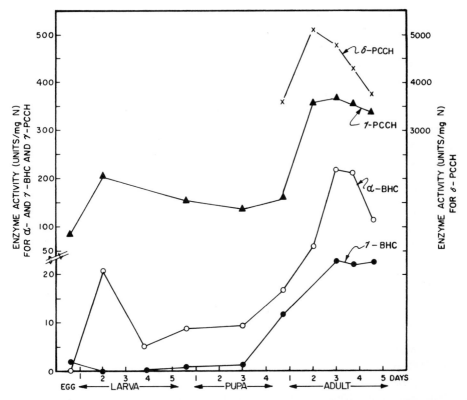

Fig. 6. Distribution of benzene hexachloride-metabolizing enzyme(s) in different stages of development of the house fly. (From Ishida and Dahm, 1965a,b.)

γ-Chlordane penetrates the fly's cuticle quite readily and accumulates unchanged in the tissues to fairly high levels within 24 hours after topical application. Thereafter, a progressive decrease in internal concentration follows but the excreta yield no detectable products. Very likely, hydrophilic metabolites are formed by insects, but no efforts have been made at their detection and identification, with the exception of the works of Korte and associates (Korte, 1967), who showed 10–25% conversion of [^{14}C]chlordane to water-soluble metabolites in *A. aegypti* larvae. No chlordane metabolism was found in larvae of *A. quadrimaculatus* (Bowman *et al.*, 1964).

In mammals, γ-chlordane is extensively metabolized to hydrophilic compounds which are largely excreted, but also localized in various tissues (Korte, 1967). A new metabolite of α- and γ-chlordane, designated as oxychlordane, has recently been isolated from the fat of several mammals (Schwemmer *et al.*, 1970; Polen *et al.*, 1971), but it was not found in chlordane-treated plants or soil.

The fate of toxaphene (chlorinated camphene containing 67–69% chlorine) in insects is even less understood. Resistant house flies which survive a dose of 10 μg/fly metabolize 74–85% of the absorbed dose in 24 hours as shown by bioassay methods (Hoffman and Lindquist, 1952) or by colorimetric analysis (Perry, 1960b).

Little difference in toxaphene metabolism is found between toxaphene-resistant and susceptible boll weevils (Lindquist et al., 1961), but this might be due to the analytical method used which does not permit differentiation between unchanged and metabolized toxaphene.

Toxaphene is broken down by enzyme preparations of the cotton leaf worm *Prodenia litura* into three components having R_b values of 0.15, 0.37, and 0.47 (Abd El Aziz et al., 1965). The enzyme requires reduced glutathione for activation, and has an optimum pH of 7.3–7.4 at 37°–39°C (Abd El Aziz et al., 1967). After long incubation periods of 2–4 hours, homogenates and acetone powders of the enzyme degrade 16–50% of the substrate. The requirement for glutathione is reminiscent of the glutathione S-aryltransferases discussed earlier in conjunction with BHC metabolism.

Resistance to toxaphene is restricted to a few insect species of agricultural importance, and due to its limited use, no purposeful studies have been made to identify the resistance mechanism.

d. Metabolism of Endosulfan. Technical endosulfan (Thiodan) consists of 90% of a mixture of two stereoisomers: a high-melting isomer (melting point, 208°–210°C), and a low-melting isomer (melting point, 108°–110°C). The remaining 10% contains endosulfan alcohol and endosulfan ether, which are relatively nontoxic. The physical and biological properties of endosulfan were investigated by Lindquist and Dahm (1957) who found the two isomers to be slightly more toxic than DDT to susceptible house flies.

Endosulfan is rapidly oxidized to endosulfan sulfate by resistant and susceptible house flies (Barnes and Ware, 1965). The oxidized product, which is as toxic as the high-melting isomer, accumulates much more rapidly in the susceptible strain, suggesting either that the S strain cannot metabolize the sulfate further or that the rate of endosulfan oxidation in the R strain is slower. Two conjugated metabolites appear in the excreta and these might be products of endosulfan sulfate metabolism. The rate of endosulfan sulfate metabolism appears to be the important factor in house fly resistance to this insecticide.

2. Resistance Due to Excretion of the Insecticide or Toxic Metabolites

Metabolism of Prolan. Prolan is one of two constituents of the insecticide Dilan, the other constituent being Bulan.

Selection pressure with Dilan against house flies results in a strain having a higher level of resistance to Prolan than to Bulan or Dilan. The reason for this is not clear, but indications are that Prolan is metabolized at a faster rate than either Bulan or Dilan by the resistant strain (Perry and Buckner, 1959).

Penetration of Prolan is fairly rapid, almost half of the applied dose (8 μg/fly) being absorbed in 24 hours. The amount of unchanged toxicant within the tissues does not exceed 0.5 μg/fly which is only slightly more than the LD_{50} for the susceptible strain. The remainder, or 87% of the absorbed dose, is excreted as (1) a neutral material soluble in common organic solvents, similar to Prolan in infrared, UV, and colorimetric absorption spectra, and as toxic to mosquito larvae as the parent compound; and (2) an acidic derivative extractable with dilute alkali, different from Prolan in photometric spectra and chemical properties (such as loss of NO_2 group from the propane moiety) and much less toxic to mosquito larvae. Rate of excretion of the toxic compound exceeds that of the acidic metabolite during the initial 24 hours after application of the toxicant. This lapse of time is well beyond the period necessary for toxicological evaluation since Prolan acts fairly rapidly. As the time interval increases beyond 24 hours, rate of excretion of the acidic metabolite surpasses that of the neutral compound.

Concerning the protective mechanism involved in Prolan resistance, it appears that excretion of the acidic metabolite is not of primary importance to the fly's survival, since it is produced in substantial quantities only after the period of greatest danger to the fly has elapsed. Excretion of the unchanged insecticide is of far greater importance in the overall protective mechanism.

In vertebrate detoxication of foreign compounds, the end-products tend to become more water-soluble than their precursors so they can be excreted (Williams, 1959). Excretion of nonpolar compounds is indeed rare. In one instance, following injection of chlorobenzene in locusts, relatively large amounts of the nonpolar chlorobenzene were found in the excreta, along with polar conjugates (Gessner and Smith, 1960). In the case of Prolan, both the unchanged fat-soluble compound and the acidic, polar metabolite are excreted simultaneously. The acidic metabolite has tentatively been identified as 3,3-bis(p-chlorophenyl)pyruvic acid (Perry and Buckner, 1959; cf. Fig. 1).

Prolan resistance has several characteristics which distinguish it from other types of resistance. For example, it exhibits cross-resistance to the DDT-type compounds and to 1,1-dianisyl neopentane which is devoid of chlorine atoms in the pentane moiety. It also shows cross-resistance to the cyclodiene compounds such as aldrin, dieldrin, heptachlor, etc., and is also highly resistant to 2-(4-chlorophenyl)-2-(2,4-dichlorophenyl)-1,1,1-trichloroethane (o-chloro-DDT). The latter compound is refractory to dehydrochlorination due to steric hindrance caused by the *ortho* chlorine atom. Most DDT-resistant strains of house flies succumb to low dosages of this compound because of their inability to detoxify it (Hennessy and O'Reilly, 1956; Hennessy *et al.*, 1961; Perry *et al.*, 1967). However, certain strains of DDT-resistant mosquitoes, *A. aegypti* and *Culex fatigans*, resist o-chloro-DDT by dehydrochlorinating it (Kimura and Brown, 1964; Kimura *et al.*, 1965) and *Culex tarsalis* tolerates this compound by other unknown mechanisms (Plapp *et al.*, 1965).

The intriguing potentialities of Prolan resistance in house flies may be summarized as follows: (1) ability to dehydrochlorinate DDT, both *in vivo* and *in vitro;* (2) detoxication of compounds such as 1,1-dianisyl neopentane, possibly via microsomal hydroxylation reactions; (3) excretion of unchanged insecticide and metabolism to acidic metabolites, such as occur with Prolan; (4) resistance to cyclodiene compounds—a resistance mechanism largely undetermined; and (5) ability to retain within the tissues without harmful effects large quantities of nondehydrochlorinatable o-chloro-DDT which, otherwise, are fatal to many other strains, both resistant and susceptible.

It is not known whether this dynamic resistance pattern is characteristic of Prolan resistance in general or is peculiar to this particular strain of house flies.

3. Resistance Due to Storage of Unchanged Insecticide or of Toxic Metabolites

Metabolic Fate of the Cyclodiene Insecticides. i. Metabolism in Vivo. Early investigations on the fate of cyclodiene compounds in insects demonstrated the biotransformation of heptachlor, aldrin, and isodrin to their epoxides, i.e., heptachlor epoxide, dieldrin, and endrin, respectively. Heptachlor epoxide, isolated and characterized from heptachlor-treated resistant house flies (Perry *et al.*, 1958), was found to be identical with the authentic compound, melting point of 159°–160.5°C, in colorimetric and infrared spectra, as well as other chemical and biological properties. Heptachlor epoxide is not metabolized further and can be recovered quantitatively when applied to either resistant or susceptible flies.

Susceptible house flies also convert heptachlor to the epoxide. The onset of symptoms of poisoning, after a latent period of 1–2 hours, coincides with the appearance of heptachlor epoxide in the tissues (Perry *et al.*, 1958). This observation led to the inference that transformation of the parent compound is an activation process not involving a detoxication mechanism. In spite of this "autointoxication" phenomenon, the resistant fly emerges unharmed. Similarly, a latent period is observed with regard to a rise in O_2 consumption in heptachlor-injected cockroaches (Harvey and Brown, 1951), presumably coinciding with the appearance of the toxic epoxide. The above hypothesis is not totally shared by other workers (Brooks, 1966; Brooks and Harrison, 1967a; Brooks et al., 1963) who showed that dihydroheptachlor which cannot be epoxidized has, nevertheless, some intrinsic toxicity of its own.

Epoxidation of heptachlor also occurs in mammals (Davidow and Radomski, 1953; Radomski and Davidow, 1953; Harris et al., 1956), in heptachlor-treated plants and soils (Bollen et al., 1958; Gannon and Bigger, 1958; Gannon and Decker, 1958; Lichtenstein and Schulz, 1959, 1960), and microorganisms (Korte, 1967; Poonwalla and Korte, 1968; Miles et al., 1969). In addition to heptachlor epoxide, soil microorganisms (Miles et al., 1969) and house flies (Brooks, 1966; Brooks and Harrison, 1964b, 1965) produce 1-hydroxychlordene, chlordene, chlordene epoxide, 1-hydroxy-2,3-epoxychlordene and one unknown metabolite, and a major metabolic product of heptachlor epoxide isolated from rat feces proved to be 1-hydroxy-2,3-epoxychlordene (Matsumura and Nelson, 1971). A schematic presentation of heptachlor metabolism in various organisms is shown in Fig. 7.

Aldrin and dieldrin have been widely used as broad-spectrum insecticides for many years, yet little is known about their mode of action in insects. More disheartening is the fact that many agricultural pests and insect vectors of disease have developed high levels of physiological resistance to these insecticides but the mechanism of this resistance is still elusive.

Metabolism of aldrin and isodrin follows much the same pattern as that

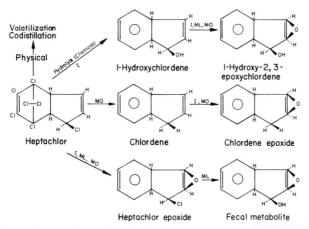

Fig. 7. Metabolism of heptachlor by insects (I), mammals (ML), and microorganisms (MO). (Adapted from Brooks, 1966; Brooks and Harrison, 1964b, 1965; Miles et al., 1969; Perry et al., 1958.)

of heptachlor. Injection of aldrin and isodrin into the American cockroach results in their partial conversion to dieldrin and endrin, respectively (Giannoti *et al.*, 1956; Giannoti, 1958), the reaction occurring chiefly in the digestive tract with the final disposition of the epoxides in various tissues, especially fat-body.

[^{14}C]Aldrin is rapidly metabolized to [^{14}C]dieldrin by S and R house flies and the onset of symptoms of poisoning, after a latent period of 2–3 hours, appears to coincide with the accumulation of dieldrin at critical sites (Perry, 1961; Perry *et al.*, 1964). However, such compounds as dihydroaldrin and dihydroisodrin which cannot be epoxidized are, nevertheless, somewhat toxic to the house fly (Brooks, 1960, 1966) but much less that aldrin and isodrin. Too, blocking the oxidation of aldrin with the synergist sesamex (Sun and Johnson, 1960) reduces the toxicity of aldrin, but not completely. These findings suggest that aldrin-type compounds might possess intrinsic toxicity independent of their epoxidation.

Aldrin epoxidation also takes place in mosquito larvae *A. aegypti* (Korte *et al.*, 1962) and *A. quadrimaculatus* (Bowman *et al.*, 1964), and in the locust *Schistocerca gregaria* (Cohen and Smith, 1961).

In the house fly, there is no evidence that the epoxides, dieldrin and endrin, are further metabolized during periods of up to 4 days after application of the toxicant (Brooks, 1960; Earle, 1963; Perry *et al.*, 1964). Similarly, there is no evidence of enhanced metabolism or excretion of dieldrin in resistant stable flies *S. calcitrans* (Mount *et al.*, 1966). In the locust, [^{36}Cl]aldrin is converted to [^{36}Cl]dieldrin at a very slow rate but no further metabolism occurs (Cohen and Smith, 1961). However, [^{36}Cl]dieldrin is slowly excreted unchanged along with small amounts of water-soluble metabolites. These findings illustrate the extreme biological stability of toxic cyclodiene epoxides in insect tissue.

There is also evidence of the production of polar metabolites of cyclodiene insecticides in other insects. Korte *et al.* (1962) and Korte and Stiasni (1964) demonstrated the presence of several hydrophilic metabolites of [^{14}C]aldrin, [^{14}C]dieldrin, and [^{14}C]telodrin in larvae of *A. aegypti*. The extensive work of Korte and associates, covering several years of investigation on the metabolism of aldrin, dieldrin, endrin, heptachlor, dihydroheptachlor, and chlordane in insects, mammals, and microorganisms was summarized by Korte (1967). In general, most organisms examined, including mosquito larvae, metabolized to some extent all the above compounds to hydrophilic metabolites, but there is no evidence that these biotransformations are in any way related to resistance.

Production of hydrophilic metabolites is also characteristic of resistant *C.p. quinquefasciatus* (Oonithan and Miskus, 1964). In this instance, treatment of adult mosquitoes with [^{14}C]dieldrin yielded an excretory product resembling 6,7-dihydroxydihydroaldrin (apparently *cis*-aldrin glycol) in its chromatographic behavior. Unfortunately, no comparison was made with a corresponding susceptible strain. Authentication of the dieldrin metabolite as *trans*-6,7-dihydroxydihydroaldrin was made by Ludwig and Korte (1965) and Tomlin (1968).

As mentioned earlier, symptoms of aldrin poisoning in the susceptible fly are not manifest until approximately 2 hours after application of the toxicant. The next few hours are characterized by rapid knockdown, at which time the insects exhibit an exhaustive burst of respiratory activity and a decrease in tissue α-glycerophosphate (Winteringham and Harrison, 1959). Therefore, the resistant insect must possess an efficient defense mechanism, not governed by detoxication or storage, which protects it during the initial critical stages of poisoning. Evidence for this comes from investigations on the vapor toxicity of aldrin (Earle, 1963) designed to bypass cuticular penetration and other barriers. The extreme tolerance of the resistant strain to such vapors might indicate that the protective mechanism resides at the site of action of the toxicant.

The site of action of the cyclodiene compounds may involve a decreased sensitivity to dieldrin of the thoracic ganglia of the resistant insect, or nerve components of the resistant strain might have less binding capacity for dieldrin than their susceptible counterparts (cf. Section III). Alternatively, perhaps one of the metabolites produced from cyclodiene degradation, such as 6,7-dihydroxydihydroaldrin, might be the neurotoxic agent, since the latter can poison the nerve cord *in situ* much more rapidly than does dieldrin (Wang and Matsumura, 1970).

A number of investigators sought differences between R and S strains of insects relevant to the resistance mechanism. Gerolt (1965) found no difference in translocation of sublethal amounts of dieldrin in R and S house fly body parts. Adult house flies and *A. aegypti* larvae converted 10% and 5%, respectively, of a sublethal dose of dieldrin to unidentified metabolites, but there was no difference between R and S insects in this respect. Similarly, the penetration, distribution, and metabolism of [14C]aldrin (Perry *et al.*, 1964) and the metabolism of isodrin (Brooks, 1960) was found to be the same in R and S house flies. Direct measurement of dieldrin penetration through the nerve sheath showed no significant differences between R and S cockroaches (Ray, 1963) and, in general, there were no large differences in rates of uptake of [14C]dieldrin by the central nervous systems of R and S house flies following injection or infusion (Schaefer and Sun, 1967). Earlier, a noninsecticidal [82]Br-labeled analogue of aldrin was found to be excreted unchanged in equal proportion by R and S house flies, and a ^{35}S-labeled analogue of dieldrin was partially metabolized to unknown derivatives by both strains (Winteringham and Harrison, 1959).

The distribution, metabolism, and localization of [^{14}C]dieldrin is practically the same in R and S house flies, except for a somewhat slower penetration of the toxicant in the R strain initially (Sellers, 1971). Although metabolism and excretion are slow in both strains, degradation is more rapid and greater quantities of water-soluble metabolites are produced in R flies. These factors alone cannot account for the dynamic resistance; hence, an unidentified mechanism must be operative in the R strain.

Brooks and Harrison (1964a,b, 1965, 1966, 1967a,b) and Brooks *et al.* (1970) extended their work on the metabolism of cyclodiene insecticides to include more than 50 related, moderately toxic cyclodiene compounds to which dieldrin-R flies show complete cross-resistance. It was found that certain compounds which are mildly toxic to S flies are markedly synergized by sesamex but the synergist has little effect on the toxicity to R flies even though sesamex inhibits the metabolism of these compounds to the same extent in both strains. The synergistic effect of sesamex implies the inhibition of oxidative metabolism of these compounds as was earlier postulated by Sun and Johnson (1960). Obviously, oxidative detoxication of these compounds is not related to the resistance mechanism resulting from selection of house flies with dieldrin since sesamex does not enhance their toxicity to the resistant strain.

The metabolic reactions occurring with aldrin and dieldrin are shown in Fig. 8. The numerous other transformations of the above compounds both *in vivo* and *in vitro* as well as their toxicities to the house fly have been

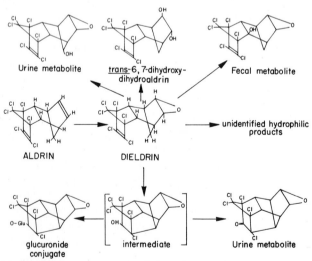

Fig. 8. Metabolism of aldrin and dieldrin by insects and mammals. (Adapted from Brooks, 1966; Brooks *et al.*, 1970; Sellers, 1971).

well documented and summarized by Brooks (1966, 1969) ; Soloway (1965) reviewed and discussed the correlation between biological activity and molecular structure of the cyclodiene insecticides.

ii. Metabolism in Vitro. Metabolism of the cyclodiene insecticides *in vitro* is mediated by the microsomal mixed-function oxidase enzyme system requiring NADPH (reduced nicotinamide adenine dinucleotide phosphate) and atmospheric oxygen.

Microsomal epoxidation of cyclodiene compounds has been investigated in several insects. Ray (1967) showed the epoxidation of aldrin by house fly microsomes and its inhibition by carbon monoxide. Lewis *et al.* (1967) studied the relationship between microsomal epoxidation of aldrin and lipid peroxidation and showed that the house fly contains an endogenous inhibitor of lipid peroxidation. Khan and Terriere (1968) and Khan (1969) examined the *in vitro* epoxidation in various genetically distinct strains of house flies and found that the same enzyme system also hydroxylates naphthalene, as was determined earlier by Schonbrod *et al.* (1968). Both reactions are carried by genes located on the second chromosome which confer resistance to a variety of insecticides.

Aldrin epoxidase, in the southern armyworm *Prodenia eridania* (Krieger and Wilkinson, 1969), in the corn earworm *Heliothis zea,* and the polyphemus moth *Antherea polyphemus,* occurs chiefly in the alimentary canal and fat-body, whereas in the house fly it occurs mainly in the abdomen. In sawfly larvae *Macremphytus varianus,* aldrin epoxidase and dihydroisodrin hydroxylase occur mostly in gut tissue in the process of active feeding, while no activity is found in fully mature larvae (Krieger *et al.,* 1970).

Among larvae of several strains of five lepidopterous species, i.e., the corn earworm, *H. zea,* tobacco budworm, *H. virescens,* pink bollworm, *Pectinophora gossypiella,* European corn borer, *Ostrinia nubilalis,* and the redbanded leaf roller, *A. velutinana,* aldrin epoxidation was highest in *H. zea* and *H. virescens* (Williamson and Schechter, 1970). There was no difference in epoxidase activity between diapausing and nondiapausing larvae. Comparative rates of aldrin epoxidation by microsomal enzymes of various insect species are shown in Table VI.

The extensive investigations by Brooks and associates on the oxidative metabolism of aldrin, isodrin, dihydroaldrin, dihydroheptachlor, and a number of related, moderately toxic, cyclodiene compounds have been summarized by Brooks (1966, 1969), and the hydration of dieldrin to *trans*-6,7,-dihydroxydihydroaldrin by the house fly has recently been reported (Brooks *et al.,* 1970) (cf. Fig. 8). The epoxide hydrase enzymes are also present in the microsomes but require no NADPH or oxgen for activation.

TABLE VI

Comparative Rates of Microsomal Aldrin Epoxidase Activity in
Various Insect Species

Enzyme source	Dieldrin formed (pmoles/min/mg protein)	Reference
Insects		
Musca domestica (whole fly)	245, 41, 197	*a, b, c*
Musca domestica (female abdomen) (S)[j]	83	*d*
Musca domestica (female abdomen) (R)[j]	713	*d*
Phormia regina (whole fly)	3	*b*
Prodenia eridania	2145	*e*
Panthea furcilla	513	*f*
Trichoplusia ni	42	*f*
Pseudaletia unipuncta	26	*f*
Danaus plexippus	8	*f*
Antheraea pernyi	147	*f*
Hyalophora cecropia	82	*f*
Acheta domesticus (male)[k]	4	*g*
Acheta domesticus (female)[k]	12	*g*
Heliothis virescens (field strain)	1090	*h*
Heliothis virescens (lab. strains)	250, 180	*h*
Ostrinia nubilalis	50	*h*
Argyrotaenia velutinana	40	*h*
Pectinophora gossypiella	20	*h*
Other species		
Rattus norvegicus (male rat liver)	84, 274	*b, i*
(female rat liver)	21	*b*
Lepus cuniculus (male rabbit liver)	344	*i*
Coturnix coturnix (male quail)	108	*b*
(female quail)	9	*b*
Salmo gairdneri (rainbow trout)	6	*b*

[a] Ray (1967).
[b] Chan *et al.* (1967).
[c] Brooks and Harrison (1969).
[d] Perry *et al.* (1971).
[e] Krieger and Wilkinson (1969).
[f] Krieger *et al.* (Cited by Hollingworth, 1971).
[g] Benke and Wilkinson (1971).
[h] Williamson and Schechter (1970).
[i] Nakatsugawa *et al.* (1965).
[j] S, Susceptible; R, resistant.
[k] Microsomes obtained from Malpighian tubules.

iii. Photoalteration of Cyclodiene Compounds. In addition to metabolism by living organisms and by enzymic reactions, several of the cyclodiene compounds undergo photolytic transformation to birdcage and half-cage structures in nature or are decomposed by irradiation at 253.7 nm (Mitchell, 1961; Roburn, 1963). Solar irradiation of aldrin and dieldrin yields the photoisomers, photoaldrin and photodieldrin, respectively (Fig. 9; Robinson *et al.*, 1966; Rosen *et al.*, 1966; Rosen and Sutherland, 1967; Henderson and Crosby, 1968). Photodieldrin is more toxic than dieldrin to albino mice and both photoisomers are more toxic than their parent compounds to susceptible house flies and mosquito larvae (Rosen *et al.*, 1966; Rosen and Sutherland, 1967). The increased toxicity and speed of action of photodieldrin might be due to enhanced penetration to the site of action. There is no difference in toxicity to resistant house flies among aldrin, dieldrin, and their corresponding photoisomers; hence, their modes of action are probably similar.

Aldrin and dieldrin can also be monodechlorinated by the action of UV light (cf. Fig. 9) (Henderson and Crosby, 1967).

The metabolism of photoaldrin and photodieldrin by house flies and mosquito larvae *A. aegypti* (Khan *et al.*, 1969) showed the presence of a metabolite, photodieldrin ketone (Fig. 10), which is identical with that found in male mice (Klein *et al.*, 1968) following administration of dieldrin. Photodieldrin ketone was not present in tissue extracts of aldrin- or dieldrin-treated flies. This metabolite is more toxic to S flies and mosquito larvae

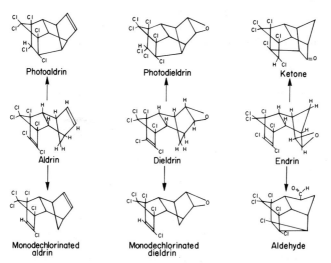

Fig. 9. Photodecomposition products of aldrin, dieldrin, and endrin in insects and mammals. (From Khan *et al.*, 1969; Rosen and Sutherland, 1967; Rosen *et al.*, 1966.)

than either photoaldrin or dieldrin. These results suggest that the greater toxicity of the photoisomers to insects might be due to the rapid formation of photodieldrin ketone which also is more rapid in its toxic action (Klein *et al.*, 1968). Mosquito larvae, however, do not form photodieldrin from photoaldrin. The possibility of a ketone formation (photoaldrin ketone, Fig. 10) has been suggested but not found (Khan *et al.*, 1969).

The cyclodiene insecticide endrin is also converted to photoisomers (Roburn, 1963) by irradiation of the deposited solid at 253.7 nm. Two of these metabolites are a ketone and an aldehyde (Fig. 9; Rosen *et al.*, 1966; Rosen and Sutherland, 1967). Photolysis of endrin in hydrocarbon solvents (Zabik *et al.*, 1971) yields a metabolite identified as the half-cage ketone, 1,8-*exo*-9,11,11-pentachloropentacyclododecan-5-one. This photolytic product of endrin is resistant to chemical degradation and its recent detection in the field might play a role in the buildup of residues in the environment. The toxicity of this compound has not been determined.

It is clear from the foregoing discussion that insect resistance to the cyclodiene compounds is independent of detoxication mechanisms whether oxidative, epoxidative, or hydrolytic.

The fact that dieldrin can bind with various components of insect nerve, and that particulate components of resistant cockroaches have less binding capacity with dieldrin than their susceptible counterparts (Matsumura and Hayashi, 1966a, 1970) supports the view that this binding property of the nervous system might be related to the poisoning effect and to the resistance mechanism. Deeper probing into nervous tissue of various insects and expanding our knowledge of insect neuropharmacology might yield fruitful results in understanding the mode of action and the mechanism of resistance of the cyclodiene insecticides.

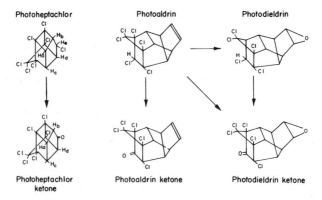

Fig. 10. Metabolism of photoaldrin, photodieldrin, and photoheptachlor by insects, mammals and UV light. (From Khan *et al.*, 1969; Klein *et al.*, 1968.)

C. BIOCHEMISTRY OF INSECT RESISTANCE TO ORGANOPHOSPHORUS INSECTICIDES

The introduction of organophosphorus (OP) insecticides for control of insect pests had entertained the hope of many entomologists that the resistance problem generated by the CH insecticides would be overcome. But history has a habit of repeating itself and, despite all our hopes to the contrary, resistance to many of the OP compounds developed rapidly in several insect species.

The great achievements in this field of organic chemistry and the dramatic consequences to the science of entomology and pest control stem largely from our more fundamental understanding of the mode of action of these compounds *vis-à-vis* the chlorohydrocarbon insecticides.

To be effective as an insecticide, an OP compound must first possess sufficient stability and suitable physicochemical properties to be absorbed and transported to the site of action. A second prerequisite is that the compound must have nucleophilic properties to act as a reactive phosphorylating agent. When these conditions are met, the OP compound will, at the site of action, bind to and inhibit a vital enzyme, cholinesterase (Casida, 1956; Fukuto, 1957; Metcalf, 1955b, 1959; O'Brien, 1960, 1967; Spencer and O'Brien, 1957) although other esterases have been implicated in the poisoning process (Casida, 1955; Hopf, 1954; Lord and Potter, 1950, 1951; Metcalf *et al.,* 1956; Oppenoorth and van Asperen, 1960, 1961; Stegwee, 1959; van Asperen and Oppenoorth, 1959, 1960), and other modes of action have been suggested (Chadwick, 1963; Mengle and Casida, 1960; Staudenmayer, 1955).

In contrast to resistance to chlorohydrocarbon (CH) insecticides, OP resistance is biochemically limited to levels peculiar to each particular insecticide, beyond which the resistance is not ordinarily augmented by further selection with the same compound. It is more specific and cross-tolerance to other OP compounds varies considerably from species to species and within strains of a single species. Furthermore, OP resistance is relatively unstable, declining rapidly following the discontinuance of selection pressure (March, 1959, 1960).

Although various insect species, notably the house fly, develop resistance to most of the OP compounds now in general use, the resistance spectrum varies greatly from strain to strain (Bell, 1968a; Busvine, 1959; Forgash and Hansens, 1962; Mengle and Casida, 1960; Oppenoorth, 1959; van den Heuvel and Cochran, 1965). Such varied patterns, along with genetic studies (Bell, 1968b; Brown and Pal, 1971; Helle, 1962; Kalra, 1970b; Kerr, 1970; Plapp and Hoyer, 1967; Tsukamoto *et al.,* 1968; Wright and Pal, 1967), indicate the presence of more than one resistance mechanism. The picture is further complicated by the fact that selection with OP compounds results, in many instances, in cross-resistance to carbamate insecticides whose mode of action is similar to that of OP compounds, and to certain CH insecticides whose mode of action (although largely undetermined) is quite dissimilar. The reverse effect, i.e., selection with CH insecticides, does not ordinarily induce resistance to OP compounds (Brown, 1958), although recent investigations indicate that where genes for oxidative metabolism of DDT and pyrethrins are present, increased tolerance to OP and carbamate insecticides is quite common (Fine, 1963; Khan and Terriere, 1968; Plapp, 1970a).

Physiological resistance to OP compounds is characterized by the degree of inhibition of cholinesterase and/or aliesterase enzymes, and by differences in rates of activation and detoxication of the OP compound. In some instances, OP resistance has been correlated with behavioristic avoidance of the insecticide, such as avoidance of malathion baits but not trichlorfon or dichlorvos baits (Kilpatrick and Schoof, 1958; Fay et al., 1958; Schmidt and LaBrecque, 1959), with lipid content of tarsi and thoracic ganglia (Reiff, 1956a) or total body lipid (Bennett and Thomas, 1963), and with rate of cuticular penetration of the OP compound (Busvine, 1957; Farnham et al., 1965; Forgash et al., 1962; Gwiazda and Lord, 1967; Hollingworth et al., 1967; Krueger et al., 1960; Matsumura and Brown, 1963a; Plapp and Hoyer, 1968).

In the living insect, the insecticide is subject to the interaction of many or all of the above mechanisms, i.e., penetration, tissue distribution, storage, activation, detoxication, and excretion. In general, death or survival of the organism will depend upon the extent of these interrelationships and the contribution of each factor to the total defense mechanism (March, 1960). However, in some instances, survival may be determined by a balance between rates of activation and detoxication, insensitivity of the OP receptor at the site of action (Mengle and Casida, 1960; Smissaert, 1964), or the production of altered specific detoxifying enzymes (van Asperen and Oppenoorth, 1959).

Metabolic Fate of OP Insecticides. The metabolism of most OP compounds follows two general types of reactivity: (a) metabolic activation of slightly or moderately toxic compounds to more active anticholinesterases, and (b) metabolic degradation leading, in most cases, to relatively innocuous derivatives. Generally, activation reactions are oxidative in nature and are catalyzed by microsomal mixed-function oxidases which reside in the endoplasmic reticulum of the cell and require NADPH and O_2 for their activity. Detoxication mechanisms may involve both oxidative reactions catalyzed by microsomal mixed-function oxidases, and hydrolytic pathways mediated by esterases of which the commonest types are the phosphatases and carboxylesterases (aliesterases).

Metabolic activation is accomplished by desulfuration of esters containing P=S groupings to form the corresponding P=O derivatives, such as occur in the conversion of malathion, parathion, methyl parathion, diazinon, coumaphos, dimethoate, Imidan, azinphosmethyl, fenthion, etc., to their corresponding phosphates (oxon derivatives); oxidation of aliphatic or aromatic thioether

$$\text{—S— to sulfoxide —}\overset{\displaystyle O}{\underset{}{\overset{\|}{S}}}\text{— and sulfone —}\overset{\displaystyle O}{\underset{\underset{\displaystyle O}{\|}}{\overset{\|}{S}}}\text{— as in demeton, disulfoton,}$$

carbophenothion, phorate, etc.

(although in some instances thioether oxidation may result in products of lower toxicity), and N-oxidation of phosphoramides as in the weakly cholinergic compounds schradan and dimefox.

Other types of activation also occur. For example, trichlorfon is converted to the more active anticholinesterase dichlorvos by dehydrochlorination and rearrangement, a process which does not involve the enzyme DDT dehydrochlorinase (Spencer and O'Brien, 1957). However, trichlorfon is enzymatically converted to dichlorvos by the digestive juice of the silkworm under a broad pH range, even at a neutral pH (Sugiyama and Shigematsu, 1969).

Metabolic degradation is brought about via several pathways including transalkylation by soluble enzymes, microsomal oxidative detoxication, microsomal S-oxidation and N-dealkylation, and hydrolytic mechanisms catalyzed by phosphatases and carboxylesterases. A schematic diagram depicting possible routes of metabolism of a hypothetical OP compound is shown in Fig. 11.

1. Types of Attack

Activation mechanisms and metabolic pathways mediated by the mixed-function oxidase system are described in detail in Chapter 10 of Volume V. It suffices to say that activation by desulfuration converts the phosphorothioate to a more active anticholinesterase compound which imparts considerably greater toxicity to the molecule (O'Brien, 1960). In the absence of degradation of the *oxon* derivative or of the parent compound, insect susceptibility to the OP insecticide remains high. For example, house fly homogenates show some conversion of acethion to acetoxon, but little or no degradative attack. This is reflected in the high susceptibility of the house fly to this compound ($LD_{50} = 9.4$ $\mu g/gm$). Cockroach gut, on the other hand, is very effective in hydrolyzing acethion; hence, its considerable

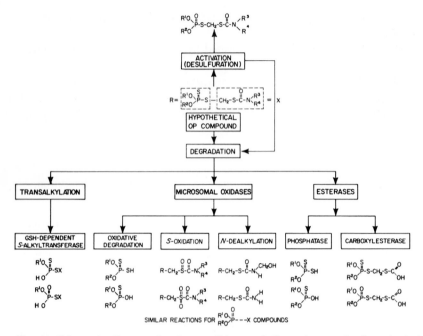

Fig. 11. Schematic diagram showing possible metabolic pathways of a hypothetical organophosphorus compound.

resistance ($LD_{50} = 375$ μg/gm) (Krueger *et al.,* 1960). The principal metabolite produced is acethion acid (O'Brien *et al.,* 1958). Similarly, the selective action of coumaphos toward insects is due to its high activation rate (O'Brien and Wolfe, 1959) and its low degradative capacity *in vivo* (Vickery and Arthur, 1960). The low toxicity of coumaphos to mammals is due to hydrolytic attack at the P—O coumarinyl link (Krueger *et al.,* 1959) and the rapid excretion of polar metabolites (Lindquist *et al.,* 1958).

The *in vivo* activation of parathion to paraoxon has been demonstrated by Metcalf and March (1953). Investigations *in vitro* with microsomal preparations from the rice stem borer (Fukami and Shishido, 1963a), cockroach fat-body (Nakatsugawa and Dahm, 1965; Vardanis and Crawford, 1964), and house fly abdomens (El Bashir and Oppenoorth, 1969) showed that desulfuration of parathion (Fig. 12) is accomplished by a mixed-function oxidase. The sulfur atom which is removed in the *in vitro* reaction is bound to the microsomes; *in vivo* it is found as inorganic sulfate (Nakatsugawa and Dahm, 1967; Nakatsugawa *et al.,* 1969).

DESULFURATION

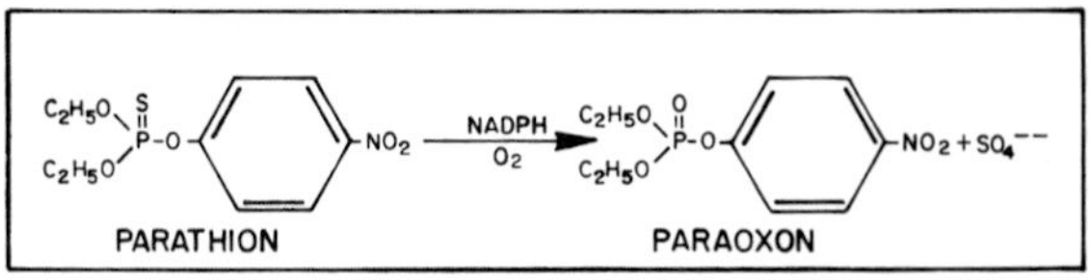

Fig. 12. Activation. Desulfuration of parathion. (From Nakatsugawa *et al.*, 1969.)

In common with other microsomal oxidases, the activating enzyme exhibits a high degree of nonspecificity and can catalyze the desulfuration of a great number of phosphorothioate insecticides (Nakatsugawa *et al.*, 1968).

2. Degradation

a. Transalkylation. Conjugation reactions utilizing glutathione are believed to be an important mechanism of detoxication of organophosphorus compounds since they yield primary metabolites. The conversion of methyl parathion, methyl paraoxon, and fenitrothion to their respective demethyl derivatives was shown to reside in the supernatant fraction and to require reduced glutathione for activation (Fukami and Shishido, 1963b, 1966; Shishido and Fukami, 1963, Fukunaga *et al.*, 1969), while deethylation of ethyl parathion proceeds much slower in both insect and mammal. However, Nolan and O'Brien (1970) made an unexpected observation that deethylation of parathion was a major metabolic pathway in the *susceptible* rather than in the resistant house fly. Enhanced dealkylation has been implicated in the mechanism of resistance to parathion and paraoxon in the rice stem borer (Kojima *et al.*, 1963), and Fukunaga *et al.* (1969) described the glutathione transferase activity in the horn beetle and silkworm larvae. Other OP compounds also undergo dealkylation reactions. Thus, *O*-dealkylation occurs with dicrotophos (Bull and Lindquist, 1966), mevinphos (Morello *et al.*, 1968), fenitrothion and methyl paraoxon (Hollingworth, 1969), dimethoate (Morikawa and Saito, 1966), bromophos (Stenersen, 1969), and diazinon (Folsom *et al.*, 1970; Lewis, 1969; Yang *et al.*, 1971b).

In the house fly, demethylation of methyl parathion and fenitrothion is an important detoxication mechanism and proceeds faster in the R strain, but its relative importance increases only slightly with an increase in dosage, apparently due to saturation of other detoxifying reactions (Hollingworth *et al.*, 1967). Conceivably, many more OP compounds can undergo dealkylation reactions of this type, but a purposeful search for this degradative process has not been applied in many instances.

The examples cited above indicate that transalkylation reactions require reduced glutathione (Fig. 13A), that the enzyme resides in the soluble frac-

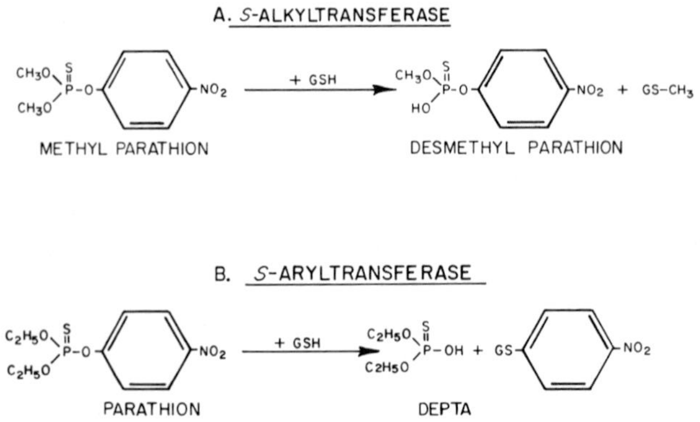

A. S-ALKYLTRANSFERASE

METHYL PARATHION   DESMETHYL PARATHION

B.   S-ARYLTRANSFERASE

PARATHION   DEPTA

Fig. 13. O-Dealkylation. (A) Dealkylation of methyl parathion. (From Fukami and Shishido, 1963b, 1966.) (B) Dearylation of parathion. DEPTA, diethylphosphorothioic acid. (From Dahm, 1970.)

tion and, most likely, belongs to the group of glutathione S-alkyl transferases (GSAT). The substrate specificity for this enzyme favors dimethyl esters of both phosphorothioates and phosphates and yields S-alkyl glutathione and mono-O-dealkylated derivatives.

Glutathione transferase reactions are also responsible for the transfer of aryl groups (Fig. 13B). Dahm (1970) reports that the P—O—aryl bond of parathion is cleaved by a soluble enzyme requiring glutathione. At low doses of injected [^{3}H]paraoxon, Nolan and O'Brien (1970) could detect small differences between S and R house flies in the time course of degradation and this was attributed to increased desarylation in the R strain. The production of diethylphosphoric acid and diethylphosphorothioic acid from diazinon in the presence of GSH and the soluble fraction (Yang et al., 1971b) also indicates the possible involvement of an S-aryltransferase in addition to microsomal oxidation to the same breakdown products (Fig. 14).

b. Oxidative N-Dealkylation. Elucidation of this type of oxidative dealkylation was first demonstrated for dicrotophos and monocrotophos by Bull and Lindquist (1964, 1966) and Menzer and Casida (1965). Dicrotophos undergoes oxidation to form the N-hydroxymethyl derivative with subsequent loss of the hydroxymethyl group to yield monocrotophos. The latter is oxidized to its N-hydroxymethyl derivative and loss of this group yields the unsubstituted amide analogue (Fig. 15). Progressive removal of methyl groups results in a progressive increase in toxicity to house flies and mice. Phosphamidon and dimethoate undergo similar dealkylation processes (Menzer and Dauterman, 1970), but no dealkylation products could be detected in dimethoate metabolism by the boll weevil and bollworm (Bull et

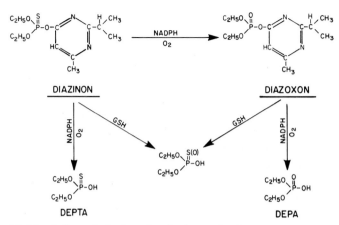

Fig. 14. Metabolism of diazinon by the house fly. DEPTA, diethylphosphorothioic acid; DEPA, diethylphosphoric acid. (From Yang *et al., 1971b.*)

al., 1963). *N*-Demethylation of famphur has been shown in cockroaches and milkweed bugs (O'Brien *et al.,* 1965), but the slow rate of degradation in the latter is well compensated by the relative insensitivity of its cholinesterase.

c. Oxidative O-Dealkylation and Dearylation. Organophosphorus triesters can undergo cleavage to diesters with the formation of relatively nontoxic metabolites. The exact site of cleavage of alkylaryl phosphates and phosphorothioates is important from the standpoint of comparative toxicity to

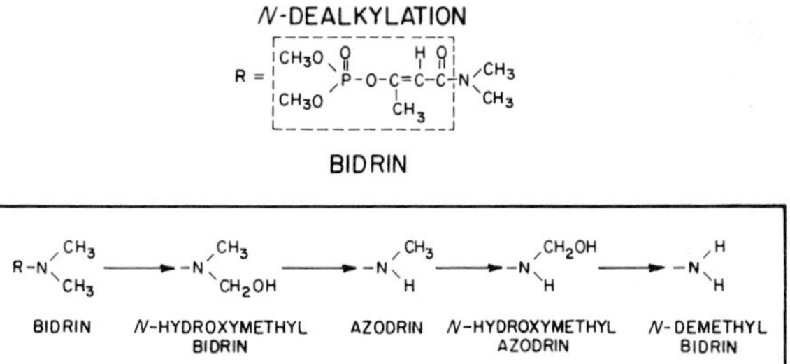

N-DEALKYLATION

$$R = \begin{bmatrix} CH_3O \\ \diagdown P-O-C=C-C-N \diagup CH_3 \\ CH_3O \diagup \quad CH_3 \quad \diagdown CH_3 \end{bmatrix}$$

BIDRIN

	BIDRIN	N-HYDROXYMETHYL BIDRIN	AZODRIN	N-HYDROXYMETHYL AZODRIN	N-DEMETHYL BIDRIN
	R-N⟨CH₃/CH₃	→ -N⟨CH₃/CH₂OH	→ -N⟨CH₃/H	→ -N⟨CH₂OH/H	→ -N⟨H/H
TOPICAL LD₅₀ mg/kg ♀ HOUSEFLY					
− sesamex	38	14	6.4	30	1.0
+ sesamex	1.0	1.2	0.8	3.4	0.9
I.P. LD₅₀ mg/kg ♀ Mouse	14	18	8	12	3

Fig. 15. N-Dealkylation. Metabolism and biological activity of Bidrin (dicrotophos), Azodrin (monocrotophos), and their metabolites. (From Menzer and Casida, 1965.)

mammals and insects. Since the isolation and identification by Plapp and Casida (1958) of metabolites resulting from cleavage at the alkyl phosphate bond, these reactions have been regarded as being hydrolytic and catalyzed by phosphatase enzymes. However, it is now recognized that such reactions may be catalyzed by mixed-function oxidase enzymes yielding O-dealkylated products. For example, diazoxon is deethylated by a microsomal preparation from resistant house flies in the presence of NADPH and O_2 (Lewis, 1969). Diazinon is not dealkylated, indicating that phosphates might be more susceptible to attack than phosphorothioates. On the other hand, parathion, but not paraoxon is dearylated by microsomal preparations from cockroach fat-body (Fig. 13B) yielding diethyl phosphorothioate and p-nitrophenol (Nakatsugawa and Dahm, 1967). This reaction is inhibited by piperonyl butoxide, giving supporting evidence of the involvement of the microsomal enzyme system.

Similarly, microsomal preparations from rat liver oxidatively dearylate parathion (Neal, 1967a,b) and diazinon, but not diazoxon (Yang *et al.,* 1971a) whereas house fly microsomes can dearylate both diazinon and diazoxon (Yang *et al.,* 1971b), the former with greater ease (Fig. 14).

In vitro studies have also shown increased dearylation as a mechanism of parathion resistance in the house fly (Nolan and O'Brien, 1970), dearylation of parathion and methyl parathion in microsomal preparations of the rice stem borer (Fukunaga *et al.,* 1969), and dearylation of isopropyl parathion by house flies, honey bees, and mice (Camp *et al.,* 1969). The production of free p-nitrophenol from methyl parathion by fat-body microsomes of the American cockroach (Vardanis and Crawford, 1964) most likely is due to oxidative dearylation rather than to hydrolysis as shown in the oxidative metabolism of [^{35}S]parathion to diethylphosphorothioic acid (DEPTA) and its inhibition by sesamex (Nakatsugawa *et al.,* 1969).

The above examples indicate that cleavage of alkyl or aryl bonds by the mixed-function oxidase system is an important detoxication mechanism in insects and mammals.

d. Sulfur Oxidation. The systemic insecticide demeton consists of two isomers, the thiono and thiol isomers. Both isomers are rapidly metabolized by the mouse and the American cockroach. The biochemical mechanism of primary importance from the toxicological standpoint involves the oxidation of the mercaptosulfur of the ethylthioethyl moiety of both isomers, first to the sulfoxide and then to the sulfone. A second important mechanism which involves only the thiono isomer is the oxidation of the thionosulfur to the corresponding phosphate with subsequent oxidation to the sulfoxide and sulfone (March *et al.,* 1955; Fukuto *et al.,* 1955, 1956). A similar pattern occurs with phorate whose chief metabolic products in *P. americana,*

O. fasciatus, and *Rhodnius prolixus* are the sulfoxide and/or the sulfone. However, there is no evidence of desulfuration activity as in demeton (Menn and Hoskins, 1962). In contrast, the southern armyworm *P. eridania* rapidly oxidizes the sulfoxide and sulfone (derived from phorate metabolism by the plant) to their corresponding phosphates (Bowman and Casida, 1957, 1958). The systemic insecticide fensulfothion is metabolized *in vitro* by cockroach gut to the sulfone, *S*-ethyl isomer, and *S*-ethyl isomer sulfone (Benjamini *et al.,* 1959a,b). Disulfoton is metabolized by boll weevils and bollworms to four oxidative products including the sulfoxide and sulfone of disulfoton and disulfoxon (Bull, 1965), but no free disulfoxon could be detected. In addition, nine hydrolytic products are produced.

Fenthion and fenoxon are oxidized to their respective sulfoxides and to a lesser extent to their sulfones (Fig. 16) by the German cockroach, the boll weevil, and the house fly (Brady and Arthur, 1961). Similar products are found in *C.p. quinquefasciatus* (Stone, 1969) and the presence of dimethyl phosphorothioic acid and dimethyl phosphoric acid are indicative of oxidative degradation rather than thionase and oxonase activity. In *S. calcitrans* and *C. lectularius,* fenthion is oxidized at the thiophosphoryl and thioether positions, the latter producing the sulfone as the predominant metabolite (Young and Berger, 1969) with subsequent hydrolysis and excretion as cresol conjugates. Oxidation of the thioether moiety is more rapid following activation of fenthion to fenoxon.

All the examples discussed above are studies of thioether oxidation *in vivo.* As yet, no data are available on the *in vitro* S-oxidation in insects. However, by analogy with similar *in vitro* metabolic pathways with car-

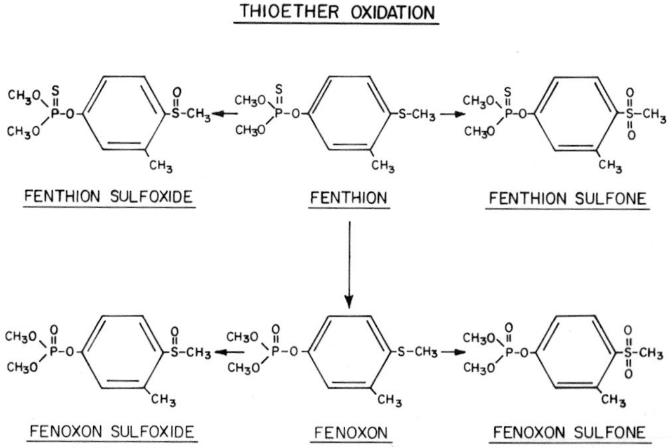

THIOETHER OXIDATION

Fig. 16. Thioether oxidation of fenthion and fenoxon in insects. (From Brady and Arthur, 1961; Stone, 1969.)

bamate insecticides (cf. Section V,D), one can assume that such reactions with OP compounds also are catalyzed by microsomal mixed-function oxidases.

3. Hydrolytic Reactions

a. Phosphatases. Historically, and until recently, degradation of insecticides was thought to occur primarily by hydrolytic routes with the commonest type being attack by phosphatases (Heath, 1961; O'Brien, 1960, 1967). These enzymes hydrolyze the phosphorus ester or the anhydride bond at P—O—C, P—S—C, P—F linkages, etc., and yield hydrolytic metabolites of the leaving group (the side chain). The weight of evidence suggests that phosphates rather than phosphorothioates are the preferred substrates for enzymic hydrolysis.

Since the isolation and identification by Plapp and Casida (1958) of hydrolytic products of ronnel, dicapthon, chlorthion, methyl parathion, parathion, and diazinon from the American cockroach, numerous other investigators have reported such hydrolytic products with other insects and with diverse OP compounds. In many instances, resistance has been attributed to more rapid hydrolysis and excretion of conjugated products of hydrolysis. Hydrolytic metabolites have been shown as products of degradation of fenthion (Brady and Arthur, 1961; Stone, 1969; Stone and Brown, 1969; Young and Berger, 1969), fenitrothion (Hollingworth et al., 1967), parathion and paraoxon (Lord and Solly, 1956; Matsumura and Hogendijk, 1964a; Mengle and Lewallen, 1966; Metcalf et al., 1956, Plapp et al., 1961), monocroptophos and dichrotophos (Bull and Lindquist, 1964, 1966), diazinon (Collins and Forgash, 1970; Farnham et al., 1965; Forgash et al., 1962), dimethoate (Brady and Arthur, 1963; Bull et al., 1963; Uchida et al., 1965; Zayed et al., 1968, 1970), trichlorfon (Arthur and Casida, 1957; Hassan et al., 1965), and famphur (O'Brien et al., 1965).

Hydrolytic degradation based solely on identification of products is no longer a valid assumption since Nakatsugawa et al. (1968) have shown that these products can be obtained by oxidation through the mediation of microsomal enzymes requiring NADPH and O_2. Thus, diethyl phosphorothioate and *p*-nitrophenol are products of oxidative metabolism of parathion rather than products of phosphatase activity. The report of *in vitro* hydrolytic cleavage of [^{32}P]parathion at the P=S bond by homogenates of R and S house flies (Matsumura and Hogendijk, 1964b) could not be confirmed by Nakatsugawa et al. (1969) or by Welling et al. (1971).

Whereas hydrolytic cleavage at the P=S bond is at present questionable, hydrolytic degradation of P=O compounds might be more of a reality. Stemming from the work of Welling et al. (1971), it appears that enhanced hydrolytic action leading to the formation of diethylphosphate and *p*-nitrophenol may account for the resistance to paraoxon in a strain of house flies, in accord with previous reports (Oppenoorth and van Asperen, 1961), but that an additional resistance factor due to increased oxidative activity (El Bashir and Oppenoorth, 1969) may be present.

Diazinon resistance has been the subject of numerous investigations and some claims of enhanced hydrolytic activity as a mechanism of resistance have been previously noted. Investigations at the subcellular level (Lewis, 1969; Lewis and Lord, 1969) indicate the presence of at least three detoxication mechanisms in several, genetically marked, strains of house flies: (1) cleavage of diazinon and diazoxon to diethylphosphorothioic acid and diethylphosphoric acid, respectively—a mechanism which resides in the microsomes and requires NADPH and O_2 (curiously, the addition of glutathione also enhances this reaction). Degradation by this route is significantly higher in the resistant strain and the reaction is inhibited by sesamex, an inhibitor of microsomal oxidases, but not by S,S,S-tributyl phosphorotrithioate, (TBTP) an inhibitor of aliesterases; (2) deethylation (transalkylation) of diazinon and diazoxon occurs entirely in the soluble fraction of strains with gene a (low aliesterase) and requires reduced glutathione for activation, with further decomposition to monoethyl phosphorothionate and monoethyl phosphate. Deethylation is inhibited by TBTP but not by sesamex. This reaction does not occur in rat liver preparations (Yang et al., 1971a); and (3) microsomal degradation of diazoxon but not of diazinon to two unidentified derivatives. This mechanism is also inhibited by sesamex but not by TBTP (Table VII). The overall metabolism of diazinon and diazoxon by the soluble fraction (GSH) and the microsomal fraction (NADPH, O_2) is at least four- to fivefold greater in R than in S strains of house flies (Folsom et al., 1970; Yang et al., 1971b).

It appears that diazinon resistance is a multifacet phenomenon with varied metabolic patterns and related to the genetic background of the insect as shown in Table IX.

A new product of diazinon metabolism, i.e., hydroxydiazinon, has recently been found in kale (Pardue et al., 1970) and in guinea pigs and sheep (Machin and Quick, 1971). This metabolite has not been reported in insects.

b. Carboxylesterases (Aliesterases). It is generally agreed that there are practically no differences between OP-resistant and OP-susceptible insects in their cholinesterase activity or in the sensitivity of this enzyme to inhibition by OP compounds (exceptions will be noted later). Some observations with resistant house flies drew attention to the possible role of aliesterases, such as enzymes hydrolyzing methyl or phenyl butyrate, in OP-resistance (van Asperen, 1958). Subsequent work showed that homogenates of six resistant

TABLE VII

Pathways of Diazinon Metabolism in Relation to Genotype Composition
of Susceptible and Resistant Strains of House Flies[a]

Strain	Degree of resistance	Type of resistance	Located on chromosome	Detoxication mechanism[b]
ocra SRS	1	None (susceptible)	—	A
ac; ar; bwb, ocra SRS	1	None (susceptible)	—	A
29	10–15	Gene a (low aliesterase)	II	A B
393	10–15	Gene a (low aliesterase)	II	A B
466.500	10	Sesamex-inhibited	V	A C
SKA	400	Gene a (low aliesterase)	II	A B C
		+ penetration factor	III	
		+ sesamex-inhibited	V	

[a] From Lewis (1969).

[b] A, cleavage of diazinon and diazoxon to diethylphosphorothioic acid and diethyl-phosphoric acid, occurring in microsomal fraction. B, deethylation of diazinon and di-azoxon (GSH-dependent S-alkyltransferase), occurring in soluble fraction. C, degrada-tion of diazoxon but not of diazinon to unidentified metabolites, occurring in microsomal fraction.

strains of different geographical origin had considerably lower aliesterase activity than their susceptible counterparts (van Asperen and Oppenoorth, 1959). This observation has been corroborated by Bigley and Plapp (1960, 1961), Collins and Forgash (1970), Forgash et al. (1962), Matsumura and Sakai (1968), and others.

There are instances, however, where aliesterase activity is not correlated with resistance in house flies. For instance, Oppenoorth and van Asperen (1961) found an OP-resistant strain with relatively high aliesterase activity, and Franco and Oppenoorth (1962) found a susceptible strain with esterase levels equal to those of the resistant strains. In contrast with previous reports on house flies, lower levels of aliesterase activity are not found in OP-re-sistant strains of either *Culex tarsalis* or *A. aegypti* (Plapp et al., 1965). Aliesterase activity in *D. melanogaster* bears no relation to insecticide re-sistance (Ogita, 1961), and in the green rice leafhopper Hayashi and Hayakawa (1962), Kojima et al. (1963), and Kasai and Ogita (1965) found more aliesterase activity in malathion-R strains than in their S-counterparts. Taken collectively, these and other reports indicate that OP-resistance is not always associated with low aliesterase activity.

Oppenoorth and van Asperen (1960) then proposed the "mutant aliester-ase" theory, which postulates the existence, in many OP-resistant strains of house flies, of a mutant allele a whose function is to produce a modified aliesterase, or OP-hydrolyzing enzyme, replacing the normal aliesterase in the susceptible strain which is under control of the wild-type gene a^+. This theory has been criticized by O'Brien (1966, 1967) on grounds that the rate of detoxication found is insufficient to account for the high degree of resistance, and that enzymatic action was not proven due to lack of identification of the hydrolytic products. In addition, the finding of Stegwee (1959) that the aliesterase of house flies could be inhibited by as much as 81% without causing ill effects strengthens the conclusion that aliesterases are of little consequence in the poisoning by OP compounds.

In spite of the uncertainties concerning the role of these enzymes, carboxylesterases have been shown to be important in the detoxication of OP compounds. In particular, one of the most studied insecticides, malathion, has attracted considerable attention because of its low mammalian toxicity and its high insecticidal activity.

The rates of metabolic attack upon different parts of the molecule vary with different organisms. Thus, the mouse and chicken degrade malathion to several metabolic products, whereas degradation is less extensive in the American cockroach (March et al., 1956). Activation to malaoxon occurs in both insects and mammals, but the balance of activation and hydrolytic activities accounts satisfactorily for the greater toxicity of malathion to the cockroach (O'Brien, 1957; Krueger and O'Brien, 1959).

Degradation of malathion involves at least two major pathways: that involving attack on P—S and S—C bonds, and that concerned with carboxylesterase hydrolysis of the diethyl succinate moiety (Fig. 17). Oxidation to malaoxon proceeds at about equal rates in R and S house flies (March, 1959; Oppenoorth, 1959), but the resistant strain degrades malaoxon more rapidly via phosphatase action. In Culex tarsalis a more important difference between R and S larvae is the much higher carboxylesterase activity of the R strain (Matsumura and Brown, 1961a), a characteristic which is genetically controlled. EPN, a carboxylesterase inhibitor, synergizes malathion against R larvae, thus establishing the importance of carboxylesterase activity. The carboxylesterase of Culex was later partially purified (Matsumura and Brown, 1963a) and found to be a low molecular weight protein of 16,000. Since the malathion-R strain cannot hydrolyze carboxymethyl malathion (which contains $COOCH_3$ in place of $COOC_2H_5$) it may be inferred that the carboxylesterase is highly specific for carboethoxy groups (Dauterman and Matsumura, 1962), a specificity also seen in malathion-resistant German cockroaches (van den Heuvel and Cochran, 1965). Like Culex and house flies, malathion-R cockroaches show little cross-resistance to other OP insecticides.

Detoxication reactions involving carboxylester hydrolysis yield the monoacid derivative of malathion (Cook and Yip, 1958). The enzyme is very

HYDROLYTIC PATHWAYS
PHOSPHATASE-CARBOXYESTERASE

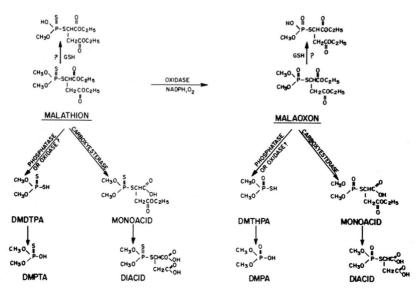

Fig. 17. Carboxyesterase and phosphatase. Metabolic pathways of malathion in insects. DMDTPA, dimethyldithiophosphoric acid; DMPTA, dimethylphosphorothioic acid; DMTHPA, dimethyl phosphorothiolate; DMPA, dimethylphosphoric acid.

active in mammals (Murphy and DuBois, 1957; Seume and O'Brien, 1960) but the activity in insects is lower (Krueger and O'Brien, 1959).

The main interstrain difference between malathion-R and malathion-S house flies (Matsumura and Hogendijk, 1964b), or blow flies *Chrysomya putoria* (Townsend and Busvine, 1969) is the superior ability of the R strains to degrade malathion to the monocarboxylic acid.

Of particular interest is the difference in malathion tolerance in two species of *Dermestes* beetles. *Dermestes maculatus* is much more resistant to malathion than *D. lardarius,* but this resistance does not extend to a number of other OP compounds investigated (Lloyd and Dyte, 1965). It appears that enhanced degradation via carboxylesterase action is the likely mechanism of resistance involved in *D. maculatus* (Dyte *et al.,* 1966) and in *Tribolium castaneum* (Dyte and Rowlands, 1968). In both works, triphenyl phosphate synergized malathion by inhibiting carboxylesterase but not phosphatase activity. Electrophoretic evidence implies the existence of at least two malathion-degrading enzymes, one of which is characteristic of the resistant insect only (Ohkawa *et al.,* 1968), and an arylesterase band has been detected in *A. aegypti* which is not inhibited by paraoxon but instead hydrolyzes it (Ziv and Brown, 1969). This band is more intense in parathion-selected than in malathion-selected strains.

Whereas most studies implicate a higher hydrolytic activity as a cause of malathion resistance, some investigators find it difficult to reconcile the small differences in oxidative activity (Lewallen and Nicholson, 1959) or in hydrolytic breakdown (Darrow and Plapp, 1960; Krueger et al., 1960; Matsumura and Brown, 1963b; Mengle and Casida, 1960) with the high malathion tolerance of the R strains. In *A. aegypti*, for example, resistance to malathion is attributed to decreased penetration and enhanced excretion of the toxicant rather than to increased hydrolysis (Matsumura and Brown, 1961b).

Since malathion is used extensively for control of agricultural pests, it is of interest to note that this insecticide is also metabolized by plants (Koivistoinen, 1961). Isolation of degradation products from plants include the monoacid derivative, indicating the presence of a carboxylesterase enzyme.

c. Amidases. Carboxyamidases also participate in the degradation of OP compounds containing a carboxyamide group, $CONR_2$ (where R is H or an alkyl group), although this type of hydrolytic cleavage has been studied more extensively in animals and plants than in insects. Among the OP compounds which are susceptible to this type of attack in insects, dimethoate takes a leading role.

Three factors appear to be involved in the considerable tolerance of fifth-instar bollworm larvae for dimethoate: (a) large amounts of dimethoate are converted to the oxygen analogue; (b) the latter and the parent compound are rapidly hydrolyzed to nontoxic derivatives; and (c) excretion of metabolic products is very efficient (Bull et al., 1963). The major site of enzymatic attack appears to be at the amide bond yielding thiocarboxy derivatives (*O,O*-dimethyl-*S*-carboxymethyl phosphorodithioate). In addition, cleavage of the S—C bond of either dimethoate or the thiocarboxy derivative yields large amounts of dithioate (*O,O*-dimethylphosphorodithioic acid). Other metabolites of interest are thioate (*O,O*-dimethylphosphorothioic acid) and dimethyl phosphate. The dimethylthiocarboxy derivative (O-methyl-*O*-hydrogen-*S*-carboxymethyl phosphorodithioate) which is a major metabolite in mammals (Chamberlain et al., 1961; Dauterman et al., 1959; Kaplanis et al., 1959; Uchida et al., 1964) is either absent or is detected only in minute quantities in bollworm larvae (Bull et al., 1963). Metabolism in boll weevils is similar to but considerably slower than in the bollworm.

In contrast, a comparative study of dimethoate metabolism in several insects (Uchida et al., 1965) indicates that total hydrolytic degradation does not correlate with sensitivity of the insect to the toxicant. Thus, the house fly, which is the most susceptible among the species tested, degrades dimethoate at the slowest rate, but the same slow rate of degradation takes place in the extremely tolerant milkweed bug. Rate of degradation of dimethoxon (the P=O analogue), as well as the small amounts of carboxyamidase products recovered, also show no correlation with toxicity. The greater susceptibility of the house fly can best be explained in terms of its much greater cholinesterase sensitivity (100-fold) to dimethoate.

Carboxyamidase activity appears to be totally absent in the pattern of dimethoate metabolism by homogenates of the house fly, American cockroach, rice borer, and green peach aphid (Morikawa and Saito, 1966), and degradation by phosphatase action is greatest by cockroach fat-body.

In cotton leafworm larvae, a major site of enzymatic attack on dimethoate occurs at the P—O—alkyl ester bonds followed by hydrolysis of the P—S bond to yield thiophosphoric acid (Zayed et al., 1968). Furthermore, the production of O,O-dimethyl-S-carboxymethyl phosphorodithioate as a metabolite indicates the presence of carboxyamidase activity. In a later study with the cotton leafworm using ^{14}C-labeled dimethoate at both —OCH$_3$ groups or at the —NHCH$_3$ group, Zayed et al. (1970) showed the presence of several oxidative and hydrolytic pathways, as well as products of carboxyamidase activity (Fig. 18). The major polar metabolite resulting from phosphatase attack at the S—C bond is believed to be α-hydroxy-N-methyl acetamide (HO—CH$_2$—CONHCH$_3$) which is excreted as a glucuronide conjugate.

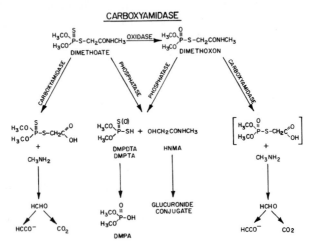

Fig. 18. Carboxyamidase. Metabolism of dimethoate in the cotton leaf worm *Prodenia litura.* DMPDTA, dimethylphophorodithioic acid; DMPTA, dimethylphosphorothioic acid; DMPA, dimethylphosphoric acid; HNMA, α-hydroxy-N-methyl acetamide. (From Zayed *et al.,* 1970.)

An interesting amidase cleavage occurs with Imidan. Mammals, insects, and plants metabolize Imidan *in vivo* via hydrolytic pathways (Fig. 19) principally to phthalamic acid and phthalic acid (Chamberlain, 1965; Ford *et al.*, 1966; McBain *et al.*, 1968; Menn and McBain, 1964). Ring hydroxylation is either absent or present in trace amounts. Imidoxon (the P=O analogue) is a product of microsomal enzyme metabolism of Imidan in the cockroach (McBain *et al.*, 1968) and in the house fly (Tsukamoto and Casida, 1967). Metabolism of the phosphate-bearing moiety has not been followed, but it is likely to involve microsomal oxidase attack yielding O,O-dimethylphosphorodithioic acid.

To recapitulate, it appears that, in general, the mechanisms responsible for insect resistance to OP compounds involve microsomal oxidations, dealkylation through mediation of glutathione, and in some instances degradation by hydrolytic pathways such as phosphatases, carboxylesterases, and amidases. However, hydrolytic reactions are subject to reinvestigation and reinterpretation inasmuch as they may, essentially, correspond to oxidative processes. Nevertheless, taken collectively, metabolic degradation per se does not fully account for the high level of resistance to OP compounds in many insect species. Alternative pathways may be involved, such as rate of penetration of the toxicant through the integument and sensitivity of the OP receptor at the site of action. Penetration is discussed separately in Section VI.

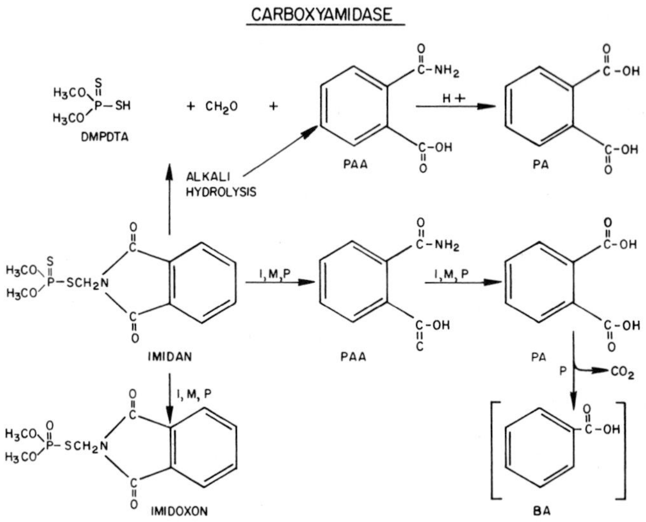

Fig. 19. Carboxyamidase. Hydrolytic cleavage of Imidan by amidases of insects (I), mammals (M), and plants (P). PAA, phthalamic acid; PA, phthalic acid; BA, benzoic acid; DMPDTA, dimethylphosphorodithioic acid. (From McBain *et al.*, 1968.)

4. Sensitivity of the OP Receptor

It is generally agreed that in most insects the cholinesterases of S and R strains do not differ quantitatively or in their intrinsic sensitivity to OP insecticides. There is evidence, however, that in a demeton-resistant strain of the spider mite *Tetranychus urticae* (Leverkusen strain) the cholinesterase is close to 150-fold less sensitive to diazoxon than that of its S counterpart (Smissaert, 1964), and the difference is even greater with paraoxon as inhibitor. These findings were confirmed by Voss and Matsumura (1964), but in another strain (Blauvelt) the resistance of the mite to malathion and parathion was attributed to its superior ability to detoxify the insecticides via phosphatase and carboxyesterase action (Matsumura and Voss, 1964).

Good correlation was found by Zahavi and Tahori (1970) between resistance and insensitivity of cholinesterase to malaoxon, malathion, dichlorvos, and phosphamidon in eight field-collected strains of the carmine spider mite *T. cinnabarinus*. The level of cholinesterase activity was not a factor in this resistance.

Genetic and biochemical studies in two parathion-R New Zealand strains of *T. urticae* and the Leverkusen strain imply that parathion resistance due to lowered cholinesterase sensitivity is governed by a single gene inheritance completely dominant over its normal allele (Ballantyne and Harrison, 1967). The resistance gene appears to be involved in determining a part or the whole structure of the aberrant cholinesterase enzyme which has a decreased sensitivity to OP inhibitors.

The strong OP resistance developed by parathion selection in a susceptible strain of *T. urticae* proved to be due to a completely dominant gene, but the cholinesterase activity of the R strain and its sensitivity to malaoxon inhibition was no different from that of the parent S strain (Herne and Brown, 1969). The selection process resulted in increased detoxication since homogenates of the R strain were considerably more active in converting malathion to water-soluble metabolites. Apparently, the gene for cholinesterase insensitivity was absent in that particular strain of mites (Niagara strain) or else the selecting agent was instrumental in determining the resistance characteristics of the progenies.

The genetic aspects of this resistance has been reviewed by Helle (1965) and the biochemical characteristics of the cholinesterases of R and S spider mites have recently been discussed by Smissaert *et al.* (1970).

Mechanisms of OP resistance in some strains of the cattle tick *Boophilus microplus* also appear to be related to decreased cholinesterase sensitivity. Thus, Lee and Batham (1966) demonstrated that the R strain possessed at least one cholinesterase which reacted more slowly with OP and carbamate compounds than did the cholinesterase of the S strain. This resistant enzyme accounted for 40% of the total activity and its susceptibility to inhibition varied with the type of inhibitor, the greatest difference occurring with the P=O analogue of carbophenothion. A strain of the blue tick which was resistant to arsenic and to chlorohydrocarbon insecticides had normal levels of cholinesterase and the enzyme was fully susceptible to the OP inhibitors.

Roulston *et al.* (1966) demonstrated that a susceptible strain of the blue tick metabolized coumaphos in the usual pattern of oxidation and hydrolysis, and that cholinesterase was the susceptible target. In a similar study with R and S strains treated with coumaphos, diazinon, and dioxathion, Schuntner *et al.* (1968) found no interstain differences in rates of penetration, oxidation to toxic metabolites, or hydrolytic reactions, even though the R strain was 12.4-fold more resistant to diazinon. However, there was a slower *in vivo* response of the R strain cholinesterase to the OP inhibitors, which is consistent with the resistant enzyme hypothesis and parallels the type of cholinesterase insensitivity found in some spider mites.

Tick resistance to insecticides, including a detailed discussion of the biochemistry of tick resistance, has recently been reviewed by Wharton and Roulston (1970).

There also are indications of cholinesterase insensitivity in mammals and insects. For example, in comparing the toxicity and metabolism of paraoxon in the frog, mouse, and cockroach, Potter and O'Brien (1963) concluded that the insensitivity of the frog to paraoxon is due to its insensitive cholinesterase and is not connected with detoxication rates. The milkweed bug which degrades famphur at a very slow rate compared to the cockroach and mouse is compensated by a relative insensitivity of its cholinesterase to famoxon (O'Brien *et al.*, 1965). Similarly, in an investigation of the penetration and metabolism of dimethoate in several insects, Uchida *et al.* (1965) found that the cholinesterase of the milkweed bug was 100-fold less sensitive to the inhibitor than that of the house fly.

D. BIOCHEMISTRY OF INSECT RESISTANCE TO CARBAMATE INSECTICIDES

The carbamate insecticides share with other classes of toxicants the now inescapable fact that, given sufficient time and selection pressure, resistance to these compounds becomes the rule rather than the exception.

Carbamate insecticides differ from OP compounds in that they are competitive rather than irreversible inhibitors of cholinesterase, i.e., the carbamate complexes with the enzyme without necessarily reacting with it chemically (Casida *et al.*, 1960; Kolbezen *et al.*, 1954; Metcalf, 1962; Metcalf and Fukuto, 1965). Evidence is also available that, in some instances, a chemical reaction occurs yielding a carbamylated enzyme (Wilson et al., 1960, 1961; O'Brien et al., 1966).

Most of the carbamates synthesized to date are patterned after the closely related, pharmacologically active compounds physostigmine (eserine) and prostigmine (neostigmine). According to Winton et al. (1958) and Metcalf et al. (1960), the requirements for toxic action of a carbamate are "(a) structural complementarity to acetylcholine, the normal substrate for the enzyme cholinesterase, (b) sufficient stability to hydrolytic attack by cholinesterase to permit the compound to act as a competitive blocking agent for cholinesterase rather than as a substrate, and (c) proper lipoid solubility and absence of a permanent electrical charge to permit penetration into the lipoid sheath surrounding the insect nerve." In general, the stronger the ionization of the carbamate the lower is its insecticidal activity (O'Brien and Matthysse, 1961; O'Brien, 1967) even though the compound might be a potent in vitro cholinesterase inhibitor. In addition to a specific acetylcholinesterase, nonspecific aliphatic esterase enzymes might be involved in the mode of action of, or resistance to the carbamates, as has been demonstrated for the OP compounds (cf. Section V,C).

Many of the commercially available carbamates are highly effective against a variety of resistant insects, but in spite of their effectiveness and their unique mode of action, the carbamates, too, are susceptible to attack by insect enzymes.

In laboratory selection experiments with a variety of carbamates (Georghiou et al., 1961; Meltzer, 1956; Moorefield, 1960; Wiesmann, 1956; Wiesmann and Kocher, 1951; and others) a high or low level of resistance was induced in several generations depending on the insect species and the selecting agent. The apparent discrepancies between in vitro inhibition of cholinesterase and the toxicity of certain compounds, e.g., Pyrolan, to the American cockroach P. americana and the forest maybeetle Melolontha melolontha (Wiesmann and Kocher, 1951), suggest that enzymatic destruction of the carbamate in vivo is responsible for the resistance. Similarly, Georghiou and Metcalf (1961b) reported that 3-isopropylphenyl N-methyl carbamate is more rapidly metabolized by R than by S house flies.

Carbaryl (Sevin) has been one of the carbamates most widely studied in insects beginning with the work of Eldefrawi and Hoskins (1961). Carbaryl-R and -S house flies metabolize [^{14}C]carbaryl to a polar metabolite which appears in the tissues and the excreta. The difference between the two strains is quantitative rather than qualitative. The milkweed bug O. fasciatus, on the other hand, metabolizes carbaryl very slowly, and this is reflected in the high susceptibility of this insect to carbaryl (LD$_{50}$ = 0.5 μg/bug). The German cockroach B. germanica presents a more complicated picture. At least six metabolites of carbaryl are formed, one of which is

1-naphthol (Eldefrawi and Hoskins, 1961). These transformations indicate a common characteristic among the species investigated; hence, the authors conclude that hydrolysis at the esteratic linkage might be the initial step in carbaryl metabolism. In S and R strains of German roaches, Ku and Bishop (1967) obtained almost complete metabolism of carbaryl to 1-naphthol and 1-naphthol conjugates.

More recent studies indicate that oxidation may be the primary pathway of carbaryl metabolism in insects and mammals. American cockroaches and house flies metabolize carbaryl via ring and side-chain hydroxylation reactions yielding 4-hydroxy-, 5-hydroxy-, N-hydroxymethyl-, and 5,6-dihydro-5,6-dihydroxy-1-naphthyl N-methylcarbamate derivatives as illustrated in Fig. 20 (Dorough and Casida, 1964; Leeling and Casida, 1966). 1-Naphthol and 1-hydroxy-5,6-dihydro-5,6-dihydroxynaphthalene are minor hydrolytic products. Some of these metabolites have been detected after topical application or injection of carbaryl to the house fly, stable fly *S. calcitrans,* boll weevil *A. grandis,* and the rice weevil *Sitophilus oryzae* (Camp and Arthur, 1967); to the boll weevil and bollworm *H. zea* larvae and adults (Andrawes and Dorough, 1967); and to the cotton leaf worm *P. litura* (Zayed *et al.,* 1966). Identical metabolites are produced *in vitro* after incubation of carbaryl with house fly abdomen homogenates or microsomes (Tsukamoto and Casida, 1967; Kuhr, 1969), fat-body and other tissues of the blow fly *Calliphora erythrocephala* (Price and Kuhr, 1969), whole tissues and tissue

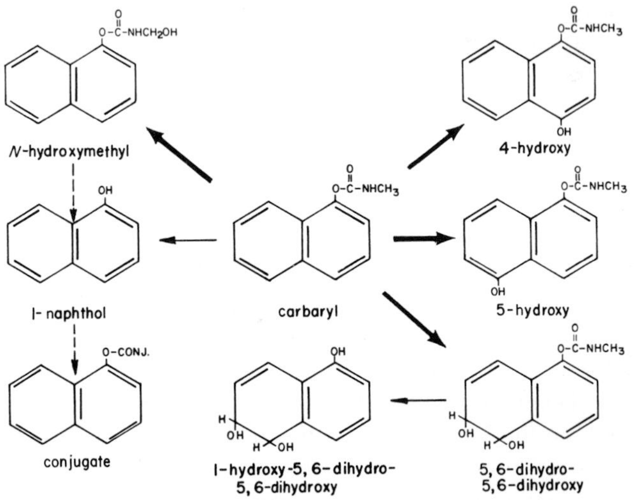

Fig. 20. Metabolism of carbaryl by insects and mammals. Major metabolic pathways in insects are designated by heavy arrows. (From Dorough and Casida, 1964; Leeling and Casida, 1966.)

homogenates of the cabbage looper *Trichoplusia ni* (Kuhr, 1970), as well as by plants (Kuhr and Casida, 1967) and mammals (Dorough and Casida, 1964; Leeling and Casida, 1966; Oonithan and Casida, 1966, 1968).

Nonpolar carbaryl metabolites are produced in varying degrees by all species examined, but water-soluble metabolites are excreted in substantial amounts. In many instances, these polar compounds are present as conjugates of precursor metabolites found in tissues. It is also evident that there is a great deal of species specificity with regard to absorption and metabolism of carbaryl and other carbamates.

Metcalf *et al.* (1967) evaluated nine carbamate insecticides labeled with ^{14}C in various parts of the molecule, for their absorption, metabolism to $^{14}CO_2$, and excretion by several R and S strains of house flies and found the following: (a) absorption and metabolism does not vary appreciably among strains; (b) rate of conversion to $^{14}CO_2$ is characteristic of the individual carbamate and the position of the ^{14}C-labeled atom in the molecule; (c) excretion of metabolites in feces is substantially higher in R flies for all the carbamates except Banol and carbaryl, due, perhaps, to the faster knockdown of S flies, since rate of metabolism is practically the same in both strains; and (d) excretion of metabolites ranges from 25 to 50% of the total radioactivity absorbed.

In general, total detoxication of carbamate insecticides to CO_2 and H_2O (Metcalf *et al.*, 1967) takes place through the following pathways:

a. Hydrolysis. Carbonyl-labeled carbamates yield phenol or oxime and N-methylcarbamic acid. The latter is metabolized to carbonic acid and eventually to CO_2 and H_2O.

$$RO^{14}\overset{\overset{O}{\|}}{C}NHCH_3 \xrightarrow{\ HOH\ } ROH \ + \ HO^{14}\overset{\overset{O}{\|}}{C}NHCH_3$$

$$HO^{14}\overset{\overset{O}{\|}}{C}NHCH_3 \longrightarrow HO^{14}\overset{\overset{O}{\|}}{C}OH \ + \ CH_3NH_2$$

$$HO^{14}\overset{\overset{O}{\|}}{C}OH \longrightarrow {}^{14}CO_2 \ + \ H_2O$$

Hydrolytic reactions occur with several carbamates but are now recognized to be of minor importance. Thus, carbaryl yields 1-naphthol (Eldefrawi *et al.*, 1961; Ku and Bishop, 1967); carbofuran (Furadan) is hydrolyzed by house flies (Dorough, 1968) and the salt marsh caterpillar *E. acrea* (Metcalf *et al.*, 1968) to the 2,3-dihydro-2,2-dimethyl-7-hydroxybenzofuran (Furadan phenol), 3-keto-Furadan phenol, and 3-hydroxy-Furadan phenol, a large part of the metabolites appearing in conjugated form (Fig. 21). Landrin hydrolysis is of minor importance in the house fly but is more prevalent in mice (Slade and Casida, 1970). Hydrolytic products of aldicarb (Temik)

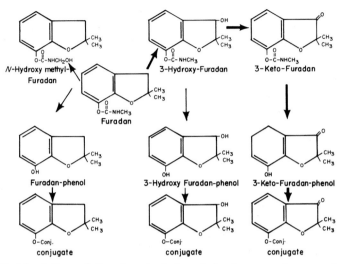

Fig. 21. Metabolism of Furadan (carbofuran) by insects and mammals. Major metabolic routes in insects are designated by heavy arrows. (From Dorough, 1968; Metcalf *et al.,* 1968.)

metabolism in the boll weevil and tobacco budworm (Bull *et al.,* 1967) include the sulfoxide and sulfone derivatives of 2-methyl-2(methylthio)-propionaldehyde oxime.

b. N-Dealkylation. This pathway yields $^{14}CO_2$ from *N*-methyl-labeled carbamates

$$ROCNH^{14}CH_3 \xrightarrow{\;[OH]\;} ROCNH^{14}CH_2OH$$

$$ROCNH^{14}CH_2OH \longrightarrow ROCNH_2 + H_2^{14}C{=}O$$

$$H_2^{14}C{=}O \xrightarrow{\;O_2\;} {}^{14}CO_2 + H_2O$$

N-Dealkylation occurs most readily with methiocarb (Mesurol) and *m*-isopropylphenyl methylcarbamate, and least with Niagara 10242, aminocarb (Matacil), and carbaryl (Metcalf *et al.,* 1967). Aminocarb is attacked on the ring methylamino group and forms the 4-methylamino, 4-amino, 4-methylformamido, and 4-formamido derivatives, and Zectran forms an analogous series of metabolites (Tsukamoto and Casida, 1967). Production of these metabolites is greater in homogenates of R flies than in those of their S counterparts. All these metabolites are less toxic to house flies than the corresponding parent compounds. On plants, the major photooxidation pathway of aminocarb and Zectran involves extensive oxidation of the di-methylamino moiety with the production of the same type metabolites, except that the methylamino and amino analogues are of relatively high toxic-ity (Abdel-Wahab and Casida, 1967). Propoxur (Baygon) is demethylated by house flies and mosquitoes via N-hydroxymethyl propoxur (Shrivastava *et al.*, 1969, 1970) but this is a minor metabolic pathway. There is evidence of some N-demethylation of aldicarb in bollweevils and tobacco budworms (Bull *et al.*, 1967).

c. O-Dealkylation. This reaction is particularly prominent with propoxur which is dealkylated at the isopropoxyphenyl moiety producing O-depropyl propoxur and $^{14}CO_2$ (Fig. 22; Metcalf *et al.*, 1967; Tsukamoto and Casida, 1967; Shrivastava *et al.*, 1969, 1970). Most of the O-dealkylated products come from the intact carbamate rather than from hydrolyzed precursors.

d. S-Dealkylation. This ordinarily takes place after sulfoxidation and sul-fonation of the carbamate, such as occurs in aldicarb.

$$H_3{}^{14}CSC(CH_3)_2CH{=}NOCNHCH_3 \xrightarrow{\text{[OH]}} \text{aldicarb sulfone}$$

$$H_3{}^{14}CSO_2C(CH_3)_2CH{=}NOCNHCH_3 \xrightarrow{\text{[OH]}} \text{S-hydroxymethyl derivative}$$

$$HO^{14}CH_2SO_2C(CH_3)_2CH{=}NOCNHCH_3 \longrightarrow \text{S-dimethyl derivative}$$

$$HOSO_2C(CH_3)_2CH{=}NOCNHCH_3 + H_2{}^{14}C{=}O$$

$$H_2{}^{14}C{=}O \longrightarrow {}^{14}CO_2 + H_2O$$

e. Ring-C Hydroxylation. This is a major pathway in carbamate detoxica-tion but there is no CO_2 formation because neither insects nor mammals can convert the ^{14}C in an aromatic ring to $^{14}CO_2$. Ring hydroxylation of propoxur to 5-hydroxy propoxur is the predominant detoxication pathway

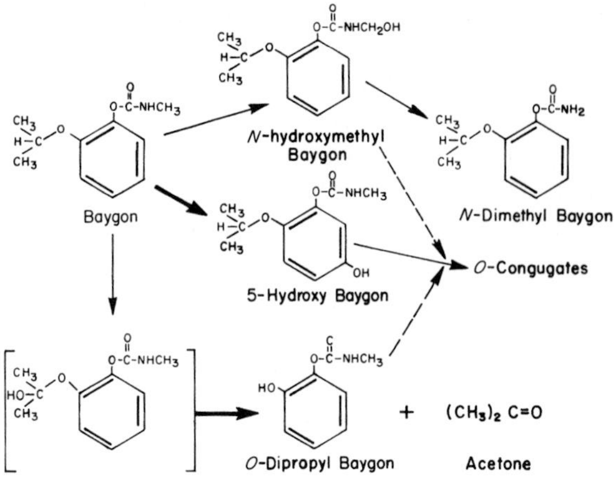

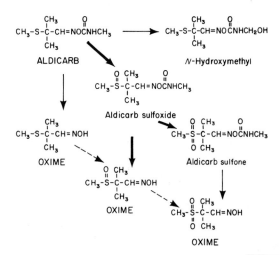

Fig. 22. Aromatic ring hydroxylation. Metabolism of Baygon (propoxur) by insects. Major degradative pathways in the house fly are designated by heavy arrows. (From Shrivastava *et al.*, 1969.)

in the house fly (Shrivastava *et al.*, 1969), and a major metabolic product in propoxur-resistant *C.p. fatigans* (Shrivastava *et al.*, 1970).

f. Sulfoxidation. In most instances, sulfoxidation of a carbamate results in the formation of more toxic metabolites (activation). For example, the major route of *in vivo* metabolism of aldicarb by house flies (Metcalf *et al.*, 1966b, 1967) and by boll weevils and tobacco budworms (Bull *et al.*,

Fig. 23. Oxidative metabolism of aldicarb (Temik) in insects. Major oxidative pathways are designated by heavy arrows. (From Bull *et al.*, 1967; Metcalf *et al.*, 1967.

1967) involves oxidation of the sulfur atom to the sulfoxide and sulfone derivatives (Fig. 23) which are more toxic than the parent compound. However, the sulfoxide and sulfone metabolites of methiocarb are less toxic to the house fly than the parent compound (Metcalf *et al.*, 1967). House fly abdomen homogenates also convert methiocarb to the sulfoxide analogue (Tsukamoto and Casida, 1967). The sulfoxide and sulfone derivatives of carbamates are stable to further degradation, but small amounts are detoxified by hydrolysis such as the conversion to oxime derivatives (Fig. 23).

g. N-Methyl Hydroxylation. This oxidative pathway is common to many carbamate insecticides, and may or may not be an important detoxication reaction depending on the compound and the insect species involved. Gut tissue and fat-body homogenates of the cabbage looper oxidize 60–76% of carbaryl to the *N*-hydroxymethyl derivative (Kuhr, 1970) and Banol is oxidized to the *N*-hydroxymethyl derivative by the cockroach *Blaberus giganteus* (Gemrich, 1967) and the house fly (Tsukamoto and Casida, 1967). The major organosoluble metabolite of propoxur in spruce budworm larvae and in *C.p. fatigans* is *N*-hydroxymethyl propoxur while it is only a minor detoxication product in the house fly (Shrivastava *et al.*, 1969, 1970). *N*-Methyl hydroxylation at the 2- and 5-*N,N*-dimethyl moieties appears to be the major factor in the detoxication of dimetilan in the German cockroach, the American cockroach, and the house fly (Fig. 24; Zubairi and Casida, 1965).

In the house fly *N*-methyl hydroxylation of propoxur, Zectran, and carbofuran constitutes only a minor detoxication pathway.

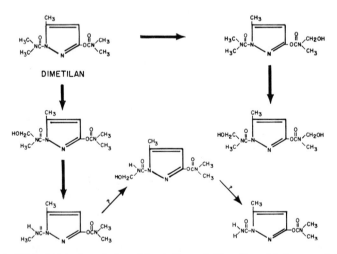

Fig. 24. N-Methyl hydroxylation. Metabolism of dimetilan. Major metabolic pathways in insects are designated by heavy arrows. (From Zubairi and Casida, 1965.)

Ring methyl hydroxylation is the major oxidative route in the metabolism of Landrin (Fig. 25) by living house flies and by the mixed-function oxidase system of house fly abdomens (Slade and Casida, 1970).

Relation of Resistance to Metabolism

Most of the studies related above show no evidence of qualitative differences in carbamate metabolism between resistant and susceptible insects. However, several factors must play important roles in the differential toxicity of carbamates to various species and among strains of the same species, including: rates of absorption, metabolism, conjugation, and excretion.

Among nine carbamates investigated (Metcalf et al., 1967) little difference in absorption was found between R and S strains of house flies. As far as metabolism to $^{14}CO_2$ is concerned, there is very little difference in total $^{14}CO_2$ production between R and S house flies. However, R flies metabolize most carbamates at a faster rate than S flies.

There is also considerable variation in carbamate metabolism to $^{14}CO_2$ depending on the structure of the carbamate and the position of the labeled ^{14}C atom. For example, ring-labeled phenylcarbamate does not produce $^{14}CO_2$, indicating the inability of the insect to metabolize the aromatic ring. Both R and S flies metabolize the isopropoxy side chain of propoxur to $^{14}CO_2$ at a rapid rate, suggesting that a major degradation pathway is via

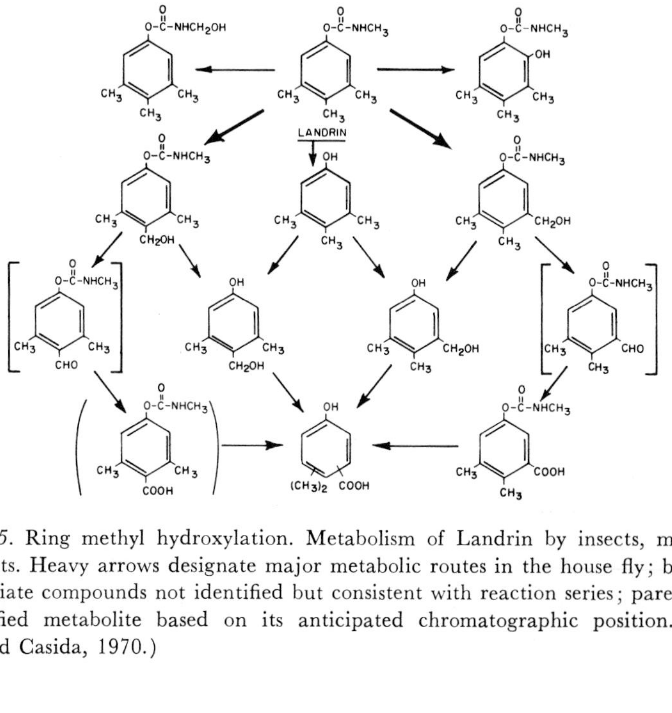

Fig. 25. Ring methyl hydroxylation. Metabolism of Landrin by insects, mammals, and plants. Heavy arrows designate major metabolic routes in the house fly; brackets: intermediate compounds not identified but consistent with reaction series; parentheses: unidentified metabolite based on its anticipated chromatographic position. (From Slade and Casida, 1970.)

O-dealkylation to form depropyl propoxur. Alternatively, ring hydroxylation yielding 5-hydroxy propoxur might be of greater importance in some strains. The relative importance of one pathway over another might be attributable to species or strain differences. An example is provided by the work of Shrivastava et al. (1969) who measured the production of [^{14}C]acetone (O-dealkylation) from propoxur-[^{14}C]isopropyl and $^{14}CO_2$ (hydrolysis) from propoxur-[^{14}C]carbonyl in 13 insect species. The ratio of carbamate ester cleavage to O-dealkylation was highest in the American cockroach and spruce budworm, indicating a predominantly hydrolytic attack, intermediate in the German cockroach and the yellow mealworm, low in the honey bee, codling moth, blow fly, and milkweed bug, and lowest in the yellow fever mosquito and the house fly, the latter pointing toward O-dealkylation as a more significant detoxication route.

Differences in rate of carbamate metabolism are well documented, especially in R and S house flies. More rapid metabolism occurs in resistant strains with carbaryl (Eldefrawi and Hoskins, 1961), 3-isopropylphenyl N-methylcarbamate (Georghiou and Metcalf, 1961b), propoxur (Shrivastava et al., 1969, 1970), aldicarb (Metcalf et al., 1966b), Isolan (Plapp et al., 1964), and ten carbamates, i.e., propoxur, carbaryl, 3-isopropylphenyl methylcarbamate, 3,5-diisopropylphenyl methylcarbamate, Banol, methiocarb, aminocarb, Zectran, Isolan, and dimetilan (Tsukamoto and Casida, 1967). Also, propoxur-resistant C.p. quinquefasciatus (Georghiou, 1965a) and C.p. fatigans (Shrivastava et al., 1970) metabolize propoxur at a faster rate than their susceptible counterparts.

Propoxur-resistant A. albimanus from El Salvador exhibit a tremendous rise in the LC_{50} upon further laboratory selection, from 0.53 ppm to >500 ppm. Cross-resistance to fenitrothion, other OP and carbamate insecticides, and DDT, is evidence of enhanced enzyme activity (Ariaratnam and Georghiou, 1971). The precise mechanisms of this high level of resistance are not well understood, especially since piperonyl butoxide and other synergists do not enhance the toxicity of propoxur except at very high insecticide concentrations. Reduced penetration might be one of the factors.

Microsomal mixed-function oxidases (MFO) of A. albimanus, A. quadrimaculatus, A. aegypti, A. triseriatus, C.p. faigans, C. peus, and C. tarsalis catalyze the hydroxylation of benzene, naphthalene, and furan rings in propoxur, carbaryl, and carbofuran, respectively; O- and N-dealkylation of propoxur; and formation of the sulfoxide and sulfone of aldicarb (Shrivastava et al., 1971). Only small amounts of hydrolysis products are produced. Enzyme activity is highest in the propoxur-selected C.p. fatigans.

The mosquito MFO system differs from that of other insects in buffer concentration, pH optimum, and incubation temperature, but has the same characteristics with respect to endogenous inhibitors. On a weight basis, adult mosquitoes show higher enzyme activity than larvae.

Resistance to carbamates is not always due to more rapid degradation of the toxicant by the resistant insect. The tobacco budworm and the bollworm are naturally tolerant to aldicarb whereas the boll weevil is quite susceptible to this carbamate (Bull *et al.,* 1967). Some of the factors which contribute to this tolerance are the slow penetration of the toxicant through the larval cuticle (although some tolerance is also manifested if the chemical is injected), and, especially, the lower sensitivity of their cholinesterase to inhibition by aldicarb and its toxic metabolite, aldicarb sulfoxide. Both compounds are effective inhibitors of boll weevil cholinesterase. The lepidopterous larval cholinesterase appears to have somewhat different properties in hydrolyzing choline esters and in its susceptibility to inhibition by excess acetylcholine (Bull *et al.,* 1967).

Rate of excretion of the parent carbamate and/or of toxic metabolites certainly plays a role in the resistance of some insects, but in many instances conjugation reactions must occur before excretion can take place. Indeed, many types of hydroxylations, especially ring hydroxylations which are very important in detoxication, undergo conjugation reactions before the compounds are excreted. Alternatively, excretion may be impeded in the susceptible insect because of a biochemical lesion causing its knockdown.

Most of the evidence supports the contention that carbamates are oxidatively metabolized primarily by microsomal enzymes. However, a partially purified soluble tyrosinase prepared from house flies catalyzes the hydroxylation of certain methyl carbamates (Abd El-Aziz *et al.,* 1969; Metcalf *et al.,* 1966a). Tyrosinase is also found in house fly microsomal preparations. According to Kuhr (1969) and Wilkinson (1968) microsomal tyrosinase does not seem to be important in carbamate metabolism in general, and the soluble fraction, which contains six times as much tyrosinase as the microsomes, fails to metabolize carbaryl or phenyl methylcarbamate. There appears to be some substrate and strain specificity in this regard.

In conclusion, carbamate insecticides are readily attacked by insect enzymes and resistance to these compounds follows much the same pattern as that of the organophosphorus insecticides. Hydrolysis at the esteratic linkage seems to be of minor importance in the overall metabolic pathway. Oxidative degradation by the mixed-function oxidase system includes *O-*, *N-*, *S-*, and *C-*dealkylation; *N-*methyl hydroxylation; ring hydroxylation; and sulfoxidation.

Rates of absorption, metabolism, conjugation, and excretion play a vital role in carbamate resistance by insects. Since the same metabolites are produced by S and R insects, the rate at which these metabolites are produced and excreted determines the ability of the insect to survive the poisoning effect.

E. METABOLIC FATE OF PYRETHRINS, ROTENONE, AND NICOTINE

1. Pyrethrins and Related Compounds

It is a well-established fact that some insects are rapidly paralyzed when exposed to pyrethrins, but that a large proportion of the individuals recover completely within 24 hours. In fly spray tests, it is usually expected that a pyrethrin formulation giving complete knockdown within 10 minutes will ordinarily show only 60–70% mortality at the end of 24 hours (Shepard, 1951). This reversal of knockdown and paralysis indicates that insects possess an efficient detoxifying mechanism capable of attacking the pyrethrin molecule at certain reactive sites.

Pyrethrins, like DDT, have a negative temperature coefficient, i.e., their toxicity diminishes with an increase in temperature (Chevalier, 1930; Hartzell and Wilcoxon, 1932; Blum and Kearns, 1956). At higher temperatures (about 35°C), hydrolysis was presumed to be so great that the insects did not accumulate a lethal dose.

Following Swingle's (1934) observation that pyrethrins were ineffective as a stomach poison, Acree et al. (1936) suggested that hydrolytic enzymes such as esterases might be involved in the breakdown of pyrethrins.

Detoxication of pyrethrins was first demonstrated in the southern armyworm P. eridania by bioassaying tissue extracts of the treated insect against mosquito larvae (Woke, 1939). In vitro degradation was highest in fat-body followed by skin and muscle, digestive tract, and hemolymph.

Most of the earlier work on pyrethrin metabolism centered around the hypothesis that hydrolysis of the ester linkage was the mechanism of detoxication. This was based chiefly on the demonstration of the production of a small amount of an unidentified acid resulting from incubation of cockroach and house fly acetone powder preparations with pyrethrins (Chamberlain, 1950), and on the incomplete identification of chrysanthemumic acid and keto-alcohols resulting from the in vivo metabolism of [14C]pyrethrins by the American cockroach (Zeid et al., 1953). The latter's findings that 3–12% of the [14C]pyrethrin dose was converted to $^{14}CO_2$ could not be confirmed later with house flies (Winteringham, 1952b; Winteringham et al., 1955), but significant amounts of [14C]pyrethrins and [14C]allethrins were shown to be metabolized to nontoxic, nonpyrethroid derivatives, and this metabolism was blocked by the synergist piperonyl cyclonene.

Bridges (1957), extending this work, found more extensive metabolism of [14C]allethrin in vivo than by lipase extracts and abdomen homogenates *in vitro,* and suggested that metabolism was not due to hydrolysis. Similarly, Hopkins and Robbins (1957) found a rapid rate of [^{14}C]allethrin metabolism in DDT-resistant house flies with approximately 44% of the absorbed dose excreted as polar metabolites and resembling allethrolone on paper chromatography. Only traces of chrysanthemumic acid and unchanged allethrin were detected. These authors pointed out that modification of the intact ester rather than hydrolysis of the ester linkage might provide a detoxication pathway.

Recently, Hayashi *et al.* (1968) showed the production *in vivo* of large quantities of allethrolone or of N-hydroxymethyl tetrahydrophthalimide in house flies exposed to [^{3}H]allethrin and [^{3}H]phthalthrin, respectively. According to Yamamoto (1970) none of these metabolic pathways have been conclusively proven.

The first breakthrough in the present concept of pyrethrin degradation was the finding by Chang and Kearns (1964) that hydrolysis plays only a minor role in pyrethrin detoxication by intact house flies. More than 96% of the absorbed dose of [^{14}C]pyrethrin I and [^{14}C]cinerin I was detoxified in 4 hours after topical application. Five unknown metabolites were detected plus a small amount of free chrysanthemumic acid which never exceeded 2.6% of the applied dose. Three of the metabolites showed an intact chrysanthemumic acid–ester linkage. The authors concluded that the detoxication process apparently is initiated on the keto-alcohol moiety leaving the acid–ester linkage intact.

The present concept of pyrethrin metabolism is that oxidative breakdown rather than hydrolytic cleavage is the main metabolic pathway. Evidence for this comes from the works of Yamamoto and Casida (1966) and Yamamoto *et al.* (1969) who first succeeded in defining the metabolic pathway in insects. The synthesis of ^{14}C-labeled pyrethroids, stereochemically pure and of high specific activity (Yamamoto and Casida, 1968), obviated the difficulties encountered in previous studies. Additional evidence of oxidative metabolism was obtained *in vitro* by using homogenates of resistant house fly abdomens fortified with NADPH, similar to the system used by Tsukamoto and Casida (1967). In this sytem [^{14}C]allethrin, labeled in either the acid or the alcohol moieties, yields more than ten, autoradiographically identical metabolites, indicating that each metabolite contains both the intact acid and alcohol moieties. A major metabolite, representing half of the total radioactivity, proved to be *trans*-allethrenoic acid or demethylallethrin II. Other metabolites are characterized as *trans*-allethrinol and *trans*-allethrinal. Low enzyme levels produce allethrinol metabolites whereas high enzyme titers oxidize the intermediates to allethrinoic acids. There is little, if any, indication of a modification of the allethrolone moiety, in contrast to earlier findings by other workers. Similar metabolic reactions occur with [^{14}C]pyrethrin I, dimethrin, and phthalthrin, and oxidative degradation of these compounds is more extensive in resistant than in susceptible house flies (Yamamoto and Casida, 1966; Yamamoto *et al.,* 1969).

In summary, the major metabolic pathway for the enzymic detoxication of the pyrethroids involves oxidation of the *trans*-methyl group of the iso-butenyl moiety to the hydroxymethyl derivative which, acted upon by other enzymes, either forms conjugates, or is further oxidized to the corresponding aldehyde and acid (Fig. 26). A minor pathway involves similar reactions at the *cis*-methyl group, and no hydrolysis or modification of the alcohol moiety is evident.

All the metabolic products obtained are less toxic to house flies than their parent compounds. Synergists containing the methylenedioxyphenyl group, particularly piperonyl butoxide, block the initial hydroxylation at the iso-butenyl–methyl group, thus stabilizing the molecule and preventing its de-toxication. On the other hand, the isobutenyl–methyl moiety is not essential for toxicity (Berteau *et al.* 1968), but compounds lacking this grouping are also synergized (Berteau and Casida, 1969), indicating, perhaps, that other sites than the isobutenylmethyl group might be susceptible to oxidative attack.

Pyrethrin I, phthalthrin, allethrin, and dimethrin deposited as thin films on glass readily undergo photodecomposition yielding at least 11 products. The photolytic pathway (Chen and Casida, 1969) follows closely that shown

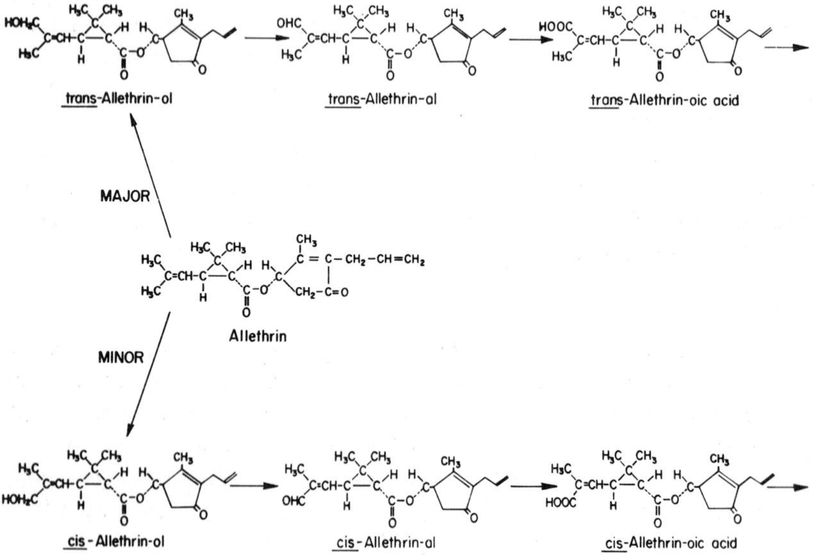

Fig. 26. Oxidative metabolism of allethrins in the house fly. (From Yamamoto and Casida, 1966; Yamamoto *et al.,* 1969.)

in the oxidative metabolism by the insect, in addition to attack on other functional groups.

Pyrethroids have very low mammalian toxicity by the oral or dermal route, but are highly toxic to insects. This selective toxicity might be due to the more efficient metabolism of pyrethroids by mammals. However, as shown above, insects, too, are capable of metabolizing these compounds. In spite of this capability, pyrethrin resistance at a level which appreciably affects control measures is not common.

The residual effectiveness of pyrethrins is not as long lasting as that of most chlorohydrocarbon insecticides, and this, perhaps, accounts for its limited use in field control operations; hence, there is less chance of attaining a high level of resistance in the field, especially if pyrethrins are used in conjunction with a synergist. There are, however, notable exceptions. Davies et al. (1958) showed an 11- to 14-fold resistance to synergized pyrethrins and a much higher resistance to allethrin (150-fold) and to synergized allethrin (45-fold) in field-collected flies. Pyrethrin-treated flies in the field also show cross-resistance to OP compounds such as dimethoate (Keiding, 1969). In the laboratory, selection of house flies with pyrethrins or DDT results in enhanced resistance to pyrethrins, synergized pyrethrins, allethrin, barthrin, cyclethrin, furethrin, in addition to DDT and other insecticides (Fine, 1961).

Pyrethrin resistance has also been reported in cockroaches B. germanica (Keller et al., 1956), bed bugs Cimex hemipterus (Busvine, 1958), blue ticks B. decoloratus (Whitehead, 1959), body lice P. humanus humanus (Cole and Clark, 1961; Nicoli and Sautet, 1955; Wright and Brown, 1957), granary weevils S. granarius (Blackith, 1953; Parkin and Lloyd, 1960), and other insects.

It is evident that under favorable conditions of selection pressure, resistance to pyrethroids may develop as readily as with other insecticides. Whether or not detoxication of these compounds in various insects is causally related to resistance has not yet been established.

2. Rotenone

It has been common knowledge for many years that rotenone depresses the oxygen uptake of many organisms. Specifically, rotenone inhibits mitochondrial respiration by blocking the enzyme system involved in the coupled oxidation of reduced NAD ($NADH_2$) and reduction of cytochrome b (Lindahl and Öberg, 1961). In insects, this type of inhibition results whether the enzyme is derived from susceptible species or from those naturally resistant to rotenone (Fukami et al., 1967). Thus, the selective toxicity of rotenone (being moderately toxic to mammals and highly toxic to fish and some insects) does not involve the primary site of action but appears to depend on the distribution pattern and rate of detoxication by various organisms (Fukami *et al.,* 1969).

Rotenone is extensively metabolized by the microsomal mixed-function oxidase system of mammals and insects. In the presence of NADPH and O_2, microsomes prepared from cockroach fat-body and midgut and from house fly abdomens metabolize rotenone principally to rotenolone I, 6',7'-dihydro-6',7'-dihydroxyrotenone and to a minor extent to 8'-hydroxy-rotenone, 8'-hydroxyrotenolone I, 6',7'-dihydro-6',7'-dihydroxyrotenolone I and II, rotenolone II derivatives, and water-soluble products. Endogenous inhibitors in cockroach fat-body and midgut soluble fractions block these reactions, but there is little or no interference from the soluble fraction of house fly homogenates. Methylenedioxyphenyl synergists inhibit rotenone metabolism (Fukami *et al.,* 1969) in accord with the mode of action of these compounds as inhibitors of microsomal oxidations.

Since rotenone metabolites are less toxic than the parent compound, hydroxylation of rotenone and conjugation of polar derivatives might provide a pathway for resistance development to this compound. So far, this metabolic pathway has not been associated with rotenone resistance in insects.

3. Nicotine

Nicotine is one of the most toxic alkaloids to insects, yet several species such as the tobacco hornworm, tobacco budworm, tobacco wireworm, green peach aphid, and others feed liberally on tobacco plants and show no symptoms of intoxication.

The ability of the green peach aphid *Myzus persicae* to feed on a plant capable of nicotine biosynthesis without suffering harmful effects is due to its ability to selectively feed in the phloem of the plant and avoid the nicotine-containing xylem (Guthrie *et al.,* 1962). On the other hand, the tobacco hornworm *Protoparce sexta* excretes and egests intact nicotine and other ingested alkaloids before a toxic dose can accumulate (Self *et al.,* 1964a).

In addition to these protective mechanisms, detoxication pathways also account for nicotine tolerance in some insects that do not feed on tobacco. Thus, *B. germanica* and *P. americana* metabolize nicotine principally to cotinine which is essentially a nontoxic metabolite, but nine other unidentified metabolites prevail in the southern armyworm *P. eridania* (Guthrie *et al.,* 1957).

Similar comparative experiments showed no metabolism of nicotine by the tobacco budworm and the cabbage looper. Only unaltered alkaloids found in the tobacco plant were present in the excreta of these insects. The tobacco wireworm, cigarette beetle, differential grasshopper, and the house fly metabolize nicotine to 1, 2, 4, and 3 other alkaloids, respectively. In all cases, at least 70% of the total metabolic products corresponds to cotinine

(Self *et al.*, 1964b). In contrast with tobacco-feeding insects, nicotine and the metabolites are not excreted by the house fly.

The possibility that nicotinophilic microorganisms associated with the tobacco hornworm or with the host plant may be capable of detoxifying nicotine was investigated by Guthrie and Apple (1961). The evidence obtained supports the contention that detoxication of nicotine by microorganisms is not a major factor in nicotine tolerance by this insect.

It is of interest to note that nicotine metabolism by mammals yields principally the same product, cotinine, and this oxidative conversion takes place in the microsomal fraction of the liver (Hucker *et al.*, 1960; Stålhandske *et al.*, 1969).

F. METABOLIC FATE OF CHEMOSTERILANTS

The first report of resistance to a chemosterilant was made by Hazard *et al.* (1964) who selected *A. aegypti* larvae with apholate and obtained a 4- to 5-fold tolerance after 11 generations of selection. In the same way, Klassen and Matsumura (1966) and Patterson *et al.* (1967) were able to induce higher levels of resistance to metepa and apholate, respectively, in the same species. Furthermore, homogenates of metepa-resistant larvae metabolized metepa more rapidly than their susceptible counterparts. George and Brown (1967) obtained a moderate tolerance to hempa in *A. aegypti* after five generations of larval selection. However, this tolerance disappeared in the sixth generation of selection due to the accumulation of inheritable recessive genetic defects resulting from inbreeding. A similar loss of tolerance to metepa occurred in house flies (Sacca and Scirocchi, 1966), perhaps for the same reason. After 10 generations of selection with metepa and 60 generations with apholate, Morgan *et al.* (1967) were unable to induce resistance to these chemosterilants in the house fly. On the other hand, Abasa and Hansens (1969) showed evidence of resistance to apholate after 15 generations of selection of a diazinon-resistant strain of house flies.

Among the reasons advanced for enhanced insect tolerance of chemosterilants, penetration and degradation are prominent factors. Working with S larvae and adult *Culex tarsalis* and with S and OP-R adult house flies, Plapp *et al.* (1962) obtained complete degradation of [^{32}P]aphoxide (metepa) by *Culex* in 48 hours, and 50% degradation by the house fly in 2 hours. Degradation rates were similar in the S and R strains and were more rapid after injection than after topical application. The principal metabolite was thought to be inorganic phosphate. The finding that R flies did not metabolize the chemical at a faster rate indicates that different enzymes might be involved in the metabolism of metepa and OP compounds. Similarly, absorption and excretion of metepa proceed rapidly in mosquitoes and house flies (Dame and Schmidt, 1964), and the difference in the doses needed to sterilize the stable fly and the screw-worm fly is related to the absorption, excretion, and metabolism of metepa by the two species (Chamberlain and Hamilton, 1964). The principal metabolite was found to be phosphoric acid and in greater quantity in the screw-worm fly.

Excretion of the chemosterilant appears to be an important factor in insect chemosterilization. Boll weevils excrete 48% of the injected dose of [^{14}C]tepa in 3 days, and 3.4% is trapped as $^{14}CO_2$, indicating that some metabolism is taking place (Hedin et al., 1967). Excretion of [^{14}C]tepa also plays an active role in the house fly (Chang et al., 1966). Excretory products are of a nonaziridinyl nature whereas some of the remaining radioactivity within the fly contains aziridinyl metabolites.

Thiotepa is rapidly oxidized to tepa by the German cockroach, house fly, and stable fly. The boll weevil oxidizes 63% of the absorbed material in 4 hours, but after that period the rate of conversion diminishes markedly (Parish and Arthur, 1965).

Because tepa is one of the most potent chemosterilants, the oxidation of thiotepa to tepa might be considered an activation process.

Hempa, a nonalkylating chemosterilant, has been the subject of several investigations. The only major metabolite found in hempa-treated male house flies and their excreta was the monodemethylated product, pentamethylphosphoric triamide (PMPT) (Chang et al., 1967). This metabolite has a much lower sterilizing activity than the parent compound.

Demethylation reactions are suggestive of microsomal mixed-function oxidase involvement. This possibility was put to test by Akov and Bořkovec (1968) and Akov et al. (1968) who compared the rates of conversion of hempa to PMPT by homogenates and microsomal preparations of susceptible and carbamate-resistant house flies. Their results (Table VIII) clearly indicate the superior ability of the resistant strains to metabolize hempa.

TABLE VIII

Metabolism of Hempa by Microsomal Preparations from
Resistant and Susceptible House Flies[a]

Fly strain	PMPT produced (nmoles)	
	Males	Females
Susceptible	0.9	1.3
Isolan-resistant	8.6	9.7
Baygon-resistant	5.2	11.0

[a] From Akov and Bořkovec (1968).

In addition to PMPT, R but not S flies contained small amounts of another metabolite, tetramethylphosphoric triamide (TMPT). These metabolites are inactive products, but the sterilizing activity of hempa is of the same magnitude to both S and R strains. This led to the assumption that oxidative demethylation of hempa to PMPT might proceed via an unstable methylol intermediate which has a sterilizing capacity equal to hempa. The important metabolic step would then be the conversion of the active methylol derivative to PMPT rather than the disappearance of hempa. Demethylation of hempa is inhibited by SKF 525-A and by methylenedioxyphenyl synergists.

It is well known that in many insect species, especially the house fly, the female has a higher capacity for insecticide metabolism than the male. Diet also plays a role in insecticide metabolism. Akov *et al.* (1968) produced evidence that female flies maintained on sugar metabolized hempa 2.5 times faster than males. Milk-fed flies of both sexes sustained increased metabolic activity, again higher in the female.

In a similar study with S and R house flies, Chang and Bořkovec (1969) determined that hempa is metabolized faster by R flies and that the latter contain a larger proportion of tetramethyl- and trimethylphosphoric triamides. Tropital (a methylenedioxyphenyl synergist) suppressed the demethylation of hempa, and the relative distribution of metabolites was substantially changed, especially in R flies, including the appearance of an unidentified new peak.

The chemosterilant hemel is metabolized by male house flies to N^2,N^2,N^4,N^6-tetramethylmelamine, N^2,N^4,N^6-trimethylmelamine, and N^2,N^2,N^4-trimethylmelamine (Chang *et al.*, 1968). Unchanged hemel occurs only in fly tissues and 11% is expired as $^{14}CO_2$. Mono- and dimethylamine metabolites are not present, but N^2,N^2,N^4,N^4-tetramethylmelamine is metabolized by male house flies to N^2,N^2,N^4-trimethylmelamine and N^2,N^4-dimethylmelamine (Chang *et al.*, 1970) with no evidence of ring cleavage.

Other chemosterilants, i.e., N,N,N',N'-tetramethyl-p-piperidinophosphonic diamide (Terranova, 1969) and N,N'-tetramethylenebis(1-aziridinecarboxamide) (Terranova and Crystal, 1970) rapidly disappear from the tissues of treated boll weevils, blow flies, and screw-worm flies, particularly when the chemicals are injected.

The biochemistry of chemosterilants in insects and mammals has been reviewed in detail by Turner (1968).

V. Supplementary Protective Mechanisms

A. STRUCTURAL ASPECTS AND INSECTICIDE PENETRATION

The first report of DDT resistance in an insect was coincidentally associated with distinct morphological characters, for Wiesmann (1947) attributed DDT resistance in the house fly to darker pigmentation of the cuticle, stiffer tarsal bristles, and thicker and more pigmented tarsi, pulvilli, and articular membranes of tarsal joints. Other morphological characteristics may segregate resistant from susceptible strains of insects (D'Allessandro *et al.*, 1949; Bigelow and LeRoux, 1954; Grayson, 1954; Mahan and Grayson, 1956) but many resistant strains also are quite indistinguishable from their susceptible counterparts (March and Lewallen, 1950). On a statistical basis, for example, 16 morphological characters did not correlate with DDT resistance in 9 house fly strains (Sokal and Hunter, 1955).

Practically every publication on insecticide metabolism *in vivo* contains information on penetration of the toxicant. Many resistant strains of insects have been shown to absorb the insecticide either at a faster rate, at a slower rate, or at an equal rate to susceptible strains. In many instances where topical application is used, absorption of the insecticide is biphasic, i.e., penetration is highest during the first few hours after application and then levels off to a slower and more constant rate, but penetration also follows first-order kinetics in some insects, such as DDT-resistant codling moths (Rose and Hooper, 1969). Permeability of the nervous system to dieldrin (Ray, 1963) or to diazoxon (Lord *et al.*, 1963) is not a factor in resistance. Hence, it might be said that permeability of the cuticle is a variable factor, well-randomized among S and R strains of varied origins.

In some instances, however, reduced rate of absorption constitues a major factor in resistance. For example, Forgash *et al.* (1962) and Farnham *et al.* (1965) found reduced permeability to be a significant factor in house fly resistance to diazinon. Similarly, reduced absorption is largely responsible for *A. aegypti* tolerance to malathion (Matsumura and Brown, 1963a), to DDT synergist-combinations in house flies (Perry, 1958) and mosquitoes (Pillai and Brown, 1965), pyrethrin-resistance in house flies (Fine *et al.*, 1967), DDT-resistance in the spotted root maggot *E. notata* (Hooper, 1965), and diazinon-selected house flies (El Bashir, 1967). Hollingworth *et al.* (1967) classify slower penetration as a secondary factor in house fly resistance to OP compounds.

Insect larvae, especially lepidopterous larvae, absorb topically applied insecticides at a slow rate. Evidently this accounts for DDT-tolerance in the corn earworm (Gast, 1961) and the tobacco budworm and bollworm (Bull *et al.*, 1967; Pate and Vinson, 1968; Vinson and Brazzel, 1966).

Observations by Plapp and Hoyer (1968) indicate that reduced absorbtion constitutes the mode of action of a gene (*R-tin*) conferring resistance to organotin compounds and acting as resistance intensifier for other insecticides. Thus, reduced absorption allows detoxication mechanisms which might be present a longer time to act, consequently enhancing the overall resistance by combining the two types of genes. A similar phenomenon occurs with the gene *pen* (for penetration delaying) which alone confers little resistance, but when introduced into a strain with a factor for deethylation of OP compounds, the interaction of the two genes enhances the resistance markedly (Sawicki, 1970).

Some resistant strains of house flies are capable of surviving repeated knockdown by intermittent exposures to DDT (Barber and Schmitt, 1949) while others are paralyzed immediately but recover when removed from contact with the insecticide (Busvine, 1951). Still others are highly resistant to knockdown (Harrison, 1952b; March and Metcalf, 1949).

Resistance to knockdown was first demonstrated by Harrison (1951) to be due to an incompletely recessive gene to which Milani (1954) gave the symbol *kdr* and assigned it to chromosome II. DDT-resistant strains possessing this gene depend on other factors than dehydrochlorination for their resistance (see reviews by Georghiou, 1965b; Oppenoorth, 1965; Plapp, 1970a). Delayed knockdown due to reduced penetration is also found in the resistant SKA strain (Sawicki and Farnham, 1968). From a biochemical standpoint, recovery from knockdown may result from detoxication of the insecticide and its removal from the site of action (Ikeshoji and Suzuki, 1959) or from incomplete and reversible enzyme inhibition such as occurs with certain carbamates (Georghiou, 1962; Pant and Self, 1968).

The above examples illustrate the fact that slow rate of penetration can confer partial protection to the insect, and in those insects which possess high levels of detoxifying enzymes, slower penetration can enhance their resistance many-fold.

As far as the writers are aware, only two instances have been recorded (Winteringham, 1952a; Gupta *et al.,* 1971) where natural immunity to DDT in an insect (*Trogoderma* larvae) is associated with complete lack of metabolism of the toxicant. Gupta *et al.* (1971) observed *Trogoderma* larvae crawling on pure DDT crystals for a week without being affected in the process. This is not due to lack of penetration, for as early as 6 hours and as late as 72 hours after application there remains unchanged in the tissues 33–40% of the amount of DDT applied, with no evidence of DDT metabolites. These amounts are even higher in various tissues of the adult, including the nervous system.

B. LIPID CONTENT AND DIETARY FACTORS

1. Lipids of Susceptible and Resistant Insects

It is well established that many insecticides, especially those of the chlorohydrocarbon type, and some of their metabolites are deposited in the adipose tissue of insects. Consequently, numerous attempts have been made to correlate the lipid content of whole insects or of individual tissues with physiological resistance to insecticides. A positive correlation has been shown in some instances. Thus, Munson and Gottlieb (1953) and Munson et al. (1954) concluded that the higher lipid content of DDT-tolerant cockroaches protects the insect from intoxication by preventing DDT from reaching the site of action. Wiesmann (1957) found more lipid in a DDT-resistant strain of house flies than in its susceptible counterpart. Similar results were obtained by Langenbuch (1954, 1955), Wiesmann and Reiff (1956), de Zulueta et al. (1957), Neri et al. (1958, 1959), and Garms et al. (1959).

DDT-R house flies have been characterized as having a higher lipid content of the tarsi, epicuticle, and thoracic ganglia; higher iodine numbers of lipids; higher cholesterol levels in various tissues (Wiesmann, 1955b; Wiesmann and Reiff, 1956; Reiff and Beye, 1960) ; as well as by histological differences from their susceptible counterparts (Wiesmann, 1957). Boll weevils surviving the toxic effects of dieldrin and toxaphene usually contain more lipid than those that succumb to the insecticides (Reiser et al., 1953; Moore et al., 1967), and the greater tolerance of the alfalfa weevil to heptachlor with increase in age correlates favorably with an increase in lipid content (Bennett and Thomas, 1963).

Dieldrin-R A. aegypti larvae have a higher neutral lipid and phospholipid content than the normal strain with close correlation between total lipid and the LC_{50} among various substrains (Khan and Brown, 1966). Since similar differences occur between individuals of a given strain it is unlikely that lipid content per se is associated with the resistance mechanism. A similar conclusion was reached by Kalra (1970a) who measured total lipids, fatty acids, and phospholipids in several R and S strains of C.p. fatigans.

Larvae of the khapra beetle Trogoderma granarium which are practically immune to DDT contain three times as much total body lipid as the more susceptible adults (Rao and Agarwal, 1969; Gupta et al., 1971). It is not clear, however, if this factor is related to resistance since the comparison involves two different stages of development.

On the other hand, a comparison of several chlordane- and dieldrin-R strains with S strains of house flies failed to show any consistent interstrain differences in lipid content (Ascher and Neri, 1961; Bridges and Cox, 1959; Doby et al., 1956; Lofgren and Cutkomp, 1956) or in the quality of lipids as revealed by infrared spectroscopy (Micks and Singh, 1958). Neither did dieldrin-R A. gambiae (Bradbury et al., 1953) nor DDT-R A. aegypti (Fast and Brown, 1962) contain more lipid than their susceptible counterparts. There are variations in total sterols among resistant and susceptible house flies, but it is the susceptible NAIDM strain which contains the largest amount of cholesterol (Enan et al., 1964).

Higher lipid content might be envisaged as solubilizing more insecticide and, perhaps, enhancing its deposition in various tissues, thereby reducing the critical concentration at the site of action. At best, this condition can afford only slight protection to the insect.

2. Dietary Factors in Insecticide Tolerance

The relation of nutritional factors to insecticide resistance follows much the same pattern as lipid content. A well-documented review (Gordon, 1961) discusses the hereditary and nonhereditary effects of nutrition on insecticide tolerance. Most of the examples given agree with the concept of vigor tolerance (Hoskins and Gordon, 1956) which delineates them from the more fundamental physiological resistance. For example, aphids and red spiders often show two- to threefold variations in susceptibility to insecticides depending on the plant they feed on (Gordon, 1961; Potter, 1956; Potter and Gilham, 1957). The two-spotted mite (Henneberry, 1964) and the carmine spider mite (Raccah and Tahori, 1970) are more susceptible to malathion if fed on bean plants grown in nutrient solutions containing high levels of phosphorus (and high levels of nitrogen in the former only). In contrast, high levels of ascorbic acid and increased lipid level in the diet (Vinson, 1967) confer resistance to DDT in tobacco budworm larvae.

House fly resistance to knockdown by DDT can be increased by substituting milk for sugar in the adult diet (Nagasawa, 1953) or by increasing the fat content of milk (Mer and Furmanska, 1953). On the other hand, milk-fed flies contain no more DDT-dehydrochlorinase than flies maintained on a sugar diet (Moorefield and Kearns, 1957).

Recent advances in mammalian and insect pharmacology have added much to clarify the effect of dietary factors on drug and insecticide metabolism. For example, a high sucrose level in the diet of rats markedly depresses microsomal hydroxylation of several drugs by the mixed-function oxidase (MFO) system (Kato, 1966). Similarly, house flies maintained only on a sucrose diet metabolize the chemosterilant hempa at a much slower rate than milk-fed flies (Akov et al., 1968), indicating an increased enzyme activity in the latter. Along the same line, Perry and Buckner (1970) demonstrated that milk-fed house flies contain more cytochrome P-450 than those fed sugar, and that the LD_{50} for various insecticides is 1.5 to 2.3 times higher for milk-fed than sugar-fed adults. Thus, the effect of dietary factors on tolerance to certain insecticides and other chemicals might well be associated with an increase in cytochrome P-450, the terminal oxidase in the microsomal enzyme system.

Other attempts have been made to correlate dietary factors with rates of MFO activity in various species. Gordon (1961) comments on larval feeding stages of polyphagous holometabolous insects as showing high and generalized tolerance to contact insecticides, this probably being the result of selection for endurance to a variety of prolonged biochemical stresses, especially exposure to secondary plant substances. This hypothesis was experimentally tested by Krieger et al. (cited by Hollingworth, 1971) who found good correlation between aldrin epoxidase activity in gut microsomes and the number of plants on which certain species are known to feed. On the other hand, Brattsten and Metcalf (1970), who examined 54 insect species from eight orders and 37 families, found a wide variation in susceptibility to carbaryl and carbaryl plus piperonyl butoxide which, in many instances, was not related to diet. The 10,000-fold range in susceptibility at the LD_{50} level and the 300-fold range in synergistic ratio should discourage any toxicologist from the ever-present temptation to generalization.

C. MISCELLANEOUS CHEMICAL DIFFERENCES

A logical extension of efforts toward understanding the mechanisms of insecticide resistance is the seeking of chemical differences which might be peculiar to either susceptible or resistant strains of a given species. Several such attempts have been reported.

Chromatographic patterns of ninhydrin-positive components of homogenates of S and R strains of *M. domestica, Fannia canicularis,* and *Phaenicia sericata* revealed a number of differences among the strains, none of which could be correlated with insecticide resistance (March and Lewallen, 1956). Instead they seemed to correlate more closely with geographical origin.

Exposure to DDT causes a significant decrease in the amino acid content of the S strain hemolymph, but not in that of the R strain of house flies (Reiff, 1956b). However, among several strains of house flies tested, the amounts of some individual free amino acids differ from one strain to another, but these differences do not appear to be related to selection by or resistance to diazinon (Lord and Solly, 1964) or to DDT (Patel et al., 1968). In DDT-poisoned cockroaches, *P. americana,* proline is practically depleted but major amino acids are not affected (Corrigan and Kearns, 1958). Clearance of injected proline after administration of DDT is considerably faster in DDT-poisoned cockroaches. This depletion may be due to excretion or to accelerated metabolism of proline caused by DDT (Hoy and Gordon, 1961).

Reversal of symptoms of DDT-poisoning results in the gradual restoration of normal proline levels in the hemolymph. Identical results are observed with TDE-poisoned cockroaches and tobacco hornworms, but other insecticides fail to produce this reversible proline depletion (Corrigan and Kearns, 1963). Furthermore, in cockroaches injected with [14C]proline, DDT increases the level of radioglutamine in the hemolymph and nerve cord while decreasing the radioproline content (Corrigan and Kearns, 1963). On the other hand, DDT, dieldrin, DFP, Baygon and n-valone, but not rotenone, reduce the free proline concentration of the central nerve cord of *P. americana* (Ray, 1964) and, in some instances, it results in the accumulation of free α-alanine and glutamine. Malathion causes a reduction in glycine, glutamate, proline, and glutamine levels of susceptible German roaches but, except for glutamine, has little effect on amino acid depletion in the resistant strain (Mansingh, 1965). An increase in free alanine occurs also in DDT-R *A. aegypti* larvae after exposure to DDT while the alanine level of the S strain remains comparatively low (Micks *et al.*, 1960).

In the house fly, DDT interferes with the metabolism of formate, glycine, and proline. After injection of [^{14}C]formate, more uric acid and allantoin and less proline are found in DDT-treated house flies than in untreated controls, but such is not the case with OP and carbamate-treated insects (Cline and Pearce, 1963). Relevant to house fly resistance to DDT is the observed DDT and thiocarbohydrazine inhibition of proline degradation and purine synthesis in the R strain while the reverse is true in the S strain (Cline and Pearce, 1966). Thus, it might be inferred that, in the R strain, the toxic action of DDT is hampered by DDT inhibition of the production of neuroactive amines or toxins, which, presumably, can be derived from proline. It is of interest, too, that in fourth-instar *T. infestans* larvae (the DDT-resistant stage) DDT causes a significant increase in free proline and in total proteins (del Villar and Mosnaim, 1967).

DDT also causes an increase in the rate of utilization of pentose phosphate cycle intermediates in *P. americana* (Silva *et al.*, 1959) and in DDT-R *T. infestans* and *M. domestica* (Agosin *et al.*, 1961a, 1963, 1966). With the introduction of genetically pure, mutant house fly strains, Plapp (1970b) was able to show a twofold increase in the level of glucose-6-phosphate dehydrogenase in DDT-R or dieldrin-R house flies. This factor is under genetic control and can be transferred from one population to another by appropriate genetic crosses.

The differences between S and R strains of insects cited above strengthens even further the conclusion that biochemical variability among species and strains is the rule rather than the exception. If biochemical defense mechanisms are rather specific in nature, as contrasted with a "multicomponent type" resistance which constitutes a summation of many independent variables (Spiller, 1958), we should then seek those mechanisms which show consistency rather than variability.

D. ECOLOGICAL AND BEHAVIORAL ASPECTS OF RESISTANCE

1. Bionomics

Numerous attempts have been made to correlate insecticide resistance with certain biological factors such as egg production, hatching period, egg viability, length of developmental stage of egg, larva, pupa, and adult; pupal and adult weights; sex ratios; and preoviposition period. A detailed discussion of these factors would appear superfluous, especially in view of the fact that most of these associations have proven to vary considerably among species and among strains of a single species (Brown, 1958; Perry, 1958). The only characteristic for which significant differences are apparent in some instances is length of the larval stage. Several DDT-R strains of house flies have been found to have a 1- to 4-day longer larval stage than S counterparts (Bruce, 1949; Pimentel et al., 1951; Norton, 1953) but in an equal number of investigations no such differences were discernible (March and Lewallen, 1950; Babers et al., 1953; Varzandeh et al., 1954). Gagliani (1952), on the other hand, found a shorter larval cycle in a DDT- and chlordane-resistant strain. Similarly, life cycle changes brought about by insecticide pressure on B. germanica appear to be the result of population selection not associated with resistance (Perkins and Grayson, 1961). Furthermore, changes in length of larval period, weight, and size brought about by external or genetic manipulation have no striking influence on DDT susceptibility in R and S D. melanogaster (Bochnig, 1960), but selection of D. melanogaster larvae for a short or a long larval period results in a lower LC_{50} for DDT in the short larval line but no change from the control in the long larval line (Hunter, 1956).

A more significant finding is that late-emerging house flies are more resistant to DDT than those emerging early (Pimentel et al., 1951). Selection of late-emerging adults for three generations (Decker and Bruce, 1952) or lengthening the larval stage by lowering the temperature (Johnston et al., 1954) confers some DDT resistance to the offspring. A six- to eightfold resistance to DDT and related compounds was obtained by selecting the late pupaters for 15 generations in the absence of a chemical (McKenzie and Hoskins, 1954). After 15 generations of selection for early and late pupation in M. domestica vicina, the LD_{50} for DDT of the early strain diminished by a factor of two, whereas that of the late strain increased sixfold (Buéi and Fukuhara, 1964). However, there was no change in the LD_{50} when BHC was used as the test compound. It is puzzling that selection for late pupation should be confined to DDT resistance and certain DDT analogues only and not extend to other type insecticides such as lindane, cyclodienes, and pyrethrins.

Selection of late pupaters also results in an increase in the level of DDT dehydrochlorinase (Kerr et al., 1957). The functional relationship between DDT resistance, DDT dehydrochlorinase, and rate of preadult development is not clear, since adult flies selected with DDT, unlike late pupaters, show no sign of retarded development.

Changes in reproductive potential following application of an insecticide have been the subject of extensive investigations. Kilpatrick and Schoof (1956) obtained a sizable increase in fly production 4 weeks after treatment of outdoor privies with aldrin, dieldrin, chlordane, and BHC, but a reduction of fly breeding with application of DDT. Knutson (1959) also observed an increase in biotic potential after 4 years of field spraying with dieldrin to control house flies, and similar results were obtained by Gratz (1966) in laboratory selection and field control of *M. domestica vicina*. In the latter case, the increased biotic potential lasted for only one or two generations after relaxation of selection pressure, suggesting that dieldrin resistance is not under genetic control. *Sitophilus granarius* weevils exposed to low concentrations of DDT produced 20% more offspring than unexposed controls (Kuenen, 1958).

Greater reproductive potential in S than in R strains of house flies is of common occurrence (Hunter *et al.*, 1958, 1959). This phenomenon shows positive correlation with ovarian size and ovarian nitrogen content (Patel *et al.*, 1968).

In some instances, sublethal doses of insecticides were found to suppress house fly fecundity and ovarian development (Beard, 1965; Georghiou, 1965c), and removal of OP-resistant *A. aegypti* from insecticide contact was found to restore the fecundity and fertility levels by stimulating protein synthesis (Inwang, 1968).

The nature of ovarian suppression by DDT involves a variety of pathological conditions as discussed and illustrated by Lineva (1962) and Beard (1965). According to Derbeneva-Ukhova *et al.* (1966) development of resistance by sublethal exposure to DDT follows three consecutive phases: (1) a short period of increased tolerance; (2) a period of susceptibility during which ovarian suppression is at its highest; and (3) a period of resistance enhancement during which disfunction of ovogenesis disappears.

Alternatively, the reproduction potential of *A. aegypti* (Sutherland *et al.*, 1967, 1970) and *C. pipiens* (Zaghloul and Brown, 1968) is increased by sublethal exposure to DDT, and in *A. albimanus* the fertility rate is higher in the dieldrin-R strain (Gilotra, 1965). Other workers find no deleterious effects on oogenesis after feeding adult house flies sublethal doses of p,p'-DDT, o,p-DDT and DDE for several generations (Walker, 1970), or after exposure of *A. aegypti* to sublethal doses of DDT (Havertz and Curtin, 1967). In fact, in the F_1 generation, treated females produced more eggs than did the controls.

2. Avoidance Mechanisms

Behavioral resistance is defined as "the ability of insects, through protective habits or behavior, to avoid lethal contact with a toxicant" (Hess, 1952). A classical example of this type is said to be the ability of certain strains of the codling moth *C. pomonella* to penetrate apples sprayed with lead arsenate, presumably due to their acquired habit of rejecting the first bite prior to penetrating the fruit (cited by Babers, 1949).

There have been reports of changes in behavioral patterns of house flies, such as fewer flies resting on DDT-sprayed walls and ceilings (King and Gahan, 1949; Bruce and Decker, 1950), or nocturnal migration from sprayed houses to nearby unsprayed vegetation (Silverman and Mer, 1952). Selective avoidance of malathion baits by certain strains of malathion-R flies have been mentioned in Section IV,C. In contrast, resistant potato flea beetles *E. cucumeris* actually feed on DDT-treated foliage more readily (Kring, 1958). Thus, the change in behavior here is in the opposite direction.

Behavioral changes in mosquitoes have been a source of considerable trouble in mosquito control, especially in connection with the worldwide malaria eradication operations of the World Health Organization. This aspect of behavior has been reviewed in detail by Mattingly (1962) and in a most recent monograph on resistance by Brown and Pal (1971).

Behavioristic resistance or avoidance is characterized by induced exophily which, in most circumstances, results from the increased irritability of adult mosquitoes contacting a residual type insecticide deposit. The results of Trapido (1952) with *A. albimanus* in Panama are a classical example of this phenomenon. Since the malaria eradication program began, there have been numerous reports linking increased irritability of anopheline mosquitoes with control failures, but one suspects that many such behavioral changes are not the result of the selective effect of the insecticide; instead they may depict the normal reaction of the insect to the toxicant. In this light, the detailed experiments of Gerold and Laarman (1967) and Gerold (1970a,b) are of particular interest. These authors attempted to show whether one could select mosquitoes for their ability to escape contact with DDT, in other words, selection for irritability. *Anopheles atroparvus* were selected for 32 generations on the basis of their escape behavior from contact with DDT deposits. The percentage escaping differed widely between males and females. Neither DDT nor the solvent used (Risella oil) was found to be essential for the escape reaction, but both factors enhanced the response. The selected mosquitoes showed greater average flight speed, more direct flight (fewer turns), and greater flight activity. Escape was not the direct result of enhanced activity. Irritation by DDT accounts for the increased flight activity, but ability to escape is thought to be a separate component of behavior. Susceptibility to DDT was similar in both strains.

Observations on the resting habits of house flies convinced Keiding (1965) that complex factors including external (environmental) and internal (physiological) may influence the activity and choice of resting places.

Do changes in behavior result in physiological resistance to insecticides? The blow fly *Phormia terrae novae* obtains a larger dose of DDT than *M. domestica* when both are exposed to a DDT deposit. This is due to the fact that *Phormia* crawls over the surface almost continuously whereas *Musca* flies up frequently and spends much time in the air (Drobozina, 1968). The house fly begins to avoid contact with the treated surface whereas the blow fly increases the duration of contact. The slow development of resistance in *Phormia* is due to its behavior with little chance of survival after contact with DDT. Resistant flies of both species are less active and less irritated by DDT. Similarly, OP-selected strains of the spotted root maggot *E. notata* survive longer periods of contact with OP compounds in plastic tubes because they spend less time on the treated papers than on the ends of the tubes (Hooper and Brown, 1965a,b). The end-resters are as susceptible to the toxicant by topical application, indicating no increase in physiological resistance. However, parallel selection with DDT and dieldrin confers true physiological resistance. The authors conclude that strains with increased irritability have greater physiological susceptibility due to decreased detoxication rate, whereas the opposite is true in the less irritable individuals.

A critical examination of the literature by Muirhead-Thomson (1960, 1968) failed to reveal any convincing evidence of the existence of behavioristic resistance in mosquitoes, i.e., instances in which irritability and avoidance have appeared or have been intensified as a result of continuous insecticide pressure. The experimental approach to measuring irritability has been hindered by the fact that behavior can readily be influenced by many external factors other than the insecticide itself, such as temperature, mosquito age, length of exposure, type of testing apparatus, mosquito species and strain, insecticide-resistance or susceptibility, etc. These factors have recently been reviewed by Kaschef (1970) and Brown and Pal (1971) and include many literature citations. Standardization of methods for measuring irritability was worked out by Coluzzi and Coluzzi (1961) and Coluzzi (1962, 1963).

VI. The Role of Synergists in Insecticide Resistance

Synergism is the phenomenon exhibited by a combination of two or more compounds such that the physiological effect produced is significantly greater than the summation of their individual effects (Bliss, 1939). As related to insect toxicology, the synergist is almost always used at high dosages but below their direct toxic level and, for this reason, most synergists have been labeled as nontoxic compounds.

The voluminous literature on this subject has been brought together and critically reviewed by Casida (1970), Fishbein and Falk (1969), Hennessy (1970), Kearns (1956), Metcalf (1955b, 1967), Metcalf *et al.* (1966a), O'Brien (1966), Sumerford (1954), Veldstra (1956), Wilkinson (1967, 1968), Wilkinson and Hicks (1969), and comprehensive reviews on the mathematical aspects of synergism have been published by Hewlett (1960) and Sakai (1960). A discussion of synergistic action from the mechanistic point of view is given in Chapter 10 of Volume V.

The practical application of synergism has been most successful with pyrethrin insecticides and related compounds and, undoubtedly, is also useful with other insecticides, but a purposeful application of this principle in pest control operations has not been widespread.

It is now well recognized that synergists enhance the activity of insecticides by inhibiting the enzymes which are responsible for their *in vivo* degradation. It is clear from the preceding discussions on insecticide metabolism that resistance to many of the commonly used insecticides is associated to a large extent with the ability of the resistant strain to detoxify the compound at a faster rate than the susceptible insect, although other factors also play a role. In most instances, these differences are quantitative in nature, so that the action of the synergist is to reduce the amount of insecticide being detoxified. The concept of interference with metabolic detoxication was provided by Lindquist *et al.* (1947) who showed that pretreating house flies with piperonyl cyclonene before application of pyrethrins resulted in enhanced knockdown and subsequent mortality. However, application of pyrethrins at an extended interval before application of the synergist had no such effect.

The biochemical significance of synergistic action has its origin in the suggestion by Wilson (1949) that piperonyl butoxide and piperonyl cyclonene most likely damage the mechanism responsible for the detoxication of pyrethrins.

Synergists may be divided into two major groups: the first group is characteristic of microsomal mixed-function oxidase inhibitors and includes diverse groups of compounds such as those containing the methylenedioxyphenyl (MDP) moiety, methylenedioxynaphthalenes, substituted phenyl-2-propynyl ethers, propynylnaphthyl ethers, benzothiadiazoles, and miscellaneous compounds such as SKF 525-A (2-diethylaminoethyl-2,2-diphenylvalerate hydrochloride), Lilly 18947 (2,4-dichloro-6-phenylphenoxyethyl diethylamine), octachlorodipropyl ether, and other experimental compounds; the second group comprises mainly analogue synergists which are structurally related to the insecticides with which they are combined.

Inhibitors of microsomal oxidations which are most important from toxicological and biochemical considerations of the resistance phenomenon are dealt with in some detail in Chapter 10 of Volume V.

With the widespread occurrence of house fly resistance to many of the chlorohydrocarbon insecticides, intensive efforts were made to find chemical adjuvants which might enhance the activity of these compounds against the resistant strains. In the course of searching for such compounds Sumerford *et al.* (1951a,b) and Speroni (1952) found 1,1-bis(*p*-chlorophenyl)ethanol (DMC), a nontoxic structural analogue of DDT, and various related compounds to effectively synergize DDT against DDT-resistant house flies. Subsequently, a large number of compounds were screened for synergistic activity with DDT, and the most active ones proved to be structural analogues of DDT with modifications mainly at the tertiary carbon moiety (Bergmann *et al.*, 1955; Blum *et al.*, 1959b; Cohen and Tahori, 1957; Kaluszymer *et al.*, 1955; March *et al.*, 1952; Reuter *et al.*, 1956; Tahori, 1955), or derivatives of diphenylamine and benzenesulfonanilides (Speroni *et al.*, 1953; Neeman *et al.*, 1956, 1957). The vast number of compounds evaluated as DDT synergists are summarized by Sumerford (1954). *N*,*N*-Dibutyl *p*-chlorobenzene sulfonamide, designated as "WARF anti-resistant," had been marketed as a DDT synergist but attained only limited success.

The synergistic activity of these compounds with DDT is undoubtedly associated with a decrease in DDE formation *in vivo*. It has been suggested (Speroni *et al.*, 1953; Perry *et al.*, 1953) that DMC may act by competing with DDT for a site on the DDT-detoxifying enzyme, pointing out the similarity between DDT dehydrochlorination and DMC dehydration. DMC is metabolized by the house fly to 1,1-bis(*p*-chlorophenyl)ethylene and then to bis(*p*-chlorophenyl)acetic acid (Perry *et al.*, 1953). A similar mechanism might be envisaged for a number of other compounds but, certainly, not all DDT synergists favor the same type of simple competitive reaction kinetics.

DDT analogues which are active *in vivo* also block the activity of the enzyme DDT dehydrochlorinase *in vitro* (Moorefield and Kearns, 1955; Lipke and Kearns, 1960) but the synergists are not attacked in the process, as they should be if the reaction were strictly of a competitive type.

Considerations of structure-activity relationships of DDT-synergists indicate that many compounds structurally related to DDT in having the bis(*p*-chlorophenyl) moiety intact are very active synergists *in vivo*. Substitution of various radicals at the *p*-Cl position reduces synergistic activity markedly (Metcalf, 1967).

Whether such structural analogues of DDT can ever be of usefulness in practical control operations against resistant insects is doubtful, since house flies selected with DDT-DMC (Moorefield and Kearns, 1955) or with DDT-WARF anti-resistant (Forgash, 1964, 1967) develop resistance to the combinations and cross-resistance to other compounds (Table IX).

In *A. aegypti*, as in house flies, DMC inhibits the dehydrochlorination

TABLE IX

EFFECT OF SELECTION WITH DDT : WARF-ANTI-RESISTANT COMBINATION ON
RESISTANCE TO INSECTICIDES[a,b]

	Strain A (Diazinon) → ↓ (A-140)	Parent strain (U) (no selection)			
		↓ (U-32)	↓ (UDS-83) DDT +WARF-	↓ (F-80)	↓ (U-95)
Insecticide	Diazinon	No selection	anti-resistant	Diazinon selected	No selection
Diazinon	70x	10x	52x	72x	6x
Malathion	6	2	4	6	2
Ronnel	15	5	15	15	4
Dimethoate	3	0.9	3	4	0.7
Isolan	5	7	13	7	4
Dimetilan	8	1.4	8	8	1.5
Pyrethrins	5	1.5	3	3	1.1
DDT	>6000	80	>6000	>6000	60
Lindane	15	30	34	32	5
Chlordane	5	—	6	—	—
Dieldrin	4	—	8	—	—
Heptachlor	4	—	7	—	—

[a] From Forgash (1967).
[b] Degree of resistance = $(LD_{50}$ R-strain$)/(LD_{50}$ S-strain$)$.

of DDT, especially in the resistant strains, thus largely restoring the toxicity of DDT (Abedi *et al.*, 1963; Kimura and Brown, 1964). Similarly, WARF antiresistant decreases the LC_{50} levels of resistant *A. aegypti* larvae by 6- to 60-fold, but has a negligible effect on susceptible larvae (Pillai *et al.*, 1963a). Unfortunately, selection of a susceptible strain with the synergist-DDT mixture results in a 20- to 40-fold increase in resistance in six generations. On the other hand, DMC does not synergize DDT against resistant *Culex fatigans* (Kalra, 1970b), but this is not due to lack of DDT dehydrochlorination by this strain. Lack of synergism by DMC was also noted in DDT-resistant *Culex tarsalis* (Kimura *et al.*, 1965; Plapp *et al.*, 1965), *T. infestans* (Fine *et al.*, 1966), and *Pediculus humanus humanus* (Perry *et al.*, 1963).

TDE labeled with ^{14}C is dehydrochlorinated to some extent by *Culex tarsalis*, but a considerable amount of FW-152 [1,1-bis(*p*-chlorophenyl)-2,2-dichloroethanol] is extracted from the larvae and from the aqueous medium (Plapp and Hennessy, 1966), indicating the involvement of a microsomal oxidase. Probably this is the reason why DMC fails to synergize DDT in this strain.

DMC also fails to synergize DDT against a pyrethrin-resistant strain of *S. granarius* which is also cross-resistant to DDT and Prolan (Lloyd, 1969). However, piperonyl butoxide, sesamex, and other MDP synergists increase the effectiveness of these insecticides against the R strain but not against the S strain. Curiously, WARF anti-R also synergizes DDT against the R strain which metabolize DDT mainly to dicofol, but this is not an isolated case since it occurs also with DDT-R *T. infestans* (Fine *et al.*, 1966).

In view of the importance of the oxidative pathway in DDT metabolism (Agosin *et al.*, 1961b), the above results are reminiscent of Oppenoorth's (1965) findings that the DDT-resistant house fly strain Fc contains little dehydrochlorinase activity which is unrelated to resistance and is not synergized by DMC, whereas the resistance is abolished by sesamex, indicating the presence of an oxidative metabolic pathway. Alternatively, in *Culex fatigans* at least, another resistance mechanism might be operative since neither DMC nor piperonyl butoxide synergize DDT against the resistant strain (Kalra, 1970b).

Autosynergism

As mentioned earlier, synergists act primarily by inhibiting the degradation of insecticides to less toxic derivatives, thus restoring to a measurable degree the original potency of the parent compound. Conceivably, such protection against degradation might be obtained by structural modifications of the toxicant at certain reactive sites with no impairment of its toxic action. Such a compound would then have a "built-in" synergist and the effect might be termed "autosynergism" (Veldstra, 1956).

Structural modifications of this type were attempted in a series of DDT analogues substituted by an additional halogen in the ortho position on one of the benzene rings (Hennessy and O'Reilly, 1956). The substituted chlorine atom in the ortho position causes steric hindrance of the hydrogen at the tertiary position, thus making the molecule refractory to dehydrochlorination.

Highest insecticidal activity against DDT-R strains of house flies was obtained with 2-(2,4-dichlorophenyl)-2-(4-chlorophenyl)-1,1,1-trichloroethane (*ortho*-chloro-DDT), but all the *o*-halogen-substituted derivatives were somewhat less toxic than *p,p'*-DDT to S strains (Hennessy *et al.*, 1961). Exceptions were noted by Perry (1964) and Hoyer and Plapp (1966) who found a Prolan-resistant strain and a DDT-resistant strain of house flies, respectively, to be almost immune to *o*-chloro-DDT. Similarly, DDT-R *A. aegypti* larvae (Abedi *et al.*, 1963) and *C. tarsalis* larvae (Plapp *et al.*, 1965) were found to be highly resistant to *o*-chloro-DDT, but resistance in *C.p. fati-gans* was only twofold (Kalra *et al.*, 1968). In neither of these cases was *o*-Cl-DDT metabolized extensively, the maximum in 24 hours being 25% in house flies and 15% in mosquitoes, and DMC proved to have only a mild or no synergistic effect with *o*-chloro-DDT (Abedi *et al.*, 1963; Kalra, 1970b; Perry *et al.*, 1967).

It is noteworthy that DDT dehydrochlorinase from resistant *A. aegypti* larvae actively dehydrochlorinates *o*-chloro-DDT to *o*-chloro-DDE, and this conversion might be viewed as conferring resistance to this strain (Kimura and Brown, 1964). These contrasting findings emphasize the danger of generalization which is frequently tempting when one tries to formulate a hypothesis correlating resistance with a biochemical factor common to all species or strains.

Following the same line of reasoning, substitution of deuterium (^{2}H) for hydrogen at the tertiary carbon in the DDT molecule could produce an isotope effect on dehydrochlorination (Barker, 1960; Dachauer *et al.*, 1963; Hennessy and O'Reilly, 1956), thereby reducing the rate of DDT breakdown and increasing its toxicity. Indeed [^{2}H]DDT was found to be highly toxic to eleven DDT-R *A. aegypti* strains (Pillai *et al.*, 1963b) and to several other resistant mosquito species (Jakob and Schoof, 1966), but not as toxic as *p,p'*-DDT to R *C. fatigans* larvae (Kalra *et al.*, 1967, 1968). On the other hand, [^{2}H]DDT proved to be only slightly more toxic than *p,p'*-DDT to S and R house flies (Moorefield *et al.*, 1962) and was shown to be metabolized with equal facility as DDT by the R strains (Barker, 1960; Perry *et al.*, 1967). Comparative toxicities of *p,p'*-DDT, *o*-chloro-DDT and [^{2}H]DDT to mosquito larvae and house flies are shown in Fig. 27.

Tritiated DDT([^{3}H]DDT) is considerably more refractory than *p,p'*-DDT to dehydrochlorination by alkali ($kH/k^3H = 12.5$) and *in vitro* by homogenates of DDT-R house flies ($kH/k^3H = 10.8$; Elliott *et al.*, 1962). Accord-

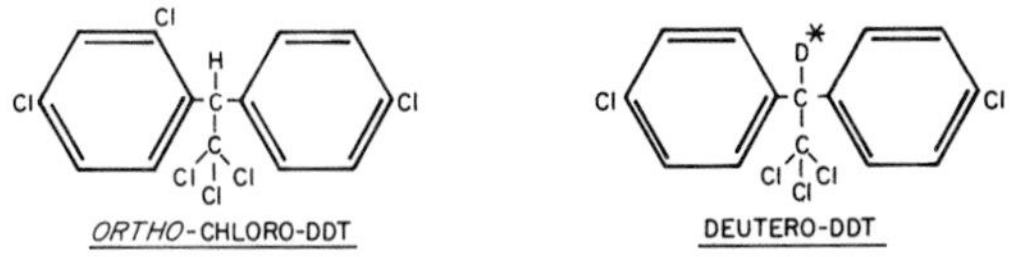

	HOUSEFLY			MOSQUITO (*A. AEGYPTI* LARVA)		
	LD50 μg/FLY			LC50 ppm		
	p,p'-DDT	*o*-Cl-DDT	DEUTERO-DDT	*p,p'*-DDT	*o*-Cl-DDT	DEUTERO-DDT
SUSCEPTIBLE	0.2	0.3	0.15	0.006	0.04	0.009
RESISTANT	>50	1.0	5.0	4.0	6.0	0.5

Fig. 27. Comparative toxicity of *p,p'*-DDT, *o*-Cl-DDT, and deutero-DDT to house flies and mosquito larvae. (From Perry, 1966.)

ingly, [^{3}H]-DDT should be more toxic than p,p'-DDT to R flies. This could not be demonstrated because of the extremely low ^{3}H content (less than 1%) of the DDT molecule.

A purposeful application of the principle of analogue synergism or of auto-synergism on a rational basis has not been practiced in the past, nor has this field been adequately explored. It is safer to predict that inhibitors of microsomal oxidations have more promise in the practical field of pest control in restoring the effectiveness of insecticides against resistant insects than those competitive type compounds synergizing DDT and its derivatives. If any practical answers to the resistance problem are to be forthcoming, the eventual solution may come sooner by a systematic exploration of different classes of synergists and a fundamental understanding of their mode of action.

VII. Summary and Conclusions

Insect resistance to insecticides is an age-old problem which has been brought into focus during the past three decades through intensification of man's efforts to control or eradicate his insect enemies.

The origin and development of resistance is biphasic: phase I is due to selection of variants in the population which carry preadaptive genes for resistance, ultimately attaining a resistance level commensurate with the gene pool originally present; in phase II, enhancement of resistance takes place by induction of preexisting detoxifying enzymes toward higher activity, thus resulting in faster breakdown of the chemical. The inducer is the insecticide itself which is used to control the insect.

Many factors are involved in insect resistance to insecticides. Among these are rate of penetration of the toxicant, rates of activation (toxication), and degradation (detoxication) of the chemical to primary metabolites, conjugation and excretion of secondary metabolites, storage of unchanged compound or of metabolites in nonsensitive tissues, sensitivity of nerve tissue and of target enzymes to the toxic agent, lipid barrier surrounding the nerve sheath, binding properties of nerve components with the chemical, etc. In addition, supplementary mechanisms such as lipid content, dietary factors, various chemical differences, biotic potential, and behavioral patterns may also play a role in resistance.

By far, the predominant factor in resistance is the ability of the resistant insect to detoxify the compound at a faster rate and in larger quantities than its susceptible counterpart. It is now generally agreed that both susceptible and resistant insects basically contain the same detoxifying enzymes, so that differences between the strains are quantitative rather than qualitative.

A high degree of correlation exists between DDT-resistance and enzymatic metabolism of DDT in the house fly and several other insects, but this relationship does not hold true for many anopheline and culicine mosquitoes with the exception, perhaps, of *Aedes aegypti* larvae. Other resistant (R) and susceptible (S) insects detoxify DDT *in vitro* with equal facility, but only the resistant strain degrades the insecticide *in vivo*. These observations indicate that other protective mechanisms than detoxication must be present. DDT and analogues are metabolized via dehydrochlorination by DDTase enzymes and via hydroxylation through the mediation of microsomal mixed-function oxidases.

Resistance to benzene hexachloride is characterized by more extensive degradation of the insecticide *in vivo* by the R strain, but little difference is noted in the enzymatic degradation *in vitro* between R and S strains.

Detoxication is not a factor in insect resistance to the cyclodiene insecticides. It appears that the resistance mechanism (although largely undetermined) might reside at the site of action of these insecticides and might involve a lower sensitivity of nerve tissue and reduced binding capacity of the toxicants with nerve components of the R strain.

The organophosphorus (OP) and carbamate insecticides share a common distinction of being particularly susceptible to attack by microsomal mixed-function oxidases as well as by soluble enzymes. The resistant insect is capable of detoxifying these compounds and excreting conjugated metabolites at a faster rate than its susceptible counterpart. This factor, coupled with reduced penetration of the toxicant, account favorably for the resistance in several species. However, in some instances it is difficult to reconcile the slight differences among S and R strains in oxidative or hydrolytic capacity, with the high levels of resistance, and only a few instances have been recorded where insensitivity of the cholinesterase enzyme to inhibition by OP and carbamate compounds can account for the resistance of the arthropod.

Other factors such as diet, higher lipid content, miscellaneous chemical differences, biotic potential, and changes in behavior cannot be considered as primary defense mechanisms but can supplement an existing, well-defined, protective mechanism such as detoxication and thereby enhance the resistance potential.

In general, so-called synergists act by inhibiting the enzymatic breakdown of insecticides, thereby restoring their potency, especially against the resistant insect. In this respect, synergists can also be used to determine if a causal relationship exists between insecticide metabolism and resistance.

While much knowledge has been gained concerning the resistance problem it is apparent that many questions remain unanswered. This is due, no doubt, to the complex nature of this phenomenon, but it also reflects on our meager knowledge of the normal physiology and biochemistry of insects, especially at the cellular level.

At the present time, chemicals are our most effective weapons in controlling insect pests, and most likely, will continue to be so for some time, but to arrest the development of resistance through increased use of insecticides is tantamount to stopping evolution. It is easier to hasten evolution. The challenge confronting entomologists is how to retard the selection and induction processes. It is here that imaginative studies and ingenious discoveries must be made through the competent and tireless efforts of those researchers who have accepted the challenge of a fundamental approach to the problem.

"Some writers warn that our widespread use of pesticides is upsetting nature's balance; indeed, there are times when nature's widespread use of pests upsets man's balance!"

VIII. Appendix

The following tabulation gives the common names and chemical names of all insecticides and other compounds referred to in the text. Proprietary names are given in parentheses. Common names are those approved by the Entomological Society of America; chemical names are in accordance with the principles of *Chemical Abstracts*. From Kenaga (1969).

Common name	Chemical name
Acethion	O,O-Diethyl-S-carboethoxymethyl phosphorodithioate
Aldicarb (Temik)	[2-Methyl-2-(methylthio) propionaldehyde O-methylcarbamoyl] oxime
Aldrin	1,2,3,4,10,10-Hexachloro-1,4,4a,5,8,8a-hexahydro-1,4-*endo-exo*-5,8-dimethanonaphthalene
Allethrin	2-Allyl-4-hydroxy-3-methyl-2-cyclopenten-1-one ester of 2,2-dimethyl-3-(2-methylpropenyl)-cyclopropanecarboxylic acid
Aminocarb (Matacil)	4-Dimethylamino-m-tolyl methylcarbamate
Apholate	2,2,4,4,6,6-Hexakis-(1-aziridinyl)-2,2,4,4,6,6-hexahydro-1,3,5,2,4,6-triazatriphosphorine
Azinphosmethyl (Guthion)	O,O-Dimethyl S [4-oxo-1,2,3-benzotriazin-3-(4H)-ylmethyl] phosphorodithioate
Banol	2-Chloro-4,5-dimethylphenyl N-methylcarbamate
Barthrin	6-Chloropiperonyl 2,2-dimethyl-3-(2-methylpropenyl) cyclopropanecarboxylate
γ-Benzene hexachloride (Lindane)	1,2,3,4,5,6-Hexachlorocyclohexane, γ-isomer
Bromophos	O-(4 Bromo-2,5-dichlorophenyl)O,O-dimethyl phosphorothioate

Common name	Chemical name
Bulan	1,1-Bis(p-chlorophenyl)-2-nitrobutane
Carbaryl (Sevin)	1-Naphthyl methylcarbamate
Carbofuran (Furadan)	2,3-Dihydro-2,2-dimethylbenzofuranyl-7-N-methylcarbamate
Carbophenothion (Trithion)	S-[(p-chlorophenylthio)methyl]O,O-diethyl phosphorodithioate
Chlordane	1,2,4,5,6,7,8,8-Octachloro-3a,4,7,7a-tetrahydro-4,7-methanoindane
Chlorthion	Dimethyl 3-chloro-4-nitrophenyl phosphorothioate
Coumaphos (Co-Ral)	O-(3-Chloro-4-methyl-2-oxo-2H-1-benzopyran-7-yl) O,O-diethyl phosphorothioate
Cyclethrin	2,2-Dimethyl-3-(2-methylpropenyl) cyclopropane carboxylic acid ester with 2-(2-cyclopenten-1-yl)-4-hydroxy-3-methyl-2-cyclopenten-1-one
DDA	Bis(p-chlorophenyl) acetic acid
DDD (See TDE)	
DDE	1,1-Dichloro-2,2-bis(p-chlorophenyl) ethylene
DDT	1,1,1-Trichloro-2,2-bis(p-chlorophenyl) ethane
Demeton (Systox)	Mixture of O,O-diethyl S-(and O)-2-[(ethylthio) ethyl)] phosphorothioates
Diazinon	O,O-Diethyl O-(2-isopropyl-4-methyl-6-pyrimidyl) phosphorothioate
Dicapthon	O-(2-Chloro-4-nitrophenyl)-O,O-dimethyl phosphorothioate
Dichlorvos (DDVP, Vapona)	2,2-Dichlorovinyl dimethyl phosphate
Dicofol (Kelthane)	4,4′-Dichloro-α-(trichloromethyl) benzhydrol
Dicrotophos (Bidrin)	3-Hydroxy-N,N-dimethyl-cis-crotonamide, dimethyl phosphate
Dieldrin	1,2,3,4,10,10-Hexachloro-6,7-epoxy-1,4,4a,5,6,7,8a-octahydro-1,4-$endo$-exo-5,8-dimethanonaphthalene
Dilan	A mixture of 1 part of Prolan and 2 parts of Bulan
Dimefox	Tetramethylphosphorodiamidic fluoride
Dimethoate (Cygon)	O,O-Dimethyl S-(N-methylcarbamoylmethyl) phosphorodithioate
Dimethrin	2,4-Dimethylbenzyl 2,2-dimethyl-3-(2-methylpropenyl) cyclopropanecarboxylate
Dimetilan	1-(Dimethylcarbamoyl)-5-methyl-3-pyrazolyl dimethylcarbamate
Dioxathion (Delnav)	S,S'-p-Dioxane-2,3-Diyl-O,O-diethyl phosphorodithioate (cis and $trans$ isomers)
Disulfoton (Di-Syston)	O,O-Diethyl S-2-[(ethylthio)ethyl] phosphorodithioate
Endosulfan (Thiodan)	6,7,8,9,10,10-Hexachloro-1,5,5a,6,9,9a-hexahydro-6,9-methano-2,4,3-benzodioxathiepin 3-oxide
Endrin	1,2,3,4,10,10-Hexachloro-6,7-epoxy-1,4,4a,5,6,7,8,8a-octahydro-1,4-$endo$-$endo$-5,8-dimethanonaphthalene

Common name	Chemical name
Famphur (Famophos)	O-[p-(Dimethylsulfamoyl)phenyl] O,O-dimethyl phosphorothioate
Fenitrothion (Sumithion)	O,O-Dimethyl O-(4-nitro-m-tolyl) phosphorothioate
Fensulfothion	O,O-Diethyl O-p-[(methylsulfinyl)phenyl] phosphorothioate
Fenthion (Baytex)	O,O-Dimethyl O-[4-(methylthio)-m-tolyl] phosphorothioate
Hemel	Hexamethylmelamine
Hempa	Hexamethylphosphoric triamide
Heptachlor	1,4,5,6,7,8,8-Heptachloro-3a,4,7,7a-tetrahydro-4,7-methanoindene
Heptachlor epoxide	1,4,5,6,7,8,8a-Heptachloro-6,7-epoxy-3a,4,7,7a-tetrahydro-4,7-methanoindene
Imidan	O,O-Dimethyl S-phthalimidomethyl phosphorodithioate
Isodrin	1,2,3,4,10,10-Hexachloro-1,4,4a,5,8,8a-hexahydro-1,4-endo-endo-5,8-dimethanonaphthalene
Isolan	1-Isopropyl-3-methyl-5-pyrazolyl dimethylcarbamate
Landrin	3,4,5-Trimethylphenyl methylcarbamate, 75%; 2,3,5-trimethylphenyl methylcarbamate, 18%
Malathion	Diethylmercaptosuccinate, S-ester with O,O-dimethyl phosphorodithioate
Metepa	Tris(2-methyl-1-aziridinyl) phosphine oxide
Methiocarb (Mesurol)	4-(Methylthio) 3,5-xylyl methylcarbamate
Methiochlor	1,1,1-Trichloro-2,2-bis(p-methylthiophenyl) ethane
Methoxychlor	1,1,1-Trichloro-2,2-bis(p-methoxyphenyl) ethane
Methyl parathion	O,O-Dimethyl O-p-nitrophenyl phosphorothioate
Monocrotophos (Azodrin)	3-Hydroxy-N-methyl cis-crotonamide dimethyl phosphate
Mevinphos (Phosdrin)	Methyl 3-hydroxy α-crotonate, dimethyl phosphate
Nicotine	1,3-(1-Methyl-2-pyrrolidyl) pyridine
Parathion	O,O-Diethyl O-p-nitrophenyl phosphorothioate
Phorate (Thimet)	O,O-Diethyl-S-[(ethylthio) methyl] phosphorodithioate
Phosphamidon (Dimecron)	2-Chloro-N,N-diethyl-3-hydroxycrotonamide, dimethyl phosphate
Phthalthrin	2,2-Dimethyl-3-(2-methylpropenyl) cyclopropanecarboxylic acid esters with N-(hydroxymethyl)-1-cyclohexene-1,2-dicarboximide
Piperonyl butoxide	α-[2-(2-Butoxyethoxy) ethoxy]-4,5-methylenedioxy-2-propyltoluene
Prolan	1,1-Bis(p-chlorophenyl)-2-nitropropane
Propoxur (Baygon)	o-Isopropoxyphenyl methyl carbamate
Pyrethrins	Pyrethrum (from plant species *Chrysanthemum cinariaefolium*), mixture of pyrethrin I and II, and cinerin I and II
Pyrolan	3-Methyl-1-phenylpyrazol-5-yl dimethylcarbamate
Ronnel, fenchlorphos	O,O-Dimethyl O-2,4,5-trichlorophenyl phosphorothioate

Common name	Chemical name
Rotenone	1,2,12,12a,Tetrahydro-2-isopropenyl-8-9-dimethoxy [1] benzopyrano-[3-4-*b*] furo [2,3-*b*] [1] benzopyran-6(6*aH*) one
Schradan, OMPA	Bis-*N*,*N*,*N'*,*N'*-tetramethylphosphorodiamidic anhydride
Sesamex	2-(2-Ethoxyethoxy) ethyl-3,4-(methylenedioxy) phenyl acetal of acetaldehyde
TDE, DDD	1,1-Dichloro-2,2-bis(*p*-chlorophenyl) ethane
Tepa	Tris(1-aziridinyl) phosphine oxide
Thiotepa	Tris(1-aziridinyl) phosphine sulfide
Toxaphene	Chlorinated camphene containing 67–69% chlorine
Trichlorfon (Dipterex)	Dimethyl (2,2,2-trichloro-1-hydroxyethyl) phosphonate
Tropital	Piperonal bis[2-(2-butoxyethoxy) ethyl] acetal
Zectran	4-Dimethylamino-3,5-xylyl methylcarbamate

Acknowledgments

The contribution by A. S. Perry was made in his private capacity and no official support or endorsement by the Public Health Service is intended, or should be inferred. The contribution by M. Agosin was supported in part by U.S. Public Health Service Grants AI-2300-10, AI-09902-01, and AI-09902-02.

References

Abasa, R. O., and Hansens, E. J. (1969). *J. Econ. Entomol.* **62,** 334.

Abd El-Aziz, S. A., Shafik, M. T., and El-Khishen, S. A. (1965). *Alexandria J. Agr. Res.* **13,** 37.

Abd El-Aziz, S. A., Shafik, M. T., and El-Khishen, S. A. (1967). *Alexandria J. Agr. Res.* **14,** 13.

Abd El-Aziz, S. A., Metcalf, R. L., and Fukuto, T. R. (1969). *J. Econ. Entomol.* **62,** 318.

Abdel-Wahab, A. M., and Casida, J. E. (1967). *J. Agr. Food Chem.* **15,** 479.

Abedi, Z. H., and Brown, A. W. A. (1961). *Ann. Entomol. Soc. Amer.* **54,** 539.

Abedi, Z. H., Duffy, J. R., and Brown, A. W. A. (1963). *J. Econ. Entomol.* **56,** 511.

Acree, F., Jr., Shaffer, P., and Haller, H. L. (1936). *J. Econ. Entomol.* **29,** 601.

Adams, C. H., and Cross, W. H. ('1967). *J. Econ. Entomol.* **60,** 1016.

Agosin, M. (1963). *Bull. W.H.O.* **29** Suppl., 69.

Agosin, M., Scaramelli, N., and Neghme, A. (1961a). *Comp. Biochem. Physiol.* **2,** 143.

Agosin, M., Michaeli, D., Miskus, R., Nagasawa, S., and Hoskins, W. M. (1961b). *J. Econ. Entomol.* **54,** 340.

Agosin, M., Scaramelli, N., Dinamarca, M. L., and Aravena, L. (1963). *Comp. Biochem. Physiol.* **8,** 311.

Agosin, M., Morello, A., and Scaramelli, N. (1964). *J. Econ. Entomol.* **64,** 974.

Agosin, M., Fine, B., Scaramelli, N., Ilivicky, J., and Aravena, L. (1966). *Comp. Biochem. Physiol.* **19,** 339.

Agosin, M., Scaramelli, N., Gil, L., and Letelier, M. E. (1969). *Comp. Biochem. Physiol.* **29,** 785.

Akov, S., and Bořkovec, A. B. (1968). *Life Sci., Part II* **7,** 1215.

Akov, S., Oliver, J. E., and Bořkovec, A. B. (1968). *Life Sci., Part II* **7,** 1207.

Andrawes, N. R., and Dorough, H. W. (1967). *J. Econ. Entomol.* **60,** 453.

Ariaratnam, V., and Georghiou, G. P. (1971). *Nature (London)* **232,** 642.

Arias, R. O., and Terriere, L. C. (1962). *J. Econ. Entomol.* **55,** 925.

Arthur, B. W., and Casida, J. E. (1957). *J. Agr. Food Chem.* **5,** 186.

Ascher, K. R. S., and Neri, I. (1961). *Entomol. Exp. Appl.* **4,** 7.

Atallah, Y. H., and Nettles, W. C., Jr. (1966). *J. Econ. Entomol.* **59,** 560.

Babers, F. H. (1949). *U.S. Dep. Agr., Bur. Entomol. Plant Quantine* **E-776.**

Babers, F. H., and Pratt, J. J., Jr. (1951). *U.S. Dep. Agr., Bur. Entomol. Plant Quarantine* **E-818.**

Babers, F. H., and Roan, C. C. (1953). *J. Econ. Entomol.* **46,** 1105.

Babers, F. H., Pratt, J. J., Jr., and Williams, M. (1953). *J. Econ. Entomol.* **46,** 914.

Balabaskaran, S., Clark, A. G., Cundell, A., and Smith, J. N. (1968). *Australas. J. Pharm.* [N.S.] **49,** Suppl., 66.

Balazs, I., and Agosin, M. (1968). *Biochim. Biophys. Acta* **157,** 1.

Ballantyne, G. H., and Harrison, R. A. (1967). *Entomol. Exp. Appl.* **10,** 231.

Bami, H. L., Sharma, M. I. D., and Kalra, R. L. (1957). *Bull. Nat. Soc. Ind. Malar.* **5,** 246.

Barber, G. W., and Schmitt, J. B. (1949). *J. Econ. Entomol.* **42,** 287.

Barker, R. J. (1960). *J. Econ. Entomol.* **53,** 35.

Barnes, W. W., and Ware, G. W. (1965). *J. Econ. Entomol.* **58,** 286.

Barton-Browne, L., and Kerr, R. W. (1967). *Entomol. Exp. Appl.* **10,** 337.

Beament, J. W. L. (1958). *J. Insect Physiol.* **2,** 199.

Beard, R. L. (1952). *J. Econ. Entomol.* **45,** 561.

Beard, R. L. (1965). *Entomol. Exp. Appl.* **8,** 193.

Becht, G. (1958). *Nature (London)* **181,** 777.

Bell, J. D. (1968a). *Bull. Entomol. Res.* **58,** 137.

Bell, J. D. (1968b). *Bull. Entomol. Res.* **58,** 191.

Benjamini, E., Metcalf, R. L., and Fukuto, T. R. (1959a). *J. Econ. Entomol.* **52,** 94.

Benjamini, E., Metcalf, R. L., and Fukuto, T. R. (1959b). *J. Econ. Entomol.* **52,** 99.

Benke, G. M., and Wilkinson, C. F. (1971). *Pestic. Biochem. Physiol.* **1,** 19.

Bennett, S. E., and Thomas, C. A., Jr. (1963). *J. Econ. Entomol.* **56,** 239.

Bergmann, E. D., Tahori, A. S., Kaluszyner, A., and Reuter, S. (1955). *Nature (London)* **176,** 266.

Berteau, P. E., and Casida, J. E. (1969). *J. Agr. Food Chem.* **17,** 931.

Berteau, P. E., Casida, J. E., and Narahashi, T. (1968). *Science* **161,** 1151.

Bigelow, R. S., and LeRoux, E. J. (1954). *Can. Entomol.* **86,** 78.

Bigley, W. S., and Plapp, F. W., Jr. (1960). *Ann. Entomol. Soc. Amer.* **53,** 360.

Bigley, W. S., and Plapp, F. W., Jr. (1961). *J. Econ. Entomol.* **54,** 904.

Binning, A., Darby, F. J., Heenan, M. P., and Smith, J. N. (1967). *Biochem. J.* **103,** 42.

Blackith, R. E. (1953). *Ann. Appl. Biol.* **40,** 106.

Bliss, C. I. (1939). *Ann. Appl. Biol.* **26,** 585.

Blum, M. S., and Kearns, C. W. (1956). *J. Econ. Entomol.* **49,** 862.

Blum, M. S., Earle, N. W., and Roussel, J. S. (1959a). *J. Econ. Entomol.* **52**, 17.

Blum, M. S., Pratt, J. J., Jr., and Bornstein, J. (1959b). *J. Econ. Entomol.* **52**, 626.

Bochnig, V. (1960). *Z. Angew. Entomol.* **47**, 323.

Bollen, W. B., Roberts, J. E., and Morrison, H. E. (1958). *J. Econ. Entomol.* **51**, 214.

Bowman, J. S., and Casida, J. E. (1957). *J. Agr. Food Chem.* **5**, 192.

Bowman, J. S., and Casida, J. E. (1958). *J. Econ. Entomol.* **51**, 838.

Bowman, M. C., Acree, F., Jr., Lofgren, C. S., and Beroza, M. (1964). *Science* **146**, 1480.

Bradbury, F. R. (1957). *J. Sci. Food Agr.* **8**, 90.

Bradbury, F. R., and Standen, H. (1955). *J. Sci. Food Agr.* **6**, 909.

Bradbury, F. R., and Standen, H. (1956a). *J. Sci. Food Agr.* **7**, 389.

Bradbury, F. R., and Standen, H. (1956b). *Nature (London)* **178**, 1053.

Bradbury, F. R., and Standen, H. (1958). *J. Sci. Food Agr.* **9**, 203.

Bradbury, F. R., and Standen, H. (1959). *Nature (London)* **183**, 983.

Bradbury, F. R., and Standen, H. (1960). *J. Sci. Food Agr.* **11**, 92.

Bradbury, F. R., Nield, P., and Newman, J. F. (1953). *Nature (London)* **172**, 1052.

Brady, U. E., Jr., and Arthur, B. W. (1961). *J. Econ. Entomol.* **54**, 1232.

Brady, U. E., Jr., and Arthur, B. W. (1963). *J. Econ. Entomol.* **56**, 477.

Brattsten, L. B., and Metcalf, R. L. (1970). *J. Econ. Entomol.* **63**, 101.

Bridges, P. M. (1957). *Biochem. J.* **66**, 316.

Bridges, R. G. (1959). *Nature (London)* **184**, 1337.

Bridges, R. G., and Cox, J. T. (1959). *Nature (London)* **184**, 1740.

Brooks, G. T. (1960). *Nature (London)* **186**, 96.

Brooks, G. T. (1966). *World Rev. Pest Contr.* **5**, 62.

Brooks, G. T. (1969). *Residue Rev.* **27**, 81.

Brooks, G. T., and Harrison, A. (1964a). *Biochem. Pharmacol.* **13**, 827.

Brooks, G. T., and Harrison, A. (1964b). *J. Insect Physiol.* **10**, 633.

Brooks, G. T., and Harrison, A. (1965). *Nature (London)* **205**, 1031.

Brooks, G. T., and Harrison, A. (1966). *Life Sci.* **5**, 2315.

Brooks, G. T., and Harrison, A. (1967a). *Life Sci.* **6**, 681.

Brooks, G. T., and Harrison, A. (1967b). *Life Sci.* **6**, 1439.

Brooks, G. T., and Harrison, A. (1969). *Biochem. Pharmacol.* **18**, 557.

Brooks, G. T., Harrison, A., and Cox, J. T. (1963). *Nature (London)* **197**, 311.

Brooks, G. T., Harrison, A., and Lewis, S. E. (1970). *Biochem. Pharmacol.* **19**, 255.

Brown, A. W. A. (1958). "Insecticide Resistance in Arthropods," Monogr. Ser. No. 38, World Health Organ., Geneva.

Brown, A. W. A. (1960). *Bull. Entomol. Soc. Amer.* **7**, 6.

Brown, A. W. A. (1963). *Farm Chem.* **126**, 21.

Brown, A. W. A. (1964). *In* "Handbook of Physiology" Amer. Physiol. Soc., (J. Field, ed.), Sect. 4, Chapter 48, pp. 773–793. Williams & Wilkins, Baltimore, Maryland.

Brown, A. W. A., and Pal, R. (1971). "Insecticide Resistance in Arthropods," 2nd ed., Monogr. Ser. No. 38. World Health Organ., Geneva.

Brown, A. W. A., and Perry, A. S. (1956). *Nature (London)* **178**, 368.

Bruce, W. N. (1949). *Pest Contr.* **17**, 7.

Bruce, W. N., and Decker, G. C. (1950). *Soap Sanit. Chem.* **26**, 122.

Buéi, K., and Fukuhara, Y. (1964). *Botyu-Kagaku* **29**, 9.

Bull, D. L. (1965). *J. Econ. Entomol.* **58**, 249.

Bull, D. L., and Adkisson, P. H. (1963). *J. Econ. Entomol.* **56**, 641.

Bull, D. L., and Lindquist, D. A. (1964). *J. Agr. Food Chem.* **12**, 310.

Bull, D. L., and Lindquist, D. A. (1966). *J. Agr. Food Chem.* **14**, 105.

Bull, D. L., Lindquist, D. A., and Hacskaylo, J. (1963). *J. Econ. Entomol.* **56**, 129.

Bull, D. L., Lindquist, D. A., and Coppedge, J. R. (1967). *J. Agr. Food Chem.* **15**, 610.

Busvine, J. R. (1951). *Nature (London)* **168**, 193.

Busvine, J. R. (1954). *Nature (London)* **174**, 783.

Busvine, J. R. (1957). *Trans. Roy. Soc. Trop. Med. Hyg.* **51**, 11.

Busvine, J. R. (1958). *Bull. W.H.O.* **19**, 1041.

Busvine, J. R. (1959). *Entomol. Exp. Appl.* **2**, 58.

Busvine, J. R., and Townsend, M. G. (1963). *Bull. Entomol. Res.* **53**, 763.

Butts, J. S., Chang, S. C., Christensen, B. E., and Wang, C. H. (1953). *Science* **117**, 699.

Camp, H. B., and Arthur, B. W. (1967). *J. Econ. Entomol.* **60**, 803.

Camp, H. B., Fukuto, T. R., and Metcalf, R. L. (1969). *J. Agr. Food Chem.* **17**, 249.

Casida, J. E. (1955). *Biochem. J.* **60**, 487.

Casida, J. E. (1956). *J. Agr. Food Chem.* **4**, 772.

Casida, J. E. (1963). *Annu. Rev. Entomol.* **8**, 39.

Casida, J. E. (1970). *J. Agr. Food Chem.* **18**, 753.

Casida, J. E., Augustinsson, K. B., and Jonsson, G. (1960). *J. Econ. Entomol.* **53**, 205.

Chadwick, L. E. (1957). *Bull. W.H.O.* **16**, 1203.

Chadwick, L. E. (1963). *In* "Handbuch der experimentellen Pharmakologie" (G. B. Koelle, ed.), Vol. 15, Chapter 16, p. 741. Springer-Verlag, Berlin and New York.

Chamberlain, R. (1950). *Amer. J. Hyg.* **52**, 153.

Chamberlain, W. F. (1965). *J. Econ. Entomol.* **58**, 51.

Chamberlain, W. F., and Hamilton, E. W. (1964). *J. Econ. Entomol.* **57**, 800.

Chamberlain, W. F., Gatterdam, P. E., and Hopkins, D. E. (1961). *J. Econ. Entomol.* **54**, 733.

Chan, T. M., Gillett, J. W., and Terriere, L. C. (1967). *Comp. Biochem. Physiol.* **20**, 731.

Chang, S. C., and Bořkovec, A. B. (1969). *J. Econ. Entomol.* **62**, 1417.

Chang, S. C., and Kearns, C. W. (1964). *J. Econ. Entomol.* **57**, 397.

Chang, S. C., Bořkovec, A. B., and Woods, C. W. (1966). *J. Econ. Entomol.* **59**, 937.

Chang, S. C., Terry, P. H., Woods, C. W., and Bořkovec, A. B. (1967). *J. Econ. Entomol.* **60**, 1623.

Chang, S. C., DeMillo, A. B., Woods, C. W., and Bořkovec, A. B. (1968). *J. Econ. Entomol.* **61**, 1357.

Chang, S. C., Woods, C. W., and Bořkovec, A. B. (1970). *J. Econ. Entomol.* **63**, 1510.

Chattoraj, A. N., and Brown, A. W. A. (1960). *J. Econ. Entomol.* **53**, 1049.

Chattoraj, A. N., and Kearns, C. W. (1958). *Bull. Entomol. Soc. Amer.* **4**, 95.

Chen, Y. L., and Casida, J. E. (1969). *J. Agr. Food Chem.* **17**, 208.

Chevalier, J. (1930). *Bull. Sci. Pharmacol.* **37**, 154.

Clark, A. G., Darby, F. J., and Smith, J. N. (1967). *Biochem. J.* **103**, 49.

Clark, A. G., Murphy, S., and Smith, J. N. (1969). *Biochem. J.* **113**, 89.

Cline, R., and Pearce, G. W. (1963). *Biochemistry* **2**, 657.

Cline, R., and Pearce, G. W. (1966). *J. Insect Physiol.* **12**, 153.

Cochran, D. G. (1956). *J. Econ. Entomol.* **49**, 43.

Cohen, A. J., and Smith, J. N. (1961). *Nature (London)* **189**, 600.

Cohen, S., and Tahori, A. S. (1957). *J. Agr. Food Chem.* **5**, 519.

Cole, M. M., and Clark, P. H. (1961). *J. Econ. Entomol.* **54**, 649.

Colhoun, E. H. (1958). *Science* **127**, 25.

Colhoun, E. H. (1959a). *Can. J. Biochem. Physiol.* **37**, 259.

Colhoun, E. H. (1959b). *Can. J. Biochem. Physiol.* **37**, 1127.

Collins, W. J., and Forgash, A. J. (1970). *J. Econ. Entomol.* **63**, 394.

Coluzzi, A., and Coluzzi, M. (1961). *Riv. Malariol.* **40**, 35.

Coluzzi, M. (1962). World Health Organ., WHO/Mal. 329, WHO/Insecticides/130, 9 pp.

Coluzzi, M. (1963). *Riv. Malariol.* **42**, 189.

Cook, B. J. (1967). *Biol. Bull.* **133**, 526.

Cook, B. J., de la Cuesta, M., and Pomonis, J. G. (1969). *J. Insect Physiol.* **15**, 963.

Cook, J. W., and Yip, G. (1958). *J. Ass. Offic. Agr. Chem.* **41**, 407.

Corrigan, J. J., and Kearns, C. W. (1958). *Bull. Entomol. Soc. Amer.* **4**, 95.

Corrigan, J. J., and Kearns, C. W. (1963). *J. Insect Physiol.* **9**, 1.

Cristol, S. (1947). *J. Amer. Chem. Soc.* **69**, 338.

Crow, J. F. (1957). *Annu. Rev. Entomol.* **2**, 227.

Crow, J. F. (1966). *Nat. Acad. Sci—Nat. Res. Counc., Publ.* **1402**, 263–275.

Dachauer, A. C., Cocheo, B., Solomon, M. G., and Hennessy, D. J. (1963). *J. Agr. Food Chem.* **11**, 47.

Dahm, P. A. (1970). *In* "Biochemical Toxicology of Insecticides" (R. D. O'Brien and I. Yamamoto, eds.), pp. 51–63. Academic Press, New York.

D'Allessandro, G., Catalano, G., Mariani, M., Scerrino, E., Smiraglia, C., and Valguarnera, G. (1949). *Sicil. Med.* **6**, 15.

Dame, D. A., and Schmidt, C. H. (1964). *J. Econ. Entomol.* **57**, 77.

Darrow, D. I., and Plapp, F. W., Jr. (1960). *J. Econ. Entomol.* **53**, 777.

Dauterman, W. C., and Matsumura, F. (1962). *Science* **138**, 694.

Dauterman, W. C., Casida, J. E., Knaak, J. B., and Kowalczyk, T. (1959). *J. Agr. Food Chem.* **7**, 188.

Davidow, B. (1950). *J. Ass. Offic. Agr. Chem.* **33**, 886.

Davidow, B., and Frawley, J. P. (1951). *Proc. Soc. Exp. Biol. Med.* **76**, 780.

Davidow, B., and Radomski, J. L. (1953). *J. Pharmacol. Exp. Ther.* **107**, 259.

Davies, M. E., Keiding, J., and von Hofsten, C. G. (1958). *Nature (London)* **182**, 1816.

Decker, G. C. (1958). *J. Agr. Food Chem.* **6**, 98.

Decker, G. C., and Bruce, W. N. (1952). *Amer. J. Trop. Med. Hyg.* **1**, 395.

del Villar, E., and Mosnaim, D. (1967). *Exp. Parasitol.* **21**, 186.

Derbeneva-Ukhova, V. P., Lineva, V. A., and Drobozina, V. P. (1966). *Bull. WHO* **34**, 939.

de Zulueta, J. Jolivet, D., Thymakis, K., and Caprari, P. (1957). *Bull. W.H.O.* **16**, 475.

Dinamarca, M. L., Agosin, M., and Neghme, A. (1962). *Exp. Parasitol.* **12**, 61.

Dinamarca, M. L., Saavedra, I., and Valdés, E. (1969). *Comp. Biochem. Physiol.* **31**, 269.

Dinamarca, M. L., Levenbook, L., and Valdés, E. (1971). *Arch. Biochem. Biophys.* **147**, 374.

Doby, J., Deblock, S., and Gaeremynck, L. (1956). *Bull. Soc. Pathol. Exot.* **49**, 56.

Dorough, H. W. (1968). *J. Agr. Food Chem.* **16**, 319.

Dorough, H. W., and Casida, J. E. (1964). *J. Agr. Food Chem.* **12**, 294.

Drobozina, V. P. (1968). *Med. Parazitol. Parazit. Bolez.* **37**, 36.

Dyte, C. E., and Rowlands, D. G. (1968). *J. Stored Prod. Res.* **4**, 157.

Dyte, C. E., Ellis, V. J., and Lloyd, C. J. (1966). *J. Stored Prod. Res.* **1**, 223.

Earle, N. W. (1963). *J. Agr. Food Chem.* **11**, 281.

Eaton, J. L., and Sternburg, J. (1967). *J. Econ. Entomol.* **60**, 1699.

El Bashir, S. (1967). *Entomol. Exp. Appl.* **10**, 111.

El Bashir, S., and Oppenoorth, F. J. (1969). *Nature (London)* **223**, 210.

Eldefrawi, M. E., and Hoskins, W. M. (1961). *J. Econ. Entomol.* **54**, 401.

Elliott, R. (1959). *Bull. W.H.O.* **20**, 777.

Elliott, R. D., Miles, J. W., and Pearce, G. W. (1962). *Abstr., Amer. Chem. Soc. Meet., Washington, D.C.* No. 27, p. 10A.

Enan, O., Miskus, R., and Craig, R. (1964). *J. Econ. Entomol.* **57**, 364.

Farnham, A. W., Lord, K. A., and Sawicki, R. M. (1965). *J. Insect Physiol.* **11**, 1475.

Fast, P. G., and Brown, A. W. A. (1962). *Ann. Entomol. Soc. Amer.* **55**, 663.

Fay, R. W., Kilpatrick, J. W., and Morris, G. C., III. (1958). *J. Econ. Entomol.* **51**, 452.

Ferguson, W. C., and Kearns, C. W. (1949). *J. Econ. Entomol.* **42**, 810.

Fir̄... ̄. C. (1961). *Nature (London)* **191**, 884.

...(1963). *Pyrethrum Post* **7**, 18.

..., M. E., and Agosin, M. (1966). *Exp. Parasitol.* **19**, 304.

... P. J., Thain, E. M., and Marks, T. B. (1967). *J. Sci. Food*

... H. L. (1969). *Environ. Res.* **2**, 297.

... *Trans. Roy. Soc. Trop. Med. Hyg.* **46**, 6.

... L. G., Philpot, R. M., Yang, R. S. H., Dauterman, W. C., ... *Life Sci., Part II* **9**, 869.

... nd Meyding, G. D. (1966). *J. Agr. Food Chem.* **14**, 83.

... *n. Entomol.* **57**, 644.

... *Entomol.* **60**, 1750.

... J. (1962). *J. Econ. Entomol.* **55**, 679.

... Riley, R. C. (1962). *J. Econ. Entomol.* **55**, 544.

... T. (1962). *Entomol. Exp. Appl.* **5**, 119.

... *Riv. Parassitol.* **20**, 107.

... 3a). *Botyu-Kagaku* **28**, 63.

...). *Botyu-Kagaku* **28**, 77.

... *J. Econ. Entomol.* **59**, 1338.

... J. E. (1967). *Science* **155**, 713.

..., and Casida, J. E. (1969). *J. Agr. Food*

... 1969). *Residue Rev.* **25**, 223.

... **1**, 147.

... and Maxon, M. G. (1955). *J. Econ.*

... d March, R. B. (1956). *J. Econ.*

... 26.

... *ol.* **51**, 1.

... *ol.* **51**, 3.

... *penmed. Parasitol.* **10**, 48.

108   ALBERT S. PERRY AND MOISES AGOSIN

George, J. A., and Brown, A. W. A. (1967). *J. Econ. Entomol.* **60**, 974.

Georghiou, G. P. (1962). *Mosquito News* **22**, 260.

Georghiou, G. P. (1965a). *Nature (London)* **207**, 883.

Georghiou, G. P. (1965b). *Advan. Pest Contr. Res.* **6**, 171.

Georghiou, G. P. (1965c). *J. Econ. Entomol.* **58**, 58.

Georghiou, G. P., and Metcalf, R. L. (1961a). *J. Econ. Entomol.* **54**, 150.

Georghiou, G. P., and Metcalf, R. L. (1961b). *J. Econ. Entomol.* **54**, 231.

Georghiou, G. P., Metcalf, R. L., and March, R. B. (1961). *J. Econ. Entomol.* **54**, 132.

Gerold, J. L. (1970a). *World Health Organ., Inform. Circ. Resistance* **VBC/IRG/70.9**, 34.

Gerold, J. L. (1970b). *World Health Organ., Inform. Circ. Resistance* **VBC/IRG/70.11**, 25.

Gerold, J. L., and Laarman, J. J. (1967). *Nature (London)* **215**, 518.

Gerolt, P. (1965). *J. Econ. Entomol.* **58**, 849.

Gessner, T., and Smith, J. N. (1960). *Biochem. J.* **75**, 165.

Giannotti, O. (1958). *Arq. Inst. Biol. (Sao Paulo)* **25**, 253.

Giannotti, O., Metcalf, R. L., and March, R. B. (1956). *Ann. Entomol. Soc. A* **49**, 588.

Gil, L., Fine, B. C., Dinamarca, M. L., Balazs, I., Busvine, J. R., and A M. (1968). *Entomol. Exp. Appl.* **11**, 15.

Gilotra, S. K. (1965). *Amer. J. Trop. Med. Hyg.* **14**, 165.

Gjullin, C. M., Lindquist, A. W., and Butts, J. S. (1952). *Mosquito News* **12**,

Goodchild, B., and Smith, J. N. (1970). *Biochem. J.* **117**, 1005.

Gordon, H. T. (1961). *Annu. Rev. Entomol.* **6**, 27.

Gordon, H. T., and Welsh, J. H. (1948). *J. Cell. Comp. Physiol.* **31**, 395.

Gratz, N. G. (1966). *Acta Trop.* **23**, 108.

Grayson, J. M. (1951). *J. Econ. Entomol.* **44**, 315.

Grayson, J. M. (1953). *J. Econ. Entomol.* **46**, 124.

Grayson, J. M. (1954). *J. Econ. Entomol.* **47**, 253.

Grigolo, A., and Oppenoorth, F. J. (1966). *Genetica* **37**, 159.

Grover, P. L., and Sims, P. (1965). *Biochem. J.* **96**, 521.

Gunther, F., and Blinn, R. (1947). *J. Amer. Chem. Soc.* **69**, 1215.

Gupta, B., Agarwal, H. C., and Pillai, M. K. K. (1971). *Pestic. Bio* **1**, 180.

Guthrie, F. E., and Apple, J. L. (1961). *J. Insect Pathol.* **3**, 426.

Guthrie, F. E., Ringler, R. L., and Bowery, T. G. (1957). *J.* **50**, 821.

Guthrie, F. E., Campbell, W. V., and Baron, R. L. (1962). *An Amer.* **55**, 42.

Gwiazda, M., and Lord, K. A. (1967). *Ann. Appl. Biol.* **59**, 221.

Hadaway, A. B. (1956). *Nature (London)* **178**, 149.

Harris, J. R., Stoddard, G. E., Bateman, G. Q., Shupe, G. L., Harris, L. E., Bahler, T. L., and Lieberman, F. V. (1956) **62**, 1334.

Harrison, C. M. (1951). *Nature (London)* **167**, 855.

Harrison, C. M. (1952a). *Bull. Entomol. Res.* **42**, 761.

Harrison, C. M. (1952b). *Trans. Roy. Soc. Trop. Med. Hyg.* **46**,

Hartzell, A., and Wilcoxon, F. (1932). *Contrib. Boyce Thompso*

Harvey, G. T., and Brown, A. W. A. (1951). *Can. J. Zool.* **29**,

Hassan, A., Zayed, S. M. A. D., and Abdel-Hamid, F. M. (1965). *Biochem. Pharmacol.* **14**, 1577.

Hatanaka, A., Hilton, B. D., and O'Brien, R. D. (1967). *J. Agr. Food Chem.* **15**, 854.

Havertz, D. S., and Curtin, T. J. (1967). *J. Med. Entomol.* **4**, 143.

Hawkins, W. B., and Sternburg, J. (1964). *J. Econ. Entomol.* **57**, 241.

Hayashi, A., Saito, T., and Iyatomi, K. (1968). *Botyu-Kagaku* **33**, 90.

Hayashi, M., and Hayakawa, M. (1962). *Jap. J. Appl. Entomol. Zool.* **6**, 250.

Hazard, E. I., Lofgren, C. S., Woodard, D. B., Ford, H. R., and Glancey, B. M. (1964). *Science* **145**, 500.

Heath, D. F. (1961). "Organophosphorus Poisons." Pergamon, Oxford.

Hedin, P. A., Wiygul, G., and Mitlin, N. (1967). *J. Econ. Entomol.* **60**, 215.

Helle, W. (1962). *Tijdschr. Plantenziekten* **68**, 155.

Helle, W. (1965). *Recent Advan. Acarol.* **2**, 71.

Henderson, G. L., and Crosby, D. G. (1967). *J. Agr. Food Chem.* **15**, 888.

Henderson, G. L., and Crosby, D. G. (1968). *Bull. Environ. Contam. Toxicol.* **3**, 131.

Henneberry, T. J. (1964). *J. Econ. Entomol.* **57**, 674.

Hennessy, D. J. (1970). *In* "Biochemical Toxicology of Insecticides" (R. D. O'Brien and I. Yamamoto, eds.), p. 105. Academic Press, New York.

Hennessy, D. J., and O'Reilly, M. G. (1956). *Abstr., Amer. Chem. Soc. Meet., Atlantic City, N.J.* Pap. 130, p. 17A.

Hennessy, D. J., Fratantoni, J., Hartigan, J., Moorefield, H. H., and Weiden, M. H. J. (1961). *Nature (London)* **190**, 341.

Herne, D. H. C., and Brown, A. W. A. (1969). *J. Econ. Entomol.* **62**, 205.

Heslop, J. P., and Ray, J. W. (1959). *J. Insect Physiol.* **3**, 395.

Hess, A. D. (1952). *Amer. J. Trop. Med. Hyg.* **1**, 371.

Hewlett, P. S. (1960). *Advan. Pest Contr. Res.* **3**, 27.

Hodgson, E., and Geldiay, S. (1959). *Biol. Bull.* **117**, 275.

Hodgson, E. S., and Smyth, T., Jr. (1955). *Ann. Entomol. Soc. Amer.* **48**, 507.

Hoffman, R. A., and Lindquist, A. W. (1952). *J. Econ. Entomol.* **45**, 233.

Holan, G. (1969). *Nature (London)* **221**, 1025.

Hollingworth, R. M. (1969). *J. Agr. Food Chem.* **17**, 987.

Hollingworth, R. M. (1971). *Bull. W.H.O.* **44**, 155.

Hollingworth, R. M., Metcalf, R. L., and Fukuto, T. R. (1967). *J. Agr. Food Chem.* **15**, 250.

Hooper, G. S. H. (1965). *J. Econ. Entomol.* **58**, 608.

Hooper, G. S. H. (1967). *Proc. Roy. Soc. Queensl.* **79**, 9.

Hooper, G. S. H. (1968). *J. Econ. Entomol.* **61**, 490.

Hooper, G. S. H. (1969). *J. Econ. Entomol.* **62**, 846.

Hooper, G. S. H., and Brown, A. W. A. (1965a). *Bull. W.H.O.* **32**, 131.

Hooper, G. S. H., and Brown, A. W. A. (1965b). *Entomol. Exp. Appl.* **8**, 263.

Hopf, H. S. (1954). *Ann. Appl. Biol.* **41**, 248.

Hopkins, T. L., and Robbins, W. E. (1957). *J. Econ. Entomol.* **50**, 684.

Hoskins, W. M. (1964). *World Rev. Pest Contr.* **3**, 85.

Hoskins, W. M., and Gordon, H. T. (1956). *Annu. Rev. Entomol.* **1**, 89.

Hoskins, W. M., and Witt, J. M. (1958). *Proc. Int. Congr. Entomol., 10th, 1956,* Vol. 2, p. 151.

Hoskins, W. M., Miskus, R., and Eldefrawi, M. E. (1958). "Seminar on Susceptibility of Insects to Insecticides, Panama" (World Health Organ. Report), p. 239.

Hoy, W., and Gordon, H. T. (1961). *J. Econ. Entomol.* **54,** 198.

Hoyer, R. F., and Plapp, F. W., Jr. (1966). *J. Econ. Entomol.* **59,** 495.

Hoyle, G. (1953). *J. Exp. Biol.* **30,** 121.

Huang, E. A., Lu, J. Y., and Chung, R. A. (1970). *Biochem. Pharmacol.* **19,** 637.

Hucker, H. B., Gillette, J. R., and Brodie, B. B. (1960). *J. Pharmacol. Exp. Ther.* **137,** 103.

Hunter, P. E. (1956). *J. Econ. Entomol.* **49,** 671.

Hunter, P. E., Cutkomp, L. K., and Kolkaila, A. M. (1958). *J. Econ. Entomol.* **51,** 579.

Hunter, P. E., Cutkomp, L. K., and Kolkaila, A. M. (1959). *J. Econ. Entomol.* **52,** 765.

Ikeshoji, T., and Suzuki, T. (1959). *Jap. J. Exp. Med.* **29,** 481.

Inagami, K. (1955). *Nippon Nogei Kagaku Kaishi* **29,** 918.

Inwang, E. E. (1968). *Can. J. Zool.* **46,** 15.

Ishaaya, J., and Chefurka, W. (1968). *Riv. Parassitol.* **29,** 289.

Ishida, M. (1968). *Agr. Biol. Chem.* **32,** 947.

Ishida, M., and Dahm, P. A. ('1965a). *J. Econ. Entomol.* **58,** 383.

Ishida, M., and Dahm, P. A. (1965b). *J. Econ. Entomol.* **58,** 602.

Jakob, W. L., and Schoof, H. F. (1966). *Mosquito News* **26,** 78.

Johnston, E. F., Bogart, R., and Lindquist, A. W. (1954). *J. Hered.* **45,** 177.

Kalra, R. L. (1970a). *Bull. W.H.O.* **42,** 623.

Kalra, R. L. (1970b). *Botyu-Kagaku* **35,** 33.

Kalra, R. L., Perry, A. S., and Miles, J. W. (1967). *Bull. W.H.O.* **37,** 651.

Karla, R. L., Perry, A. S., and Miles, J. W. (1968). *Indian J. Exp. Biol.* **6,** 37.

Kaluszyner, A., Reuter, S., and Bergmann, E. D. (1955). *J. Amer. Chem. Soc.* **77,** 4164.

Kaplanis, J. N., Robbins, W. E., Darrow, D. I., Hopkins, D. E., Monroe, R. E., and Treiber, G. (1959). *J. Econ. Entomol.* **52,** 1190.

Kapoor, I. P., Metcalf, R. L., Nystrom, R. F., and Sangha, G. K. (1970). *J. Agr. Food Chem.* **18,** 1145.

Kasai, T., and Ogita, Z. (1965). *SABCO (Soc. Areas Biol. Chem. Overlap).* *J.* **1,** 130.

Kaschef, A. H. (1970). *Bull. W.H.O.* **42,** 917.

Kato, R. (1966). *J. Biochem. (Tokyo)* **59,** 574.

Kauer, K. C., Duvall, R. B., and Alquist, F. N. (1947). *Ind. Eng. Chem.* **39,** 1335.

Kearns, C. W. (1955). *In* "Origins of Resistance to Toxic Agents" (M. G. Sevag, R. D. Reid, and O. E. Reynolds, eds.), pp. 148–58. Academic Press, New York.

Kearns, C. W. (1956). *Annu. Rev. Entomol.* **1,** 123.

Kearns, C. W. (1957). *World Health Organ., Inform. Circ. Resistance Problem* No. 8.

Keiding, J. (1965). *Riv. Parassitol.* **26,** 45.

Keiding, J. (1967). *World Rev. Pest Contr.* **6,** 115.

Keiding, J. (1969). Annual Report. Government Pest Infestation Laboratory, Lyngby, Denmark.

Keller, J. C., Clark, P. H., and Lofgren, C. S. (1956). *Pest Contr.* **24,** 14.

Kenaga, E. E. (1969). *Bull. Entomol. Soc. Amer.* **15,** 85.

Kerr, R. W. (1963). *J. Aust. Inst. Agr. Sci.* **29,** 31.

Kerr, R. W. (1970). *Aust. J. Biol. Sci.* **23,** 377.

Kerr, R. W., Venables, D. G., Roulston, W. J., and Schnitzerling, H. J. (1957). *Nature (London)* **180,** 1132.

Dyte, C. E., Ellis, V. J., and Lloyd, C. J. (1966). *J. Stored Prod. Res.* 1, 223.
Earle, N. W. (1963). *J. Agr. Food Chem.* 11, 281.
Eaton, J. L., and Sternburg, J. (1967). *J. Econ. Entomol.* 60, 1699.
El Bashir, S. (1967). *Entomol. Exp. Appl.* 10, 111.
El Bashir, S., and Oppenoorth, F. J. (1969). *Nature (London)* 223, 210.
Eldefrawi, M. E., and Hoskins, W. M. (1961). *J. Econ. Entomol.* 54, 401.
Elliott, R. (1959). *Bull. W.H.O.* 20, 777.
Elliott, R. D., Miles, J. W., and Pearce, G. W. (1962). *Abstr., Amer. Chem. Soc. Meet., Washington, D.C.* No. 27, p. 10A.
Enan, O., Miskus, R., and Craig, R. (1964). *J. Econ. Entomol.* 57, 364.
Farnham, A. W., Lord, K. A., and Sawicki, R. M. (1965). *J. Insect Physiol.* 11, 1475.
Fast, P. G., and Brown, A. W. A. (1962). *Ann. Entomol. Soc. Amer.* 55, 663.
Fay, R. W., Kilpatrick, J. W., and Morris, G. C., III. (1958). *J. Econ. Entomol.* 51, 452.
Ferguson, W. C., and Kearns, C. W. (1949). *J. Econ. Entomol.* 42, 810.
Fine, B. C. (1961). *Nature (London)* 191, 884.
Fine, B. C. (1963). *Pyrethrum Post* 7, 18.
Fine, B. C., Letelier, M. E., and Agosin, M. (1966). *Exp. Parasitol.* 19, 304.
Fine, B. C., Godin, P. J., Thain, E. M., and Marks, T. B. (1967). *J. Sci. Food Agr.* 18, 220.
Fishbein, L., and Falk, H. L. (1969). *Environ. Res.* 2, 297.
Fletcher, T. E. (1952). *Trans. Roy. Soc. Trop. Med. Hyg.* 46, 6.
Folsom, M. D., Hansen, L. G., Philpot, R. M., Yang, R. S. H., Dauterman, W. C., and Hodgson, E. (1970). *Life Sci., Part II* 9, 869.
Ford, I. M., Menn, J. J., and Meyding, G. D. (1966). *J. Agr. Food Chem.* 14, 83.
Forgash, A. J. (1964). *J. Econ. Entomol.* 57, 644.
Forgash, A. J. (1967). *J. Econ. Entomol.* 60, 1750.
Forgash, A. J., and Hansens, E. J. (1962). *J. Econ. Entomol.* 55, 679.
Forgash, A. J., Cook, B. J., and Riley, R. C. (1962). *J. Econ. Entomol.* 55, 544.
Franco, M., and Oppenoorth, F. J. (1962). *Entomol. Exp. Appl.* 5, 119.
Frontali, N., and Carta, S. (1959). *Riv. Parassitol.* 20, 107.
Fukami, J. I., and Shishido, T. (1963a). *Botyu-Kagaku* 28, 63.
Fukami, J. I., and Shishido, T. (1963b). *Botyu-Kagaku* 28, 77.
Fukami, J. I., and Shishido, T. (1966). *J. Econ. Entomol.* 59, 1338.
Fukami, J. I., Yamamoto, I., and Casida, J. E. (1967). *Science* 155, 713.
Fukami, J. I., Shishido, T., Fukunaga, K., and Casida, J. E. (1969). *J. Agr. Food Chem.* 17, 1217.
Fukunaga, K., Fukami, J., and Shishido, T. (1969). *Residue Rev.* 25, 223.
Fukuto, T. R. (1957). *Advan. Pest Contr. Res.* 1, 147.
Fukuto, T. R., Metcalf, R. L., March, R. B., and Maxon, M. G. (1955). *J. Econ. Entomol.* 48, 347.
Fukuto, T. R., Wolf, J. P., Metcalf, R. L., and March, R. B. (1956). *J. Econ. Entomol.* 49, 147.
Gagliani, M. (1952). *Boll. Soc. Ital. Biol. Sper.* 26, 326.
Gannon, N., and Bigger, J. H. (1958). *J. Econ. Entomol.* 51, 1.
Gannon, N., and Decker, G. C. (1958). *J. Econ. Entomol.* 51, 3.
Garms, R. (1961). *Z. Tropenmed. Parasitol.* 11, 353.
Garms, R., Weyer, F., and Rehm, W. F. (1959). *Z. Tropenmed. Parasitol.* 10, 48.
Gast, R. T. (1961). *J. Econ. Entomol.* 54, 1203.
Gemrich, E. G. (1967). *J. Agr. Food Chem.* 15, 617.

George, J. A., and Brown, A. W. A. (1967). *J. Econ. Entomol.* **60,** 974.

Georghiou, G. P. (1962). *Mosquito News* **22,** 260.

Georghiou, G. P. (1965a). *Nature (London)* **207,** 883.

Georghiou, G. P. (1965b). *Advan. Pest Contr. Res.* **6,** 171.

Georghiou, G. P. (1965c). *J. Econ. Entomol.* **58,** 58.

Georghiou, G. P., and Metcalf, R. L. (1961a). *J. Econ. Entomol.* **54,** 150.

Georghiou, G. P., and Metcalf, R. L. (1961b). *J. Econ. Entomol.* **54,** 231.

Georghiou, G. P., Metcalf, R. L., and March, R. B. (1961). *J. Econ. Entomol.* **54,** 132.

Gerold, J. L. (1970a). *World Health Organ., Inform. Circ. Resistance* **VBC/IRG/70.9,** 34.

Gerold, J. L. (1970b). *World Health Organ., Inform. Circ. Resistance* **VBC/IRG/70.11,** 25.

Gerold, J. L., and Laarman, J. J. (1967). *Nature (London)* **215,** 518.

Gerolt, P. (1965). *J. Econ. Entomol.* **58,** 849.

Gessner, T., and Smith, J. N. (1960). *Biochem. J.* **75,** 165.

Giannotti, O. (1958). *Arq. Inst. Biol. (Sao Paulo)* **25,** 253.

Giannotti, O., Metcalf, R. L., and March, R. B. (1956). *Ann. Entomol. Soc. Amer.* **49,** 588.

Gil, L., Fine, B. C., Dinamarca, M. L., Balazs, I., Busvine, J. R., and Agosin, M. (1968). *Entomol. Exp. Appl.* **11,** 15.

Gilotra, S. K. (1965). *Amer. J. Trop. Med. Hyg.* **14,** 165.

Gjullin, C. M., Lindquist, A. W., and Butts, J. S. (1952). *Mosquito News* **12,** 201.

Goodchild, B., and Smith, J. N. (1970). *Biochem. J.* **117,** 1005.

Gordon, H. T. (1961). *Annu. Rev. Entomol.* **6,** 27.

Gordon, H. T., and Welsh, J. H. (1948). *J. Cell. Comp. Physiol.* **31,** 395.

Gratz, N. G. (1966). *Acta Trop.* **23,** 108.

Grayson, J. M. (1951). *J. Econ. Entomol.* **44,** 315.

Grayson, J. M. (1953). *J. Econ. Entomol.* **46,** 124.

Grayson, J. M. (1954). *J. Econ. Entomol.* **47,** 253.

Grigolo, A., and Oppenoorth, F. J. (1966). *Genetica* **37,** 159.

Grover, P. L., and Sims, P. (1965). *Biochem. J.* **96,** 521.

Gunther, F., and Blinn, R. (1947). *J. Amer. Chem. Soc.* **69,** 1215.

Gupta, B., Agarwal, H. C., and Pillai, M. K. K. (1971). *Pestic. Biochem. Physiol.* **1,** 180.

Guthrie, F. E., and Apple, J. L. (1961). *J. Insect Pathol.* **3,** 426.

Guthrie, F. E., Ringler, R. L., and Bowery, T. G. (1957). *J. Econ. Entomol.* **50,** 821.

Guthrie, F. E., Campbell, W. V., and Baron, R. L. (1962). *Ann. Entomol. Soc. Amer.* **55,** 42.

Gwiazda, M., and Lord, K. A. (1967). *Ann. Appl. Biol.* **59,** 221.

Hadaway, A. B. (1956). *Nature (London)* **178,** 149.

Harris, J. R., Stoddard, G. E., Bateman, G. Q., Shupe, G. L., Greenwood, D. A., Harris, L. E., Bahler, T. L., and Lieberman, F. V. (1956). *J. Econ. Entomol.* **62,** 1334.

Harrison, C. M. (1951). *Nature (London)* **167,** 855.

Harrison, C. M. (1952a). *Bull. Entomol. Res.* **42,** 761.

Harrison, C. M. (1952b). *Trans. Roy. Soc. Trop. Med. Hyg.* **46,** 255.

Hartzell, A., and Wilcoxon, F. (1932). *Contrib. Boyce Thompson Inst.* **4,** 107.

Harvey, G. T., and Brown, A. W. A. (1951). *Can. J. Zool.* **29,** 42.

Hassan, A., Zayed, S. M. A. D., and Abdel-Hamid, F. M. (1965). *Biochem. Pharmacol.* **14**, 1577.

Hatanaka, A., Hilton, B. D., and O'Brien, R. D. (1967). *J. Agr. Food Chem.* **15**, 854.

Havertz, D. S., and Curtin, T. J. (1967). *J. Med. Entomol.* **4**, 143.

Hawkins, W. B., and Sternburg, J. (1964). *J. Econ. Entomol.* **57**, 241.

Hayashi, A., Saito, T., and Iyatomi, K. (1968). *Botyu-Kagaku* **33**, 90.

Hayashi, M., and Hayakawa, M. (1962). *Jap. J. Appl. Entomol. Zool.* **6**, 250.

Hazard, E. I., Lofgren, C. S., Woodard, D. B., Ford, H. R., and Glancey, B. M. (1964). *Science* **145**, 500.

Heath, D. F. (1961). "Organophosphorus Poisons." Pergamon, Oxford.

Hedin, P. A., Wiygul, G., and Mitlin, N. (1967). *J. Econ. Entomol.* **60**, 215.

Helle, W. (1962). *Tijdschr. Plantenziekten* **68**, 155.

Helle, W. (1965). *Recent Advan. Acarol.* **2**, 71.

Henderson, G. L., and Crosby, D. G. (1967). *J. Agr. Food Chem.* **15**, 888.

Henderson, G. L., and Crosby, D. G. (1968). *Bull. Environ. Contam. Toxicol.* **3**, 131.

Henneberry, T. J. (1964). *J. Econ. Entomol.* **57**, 674.

Hennessy, D. J. (1970). *In* "Biochemical Toxicology of Insecticides" (R. D. O'Brien and I. Yamamoto, eds.), p. 105. Academic Press, New York.

Hennessy, D. J., and O'Reilly, M. G. (1956). *Abstr., Amer. Chem. Soc. Meet., Atlantic City, N.J.* Pap. 130, p. 17A.

Hennessy, D. J., Fratantoni, J., Hartigan, J., Moorefield, H. H., and Weiden, M. H. J. (1961). *Nature (London)* **190**, 341.

Herne, D. H. C., and Brown, A. W. A. (1969). *J. Econ. Entomol.* **62**, 205.

Heslop, J. P., and Ray, J. W. (1959). *J. Insect Physiol.* **3**, 395.

Hess, A. D. (1952). *Amer. J. Trop. Med. Hyg.* **1**, 371.

Hewlett, P. S. (1960). *Advan. Pest Contr. Res.* **3**, 27.

Hodgson, E., and Geldiay, S. (1959). *Biol. Bull.* **117**, 275.

Hodgson, E. S., and Smyth, T., Jr. (1955). *Ann. Entomol. Soc. Amer.* **48**, 507.

Hoffman, R. A., and Lindquist, A. W. (1952). *J. Econ. Entomol.* **45**, 233.

Holan, G. (1969). *Nature (London)* **221**, 1025.

Hollingworth, R. M. (1969). *J. Agr. Food Chem.* **17**, 987.

Hollingworth, R. M. (1971). *Bull. W.H.O.* **44**, 155.

Hollingworth, R. M., Metcalf, R. L., and Fukuto, T. R. (1967). *J. Agr. Food Chem.* **15**, 250.

Hooper, G. S. H. (1965). *J. Econ. Entomol.* **58**, 608.

Hooper, G. S. H. (1967). *Proc. Roy. Soc. Queensl.* **79**, 9.

Hooper, G. S. H. (1968). *J. Econ. Entomol.* **61**, 490.

Hooper, G. S. H. (1969). *J. Econ. Entomol.* **62**, 846.

Hooper, G. S. H., and Brown, A. W. A. (1965a). *Bull. W.H.O.* **32**, 131.

Hooper, G. S. H., and Brown, A. W. A. (1965b). *Entomol. Exp. Appl.* **8**, 263.

Hopf, H. S. (1954). *Ann. Appl. Biol.* **41**, 248.

Hopkins, T. L., and Robbins, W. E. (1957). *J. Econ. Entomol.* **50**, 684.

Hoskins, W. M. (1964). *World Rev. Pest Contr.* **3**, 85.

Hoskins, W. M., and Gordon, H. T. (1956). *Annu. Rev. Entomol.* **1**, 89.

Hoskins, W. M., and Witt, J. M. (1958). *Proc. Int. Congr. Entomol., 10th, 1956,* Vol. 2, p. 151.

Hoskins, W. M., Miskus, R., and Eldefrawi, M. E. (1958). "Seminar on Susceptibility of Insects to Insecticides, Panama" (World Health Organ. Report), p. 239.

Hoy, W., and Gordon, H. T. (1961). *J. Econ. Entomol.* **54**, 198.

Hoyer, R. F., and Plapp, F. W., Jr. (1966). *J. Econ. Entomol.* **59**, 495.

Hoyle, G. (1953). *J. Exp. Biol.* **30**, 121.

Huang, E. A., Lu, J. Y., and Chung, R. A. (1970). *Biochem. Pharmacol.* **19**, 637.

Hucker, H. B., Gillette, J. R., and Brodie, B. B. (1960). *J. Pharmacol. Exp. Ther.* **137**, 103.

Hunter, P. E. (1956). *J. Econ. Entomol.* **49**, 671.

Hunter, P. E., Cutkomp, L. K., and Kolkaila, A. M. (1958). *J. Econ. Entomol.* **51**, 579.

Hunter, P. E., Cutkomp, L. K., and Kolkaila, A. M. (1959). *J. Econ. Entomol.* **52**, 765.

Ikeshoji, T., and Suzuki, T. (1959). *Jap. J. Exp. Med.* **29**, 481.

Inagami, K. (1955). *Nippon Nogei Kagaku Kaishi* **29**, 918.

Inwang, E. E. (1968). *Can. J. Zool.* **46**, 15.

Ishaaya, J., and Chefurka, W. (1968). *Riv. Parassitol.* **29**, 289.

Ishida, M. (1968). *Agr. Biol. Chem.* **32**, 947.

Ishida, M., and Dahm, P. A. (1965a). *J. Econ. Entomol.* **58**, 383.

Ishida, M., and Dahm, P. A. (1965b). *J. Econ. Entomol.* **58**, 602.

Jakob, W. L., and Schoof, H. F. (1966). *Mosquito News* **26**, 78.

Johnston, E. F., Bogart, R., and Lindquist, A. W. (1954). *J. Hered.* **45**, 177.

Kalra, R. L. (1970a). *Bull. W.H.O.* **42**, 623.

Kalra, R. L. (1970b). *Botyu-Kagaku* **35**, 33.

Kalra, R. L., Perry, A. S., and Miles, J. W. (1967). *Bull. W.H.O.* **37**, 651.

Karla, R. L., Perry, A. S., and Miles, J. W. (1968). *Indian J. Exp. Biol.* **6**, 37.

Kaluszyner, A., Reuter, S., and Bergmann, E. D. (1955). *J. Amer. Chem. Soc.* **77**, 4164.

Kaplanis, J. N., Robbins, W. E., Darrow, D. I., Hopkins, D. E., Monroe, R. E., and Treiber, G. (1959). *J. Econ. Entomol.* **52**, 1190.

Kapoor, I. P., Metcalf, R. L., Nystrom, R. F., and Sangha, G. K. (1970). *J. Agr. Food Chem.* **18**, 1145.

Kasai, T., and Ogita, Z. (1965). *SABCO (Soc. Areas Biol. Chem. Overlap).* *J.* **1**, 130.

Kaschef, A. H. (1970). *Bull. W.H.O.* **42**, 917.

Kato, R. (1966). *J. Biochem. (Tokyo)* **59**, 574.

Kauer, K. C., Duvall, R. B., and Alquist, F. N. (1947). *Ind. Eng. Chem.* **39**, 1335.

Kearns, C. W. (1955). *In* "Origins of Resistance to Toxic Agents" (M. G. Sevag, R. D. Reid, and O. E. Reynolds, eds.), pp. 148–58. Academic Press, New York.

Kearns, C. W. (1956). *Annu. Rev. Entomol.* **1**, 123.

Kearns, C. W. (1957). *World Health Organ., Inform. Circ. Resistance Problem* No. 8.

Keiding, J. (1965). *Riv. Parassitol.* **26**, 45.

Keiding, J. (1967). *World Rev. Pest Contr.* **6**, 115.

Keiding, J. (1969). Annual Report. Government Pest Infestation Laboratory, Lyngby, Denmark.

Keller, J. C., Clark, P. H., and Lofgren, C. S. (1956). *Pest Contr.* **24**, 14.

Kenaga, E. E. (1969). *Bull. Entomol. Soc. Amer.* **15**, 85.

Kerr, R. W. (1963). *J. Aust. Inst. Agr. Sci.* **29**, 31.

Kerr, R. W. (1970). *Aust. J. Biol. Sci.* **23**, 377.

Kerr, R. W., Venables, D. G., Roulston, W. J., and Schnitzerling, H. J. (1957). *Nature (London)* **180**, 1132.

Khan, M. A. Q. (1969). *J. Econ. Entomol.* **62**, 388.

Khan, M. A. Q., and Brown, A. W. A. (1966). *J. Econ. Entomol.* **59**, 1512.

Khan, M. A. Q., and Terriere, L. C. (1968). *J. Econ. Entomol.* **61**, 732.

Khan, M. A. Q., Rosen, J. D., and Sutherland, D. J. (1969). *Science* **164**, 318.

Kilpatrick, J. W., and Schoof, H. F. (1956). *Pub. Health Rep.* **71**, 787.

Kilpatrick, J. W., and Schoof, H. F. (1958). *J. Econ. Entomol.* **51**, 18.

Kimura, T., and Brown, A. W. A. (1964). *J. Econ. Entomol.* **57**, 710.

Kimura, T., Duffy, J. R., and Brown, A. W. A. (1965). *Bull. W.H.O.* **32**, 557.

King, W. V., and Gahan, J. B. (1949). *J. Econ. Entomol.* **42**, 405.

Klassen, W., and Matsumura, F. (1966). *Nature (London)* **209**, 1155.

Klein, A. K., Link, J. D., and Ives, N. F. (1968). *J. Ass. Offic. Anal. Chem.* **51**, 805.

Knutson, H. (1959). *Entomol. Soc. Amer., Misc. Publ.* **1**, 27.

Koivistoinen, P. (1961). *Ann. Acad. Sci. Fenn., Ser. A4* **51**, 1–91.

Kojima, K., Nagae, Y., Ishizuka, T., and Shiino, A. (1958). *Botyu-Kagaku* **23**, 12.

Kojima, K., Ishizuka, T., Shiino, A., Kitakata, S. (1963). *Jap. J. Appl. Zool.* **7**, 63.

Kolbezen, M. J., Metcalf, R. L., and Fukuto, T. R. (1954). *J. Agr. Food Chem.* **2**, 864.

Koransky, W., Portig, J., Vohland, H. W., and Klempau, I. (1964). *Naunyn-Schmiedbergs Arch. Exp. Pathol. Pharmakol.* **247**, 49.

Korte, F. (1967). *Botyu-Kagaku* **32**, 46.

Korte F., and Stiasni, M. (1964). *Justus Liebigs Ann. Chem.* **673**, 146.

Korte, F., Ludwig, G., and Vogel, J. (1962). *Justus Liebigs Ann. Chem.* **656**, 135.

Krieger, R. I., and Wilkinson, C. F. (1969). *Biochem. Pharmacol.* **18**, 1403.

Krieger, R. I., Gilbert, M. D., and Wilkinson, C. F. (1970). *J. Econ. Entomol.* **63**, 1322.

Kring, J. B. (1958). *J. Econ. Entomol.* **51**, 823.

Krueger, H. R., and O'Brien, R. D. (1959). *J. Econ. Entomol.* **52**, 1063.

Krueger, H. R., Casida, J. E., and Niedermeier, R. P. (1959). *J. Agr. Food Chem.* **7**, 182.

Krueger, H. R., O'Brien, R. D., and Dauterman, W. C. (1960). *J. Econ. Entomol.* **53**, 25.

Ku, T. Y., and Bishop, J. L. (1967). *J. Econ. Entomol.* **60**, 1328.

Kuenen, D. J. (1958). *Entomol. Exp. Appl.* **1**, 147.

Kuhr, R. J. (1969). *J. Agr. Food Chem.* **17**, 112.

Kuhr, R. J. (1970). *J. Agr. Food Chem.* **18**, 1023.

Kuhr, R. J., and Casida, J. E. (1967). *J. Agr. Food Chem.* **15**, 814.

Langenbuch, R. (1954). *Naturwissenschaften* **41**, 70.

Langenbuch. R. (1955). *Z. Pflanzenkrankh. (Pflanzenpathol.) Pflanzenschutz* **62**, 564.

Lee, R. M., and Batham, P. (1966). *Entomol. Exp. Appl.* **9**, 13.

Leeling, N. C., and Casida, J. E. (1966). *J. Agr. Food Chem.* **14**, 281.

LeRoux, E. J., and Morrison, F. O. (1954). *J. Econ. Entomol.* **47**, 1058.

Lewallen, L. L., and Nicholson, L. M. (1959). *Ann. Entomol. Soc. Amer.* **52**, 767.

Lewis, J. B. (1969). *Nature (London)* **224**, 917.

Lewis, J. B., and Lord, K. A. (1969). *Proc. Brit. Insect. Fungic. Conf., 5th,* 1969 p. 465.

Lewis, S. E., Wilkinson, C. F., and Ray, J. W. (1967). *Biochem. Pharmacol.* **16**, 1195.

Lichtenstein, E. P., and Schulz, K. R. (1959). *J. Econ. Entomol.* **52**, 118.
Lichtenstein, E. P., and Schulz, K. R. (1960). *J. Econ. Entomol.* **53**, 192.
Lindahl, P. E., and Öberg, K. E. (1961). *Exp. Cell Res.* **23**, 221.
Lindquist, A. W., Madden, A., and Wilson, H. (1947). *J. Econ. Entomol.* **40**, 426.
Lindquist, A. W., Yates, W., Roth, A. R., Hoffman, R. A., and Butts, J. S. (1951a). *J. Econ. Entomol.* **44**, 167.
Lindquist, A. W., Roth, A. R., and Hoffman, R. A. (1951b). *J. Econ. Entomol.* **44**, 931.
Lindquist, D. A., and Dahm, P. A.(1956). *J. Econ. Entomol.* **49**, 579.
Lindquist, D. A., and Dahm, P. A. (1957). *J. Econ. Entomol.* **50**, 483.
Lindquist, D. A., Burns, E. C., Pant, C. P., and Dahm, P. A. (1958). *J. Econ. Entomol.* **51**, 204.
Lindquist, D. A., Brazzel, J. R., and Davich, T. B. (1961). *J. Econ. Entomol.* **54**, 299.
Lineva, V. A. (1962). *J. Hyg., Epidemiol., Microbiol., Immunol.* **6**, 271.
Lipke, H. (1960). *J. Econ. Entomol.* **53**, 31.
Lipke, H., and Chalkley, J. (1964). *Bull. W.H.O.* **30**, 57.
Lipke, H., and Kearns, C. W. (1959a). *J. Biol. Chem.* **234**, 2123.
Lipke, H., and Kearns, C. W. (1959b). *J. Biol. Chem.* **234**, 2129.
Lipke, H., and Kearns, C. W. (1960). *Advan. Pest Contr. Res.* **3**, 253.
Litvak, S., and Agosin, M. (1968). *Biochemistry* **7**, 1560.
Lloyd, C. J. (1969). *J. Stored Prod. Res.* **5**, 357.
Lloyd, C. J., and Dyte, C. E. (1965). *J. Stored Prod. Res.* **1**, 159.
Lofgren, C. S., and Cutkomp, L. K. (1956). *J. Econ. Entomol.* **49**, 167.
Lord, K. A., and Potter, C. (1950). *Nature (London)* **166**, 893.
Lord, K. A., and Potter, C. (1951). *Ann. Appl. Biol.* **38**, 495.
Lord, K. A., and Solly, S. R. B. (1956). *Chem. Ind. (London)* p. 1352.
Lord, K. A., and Solly, S. R. B. (1964). *Biochem. Pharmacol.* **13**, 1341.
Lord, K. A., Molloy, F. M., and Potter, C. (1963). *Bull. Entomol. Res.* **54**, Part 2, 189.
Lovell, J. B., and Kearns, C. W. (1959). *J. Econ. Entomol.* **52**, 931.
Ludwig, G., and Korte, F. (1965). *Life Sci.* **4**, 2027.
McBain, J. B., Menn, J. J., and Casida, J. E. (1968). *J. Agr. Food Chem.* **16**, 813.
McKenzie, R. E., and Hoskins, W. M. (1954). *J. Econ. Entomol.* **47**, 984.
Machin, A. F., and Quick, M. P. (1971). *Bull. Environ. Contam. Toxicol.* **6**, 26.
Mahan, J. G., and Grayson, J. M. (1956). *Va. J. Sci.* **7**, 166.
Mansingh, A. (1965). *J. Insect Physiol.* **11**, 1389.
March, R. B. (1952a). *Nat. Acad. Sci.—Nat. Res. Counc., Publ.* **219**, 45.
March, R. B. (1952b). *J. Econ. Entomol.* **45**, 452.
March, R. B. (1959). *Entomol. Soc. Amer., Misc. Publ.* **1**, 13.
March, R. B. (1960). *Entomol. Soc. Amer., Misc. Publ.* **2**, 139.
March, R. B., and Lewallen, L. L. (1950). *J. Econ. Entomol.* **43**, 721.
March, R. B., and Lewallen, L. L. (1956). *Ann. Entomol. Soc. Amer.* **49**, 571.
March, R. B., and Metcalf, R. L. (1949). *Calif., Dep. Agr., Bull.* **38**, 93.
March, R. B., Metcalf, R. L., and Lewallen, L. L. (1952). *J. Econ. Entomol.* **45**, 851.
March, R. B., Metcalf, R. L., Fukuto, T. R., and Maxon, M. G. (1955). *J. Econ. Entomol.* **48**, 355.
March, R. B., Fukuto, T. R., Metcalf, R. L., and Maxon, M. G. (1956). *J. Econ. Entomol.* **49**, 185.

Matsumura, F., and Brown, A. W. A. (1961a). *J. Econ. Entomol.* **54,** 1176.

Matsumura, F., and Brown, A. W. A. (1961b). *Mosquito News* **21,** 192.

Matsumura, F., and Brown, A. W. A. (1963a). *Mosquito News* **23,** 26.

Matsumura, F., and Brown, A. W. A. (1963b). *J. Econ. Entomol.* **56,** 381.

Matsumura, F., and Hayashi, M. (1966a). *Science* **153,** 757.

Matsumura, F., and Hayashi, M. (1966b). *Mosquito News* **26,** 190.

Matsumura, F., and Hayashi, M. (1970). *J. Agr. Food Chem.* **17,** 231.

Matsumura, F., and Hogendijk, C. J. (1964a). *J. Agr. Food Chem.* **12,** 447.

Matsumura, F., and Hogendijk, C. J. (1964b). *Entomol. Exp. Appl.* **7,** 179.

Matsumura, F., and Nelson, J. O. (1971). *Bull. Environ. Contam. Toxicol.* **5,** 489.

Matsumura, F., and O'Brien, R. D. (1966). *J. Agr. Food Chem.* **14,** 36.

Matsumura, F., and Sakai, K. (1968). *J. Econ. Entomol.* **61,** 598.

Matsumura, F., and Voss, G. (1964). *J. Econ. Entomol.* **57,** 911.

Mattingly, P. F. (1962). *Annu. Rev. Entomol.* **7,** 419.

Melander, A. L. (1914). *J. Econ. Entomol.* **7,** 167.

Meltzer, J. (1956). *Meded. Landbouwhogesch. Opzoekingssta. Staat Gent* **21,** 459.

Mengle, D. E., and Casida, J. E. (1960). *J. Agr. Food Chem.* **8,** 431.

Mengle, D. E., and Lewallen, L. L. (1966). *J. Econ. Entomol.* **59,** 743.

Menn, J. J., and Hoskins, W. M. (1962). *J. Econ. Entomol.* **55,** 90.

Menn, J. J., and McBain, J. B. (1964). *J. Agr. Food Chem.* **12,** 162.

Menn, J. J., Benjamini, E., and Hoskins, W. M. (1957). *J. Econ. Entomol.* **50,** 67.

Menzel, D. B., Smith, S. M., Miskus, R., and Hoskins, W. M. (1961). *J. Econ. Entomol.* **54,** 9.

Menzer, R. E., and Casida, J. E. (1965). *J. Agr. Food Chem.* **13,** 102.

Menzer, R. E., and Dauterman, W. C. (1970). *J. Agr. Food Chem.* **18,** 1031.

Mer, G. G., and Furmanska, W. (1953). *Riv. Parassitol.* **14,** 49.

Merrell, D. J., and Underhill, J. C. (1956). *J. Econ. Entomol.* **49,** 300.

Metcalf, R. L. (1955a). *Physiol. Rev.* **35,** 197.

Metcalf, R. L. (1955b). "Organic Insecticides. Their Chemistry and Mode of Action." Wiley (Interscience), New York.

Metcalf, R. L. (1959). *Bull. Entomol. Soc. Amer.* **5,** 3.

Metcalf, R. L. (1962). *Pest Contr.* **30,** 20.

Metcalf, R. L. (1967). *Annu. Rev. Entomol.* **12,** 229.

Metcalf, R. L., and Fukuto, T. R. (1965). *J. Agr. Food Chem.* **13,** 220.

Metcalf, R. L., and Fukuto, T. R. (1968). *Bull. W.H.O.* **38,** 633.

Metcalf, R. L., and March, R. B. (1953). *Ann. Entomol. Soc. Amer.* **46,** 63.

Metcalf, R. L., Maxon, M. G., Fukuto, R. T., and March, R. B. (1956). *Ann. Entomol. Soc. Amer.* **49,** 274.

Metcalf, R. L., Fukuto, T. R., and Winton, M. Y. (1960). *J. Econ. Entomol.* **53,** 828.

Metcalf, R. L., Fukuto, T. R., Wilkinson, C. F., Fahmy, M. H., Abd El-Aziz, S. A., and Metcalf, E. R. (1966a). *J. Agr. Food Chem.* **14,** 555.

Metcalf, R. L., Fukuto, T. R., Collins, C., Borck, K., Burk, J., Reynolds, H. T., and Osman, M. F. (1966b). *J. Agr. Food Chem.* **14,** 579.

Metcalf, R. L., Osman, M. F., and Fukuto, T. R. (1967). *J. Econ. Entomol.* **60,** 445.

Metcalf, R. L., Fukuto, T. R., Collins, C., Borck, K., Abd El-Aziz, S. A., Munoz, R., and Cassil, C. C. (1968). *J. Agr. Food Chem.* **16,** 300.

Micks, D. W. (1960). *Bull. W.H.O.* **22,** 519.

Micks, D. W., and Singh, K. R. P. (1958). *Tex. Rep. Biol. Med.* **16,** 355.

Micks, D. W., Ferguson, M. J., and Singh, K. R. P. (1960). *Science* **131,** 1615.

Milani, R. (1954). *Riv. Parassitol.* **15**, 513.
Milburn, N., Weiant, E. A., and Roeder, K. D. (1960). *Biol. Bull.* **118**, 111.
Miles, J. R. W., Tu, C. M., and Harris, C. R. (1969). *J. Econ. Entomol.* **62**, 1334.
Miller, S., and Perry, A. S. (1964). *J. Agr. Food Chem.* **12**, 167.
Mitchell, L. C. (1961). *J. Ass. Offic. Anal. Chem.* **44**, 643.
Miyake, S. S., Kearns, C. W., and Lipke, H. (1957). *J. Econ. Entomol.* **50**, 359.
Moore, R. F., Jr., Hopkins, A. R., Taft, H. M., and Anderson, L. L. (1967). *J. Econ. Entomol.* **60**, 64.
Moorefield, H. H. (1956). *Contrib. Boyce Thompson Inst.* **18**, 303.
Moorefield, H. H. (1958). *Contrib. Boyce Thompson Inst.* **19**, 501.
Moorefield, H. H. (1960). *Entomol. Soc. Amer., Misc. Publ.* **2**, 145.
Moorefield, H. H., and Kearns, C. W. (1955). *J. Econ. Entomol.* **48**, 403.
Moorefield, H. H., and Kearns, C. W. (1957). *J. Econ. Entomol.* **50**, 11.
Moorefield, H. H., Weiden, M. H. J., and Hennessy, D. J. (1962). *Contrib. Boyce Thompson Inst.* **21**, 481.
Morello, A. (1964). *Nature (London)* **203**, 785.
Morello, A., Vardanis, A., and Spencer, E. Y. (1968). *Can. J. Biochem.* **46**, 885.
Morgan, P. B., LaBrecque, G. C., Smith, C. N., Meifert, D. W. and Murvosh, C. M. (1967). *J. Econ. Entomol.* **60**, 1064.
Morikawa, O., and Saito, T. (1966). *Botyu-Kagaku* **31**, 130.
Mount, G. A., Lofgren, C. S., Bowman, M. C., and Acree, F., Jr. (1966). *J. Econ. Entomol.* **59**, 1352.
Muirhead-Thomson, R. C. (1960). *Bull. W.H.O.* **22**, 721.
Muirhead-Thomson, R. C. (1968). "Ecology of Insect Vector Populations." Academic Press, New York.
Mullins, L. J. (1954). *Chem. Rev.* **54**, 289.
Mullins, L. J. (1955). *Science* **122**, 118.
Munson, S. C., and Gottlieb, M. I. (1953). *J. Econ. Entomol.* **46**, 798.
Munson, S. C., Padilla, G. M., and Weissmann, M. L. (1954). *J. Econ. Entomol.* **47**, 578.
Murphy, S. D., and DuBois, K. P. (1957). *Proc. Soc. Exp. Biol. Med.* **96**, 813.
Nagasawa, S. (1953). *Botyu-Kagaku* **18**, 22.
Nakatsugawa, T., and Dahm, P. A. (1965). *J. Econ. Entomol.* **58**, 500.
Nakatsugawa, T., and Dahm, P. A. (1967). *Biochem. Pharmacol.* **16**, 25.
Nakatsugawa, T., Ishida, M., and Dahm, P. A. (1965). *Biochem. Pharmacol.* **14**, 1853.
Nakatsugawa, T., Tolman, N. M., and Dahm, P. A. (1968). *Biochem. Pharmacol.* **17**, 1517.
Nakatsugawa, T., Tolman, N. M., and Dahm, P. A. (1969). *Biochem. Pharmacol.* **18**, 1103.
Narahashi, T. (1964). *Jap. J. Med. Sci. Biol.* **17**, 46.
Neal, R. A. (1967a). *Biochem. J.* **103**, 183.
Neal, R. A. (1967b). *Biochem. J.* **105**, 289.
Neeman, M., Modiano, A., Mer, G. G., and Cwilich, R. (1956). *Nature (London)* **177**, 800.
Neeman, M., Mer, G. G., Cwilich, R., Modiano, A., and Zacks, S. (1957). *J. Sci. Food Agr.* **8**, 55.
Neri, I., Ascher, K. R. S., and Mosna, E. (1958). *Indian J. Malariol.* **12**, 565.
Neri, I., Ascher, K. R. S., and Mosna, E. (1959). World Health Organ., WHO/Mal/249, WHO/Insecticides 105 8 pp.

Nicoli, R. M., and Sautet, J. (1955). *Inst. Nat. Hyg. (Paris) Monogr.* **8**, 1–78.
Nolan, J., and O'Brien, R. D. (1970). *J. Agr. Food Chem.* **18**, 802.
Norton, R. (1953). *Contrib. Boyce Thompson Inst.* **17**, 105.
O'Brien, R. D. (1957). *J. Econ. Entomol.* **50**, 159.
O'Brien, R. D. (1960). "Toxic Phosphorus Esters." Academic Press, New York.
O'Brien, R. D. (1966). *Annu. Rev. Entomol.* **11**, 369.
O'Brien, R. D. (1967). "Insecticides: Action and Metabolism." Academic Press, New York.
O'Brien, R. D., and Matsumura, F. (1964). *Science* **146**, 657.
O'Brien, R. D., and Matthysse, J. G. (1961). *Agr. Chem.* **16**, 16.
O'Brien, R. D., and Wolfe, L. S. (1959). *J. Econ. Entomol.* **52**, 692.
O'Brien, R. D., Thorn, G. D., and Fisher, R. W. (1958). *J. Econ. Entomol.* **51**, 714.
O'Brien, R. D., Kimmel, E. C., and Sferra, P. R. (1965). *J. Agr. Food Chem.* **13**, 366.
O'Brien, R. D., Hilton, B. D., and Gilmour, L. (1966). *Mol. Pharmacol.* **2**, 593.
Ogita, Z. (1961). *Botyu-Kagaku* **26**, 93.
Ohkawa, H., Eto, M., Oshima, Y., Tanaka, F., and Umeda, K. (1968). *Botyu-Kagaku* **33**, 139.
Oonithan, E. S., and Casida, J. E. (1966). *Bull. Environ. Contam. Toxicol.* **1**, 59.
Oonithan, E. S., and Casida, J. E. (1968). *J. Agr. Food Chem.* **16**, 28.
Oonithan, E. S., and Miskus, R. (1964). *J. Econ. Entomol.* **57**, 425.
Oppenoorth, F. J. (1954). *Nature (London)* **173**, 1001.
Oppenoorth, F. J. (1955). *Nature (London)* **175**, 124.
Oppenoorth, F. J. (1956). *Arch. Neer. Zool.* **12**, 1.
Oppenoorth, F. J. (1959). *Entomol. Exp. Appl.* **2**, 304.
Oppenoorth, F. J. (1965a). *Proc. Int. Congr. Entomol. 12th, 1964* p. 240.
Oppenoorth, F. J. (1965b). *Annu. Rev. Entomol.* **10**, 185.
Oppenoorth, F. J., and Houx, N. W. H. (1968). *Entomol. Exp. Appl.* **11**, 81.
Oppenoorth, F. J., and van Asperen, K. (1960). *Science* **132**, 298.
Oppenoorth, F. J., and van Asperen, K. (1961). *Entomol. Exp. Appl.* **4**, 311.
Oppenoorth, F. J., and Voerman, S. (1965). *Entomol. Exp. Appl.* **8**, 293.
Ozbas, S., and Hodgson, E. S. (1958). *Proc. Nat. Acad. Sci. U.S.* **44**, 825.
Pal, R., and Kalra, R. L. (1965). World Health Organ./Vector Control/120.65. 51 pp.
Pant, C. P., and Self, L. S. (1968). *Mosquito News* **28**, 630.
Pardue, J. R., Hansen, E. A., Barron, R. P., and Chen, J. Y. (1970). *J. Agr. Food Chem.* **18**, 405.
Parish, J. C., and Arthur, B. W. (1965). *J. Econ. Entomol.* **58**, 976.
Parkin, E. A., and Lloyd, C. J. (1960). *J. Sci. Food Agr.* **11**, 471.
Pate, T. L., and Vinson, S. B. (1968). *J. Econ. Entomol.* **61**, 1135.
Patel, N. G., and Cutkomp, L. K. (1967). *J. Econ. Entomol.* **60**, 783.
Patel, N. G., and Cutkomp, L. K. (1968). *J. Econ. Entomol.* **61**, 931.
Patel, N. G., Cutkomp, L. K., and Ikeshoji, T. (1968). *J. Econ. Entomol.* **61**, 1079.
Patterson, R. S., Lofgren, C. S., and Boston, M. D. (1967). *J. Econ. Entomol.* **60**, 1673.
Perkins, B. D., and Grayson, J. M. (1961). *J. Econ. Entomol.* **54**, 747.
Perry, A. S. (1953). "Technology—Branch Summary of Investigations," vol. 1, p. 45. U.S. Pub. Health Serv., Center for Disease Control, Savannah, Georgia.
Perry, A. S. (1958). *Proc. Int. Congr. Entomol. 10th, 1956* Vol. 2, p. 157.
Perry, A. S. (1960a). *Bull. W.H.O.* **22**, 743.

Perry, A. S. (1960b). *Entomol. Soc. Amer., Misc. Publ.* 2, 119.

Perry, A. S. (1960c). *J. Agr. Food Chem.* 8, 266.

Perry, A. S. (1961). *Bull. Entomol. Soc. Amer.* 7, 167.

Perry, A. S. (1964). *In* "The Physiology of Insecta" (M. Rockstein, ed.), Vol. 3, pp. 285–378. Academic Press, New York.

Perry, A. S. (1966). *Mosquito News* 26, 302.

Perry, A. S., and Buckner, A. J. (1958). *Amer. J. Trop. Med. Hyg.* 7, 620.

Perry, A. S., and Buckner, A. J. (1959). *J. Econ. Entomol.* 52, 997.

Perry, A. S., and Buckner, A. J. (1970). *Life Sci., Part II* 9, 335.

Perry, A. S., and Hoskins, W. M. (1950). *Science* 111, 600.

Perry, A. S., and Hoskins, W. M. (1951). *J. Econ. Entomol.* 44, 850.

Perry, A. S., Dale, E. W., and Buckner, A. J. (1971). *Pestic. Biochem. Physiol.* 1, 131.

Perry, A. S., Hennessy, D. J., and Miles, J. W. (1967). *J. Econ. Entomol.* 60, 568.

Perry, A. S., Mattson, A. M., and Buckner, A. J. (1953). *Biol. Bull.* 104, 426.

Perry, A. S., Mattson, A. M., and Buckner, A. J. (1958). *J. Econ. Entomol.* 51, 346.

Perry, A. S., Miller, S., and Buckner, A. J. (1963). *J. Agr. Food Chem.* 11, 457.

Perry, A. S., Pearce, G. W., and Buckner, A. J. (1964). *J. Econ. Entomol.* 57, 867.

Pillai, M. K. K., and Brown, A. W. A. (1965). *J. Econ. Entomol.* 58, 255.

Pillai, M. K. K., Abedi, Z. H., and Brown, A. W. A. (1963a). *Mosquito News* 23, 112.

Pillai, M. K. K., Hennessy, D. J., and Brown, A. W. A. (1963b). *Mosquito News* 23, 118.

Pilou, D. P., and Glasser, R. E. (1951). *Can. J. Zool.* 29, 90.

Pimentel, D., Dewey, J. E., and Schwardt, H. H. (1951). *J. Econ. Entomol.* 44, 477.

Plapp, F. W., Jr. (1970a). *In* "Biochemical Toxicology of Insecticides" (R. D. O'Brien and I. Yamamoto, eds.), pp. 179–192. Academic Press, New York.

Plapp, F. W., Jr. (1970b). *J. Econ. Entomol.* 63, 1768.

Plapp, F. W., Jr., and Casida, J. E. (1958). *J. Econ. Entomol.* 51, 800.

Plapp, F. W., Jr., and Casida, J. E. (1969). *J. Econ. Entomol.* 62, 1174.

Plapp, F. W., Jr., and Hennessy, D. J. (1966). *Mosquito News* 26, 527.

Plapp, F. W., Jr., and Hoyer, R. F. (1967). *J. Econ. Entomol.* 60, 768.

Plapp, F. W., Jr., and Hoyer, R. F. (1968). *J. Econ. Entomol.* 61, 1298.

Plapp, F. W., Jr., Bigley, W. S., Darrow, D. I., and Eddy, G. W. (1961). *J. Econ. Entomol.* 51, 800.

Plapp, F. W., Jr., Bigley, W. S., Chapman, G. A., and Eddy, G. W. (1962). *J. Econ. Entomol.* 55, 607.

Plapp, F. W., Jr., Chapman, G. A., and Bigley, W. S. (1964). *J. Econ. Entomol.* 57, 692.

Plapp, F. W., Jr., Chapman, G. A., and Morgan, J. W. (1965). *J. Econ. Entomol.* 58, 1064.

Polen, P. B., Hester, M., and Benziger, J. (1971). *Bull. Environ. Contam. Toxicol.* 5, 521.

Poonwalla, N. H., and Korte, F. (1968). *J. Agr. Food Chem.* 16, 15.

Potter, C. (1956). *Chem. Ind.* (*London*) 42, 1178.

Potter, C., and Gilham, E. M. (1957). *Bull. Entomol. Res.* 48, 317.

Potter, J. L., and O'Brien, R. D. (1963). *Entomol. Exp. Appl.* 6, 319.

Pratt, J. J., Jr., and Babers, F. H. (1953). *J. Econ. Entomol.* 46, 700.

Price, G. M., and Kuhr, R. J. (1969). *Biochem. J.* **112**, 133.
Quarterman, K. D., and Schoof, H. F. (1958). *Amer. J. Trop. Med. Hyg.* **7**, 74.
Raccah, B., and Tahori, A. S. (1970). *J. Econ. Entomol.* **63**, 567.
Radomski, J. L., and Davidow, B. (1953). *J. Pharmacol. Exp. Ther.* **107**, 266.
Rao, K. D. P., and Agarwal, H. C. (1969). *Comp. Biochem. Physiol.* **30**, 161.
Ray, J. W. (1963). *Nature (London)* **197**, 1226.
Ray, J. W. (1964). *J. Insect Physiol.* **10**, 587.
Ray, J. W. (1967). *Biochem. Pharmacol.* **16**, 99.
Reed, W. T., and Forgash, A. J. (1968). *Science* **160**, 1232.
Reed, W. T., and Forgash, A. J. (1969). *J. Agr. Food Chem.* **17**, 896.
Reed, W. T., and Forgash, A. J. (1970). *J. Agr. Food Chem.* **18**, 475.
Reiff, M. (1956a). *Rev. Suisse Zool.* **63**, 317.
Reiff, M. (1956b). *Verh. Naturforsch. Ges. Basel* **67**, 133.
Reiff, M., and Beye, F. (1960). *Acta Trop.* **17**, 1.
Reiser, R., Chadbourne, D. S., Kuiken, K. A., Rainwater, C. F., and Ivy, E. G. (1953). *J. Econ. Entomol.* **46**, 337.
Reuter, S., Cohen, S., Mechoulam, R., Kaluszyner, A., and Tahori, A. S. (1956). *Riv. Parassitol.* **17**, 125.
Richards, A. G. (1943). *J. N.Y. Entomol. Soc.* **51**, 55.
Richards, A. G. (1944). *J. N.Y. Entomol. Soc.* **52**, 285.
Robbins, W. E., and Dahm, P. A. (1955). *J. Agr. Food Chem.* **3**, 500.
Robertson, J. G. (1957). *Can. J. Zool.* **35**, 629.
Robinson, J., Richardson, A., Bush, B., and Edgar, K. (1966). *Bull. Environ. Contam. Toxicol.* **1**, 127.
Roburn, J. (1963). *Chem. Ind. (London)* No. 38, 1555.
Roeder, K. D., and Weiant, E. A. (1948). *J. Cell. Comp. Physiol.* **32**, 175.
Roeder, K. D., and Weiant, E. A. (1951). *Ann. Entomol. Soc. Amer.* **44**, 373.
Rose, H. A., and Hooper, G. S. H. (1969). *J. Econ. Entomol.* **62**, 857.
Rosen, J. D., and Sutherland, D. J. (1967). *Bull. Environ. Contam. Toxicol.* **2**, 1.
Rosen, J. D., Sutherland, D. J., and Lipton, G. R. (1966). *Bull. Environ. Contam. Toxicol.* **1**, 133.
Roulston, W. J., Schuntner, C. A., and Schnitzerling, H. J. (1966). *Aust. J. Biol. Sci.* **19**, 619.
Rowlands, D. G. (1968). *J. Stored Prod. Res.* **4**, 183.
Rowlands, D. G., and Lloyd, C. J. (1969). *J. Stored Prod. Res.* **5**, 413.
Sacca, G., and Scirocchi, A. (1966). World Health Organ., Mimeo Publ. WHO/Vector Control/66.192.
Sakai, S. (1960). "Joint Action of Insecticides." Yashima Chem. Co., Kanagawa, Japan.
Sawicki, R. M. (1970). *Pestic. Sci.* **1**, 84.
Sawicki, R. M., and Farnham, A. W. (1968). *Entomol. Exp. Appl.* **11**, 133.
Schaefer, C. H., and Sun, Y. P. (1967). *J. Econ. Entomol.* **60**, 1580.
Schechter, M. S., Soloway, S. B., Hayes, R. A., and Haller, H. L. (1945). *Ind. Eng. Chem., Anal. Ed.* **17**, 704.
Schmidt, C. H., and LaBrecque, G. C. (1959). *J. Econ. Entomol.* **52**, 345.
Schonbrod, R. D., Philleo, W. W., and Terriere, L. C. (1965). *J. Econ. Entomol.* **58**, 74.
Schonbrod, R. D., Khan, M. A. Q., Terriere, L. C., and Plapp, F. W., Jr. (1968). *Life Sci.* **7**, 681.
Schuntner, C. A., Roulston, W. J., and Schnitzerling, H. J. (1968). *Aust. J. Biol. Sci.* **21**, 97.

Schwemmer, B., Cochrane, W. P., and Polen, P. B. (1970). *Science* **169**, 1087.

Self, L. S., Guthrie, F. E., and Hodgson, E. (1964a). *J. Insect Physiol.* **10**, 907.

Self, L. S., Guthrie, F. E., and Hodgson, E. (1964b). *Nature (London)* **204**, 300.

Sellers, L. G. (1971). Ph.D. Thesis, Department of Entomology, North Carolina State University, Raleigh.

Sellers, L. G., and Guthrie, F. E. (1971). *J. Econ. Entomol.* **64**, 352.

Seume, F. W., and O'Brien, R. D. (1960). *J. Agr. Food Chem.* **8**, 36.

Shankland, D. L., and Kearns, C. W. (1959). *Ann. Entomol. Soc. Amer.* **52**, 386.

Shepard, H. H. (1951). "The Chemistry and Action of Insecticides." McGraw-Hill, New York.

Sherman, M. V., Evans, R., Nesyto, E., and Radlowski, C. (1971). *Nature (London)* **232**, 118.

Shishido, T., and Fukami, J. I. (1963). *Botyu-Kagaku* **28**, 69.

Shrivastava, S. P., Tsukamoto, M., and Casida, J. E. (1969). *J. Econ. Entomol.* **62**, 483.

Shrivastava, S. P., Georghiou, G. P., Metcalf, R. L., and Fukuto, T. R. (1970). *Bull. W.H.O.* **42**, 931.

Shrivastava, S. P., Georghiou, G. P., and Fukuto, T. R. (1971). *Entomol. Exp. Appl.* **14**, 333.

Silva, G. M., Doyle, W. P., and Wang, C. H. (1959). *Arq. Port. Bioquim.* **3**, 298.

Silverman, P. H., and Mer, G. G. (1952). *Riv. Parassitol.* **13**, 123.

Sims, P., and Grover, P. L. (1965). *Biochem. J.* **95**, 156.

Singh, J., and Malaiyandi, M. (1969). *Bull. Environ. Contam. Toxicol.* **4**, 337.

Slade, M., and Casida, J. E. (1970). *J. Agr. Food Chem.* **18**, 467.

Smissaert, H. R. (1964). *Science* **143**, 129.

Smissaert, H. R., Voerman, S., Oostenbrugge, L., and Renooy, N. (1970). *J. Agr. Food Chem.* **18**, 66.

Smith, J. N. (1955). *Biol. Rev. Cambridge Phil. Soc.* **30**, 455.

Smith, J. N. (1962). *Annu. Rev. Entomol.* **7**, 465.

Smith, J. N. (1964). *Comp. Biochem.* **6**, 403–457.

Smith, J. N., and Turbert, H. B. (1961). *Nature (London)* **189**, 600.

Smith, J. N., and Turbert, H. B. (1964). *Biochem. J.* **92**, 127.

Smyth, T., Jr., and Roys, C. C. (1955). *Biol. Bull.* **108**, 66.

Sokal, R. R., and Hunter, P. E. (1955). *Ann. Entomol. Soc. Amer.* **48**, 499.

Soloway, S. B. (1965). *Advan. Pest Contr. Res.* **6**, 85.

Spencer, E. Y., and O'Brien, R. D. (1957). *Annu. Rev. Entomol.* **2**, 261.

Speroni, G. (1952). *Chim. Ind. (Milan)* **34**, 309.

Speroni, G., Losco, G., Santi, R., and Peri, C. (1953). *Chim. Ind. (Paris)* **69**, 658.

Spiller, D. (1958). *N.Z. Entomol.* **2**, 1.

Stålhandske, T., Slanina, P., Tjälve, H., Hansson, E., and Schmiterlöw, C. G. (1969). *Acta Pharmacol. Toxicol.* **27**, 363.

Staudenmayer, T. (1955). *Z. Vergl. Physiol.* **37**, 416.

Stegwee, D. (1959). *Nature (London)* **184**, 1253.

Stenersen, J. H. V. (1965). *Nature (London)* **207**, 660.

Stenersen, J. (1969). *J. Econ. Entomol.* **62**, 1043.

Sternburg, J. (1960). *J. Agr. Food Chem.* **8**, 257.

Sternburg, J. (1963). *Annu. Rev. Entomol.* **8**, 19.

Sternburg, J., and Kearns, C. W. (1950). *Ann. Entomol. Soc. Amer.* **43**, 444.

Sternburg, J., and Kearns, C. W. (1952a). *J. Econ. Entomol.* **45**, 497.

Sternburg, J., and Kearns, C. W. (1952b). *Science* **116**, 144.

Sternburg, J., and Kearns, C. W. (1956). *J. Econ. Entomol.* **49**, 548.

Sternburg, J., Kearns, C. W., and Bruce, W. N. (1950). *J. Econ. Entomol.* **43**, 214.

Sternburg, J., Vinson, E. B., and Kearns, C. W. (1953). *J. Econ. Entomol.* **46**, 513.

Sternburg, J., Kearns, C. W., and Moorefield, H. (1954). *J. Agr. Food Chem.* **2**, 1125.

Sternburg, J., Chang, S. C., and Kearns, C. W. (1959). *J. Econ. Entomol.* **52**, 1070.

Stone, B. F. (1969). *J. Econ. Entomol.* **62**, 977.

Stone, B. F., and Brown, A. W. A. (1969). *Bull. W.H.O.* **40**, 401.

Sugiyama, H., and Shigematsu, H. (1969). *Botyu-Kagaku* **34**, 79.

Sumerford, W. T. (1954). *J. Agr. Food Chem.* **2**, 310.

Sumerford, W. T., Goette, M. B., Quarterman, K. D., and Schenck, S. L. (1951a). *Science* **114**, 6.

Sumerford, W. T., Fay, R. W., Goette, M. B., and Allred, A. M. (1951b). *J. Nat. Malar. Soc.* **10**, 345.

Sun, Y. P., and Johnson, E. R. (1960). *J. Agr. Food Chem.* **8**, 261.

Sutherland, D. J., Beam, F. D., and Gupta, A. P. (1967). *Mosquito News* **27**, 316.

Sutherland, D. J., Siewierski, M., Marei, A. H., and Helrich, K. (1970). *Mosquito News* **30**, 8.

Swift, F. C., and Forgash, A. J. (1959). *Entomol. Soc. Amer., Cotton States Br., 33rd Annu. Meet.* p. 4.

Swingle, M. C. (1934). *J. Econ. Entomol.* **28**, 1101.

Tahori, A. S. (1955). *J. Econ. Entomol.* **48**, 638.

Tahori, A. S., and Hoskins, W. M. (1953). *J. Econ. Entomol.* **46**, 302 and 829.

Telford, J. N., and Matsumura, F. (1971). *J. Econ. Entomol.* **64**, 230.

Terranova, A. C. (1969). *J. Econ. Entomol.* **62**, 821.

Terranova, A. C., and Crystal, M. M. (1970). *J. Econ. Entomol.* **63**, 455.

Terriere, L. C., and Schonbrod, R. D. (1955). *J. Econ. Entomol.* **48**, 736.

Terriere, L. C., Boose, R. B., and Roubal, W. T. (1961). *Biochem. J.* **79**, 620.

Tombes, A. S., and Forgash, A. J. (1961). *J. Insect Physiol.* **7**, 216.

Tomlin, A. D. (1968). *J. Econ. Entomol.* **61**, 855.

Townsend, M. G., and Busvine, J. R. (1969). *Entomol. Exp. Appl.* **12**, 243.

Trapido, H. (1952). *Amer. J. Trop. Med. Hyg.* **1**, 853.

Treherne, J. E. (1962). *Times Sci. Rev.* **43**, 5.

Tsukamoto, M. (1959). *Botyu-Kagaku* **24**, 141.

Tsukamoto, M. (1960). *Botyu-Kagaku* **25**, 156.

Tsukamoto, M. (1961). *Botyu-Kagaku* **26**, 74.

Tsukamoto, M., and Casida, J. E. (1967). *Nature (London)* **213**, 49.

Tsukamoto, M., and Ogaki, M. (1953). *Botyu-Kagaku* **18**, 39.

Tsukamoto, M., and Suzuki, R. (1964). *Botyu-Kagaku* **29**, 76.

Tsukamoto, M., Narahashi, T., and Yamasaki, T. (1965). *Botyu-Kagaku* **30**, 128.

Tsukamoto, M., Shrivastava, S. P., and Casida, J. E. (1968). *J. Econ. Entomol.* **61**, 50.

Turner, R. B. (1968). *In* "Principles of Insect Chemosterilization" (G. C. LaBrecque and C. N. Smith, eds.), pp. 161–274. Appleton, New York.

Twarog, B. M., and Roeder, K. D. (1956). *Biol. Bull.* **111**, 278.

Twarog, B. M., and Roeder, K. D. (1957). *Ann. Entomol. Soc. Amer.* **50**, 231.

Uchida, T., Dauterman, W. C., and O'Brien, R. D. (1964). *J. Agr. Food Chem.* **12**, 48.

Uchida, T., Rahmati, H. S., and O'Brien, R. D. (1965). *J. Econ. Entomol.* **58**, 831.

van Asperen, K. (1954). *Arch. Int. Pharmacodyn. Ther.* **99**, 368.

van Asperen, K. (1958). *Nature (London)* **181**, 355.

van Asperen, K., and Oppenoorth, F. J. (1954). *Nature (London)* **173**, 1000.

van Asperen, K., and Oppenoorth, F. J. (1959). *Entomol. Exp. Appl.* **2**, 48.

van Asperen, K., and Oppenoorth, F. J. (1960). *Entomol. Exp. Appl.* **3**, 68.

van den Heuvel, M. J., and Cochran, D. G. (1965). *J. Econ. Entomol.* **58**, 872.

Vardanis, A., and Crawford, L. G. (1964). *J. Econ. Entomol.* **57**, 136.

Varzandeh, M., Bruce, W. N., and Decker, G. C. (1954). *J. Econ. Entomol.* **47**, 129.

Veldstra, H. (1956). *Pharmacol. Rev.* **8**, 339.

Vickery, D. S., and Arthur, B. W. (1960). *J. Econ. Entomol.* **53**, 1037.

Vinson, E. B., and Kearns, C. W. (1952). *J. Econ. Entomol.* **45**, 484.

Vinson, S. B. (1967). *J. Econ. Entomol.* **60**, 565.

Vinson, S. B., and Brazzell, J. R. (1966). *J. Econ. Entomol.* **59**, 600.

Voss, G., and Matsumura, F. (1964). *Nature (London)* **202**, 319.

Walker, T. F. (1970). *Bull. Entomol. Res.* **60**, 291.

Wang, C. M., and Matsumura, F. (1970). *J. Econ. Entomol.* **63**, 1731.

Weiant, E. A. (1955). *Ann. Entomol. Soc. Amer.* **48**, 489.

Welling, W., Blaakmeer, P., Vink, G. J., and Voerman, S. (1971). *Pestic. Biochem. Physiol.* **1**, 61.

Welsh, J. H., and Gordon, H. T. (1947). *J. Cell. Comp. Physiol.* **30**, 147.

Wharton, R. H., and Roulston, W. J. (1970). *Annu. Rev. Entomol.* **15**, 381.

Whitehead, G. B. (1959). *Nature (London)* **184**, 378.

Wiesmann, R. (1947). *Mitt. Schweiz. Entomol. Ges.* **20**, 484.

Wiesmann, R. (1955a). *Mitt. Schweiz. Entomol. Ges.* **28**, 251.

Wiesmann, R. (1955b). *Congr. Int. Phytopharm., Mondorf les Baines, J. Trav.* p. 69.

Wiesmann, R. (1956). *World Health Organ., Inform Circ. Resistance* **3**, 1.

Wiesmann, R. (1957). *J. Insect Physiol.* **1**, 187.

Wiesmann, R., and Kocher, C. (1951). *Z. Angew. Entomol.* **33**, 297.

Wiesmann, R., and Reiff, M. (1956). *Verh. Naturforsch. Ges. Basel* **67**, 311.

Wigglesworth, V. B. (1956). *J. Roy. Soc. Arts* **104**, 426.

Wilkinson, C. F. (1967). *J. Agr. Food Chem.* **15**, 139.

Wilkinson, C. F. (1968). *World Rev. Pest Contr.* **7**, 155.

Wilkinson, C. F. (1968). *In* "Enzymatic Oxidations of Toxicants" (E. Hodgson, ed.), pp. 113–149. North Carolina State Univ., Raleigh.

Wilkinson, C. F., and Hicks, L. J. (1969). *J. Agr. Food Chem.* **17**, 829.

Williams, R. T. (1959). "Detoxication Mechanisms." Wiley, New York.

Williamson, R. L., and Schechter, M. S. (1970). *Biochem. Pharmacol.* **19**, 1719.

Wilson, C. S. (1949). *J. Econ. Entomol.* **42**, 423.

Wilson, I. B., Hatch, M. A., and Ginsburg, S. (1960). *J. Biol. Chem.* **235**, 2312.

Wilson, I. B., Harrison, M. A., and Ginsburg, S. (1961). *J. Biol. Chem.* **236**, 1498.

Wilson, W. E., Fishbein, L., and Clements, S. T. (1971). *Science* **171**, 180.

Winteringham, F. P. W. (1952a). *Nat. Acad. Sci.—Nat. Res. Counc., Publ.* **219**, 61.

Winteringham, F. P. W. (1952b). *Science* **116**, 482.

Winteringham, F. P. W., and Barnes, J. M. (1955). *Physiol. Rev.* **35**, 701.

Winteringham, F. P. W., and Harrison, A. (1959). *Nature (London)* **184**, 608.

Winteringham, F. P. W., Loveday, P. M., and Harrison, A. (1951). *Nature (London)* **167**, 106.

Winteringham, F. P. W., Harrison, A., and Bridges, P. M. (1955). *Biochem. J.* **61**, 359.

Winton, M. Y., Metcalf, R. L., and Fukuto, T. R. (1958). *Ann. Entomol. Soc. Amer.* **51**, 436.

Woke, P. (1939). *J. Agr. Res.* **58**, 289.

Wright, J. W., and Brown, A. W. A. (1957). *Bull. W.H.O.* **16**, 9.

Wright, J. W., and Pal, R. (1967). "Genetics of Insect Vectors of Disease." Elsevier, Amsterdam.

Yamamoto, I. (1970). *In* "Biochemical Toxicology of Pesticides" (R. D. O'Brien and I. Yamamoto, eds.), pp. 193–200. Academic Press, New York.

Yamamoto, I., and Casida, J. E. (1966). *J. Econ. Entomol.* **59**, 1542.

Yamamoto, I., and Casida, J. E. (1968). *Agr. Biol. Chem.* **32**, 1382.

Yamamoto, I., Kimmel, E. C., and Casida, J. E. (1969). *J. Agr. Food Chem.* **17**, 1227.

Yamasaki, T., and Narahashi, T. (1958). *Botyu-Kagaku* **23**, 146.

Yang, R. S. H., Hodgson, E., and Dauterman, W. C. (1971a). *J. Agr. Food Chem.* **19**, 10.

Yang, R. S. H., Hodgson, E., and Dauterman, W. C. (1971b). *J. Agr. Food Chem.* **19**, 14.

Young, S. Y., and Berger, R. S. (1969). *J. Econ. Entomol.* **62**, 727.

Zabik, M. J., Schuetz, R. D., Burton, W. L., and Pape, B. E. (1971). *J. Agr. Food Chem.* **19**, 308.

Zaghloul, T. M. A., and Brown, A. W. A. (1968). *Bull. W.H.O.* **38**, 459.

Zahavi, M., and Tahori, A. S. (1970). *Biochem. Pharmacol.* **19**, 219.

Zayed, S. M. A. D., Hassan, A., and Hussein, T. M. (1965). *Z. Naturforsch. B* **20**, 587.

Zayed, S. M. A. D., Hassan, A., and Hussein, T. M. (1966). *Biochem. Pharmacol.* **15**, 2057.

Zayed, S. M. A. D., Hassan, A., and Fakhr, I. M. I. (1968). *Biochem. Pharmacol.* **17**, 1339.

Zayed, S. M. A. D., Hassan, A., Fakhr, I. M. I., and Bahig, M. R. E. (1970). *Biochem. Pharmacol.* **19**, 17.

Zeid, M., Dahm, P., Hein, R., and McFarland, R. (1953). *J. Econ. Entomol.* **46**, 324.

Ziv, M., and Brown, A. W. A. (1969). *Mosquito News* **29**, 456.

Zubairi, M. Y., and Casida, J. E. (1965). *J. Econ. Entomol.* **58**, 403.

Chapter 2

THE STRUCTURE AND FORMATION OF THE INTEGUMENT IN INSECTS

Michael Locke

123

I. Introduction

The cuticle covering the body determines the characteristics of insects in a much more profound way than the skin of other groups of animals. It is skin, skeleton, and a food reserve. It also takes part in the formation of nearly all sense organs and its properties have markedly influenced the design of organs for respiration. Like the skin, its hardness makes it resistant to abrasion and has resulted in the evolution of minute surface structures important in protection, heat conservation, color, and surface properties. Particular layers and components form barriers to water, ions, pathogens, and insecticides. Like the skeleton, it is the only structural tissue determining the form of the body. Tough rigid components form the articular skeleton, wings, and appendages, with flexible regions for joints and an elastic component to store energy for economic movement and to conserve hydrostatic pressure. The mouthparts are formed from specially hardened cuticle. The inner layers are a food reserve resorbed during starvation and at molting. Lastly, the design of all mechanoreceptors, chemoreceptors, and organs of vision is intimately bound up with the properties of the cuticle. It is for these reasons that studies of the integument occupy a central position in insect biology.

There is another equally important reason for studies of the integument. It provides suitable experimental material for the solution of fundamental problems of general interest. For the biophysicist, there is the problem of the orientation of macromolecules and fibrous polymers. For the biochemist, there is the synthesis of an array of biologically uncommon molecules. For the physiologist, there are membranes of selective and polarized permeability. For the endocrinologist, there are morphological changes readily induced by slight changes in hormonal milieu, and for the developmental biologist there is an epithelium, one cell thick, whose differentiation in space and time is visibly recorded by the synthesis of layers of cuticle.

A. Strength, Size, and Shape

Many derivatives of the integument, and even whole insects, are supported by layers of cuticle which are not very much thicker than a cell membrane and yet are mechanically stable. There are three reasons for this: the cuticle differs from a cell membrane in its mechanical properties, the structures concerned are often little larger than large protozoa or some single metazoan cells, and the skeleton is arranged as a shell.

1. Strength

Until recently there have been no direct measurements of the mechanical properties of cuticle as it functions on the animal. Jensen and Weis-Fogh (1962) have measured the elastic modulus and tensile strength of both hard tibial cuticle and elastic wing hinges in *Schistocerca*. It has also been possible to obtain a minimal value for the elastic modulus of hairs in compression in the plastron of *Aphelocheirus* from the data of Thorpe and Crisp (1947). Typical values for cuticle, skeletal materials, organic fibers, noncrystalline polymers, and metals are given in Table I.

From this we can see that hard cuticle is of medium rigidity and rather low tensile strength. Weight for weight, most of the organic materials compare quite favorably with metals. No values are obtainable for many of the cuticles most commonly met with, for example, those of caterpillars and blood-sucking bugs. Cuticles from these animals are plastic; stress results in permanent deformation.

Cuticle can also be exceedingly hard, approximately 3 on Moh's scale. The mandibles can scratch calcite and can bore through tin, copper, zinc, or silver (Bailey, 1954).

2. Size

It is not generally realized how thin a skeleton in the form of a shell can be for adequate support, provided it is small and has the appropriate mechanical properties. Shell is used here in the engineering sense to mean a surface-supporting structure, which is thin in relation to total size. Most insect skeletal structures, whether abdominal segments, appendages, or hairs, are more or less tubular in shape. Other parts approximate in shape to a sphere or to curved panels. The critical stress has been calculated (see accompanying tabulation) for shells with these shapes approximating in size to an insect or its appendages.

Structure	Stress	Critical stress given by	Critical stress
Cylinder or curved panel	Axial compression	$\dfrac{Eh}{a\sqrt{3(1-v^2)}}$	100 gm/mm²
Sphere	Uniform compression	(Same as above)	(Same as above)

TABLE I

The Mechanical Properties of Cuticle as Compared with Other Skeletal Materials, Organic Fibers, Noncrystalline Polymers, and Metals[a]

Material and source	Elastic modulus E in extension (kg/mm²)	Tensile strength (kg/mm²)	Compressive strength (kg/mm²)	Extension at break (%)	Specific gravity (gm/ml)	Remarks
Cuticle and chitin						
Solid cuticle	960	9.6	—	2–3	1.2	Jensen and Weis-Fogh (1962)
"Chitin" (dry)	4500	58	—	1.3	—	Herzog (1926)
Purified chitin (dry)	—	9.5	—	—	—	Thor and Henderson (1940)
Purified chitin (wet)	—	1.8	—	—	—	Thor and Henderson (1940)
Pure resilin, dragonfly	0.2	0.3	—	—	—	Weis-Fogh (1961c)
Rubberlike ligament, locust	0.2	0.3	0.3	300	—	E normal to lamellae; Weis-Fogh (1961c)
Solid cuticle, locust	960	9.6	—	2–3	—	Laminate; Jensen and Weis-Fogh (1962)
"Balkenlage," beetle	4500	5.8	—	1.3	—	Fibrous; Herzog (1926)
Aphelocheirus hair pile	340	—	—	—	—	E in compression, minimal value calculated from Thorpe and Crisp's data (1947)[b]
Cell membranes						
Sea urchin eggs	2×10^{-4}	—	—	—	—	Mitchison and Swann (1954)
Vertebrate skeletal materials						
Collagen	140	56	18	—	1.3	Currey (1962)
Bone	1800	10.5	2.8	—	1.8	Currey (1962); collagen and apatite
Cartilage	—	1.4	—	—	—	Koch (1917)
Natural fibrous materials						
Cellulose, ramie	2500–500	90–110	—	4	1.6	Highly oriented, polycrystalline, no primary interlockings
Cellulose, cotton	600–1100	50–60	—	6–8	1.6	
Protein, silk fibroin	700–1000	35–60	—	20–25	1.3	
Protein, zein	250	12	—	20–25	1.3	Mainly noncrystalline
Protein, wool keratin	100–300	15–20	—	30–40	1.3	Complex α to β transformation

Wood						
Mahogany	880	11	3.5	—	0.8	Kaye and Laby (1948)
Rubbers (*Hevea* rubber)						
Hevea, unvulcanized	0.05	0.1–0.3	—	up to 1200	0.9	No interlockings
Hevea, soft vulcanized	up to 0.5	3	—	700	—	Few interlockings
Hevea, hard vulcanized (ebonite)	350	7	8	4	—	Filled space networks
Synthetic materials						
Polyesters, Terylene	1000–1800	80	—	10	1.4	Highly oriented, polycrystalline
Polyamides, nylon 66	200–500	40–60	—	20–30	1.1	Oriented, mainly crystalline
Polystyrene fibers	250–400	5	—	2.5	1.1	Noncrystalline
Resins and laminates						
Protein, casein plastics	350	8	25	2.5	—	Formaldehyde tanned
Phenol–formaldehyde: cast, unfilled	280	7	15	1–5	—	Space network
Cellulose-filled	1000	6	15	1–5	—	Space network
Paper laminate	1000	9	20	1–5	—	Shear strength normal to lamina
Cotton laminate	850	8	23	1–5	—	10 kg/mm², in plane 3 kg/mm²
Metals						
Steel	20,000	100	—	—	7.8	Kaye and Laby (1948)
Aluminum	7000	10	—	—	2.7	Kaye and Laby (1948)

[a] Data taken mainly from Currey (1962), Jensen and Weis-Fogh (1962), Kaye and Laby (1948), and Roff (1956).

[b] Calculated from the data of Thorpe and Crisp (1947) on the resistance of the plastron hairs to pressure. The hairs are 6 μm long by 0.18 μm in diameter and are spaced evenly with a density of 2.5×10^6 per mm². They prevent the entry of water and maintain an air–water interface at the tips of the hairs at pressures up to 4 atm. At higher pressures, water penetrates between the hairs. Thus, 4 atm is a minimal value for resistance of the hair pile to collapse and is within the elastic limits. Also, 4 atm is equivalent to 4×10^6 dynes per cm² and it would operate on 2.5×10^8 hairs. Each hair can be thought of as a centrally compressed bar supporting a stress of at least 1.6×10^{-2} dynes per cm². The critical stress P, for a centrally compressed bar is given by $P = \pi EAk^2/l^2$ where E = elastic modulus; A = cross-sectional area of the bar; l = length of the bar; and k = radius of gyration. Thus, if the material obeys Hooke's law, 340 kg/mm² is a minimal value for the elastic modulus in compression.

For comparison with cuticular structures, the wall thickness h has been made 500 Å; the diameter $2a$, 1/2 mm; $E = 3 \times 10^{10}$ dynes/cm^2; and v, Poisson's ratio, is 1:2, similar to rubber and most biological materials.

The appendage of a *Rhodnius* larva might have a diameter of 1/2 mm with the cuticle 100 μm thick, giving a critical stress for axial compression of 80 kg/mm^2. Such a larva would weigh about 30 mg so that the stresses, if all the load is placed on one leg, would only be a very small fraction of the critical load. In a human femur under a static load of 100 pounds, the compression in the neck may reach about 1 kg/mm^2. Bone can take a compressive stress of about 17 kg/mm^2, so the standing load is rather more than 5% of the breaking strength (Koch, 1917). The safety factors for bearing static loads in insect skeletons compare favorably with vertebrates and give them a margin of safety for performing dynamic feats inappropriate for larger organisms.

In many distortions of shells (using shell here in the engineering sense), the critical stress is related to some power of the ratio of thickness to diameter. For example, in a cylinder subjected to torsion, the critical stress varies with the ratio $(h/a)^{3/2}$ and in a compressed circular ring the critical stress is proportional to $(h/a)^2$. Thus, stiff, superficial membranes are mechanically appropriate only for organisms and structures in the small size range occupied by insects. Unless the shell retains the same relative proportions in large animals, the mechanical advantages of a thin shell are lost.

3. Shape

The mechanical advantages of a skeleton on the outside, a tubular shell, were pointed out many years ago by Chetverikov (1915), but the argument deserves to be repeated here. The modulus of resistance to bending in a solid cylinder and in a hollow one is expressed by

$$W = \pi D^3/32$$

and

$$W_1 = \pi(D_1{}^4 - d^4)/32D_1$$

where W and W_1 are the respective moduli, D is the diameter of the solid cylinder, D_1 the outer, and d the inner diameter of the hollow cylinder. If we assume that an appendage with a central skeleton has the same cross section as one with a shell, and the cross sections of the skeletons are equal in area (i.e., they both use the same amount of skeletal material), then the shell is three times stronger than the solid rod. If the skeletons are to be of equal strength, the rod must occupy 84% of the cross-sectional area.

These calculations involve many assumptions and greatly oversimplify the problems, but they serve to demonstrate that in the size range occupied by insects the strength of cuticle arranged in shells is adequate for most mechanical purposes. Design engineers interested in miniaturization might perhaps learn something from insect structures.

B. GROWTH AND ECDYSIS

Perhaps the most important consequence of having a hard integument is the limitation upon growth, necessitating ecdysis. In a few insects (many holometabolous larvae, queen termites, blood-sucking bugs), much of the integument is plastic throughout an instar and stretches to the limit of the epicuticle; but in most, plasticity and increase in size are limited to brief periods before hardening immediately after eclosion from the exuvium.* During a stadium such insects are restricted in growth to expansion at flexible intersegmental membranes. This need for ecdysis has resulted in cyclical activity of the tissues concerned in cuticle formation and in their control systems. The side effects upon the rest of the animal during the formation of a new integument and at molting are so far reaching that insects are commonly, but incorrectly, said to grow discontinuously. This error emphasizes the importance of the integument. However, growth, in the sense of increase in living tissue as a result of the assimilation of food, takes place mainly in the intermolt period. Nor is cell division only associated with the molt cycle; in the gut and nervous system at least there may be a closer correlation with feeding than molting. Growth in the nervous system of *Drosophila* is continuous throughout larval and pupal life (Power, 1952). We should think of insects, not as organisms with discontinuous growth, but as organisms in which a set of syntheses, predominantly in a few cell types, are cyclically hypertrophied, thus overshadowing normal growth. A study of molting processes then becomes an exciting study of growth with phases of synthesis discretely grouped in space and time for the convenience of the experimenter.

C. THE BARRIER TO THE ENVIRONMENT

The cuticle may compose up to half the dry weight of an insect and represents a major investment of raw materials. It is a triumph of design that only a small fraction of this is outside the body metabolic pool. In a vertebrate everything secreted by the epidermis is lost. The cuticle of annelids, which is often compared to that of arthropods because, like endocuticle, it is a mucopolysaccharide, is also functionally outside the animal; it is lost

---

* *Exuviae-arum* f. is a Latin plural having no singular. In spite of this, the word exuvium is in common use as a singular, meaning a skin which is laid aside. In using exuvium in this sense, I have followed current practice.

from the organism's metabolic pool from the moment of secretion (Millard and Rudall, 1960). In an insect, only the outer layers of the cuticle which are chemically most inert and mechanically most rigid make up the exuvium. This economy has been made possible by the evolution of the epicuticle as the barrier to the environment: proximal to the epicuticle the integument is living, in the sense that the epidermis can exchange and draw upon its components either cyclically at molting or continuously during starvation. The functional difference between vertebrate, arthropod, and most other types of invertebrate integument is described below (Fig. 1).

The extreme thinness of the primary barrier to the environment is both a virtue in its economy and a limitation in its fragility. The epicuticle is damaged by the most minute abrasions, and the separation from the epidermis by the endocuticle creates a problem in repair. Quite apart from the demands of growth, molting is a necessity for the periodic replacement of the epicuticle.

II. The Structure of the Integument

The following account of the structure of the integument is intended to provide a basis for the discussion (see Section III) upon the control of the

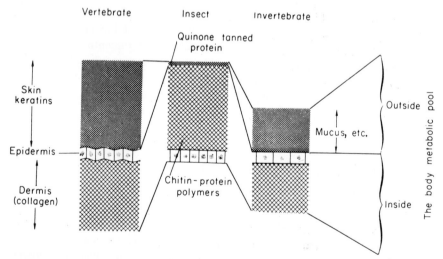

Fig. 1. The limits of epidermal control in vertebrate, insect, and in most nonarthropod invertebrate integuments. In nonarthropod groups, inert "dead" material begins immediately outside the epidermis; the region within the body metabolic pool is defined by the cells. In insects and other arthropods, the limits of the metabolic pool are set by the exocuticle.

formation of its many layers. For a detailed account of earlier work, see the reviews of Wigglesworth (1948a, 1953a, 1957) and Richards (1951, 1958b).

A. THE BASEMENT MEMBRANE

The epidermis is separated from the hemocoel by a layer of neutral mucopolysaccharide secreted by a type of hemocyte (Wigglesworth, 1956). With the electron microscope, it appears as an amorphous granular layer up to $1/2$ μm thick. Other cells may take part in the formation of the basement membrane in other parts of the body. Most is known about the basement membrane round the nervous system where it contains collagen fibrils (Smith and Wigglesworth, 1959; Hess, 1958). Other studies are those of Ashurst (1959), Ashurst and Chapman (1961), Baccetti (1955, 1956a,b), Pipa and Cook (1958), and Richards and Schneider (1958).

Nerves and tracheoles run between the basement membrane and the cells before ramifying further. The tracheoles may give the appearance of penetrating the epidermal cells but they always carry their own plasma membrane with them. There are many peripheral nerves but connections with the epidermal cells have not been described. If the basement membrane of the epidermal cells is comparable to the neural lamella in its permeability the cells may exist in an environment very different in composition from the blood.

B. THE EPIDERMIS

The epidermis forms a continuous sheet of polygonal cells below the cuticle. In *Locusta* the density of cells increases during each stadium until the molt, but at comparable times the density is similar in different stadia (Strich-Halbwachs, 1959). Epidermal cells are often polyploid. Those in *Calpodes* contain numerous nucleoli. In many Hymenoptera and probably in all Diptera Cyclorrhapha, the epidermal cells do not divide but grow in size as the larva grows. In *Calliphora erythrocephala* the epidermal cells are about eleven times as wide in the third as in the first instar. On the other hand in many Diptera Nematocera, the epidermal cells divide as the larva grows, although there is also an increase in the average size of the cells (Hinton, 1961). The form and contents of the cells vary greatly with their activity. At some stages during cuticle deposition the outer plasma membrane is folded, increasing the surface area for the absorption of compounds from the blood. There are numerous connections between adjacent cells (Locke, 1961) having the form of the septate desmosomes described by Wood (1959) in *Hydra*. The plasma membrane next to the cuticle is microvillate during cuticle deposition (Fig. 2) and lobed with much smooth endoplasmic reticulum during the secretion of wax (Fig. 3). The axial filaments of the pore canals pass through the plasma membrane into the cell.

Many epidermal cells contain characteristic orange to red pigment granules. In osmium-fixed tissue they appear under the electron microscope as concentric shells of dense and less dense material. Some newly emerged insects—for example, *Rhodnius* and *Triatoma*—appear bright orange-red before the color of these granules is obscured by pigment patterns in the cuticle. In other insects (*Calpodes*) the color and number of granules appear to vary genetically. The granules disappear from cicatrices in wounded areas. These granules are probably the same as those described in the epidermis of gryllids and phasmids by Fuzeau-Braesch (1957) which are composed of a type of insectorubin capable of undergoing oxidation–reduction changes. They are unimportant in contributing to the gross coloration in gryllids but in phasmids they undergo a hormonally controlled migration with a diurnal rhythm (Raabe, 1959). These granules have also been studied in acridiids by Nickerson (1956).

C. The Endocuticle

There have been no chemical assays of endocuticle alone, but since it constitutes the bulk of the integument the analyses of whole cuticles are probably nearly the same. One of the most recent analyses is that of DeHass *et al.* (1957) on *Anabrus simplex*. For other analyses and the general chemistry see Hackman (Chapter 3, this volume).

The inner face of the endocuticle next to the cells is frequently different from the more peripheral endocuticle; Schmidt (1956) called it the subcuticular layer and thought of it as a glue holding the cuticle to the cells, but is is more probable that it is newly secreted endocuticle in which the fibers are not ordered and have different chemical reactions. This layer shows up characteristically after oxidation and staining with paraldehyde fuchsin. It appears granular under the electron microscope but the granules already have the alignment characteristic of the microfibers of the endocuticle (Fig. 2). In some cuticles the region next to the cell may be filled with lipid-water liquid crystals which later find their way through the pore canals to the surface. Figure 4 shows this condition in the wax-secreting cuticle of the honey bee.

The endocuticle is a layer from 10 to 200 μm thick composed of numerous lamellae. The electron microscope resolves these lamellae as patterns of microfibers arranged in sheets corresponding to a dense part of a lamella but which curve out at right angles between the sheets (Figs. 5, 6, and 7). So far, the microfibers have always been observed to have the same orientation

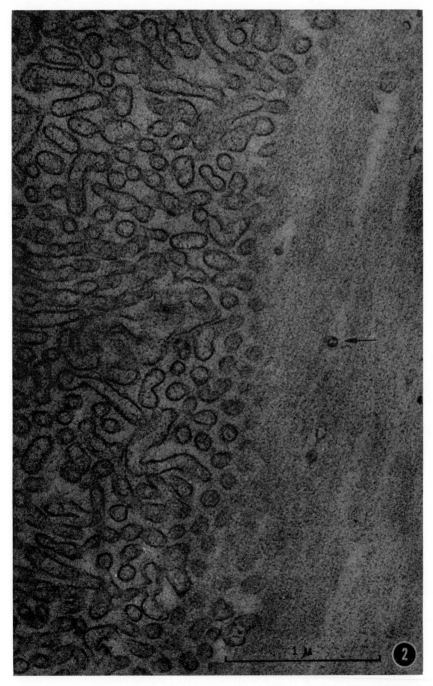

Fig. 2. Section in the plane of the cuticle where it adjoins the epidermis in a *Tenebrio* larva 24 hours after molting while endocuticle is still being deposited. The surface of the epidermis is microvillate and the newly secreted endocuticle (Schmidt's subcuticular layer) is granular. The arrow points to a pore canal filament. × 42,000.

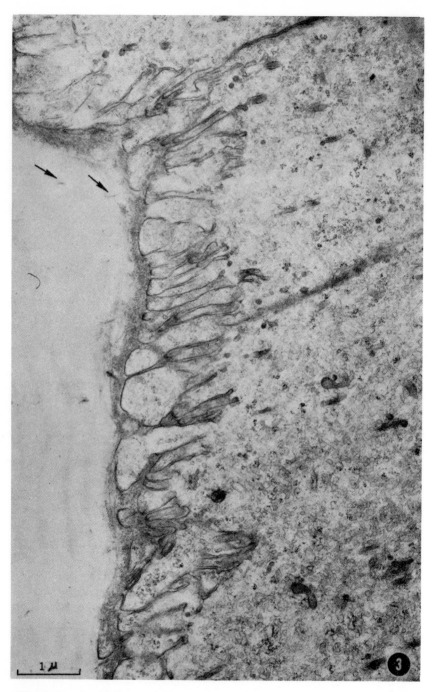

Fig. 3. Transverse section of the cuticle and epidermal edge in a *Calpodes* larva during the secretion of wax. The cells are filled with smooth endoplasmic reticulum, and the plasma membrane is lobed. Some middle phase lipid-water liquid crystals (arrows) penetrate the lamellae of the endocuticle. $\times$ 17,300.

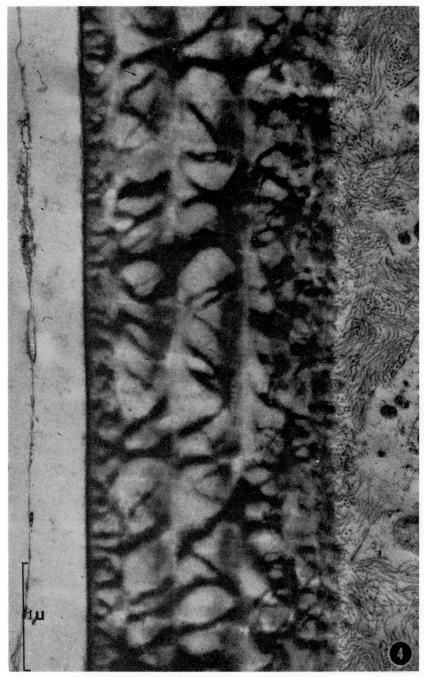

Fig. 4. Transverse section of the wax-secreting cuticle of a honey bee. Lipid-water liquid crystals accumulate on the inner face and pass through the pore canals. × 28,700.

Fig. 5. Slightly oblique transverse section of the cuticle of a *Tenebrio* larva. The epicuticle is toward the bottom left corner. The pore canals contain lipids and follow the pattern of the microfibers in the lamellae. They are not helical. See Figs. 7 and 10. × 6000. Scale = 1 μm.

Fig. 6. Transverse section of the endocuticle of a *Galleria* larva. This is a thick section to show the pattern of microfibrils making up lamellae. Compare with Figs. 7, 8, and 9. ×50,000. Scale = 1 μm.

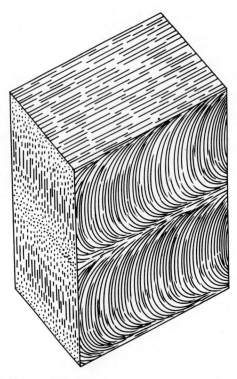

Fig. 7. Diagram of the apparent arrangement of microfibers to form lamellae
in the endocuticle. The diagram shows two lamellae. The top of the cube is in
the plane of the surface of the cuticle. (From Locke, 1961.) This pattern has
been interpreted by Bouligand to be the result of overlapping images of helicoidally
arranged stacks of microfibers (Bouligand, 1965).

in adjacent sheets, but there is no reason why the orientation should not
change from layer to layer, giving structures with extra strength in particular
directions. Although there are many parallels between plant cuticles with
cellulose, and arthropod cuticles with chitin-protein polymers, the arrangement of microfibers is fundamentally different. Cellulose occurs in flat or
woven sheets almost restricted to two dimensions (Preston, 1962). Chitin
in the endocuticle occurs in this more complex three-dimensional pattern.
The chitin of peritrophic membranes, however, is much more like the cellulose of plant cell walls. In insects and other arthropods peritrophic membranes
are like loosely woven fish nets (de Mets, 1962). Perhaps this difference
from the endocuticle is due to the low chitin content. Chemically, the peritrophic membrane of several insects has been shown to be more like a vertebrate gastric mucin (de Mets and Jeuniaux, 1962; Day, 1949).

The lamellae have a characteristic pattern of fibers. In sections normal
to the surface, each lamella seems to be composed of fibers radiating out

in a fan shaped pattern (Fig. 7). The pattern is like a series of C's, the amplitude of the curve varying with the plane of section. Bouligand (1965) has shown how this appearance is an optical illusion created by overlapping planes of fibers. The C pattern results from fibers in planes or laminae parallel to the surface. The fibers in each lamina are parallel to one another but the orientation changes slightly in the same direction from one lamina to the next. Each 180° turn gives rise to a lamella, each 360° turn to two lamellae differing from one another only if the component fibers are polarized. A lamella as such only exists with respect to the plane from which it is viewed, and the C pattern only appears in sections which are slightly oblique. The pattern is symmetrical except for the direction of rotation of the planes of successive laminae. It remains to be seen whether the direction of rotation is constant. The model is particularly satisfactory in accounting for helical pore canals which presumably follow gaps in successive layers in a helical fashion. Bouligand's model also makes it easy to comprehend transitions in structure between lamellar cuticle and cuticle with layers oriented primarily in one direction. There are two problems in lamellar structure. How is the orientation of fibers in each lamina determined, and what determines the frequency of microcycles of rotation of successive laminae to make lamellae of different dimensions? (Locke, 1967).

The lamella system varies in the degree of cross linking and hence the stability of the lamellar microfiber system. Lack of cross links would be expected in cuticles needing plasticity or elasticity but not rigidity. In this connection it is of interest that the microfibers are most easily resolved in the plastic cuticles of caterpillars (*Calpodes* and *Galleria*). They are least well seen in the exocuticle and the hard endocuticle of *Tenebrio*. The orientation would also be advantageous to give uniform elasticity in cuticles heavily reinforced with resilin, because of the possibility of distortion in any direction. The elastic cuticle round the bases of hairs has the fiber pattern typical of a lamella (Noble-Nesbitt, personal communication).

Weis-Fogh (1960, 1961a,b,c) described a structural protein, resilin, which occurs in a pure state particularly in some wing hinges and muscle insertions. It shows long-range rubberlike elasticity with no permanent deformation. Unstrained, it is optically isotropic but becomes birefringent on deformation. It is probably built from randomly coiled peptide chains linked together in a stable three-dimensional network by two aromatic α-amino acids containing a phenolic group, one of which is a diaminodicarboxylic acid and the other a traiminotricarboxylic acid (Andersen, 1963).

D. THE EXOCUTICLE

The exocuticle may be defined as the region of the endocuticle adjacent to the epicuticle which is so stabilized that it is not attacked by the molting

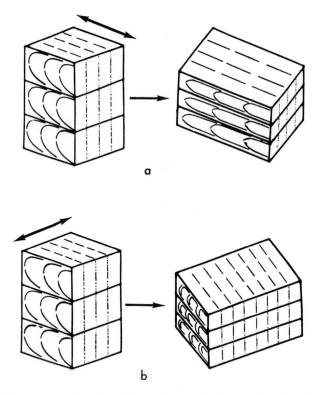

Fig. 8. Diagrams to show the effects of extension in two directions in the plane of the cuticle upon the orientation of microfibers. (a) Extension in the main axis of alignment of the microfibers would lead to the formation of laminae (as defined in the text) with the fibers predominantly in one direction in the plane of the cuticle; (b) extension at right angles to (a) would lead to laminae with mibrofibers still partly normal to the surface. In the light of Bouligand's work the above interpretation is incorrect.

fluid and is left behind with the exuvium at molting. In this sense it includes all the darkly colored, quinone-tanned cuticle. It may vary in thickness from a thin outer layer, indistinguishable from the epicuticle (as in soft-bodied larvae) to almost the whole thickness of the cuticle (as in adult beetles). In some larvae (Lepidoptera and Hymenoptera) the exocuticle contains calcium oxalate crystals (Weaver, 1958). It is uncertain whether the first-formed endocuticle destined to be stabilized as exocuticle is qualitatively different from that formed later, which remains labile. The extent of quinone tanning could be determined either by the suitability of the substrate or by the amount and time over which tanning agents are extruded. The term mesocuticle has been proposed for lightly colored but stabilized regions

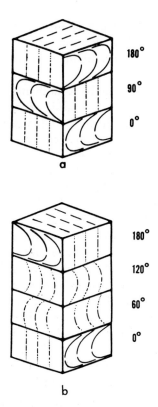

Fig. 9. Diagram of the arrangement of microfibers if successive lamellae were to be secreted with different orientations. (a) Orientation changing by 90° from one lamella to the next; (b) orientation changing by 60°. In the light of Bouligand's work the above interpretation is incorrect.

between the exo- and endocuticle. The mesocuticle differs in staining properties and presumably in chemistry from the cuticle on each side of it (see, for example, Lower, 1957a,b).

E. Pore Canals

A number of processes occur at the surface of the insect cuticle, viz., secretion and repair of wax layers, tanning of the endocuticle to form exocuticle, and the repair of surface cuticle after abrasion (Dennell, 1958; Wigglesworth, 1948a,b). For the cells to participate in these processes, they must maintain some route across the cuticle, except in the loosely knit endocuticle of some caterpillars. The pore canals form this route. They are extracellular in the sense that they are not lined by an extension of the epidermal plasma membrane but they do contain a filament from the cell (Figs. 5, 19, 20; Locke, 1961). Some pore canals seem to have several filaments, like the cilia from the rat oviduct which are intermediate in structure with microvilli (Nilsson, 1957).

The pore canal filament may be a mechanism for keeping a hole in the newly secreted cuticle until it has hardened and the canal is permanent. The filament might accomplish this by inhibiting fiber formation in its immediate vicinity. The pore canal filaments might also function as anchors to stick the epithelium to the endocuticle.

In light microscope studies (Richards, 1953), pore canals from several insects have been described as helically coiled. Helical pore canals were also reported in an electron microscope study of thick sections of cockroach cuticle (Richards and Anderson, 1942) and in *Rhodnius* the crescentic holes seen in cross section were taken to confirm their helical nature (Locke, 1957). However, another shape is more probable, and the hypothesis that pore canals are usually helical should be reconsidered (Locke, 1961).

In many sections, some pore canals have a circular outline (Figs. 19, 20). The basic shape is probably that of a cylinder bent either in a helix or in some other way. If the pore canals are indeed helical, the shape and the orientation of the holes seen in cross section will be influenced by two main considerations: (1) the phase relations of different helices, and (2) the plane of the section. The shape of the tube and the pitch and regularity of the helix do not affect the following conclusion.

a. The Phase Relations of Different Helices. The orientation of the crescentic holes seen in section is influenced by their position relative to each lamella. In any one plane the outline of a pore canal has approximately the same orientation as that of any other pore canal in a similar position with respect to the lamella. Therefore, whether or not they are helices, they are in phase, and the pattern repeats itself in each lamella.

b. The Plane of the Section. If a number of helical tubes are cut in transverse section, then the crescentic holes observed would tend to have a random orientation, as in Fig. 10a. Even if they are all in phase, the crescents will have all orientations except in the rare sections exacty in the plane of the lamella, or, when all the pore canals are equidistant from one another, in sections cut in the set of planes parallel to the pitch of the helix. Many sections have been cut for the electron microscope in all planes, but the orientation expected for helices has not been seen. The usual orientation is as in Fig. 10b.

The most probable explanation is that the pore canals follow the arrangement of fibers in the lamellae, the hole taking a curved course repeated in the same plane in each lamella. The shape and orientation of the pore

Fig. 10. Diagram of the configuration of pore canals in section. (a) The arrange-ment expected for helical pore canals when cut transversely, i.e., tangential to the surface; (b) the arrangement and shape of pore canals found in slightly oblique tangential sections. See, for example, Fig. 5. (After Locke, 1961.) The shape of pore canals has been shown to be helical in the light of Bouligand's work. Neville and Luke (1969).

canals seen in some sections can only be explained in this way. In some sections the holes are symmetrical ellipses, the long axis of each hole being parallel to its neighbors.

Helical canals, if they occur, can readily be explained by this observation that a pore canal follows the fiber pattern of the lamellae. A helical pore canal would result if the fiber pattern were to rotate about an axis normal to the surface, from one lamella to the next. This pattern has not been seen in the insects studied, but it may occur.

F. THE EPICUTICLE

The epicuticle is a composite structure involving secretions of the epider-mis and also the dermal glands which are derived from epidermal cells. In sections with the light microscope it appears as a thin refractile line. With the electron microscope it is seen to have the structure shown in Fig. 11. Very little is known about the composition of any of the constituent layers except the wax. The other layers cannot be isolated and have only been characterized by staining and the crudest chemical techniques. At the present time it is difficult to reconcile the terminology of layers deduced from physiological and light microscope studies with results obtained with the electron microscope. The difficulties may also be due to the different insects studied. The problem would be unimportant were it not for the fact that the epicuticle determines the pattern, the surface properties, and most of the impermeability of the integument; its structure therefore impinges upon insect biology in a most direct way.

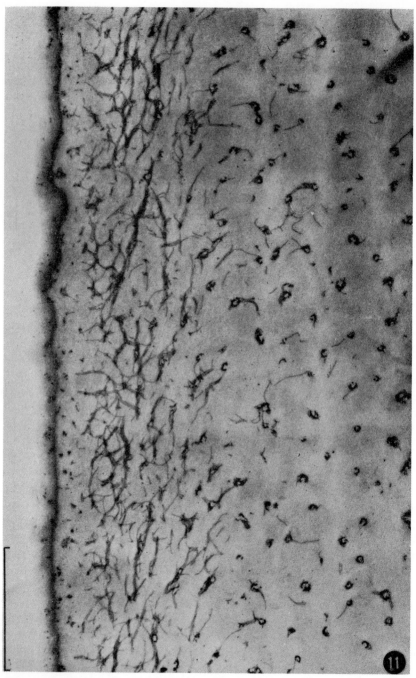

Fig. 11. Oblique section through the epicuticle of a *Tenebrio* larva. The cuticulin is cut obliquely and shows the black dots of the lipid-filled wax canals even where it is most thin. Internal to the cuticulin many middle-phase lipid–water liquid crystals ramify in the homogeneous inner epicuticle. To the right of the picture are pore canals cut almost transversely. × 33,000. Scale = 1 μm.

The traditionally accepted structure of the epicuticle is one of four layers: cement, wax, polyphenol, and cuticulin, respectively, from the outside to the inside (Wigglesworth, 1947, 1948a,c, 1953a). The electron microscope confirms that cement and wax are usually present. The cuticulin is a composite structure, cuticulin *sensu strictu* and a dense homogeneous layer, the outer part of which probably contains, or has contained, polyphenols. According to Dennell and Malek (1955a,b) and Malek (1958), the protein epicuticle in *Schistocerca* and *Periplaneta* consists of two layers, paraffin and cuticulin, distinguished by their staining although not consistently colored in different species. It is not possible to make a firm parallel between these findings and those of the electron microscope but their two layers probably correspond to regions of the dense homogeneous layer which may be chemically different, perhaps with more or less polyphenol and hydrocarbons, and which also differ in chemistry and structure between different insects.

As far as present evidence allows we can say that the whole insect is covered by the cuticulin layer *sensu strictu* except for some gaps in sensory structures. Most of it, including trachea but excluding tracheoles, parts of the lining of some glands (Mercer and Brunet, 1959) and sense organs (Slifer, 1961), is also covered by the dense homogeneous layer. Cement is present where dermal glands occur and wax is universal [although wax canals have not yet been observed in tracheae there is evidence for the presence of a lipid lining (Wigglesworth, 1953b)].

1. The Cement Layer

On the outside of many cuticles is a layer of cement secreted by the dermal glands, also called Verson's glands in Lepidoptera. These glands have been described by Kramer and Wigglesworth (1950), Malek (1958), and Wigglesworth (1953c). The nature and function of the cement probably varies in different insects but a clue to its general composition may come from the lac insect, *Laccifer lacca,* in which the cement is probably the shellac of commerce (Beament, 1955). Shellac is a mixture of laccose and lipids. The cement may also be partly tanned protein (Way, 1950), but we shall probably not be far wrong if we think of it functionally as a mixture something like varnish. It varies greatly in thickness and extent. It is thickest in the exposed cuticles of beetles and absent in many adults covered with scales although it is present in some, e.g., *Thermobia* (A. J. Watson, personal communication). In *Rhodnius* it covers most of the surface (Wigglesworth, 1947) but on the dorsal abdominal tergites of the adults it is incomplete under the wings, forming flat flowerlike plugs to the mouths of the glands. In caterpillars like *Diataraxia* the cement occupies the valleys between the tubercles (Way, 1950).

The outer surface of many insects has a fuzzy indistinct outline as seen in section with the electron microscope. The layer is much less distinct than the secretion of the dermal glands and is perhaps preecdysial in origin resulting from the products of digestion of the old cuticle. Like the cement it would be important in determining the surface properties.

There are several possible functions for the cement. In some insects it is certainly protective. The delicate patterning of the underlying layers is given a complete coat of hard transparent varnish. This is impossible in cuticles which expand during an instar, and in caterpillars and blood-sucking larvae it is a meshwork which must distort during growth. In these insects the cement might serve as a reservoir of lipids which could leak out to replace lost surface lipids. This role as a sponge to soak up mobile lipids is probably the main one in cockroaches, which are covered with very low melting point waxes (Beament, 1955). If there is a reserve of mobile wax in the cement it could play an important role in rapidly sealing over surface abrasions to prevent water loss. Only very thin layers of lipid are necessary to prevent water loss, but such films are very vulnerable to absorption and mechanical rupture (Beament, 1959).

Wigglesworth's experiments (1947) show that the wax layer is overlain by the cement; the protection of the wax may be its most important function. In small insects the high contact angle presented by a waxy surface is an essential safeguard against being trapped below a film of water. The surface of some large insects, on the other hand, is hydrophilic. The larva of *Calpodes*, for example, prefers to live in a private bath of water within a folded leaf. The wax surface may be masked by a hydrophilic cement. In *Thermobia* the cement is highly hydrophobic and also repels most organic solvents (A. J. Watson, personal communication).

2. The Wax Layers

The chemistry of cuticular waxes is well reviewed in Gilmour (1960) and will not be elaborated here. They are usually mixtures of hydrocarbons with odd and even numbers of carbon atoms in the range C_{25}–C_{31}, and esters of even-numbered fatty acids and alcohols in the range C_{24}–C_{34}. A recent and fairly complete analysis of the wax of the Mormon Cricket, *Anabrus simplex*, is given in Table II (Baker *et al.*, 1960). The hydrocarbons are not unlike those in plant waxes in which the composition is sufficiently constant for an analysis to be useful in the systematic separation of closely related species (Eglinton *et al.*, 1962a,b). It would be interesting to know if the composition of insect waxes is also species specific. Koidsumi (1957) has recorded that fatty acids, particularly caprylic or caproic acids, are important in the resistance of *Bombyx* and *Chilo* to attack by fungi.

Three regions may be recognized in the wax at the surface of insects.

TABLE II

THE COMPOSITION OF MORMON CRICKET WAX, *Anabrus simplex*[a]

Chemical class	Percent of wax	Individual component	Percent of class	Percent of wax
Hydrocarbons	48–58	C_{12}, C_{13}, C_{14}, C_{15}, C_{16}	12.5	6–7
		C_{17}, C_{18}, C_{19}, C_{20}, C_{21}	Equally	
		C_{22}, C_{23}, C_{24}, C_{25}, C_{26}	distributed	
		C_{27}–C_{28}	20.6	10–12
		C_{20}–C_{31}	55.3	27–32
		C_{32} ?	11.4	5–7
Free acids	15–18	C_{12}	0.4	1
		C_{14}	1.7	0.4
		C_{14}—1 double bond	0.5	0.1
		C_{14}—2 double bonds	0.3	0.1
		C_{14}—3 double bonds	0.6	0.1
		C_{16}	7.9	1–1.4
		C_{16}—1 double bond	4.0	0.6
		C_{16}—2 double bonds	0.3	0.1
		C_{16}—3 double bonds	0.7	0.1
		C_{18}	2.4	0.4
		C_{18}—1 double bond	48.4	7–9
		C_{18}—2 double bonds	26.8	4–5
		C_{18}—3 double bonds	6.0	1
		C_{20}	Trace	Trace
Esters	9–11	Saturated	34	3.2–3.9
		Saturated and unsaturated	28	2.7–3.3
		Saturated and/or hydroxy	15	1.4–1.7
		Unsaturated and/or hydroxy	23	2.2–2.7
Free alcohols	2–3	Cholesterol		2–3
Polymers	12–14	Acidic resins of varying molecular size		
Unidentified	2–4			2–4

[a] Data from Baker *et al.* (1960).

According to Beament (1945, 1961) there is a monolayer of lipid next to the cuticle. This is sometimes visible in electron micrographs. Most of the lipid is less well ordered and forms a much thicker layer, often permeating the cement. On the very outside in many insects, the wax forms a bloom comparable to that described on plant surfaces by Juniper (1960).

The evidence for the presence of a monolayer at the surface is reviewed by Beament (1961). The most telling evidence is perhaps the similarity in transitions of water permeability between artificial monolayers on membranes and insect cuticles. In some insects (*Rhodnius, Tenebrio, Pieris;* Beament, 1959) there are two critical temperatures for water loss (Fig. 12). He argued from this that there might be two ordered layers forming the main barrier to water loss, one at the surface of the cement and one on the outer face of the cuticulin. Recent, still unpublished, work by Thompson (1964) on the changes in electrical resistance of lipid double layers with temperature, suggests an alternative explanation of Beament's observation. Thompson showed that the electrical resistance of an artificially prepared lipid double layer goes through two maxima and three minima in the temperature range from 19° to 36°C. It seems probable that the monolayer of lipid at the surface of the cuticle postulated by Beament could go through similar changes to give the two abrupt changes in water permeability which he found. Beament's and Thompson's findings are compared in Figs. 12 and 13.

The monolayer is probably not static but is in a liquid phase being moved, lost, and replaced. C. T. Lewis (1962) used an autoradiographic technique to trace the spread of a radioiodine preparation of diiodooctadecane from the tarsi of adult *Phormia* over the body. He found that oil films were established over the whole of the integument within a few minutes of exposure.

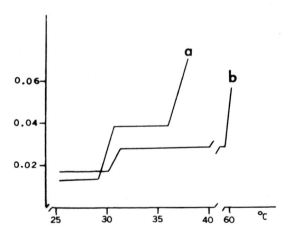

Fig. 12. The permeability of the cuticle to water. Graph showing changes in the permeability/temperature relationship with age of pupae of *Pieris:* (a) killed 24 hours after pupation; (b) killed 5 days after pupation. There are two discontinuities in permeability, the upper one varying with age. Compare the permeability to water with the electrical properties of a lipid bilayer in Fig. 13. Ordinate, permeability to water in mg/animal/mm Hg/hour. Abscissa, surface temperature in °C. (After Beament, 1959.)

Fig. 13. The electrical properties of a lipid bilayer. Graph showing the changes in specific resistance with temperature of a lipid bilayer formed at 36°C. At 36°C the resistance per cm² is 3.7×10^6 Ω. As the temperature is lowered the resistance increases linearly until 29°C when there is an abrupt decrease followed by an increase and a second abrupt decrease at about 22°C. The bilayer was created between two aqueous phases for the resistance across it to be measured. Compare the changes in resistance of a lipid bilayer with the permeability of an insect cuticle to water in Fig. 12. Ordinate, resistance in MΩ per cm². Abscissa, temperature in °C. (After Thompson, 1964.)

Five to fifteen minutes was enough to coat even the wings with a mono-molecular film in a living fly.

The bulk of the lipid at the surface of insects, enough for a layer $\frac{1}{10}$ to 1 μm thick, is probably in the less well-oriented layers permeating the cement.

Juniper (1960) was able to study the waxy bloom at the surface of many plants by viewing direct carbon replicas in the electron microscope. There has not yet been any comparable study in insects, although Holdgate and Seal (1956) give a picture of the surface of a *Tenebrio* pupa taken with an electron microscope adapted for reflexion microscopy. The pattern of wax blooms at the surface of insects is at least as complex as that in plants. The wax at the surface of several insects is shown in Figs. 14 and 15 (M. Locke and R. Anderson, unpublished observations, 1963). In these examples, although the wax has a very varied and characteristic form, it nevertheless arises from a smooth surface of cuticle. The pattern of wax must therefore result from the distribution of wax canals through the epicuticle, the nature of the wax, and the conditions under which it is deposited. In other insects the surface of the cuticle is elaborately patterned to ensure a characteristic

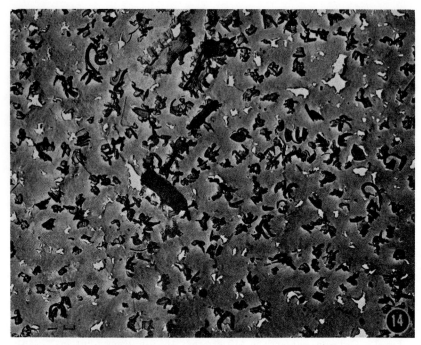

Fig. 14. Carbon replica of the early phases of formation of a wax bloom on the surface of a pupa of *Calpodes ethlius*. At a later stage, the surface is completely covered with extrusions. (After M. Locke and R. Anderson, unpublished observations, 1963.) × 3100. Scale = 1 μm.

wax form. For example, the larva of *Calpodes ethlius* has crater-shaped tubercles (Fig. 16) with wax canals situated around the rims causing the wax to emerge in the form of cylinders (Figs. 17 and 18). The molecules in the projections forming the blooms may be highly ordered. The wax cylinders from *Calpodes* larvae and the blooms on *Tenebrio* and *Calpodes* pupae are all strongly birefringent (M. Locke and R. Anderson, unpublished observations, 1963).

Although these wax blooms may look very different, they all have one feature in common. When projecting from the surface, and even more when broken up, they give wax in the form of thin sheets or plates having the maximum surface for volume of wax. The surface of the insect thus has a high contact angle for water which tends to roll up into spheres coated with wax fragments.

The function of the specialized wax secretion in *Calpodes* is an enigma. It may be concerned in the conservation of water which subsequently evaporates and cools the pupa. The wax filaments are broken up around the flimsy

Fig. 15. Carbon replica of the wax bloom on the surface of a pupa of *Tenebrio* at an early stage. The bloom is mainly in the form of crystalline flakes but there are some pieces (arrowed) which appear almost as if they have been extruded like toothpaste from a tube. (After M. Locke and R. Anderson, unpublished observations, 1963.) × 4900. Scale = 1 μm.

cocoon inside a folded leaf. At night, rain and dew collect in spheres within the leaf and evaporate in the heat of the day. Without wax, the water forms a thin film and drips from the coccoon.

3. Wax Precursors and the Liquid-Crystalline Phases of Lipid-Water Systems

Elongate filamentous structures 60 to 130 Å in diameter have been found in the cuticle of all insects examined with the electron microscope (Locke, 1960a, 1961). They occur in abundance between the cells and the cuticle in the wax-secreting integument of the honey bee, in the pore canals of *Rhodnius, Galleria,* and *Tenebrio,* and penetrate the epicuticle in all species (Figs. 18, 19, 20, 21, 22). The filaments appear more or less solid after osmium fixation and tubular after permanganate. They are similar in appearance to the liquid-crystalline phases of lipid-water systems described by Luzzati and Husson (1962) and Stoeckenius (1962).

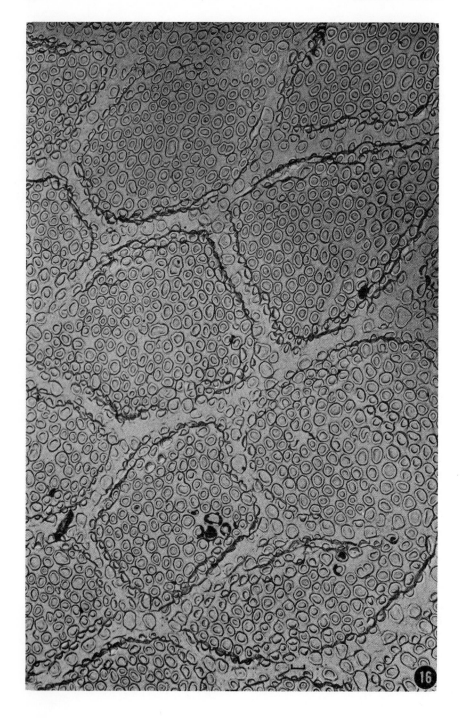

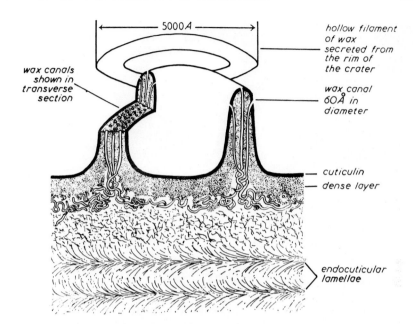

Fig. 17. The structure of the epicuticle in *Calpodes ethlius* larvae in the region where extra wax is secreted (compare Figs. 16 and 18). The epicuticle is traversed by filamentous structures in hexagonal array believed to be lipid-water liquid crystals in the middle phase. These emerge around the rim and crater-shaped tubercles. The wax is secreted in the form of a cylinder from the rim of each crater.

Luzzati and Husson used X-ray scattering techniques to deduce the structure of liquid-crystalline phases of lipid-water systems, varying both the temperature and the proportion of water. They found that a number of soaps went through a sequence of phases—neat, cubic, complex hexagonal, rectangular, deformed middle, and middle—as the temperature was lowered and the proportion of water increased. The neat phase is an alternate sequence of planar layers of lipid and water. The cubic phase is isotropic and the structure is one of spheres closely packed in a cubic lattice. In the complex hexagonal phase, the lipids form a double-layered cylindrical shell, water being both outside and inside the cylinder (Fig. 23b). The rectangular

---

Fig. 16. Carbon replica of the surface of the cuticle which secretes extra wax in the larva of *Calpodes ethlius*. The surface is covered with tiny crater-shaped tubercles. Wax canals pass through the walls of each crater in hexagonal array (Figs. 17, 18). From the rim, a cylindrical filament of wax emerges. The polygonal grouping of the craters follows the pattern of the epidermal cells which secreted the epicuticle. × 3900. Scale = 1 μm.

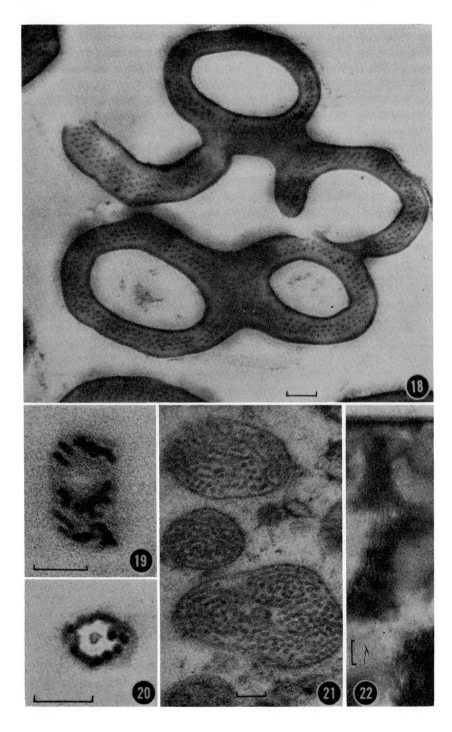

and deformed middle phases are derived from the middle phase in which the lipids are in an array of single-layered cylinders of indefinite length, the hydrocarbon chains filling the interior of the cylinders with the hydrophilic groups an water outside (Fig. 23a). The cylinders form a hexagonal array. Stoeckenius confirmed these findings with the electron microscope by direct observation of osmium-fixed soaps and phospholipids.

The waxes of insects have longer hydrocarbon chains than the lipids studied above, and the liquid crystals are correspondingly larger. Even so, in *Calpodes* the filaments are little more than 60 Å in diameter, which would correspond to two hydrocarbon chains 20 to 30 Å long arranged radially in the cylinders of the middle phase. In the epicuticle forming the wall of each crater-shaped tubercle there are 150 to 300 cylinders in a perfect hexagonal array (Figs. 17 and 18). In *Galleria* and *Tenebrio* the filaments are a little larger but probably still correspond to the cylinders of the middle phase. In honey bee cuticle, however, there are cylinders almost 200 Å in diameter which are probably in the complex hexagonal phase (Fig. 22).

On the basis of structure alone there is, therefore, a strong argument for supposing that the filamentous structures seen in cuticle are lipid-water liquid crystals. There is also the strongest circumstantial evidence. The presumed lipid-water liquid crystals are most abundant in regions where lipids would be expected, such as the cuticle where wax is secreted in the honey bee and, in *Calpodes,* the wax is secreted in a shape which follows exactly the pattern of the openings of the wax canals through the epicuticle to the exterior. It was for this reason that the liquid crystals were first called wax canal filaments, and the space they occupy in the epicuticle was referred

---

Figs. 18–22. Fig. 18, transverse section through several of the crater-shaped tubercles which secrete wax in *Calpodes ethlius* larvae. The hexagonal array of dots in the walls of each crater are believed to be sections of lipid-water liquid crystals in the middle phase. They are about 60 Å in diameter. Compare with Figs. 16 and 17. (From Locke, 1960a.) × 85,000. Scale = 1000 Å. Fig. 19, tangential section of *Tenebrio* larval cuticle about ¾ of the way through the cuticle from the cells to show one pore canal. The pore canal contains numerous filaments believed to be lipid-water liquid crystals in the middle phase. × 143,000. Scale = 1000 Å. Fig. 20, tangential section of *Tenebrio* larval cuticle just below the epicuticle showing a single pore canal with an axial filament and lipid-water filaments arranged in a ring at the periphery. × 163,000. Scale = 1000 Å. Fig. 21, transverse section of bundles of lipid-water liquid crystals enclosed by the plasma membrane of an epidermal cell below the wax-secreting cuticle in a honey bee. The filaments appear superficially to be intracellular but they are in fingerlike infoldings of the plasma membrane. They may form a layer below the cuticle (Fig. 4) before passing through the pore canals (Fig. 22). (From Locke, 1961.) × 77,000. Scale = 1000 Å. Fig. 22, transverse section of the wax-secreting cuticle of the honey bee near the surface showing numerous filamentous structure about 200 Å in diameter which may be lipid-water liquid crystals in the complex hexagonal phase. × 69,000. Scale = 1000 Å.

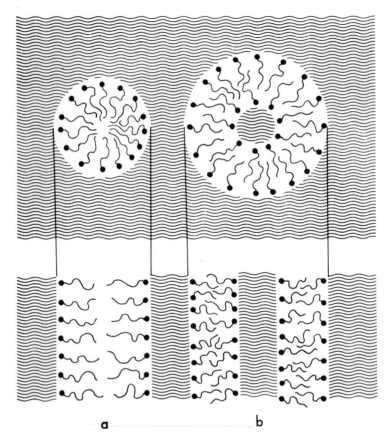

Fig. 23. Diagram of the structure of two phases of lipid-water liquid crystal. Above, transverse; below, longitudinal section. (a) The middle phase, in which the lipids form cylinders of indefinite length. Hydrocarbon chains fill the interior of the cylinders; hydrophilic groups are at the surface and water is outside; (b) the complex hexagonal phase, in which the cylinders are double-walled, with hydrophilic groups and water both inside and outside (compare with Fig. 22). Both phases tend to be in hexagonal arrays. (After Luzzati and Husson, 1962.)

to as a wax canal since it provides the route for the transport of wax to the surface.

The probable structure of these cuticular components is given in Fig. 24. There is no direct evidence (but it seems most probable) that the layer forming the cylinder of the liquid crystal is continuous with the surface monolayer. This figure suggests that the motive force driving lipids to the surface may be surface tension. On the free surface of the cuticulin, liquid lipids would be at a lower energy level than when constrained in the narrow

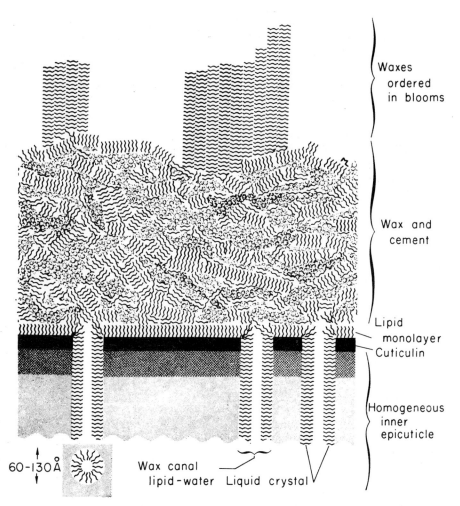

Fig. 24. The structure of the surface of an insect. Not all insects have all these structures. The cement is frequently lacking, and wax blooms would then form directly over the wax canals. In contrast to the crystalline wax blooms, the inner waxes may be liquid and mobile. There is no direct evidence that the lipids in the wax canals are continuous with the surface monolayer, but it seems most probable. The lipids in the wax canals may not always be in the middle phase as drawn here. Complex hexagonal and cubic phases have also been seen. The homogeneous inner epicuticle is frequently denser next to the cuticulin as shown, perhaps owing to its content of osmiophilic polyphenols.

cylindrical micelles; they would also flow out if lipids were to be continually drained from the surface, either by adsorption to the cement or by crystallization to form the blooms.

In addition to providing a route to the surface for waxes, the wax canals are probably also the route by which oil-soluble compounds enter an insect. The topical application of compounds with juvenile hormone activity is one of the most certain methods of inducing a systemic effect in *Rhodnius* (Wigglesworth, 1963a). The wax canals are presumably the port of entry for oil-soluble insecticides. Liquid-crystalline phases can incorporate large amounts of lipid-soluble substances (McBain, 1950). Liquid crystals in the endocuticle and even adjacent to the epidermis might be expected to come into rapid equilibrium with lipid contaminants of the surface monolayer if the lipids do form a continuous liquid system.

4. An Hypothesis for the Asymmetrical Permeability of Cuticle to Water

The problem of cuticular asymmetry is discussed in Beament (1961). In sect cuticles exposed to water on the cell side are far less permeable to water than the same cuticles when exposed to water on the outside and submitted to transpiration in the reverse direction (see, for example, Fig. 25 showing that the gain of water by live mealworms in high humidities takes place

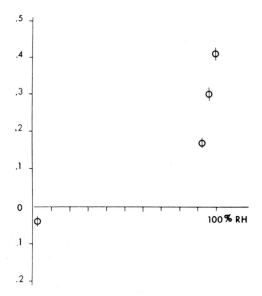

Fig. 25. The gain and loss of water by live *Tenebrio* larvae at different humidities at 30°C. Each point is the mean of about 25 animals. The cuticle behaves in a markedly asymmetrical manner. Even though the pressure drop into the animal cannot be as great as the pressure drop from body fluids to 0% RH externally, the passage of water into the animal in the direction epicuticle-endocuticle is much greater than the passage of water out in the direction endocuticle-epicuticle. Ordinate, rate of gain or loss of water in mg/cm²/hour; abscissa, relative humidity.

at a much greater rate than the maximum loss at low humidities). The previous account of the two forms of lipid-water liquid crystal suggests an hypothesis to account for this asymmetry. If both phases can exist in the epicuticle at the surface of the cuticulin we should expect high permeability to water with the complex hexagonal and very low water permeability with the middle phase. If the presence of water at the surface can induce the change, the observed asymmetry would be accounted for (Fig. 26). An objection might be that the dimensions of the spaces occupied by the liquid crystals in the epicuticle are constant, and would not permit such a change in the contained water and lipid. The change from complex hexagonal to the middle phase need only take place over a very short length of the cylinder to reduce permeability markedly. It is difficult to determine what the dimensions of the wax canals may be since we only infer their presence from the lipid they contain. Fortunately, this hypothesis can be tested by electron microscope observations, although with some difficulty.

5. Cuticulin and the Protein Epicuticle

In early studies on the epicuticle Wigglesworth observed the formation of oil droplets when he treated it with Schultze's reagent. He deduced that it must be a lipoprotein and called it cuticulin (Wigglesworth, 1948a). Since

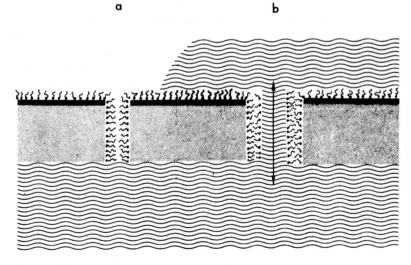

Fig. 26. Diagram to explain a possible effect of a phase change of the lipids within the wax canals upon the permeability of cuticle to water. A wax canal occupied by a lipid-water liquid crystal (a) in the middle phase and (b) in the complex hexagonal phase. If water or a high humidity in the environment can induce this phase change, then the apparent asymmetry in permeability to water would be accounted for.

it turns out that this structure has several components, the term now lacks precise meaning. It has also been used in several different senses by other authors. It is proposed here to use cuticulin to mean the dense outermost layer of the protein epicuticle which is universally present and has about the thickness of a cell membrane (or a little more). Its composition is unknown but it has been found to cover the entire surface of insects from tracheoles to gland ducts, hairs, and scales. The only thing known of its structure is that it appears as a thin line in electron micrographs. If the lining of tracheoles is the same, as it appears to be, then it may originate as a lipid double layer. In newly formed tracheoles the lining is about 85 Å thick and can be resolved as twin dense lines (Locke, 1958a). In later tracholes only a single dense line is resolvable. It is perhaps of interest that very thin membranes form spontaneously at the surface of injured cells in a number of animals (insect eggs, Jones, 1958, 1959; various invertebrates, Monné, 1960). In these examples it seems that a lipid double layer could be protected by proteins in solution which are later stabilized by quinones produced in wounding. This would be a useful first hypothesis to test for the structure and formation of the cuticulin layer.

The cuticulin is perhaps the most important layer in the cuticle. Its functions are listed below.

a. *Permeability Barrier.* During its existence it has to be permeable to the factor activating the molting gel and the products of hydrolysis of the old cuticle, but impermeable to the molting fluid and resistant to digestion by it. It is permeable to waxes and under some circumstances, as when an insect is immersed in oil (Wigglesworth, 1942), it is permeable to water. In those insects which actively take up water it also has to be permeable to water passing into the insect.

b. *The Limitation of Growth.* In those insects such as blood-sucking bugs and caterpillars which grow within a stadium by expansion of the integument, the size achieved at the end of the stadium is limited by the dimensions of the epicuticle laid down on the surface of the epidermis before ecdysis. It is for this reason that in these insects the newly molted surface is deeply pleated.

c. *The Surface Pattern.* The elaborate sculpturing of the surface, the formation of hairs, scales, and microtrichia, and the elegant patterning of tracheal and tracheoles all take place at an epidermal cell surface at the time of formation of the cuticulin layer. It is questionable which is the prime mover, the plasma membrane assuming a shape for the deposition of the cuticulin, or the cuticulin responding to forces which carry the plasma membrane along with it.

d. Surface Properties. If Beament is correct in supposing that there is a lipid monolayer in the epicuticle, the cuticulin provides the surface to support it. If the waxes are polar, then the surface of the monolayer could be lipophilic or lipophobic depending upon the properties of the cuticulin and the orientation of the lipid molecules (Fig. 27; Beament, 1961).

Immediately below the cuticulin in most cuticles is a much wider layer of homogeneous, rather dense cuticle which is the refractile line called the epicuticle in sections with light microscopy. Nothing is known of its composition apart from Wigglesworth's early work. Just before the secretion of wax Wigglesworth observed the progressive spread of argentaffin-positive material from the tips of the pore canals until it eventually covered the surface. He supposed that there was a discrete layer of polyphenols on the outside of the protein epicuticle, stabilizing both it and the underlying endocuticle (Wigglesworth, 1947). Although there is no discrete layer to be seen with the electron microscope, the region just below the cuticulin is more electron dense in osmium-fixed preparations, which would be expected if it contained polyphenols tending to bind osmium. The pore canals stop below this homogeneous dense layer in *Galleria* and *Tenebrio* but it is penetrated by the wax canals, which in *Calpodes* are much wider on the inner edge than the diameter of a middle phase lipid-water liquid crystal.

G. The Ecdysial Membrane

Passoneau and Williams (1953) first described a membrane between the pupa and adult cuticles of developing *Cecropia* which they called the "ecdy-

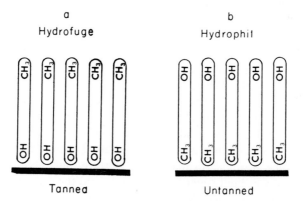

Fig. 27. Diagram showing how the condition of the cuticulin might affect the surface properties of cuticle through the orientation of the surface lipids. (a) Hydrofuge outer surface produced by mounting a normal monolayer of long-chain alcohol on a tanned cuticulin substrate; (b) hydrophil outer surface produced by mounting a reverse monolayer on an untanned cuticulin substrate. (After Beament, 1961.)

sial membrane." Since then, the ecdysial membrane has been studied in *Schistocerca* by Malek (1956, 1958, 1959), in *Rhodnius* by Locke (1958a), in Lepidoptera by Richards (1955, 1958a) and Lower (1957a,b), and in *Sialis* by Selman (1960). In his earlier work on molting in *Rhodnius*, Wigglesworth figures a layer probably corresponding to the ecdysial membrane, but he included it with the debris of endocuticle attacked by molting fluid (see Wigglesworth, 1957).

In *Schistocerca*, the ecdysial membrane is formed from the inner few lamellae of the endocuticle which are resistant to the action of the molting fluid as well as to concentrated mineral acids. These laminae separate from the old cuticle when it is dissolved by the molting fluid and give rise to a discrete membrane which persists throughout the premolt period eventually being cast as the inner lining of the larval exuvium.

In *Calpodes* larvae the epidermal cells withdraw from the endocuticle and secrete numerous dense osmiophilic droplets, some of which run together to form a membrane on the inner face of the endocuticle. At a slightly later stage both droplets and complete membrane can be seen. Similar events take place during the formation of the tracheal cuticle in *Rhodnius*. There is a distinct stage after the deposition of the cuticulin but before the activation of the molting fluid when the membrane is a reticulum of coalescing droplets. The membrane, or the remains of it, is shed with the exuvium at ecdysis.

H. Tracheal and Tracheolar Cuticle

For descriptions of the pattern of branching in the tracheal system in *Sciara*, see Keister (1948); in various Diptera, Whitten (1959, 1960); in *Locusta*, Albrecht (1953); and in *Rhodnius*, Wigglesworth (1954a) and Locke (1958b). The tracheae arise as segmental ingrowths of the integument which branch over the internal organs and unite laterally to form longitudinal tracheal trunks. Where tracheal cells from different segments meet they secrete cuticle with a characteristic randomly buckled pattern different from the usual helical or annular pattern of the taenidia. These positions, called nodes, are the tracheal homologue of the intersegmental membranes.

The structure of a trachea is given in Fig. 28 (Locke, 1957). The cuticle forming the tube is similar to that on the surface, but there is no cement, no detectable wax or lipid-water liquid crystals, and the exo- and endocuticle are reduced to the taenidial thickenings. In most insects there is no cuticle between the taenidia and the epithelium, so that the trachea is a freely extensible tube resisting lateral compression by virtue of the taenidia and extending to the limit of the lining epicuticle. The axial orientation of the chitin micelles in the epicuticle and the circumferential orientation in the taenidia would function to prevent overextension and lateral compression, respectively. In a few insects (e.g., *Phlegethontius* larvae, Lepidoptera, Sphingidae) there may be a continuous endocuticle in the largest tracheae which therefore resist both axial and lateral deformation.

The tracheoles arise from single tracheal cells and unlike the tracheal they do not shed and reform their lining membrane but remain unchanged from instar to instar (Wigglesworth, 1954a, 1961b, 1962). They are similar in structure to a simplified trachea but with a different buckling frequency consistent with their dimensions (Locke, 1958a); like the tracheal cuticle, the tracheolar cuticle is secreted at the surface of a plasma membrane. The cuticle membrane is uniform and sharply delineated like the cuticulin layer which lines tracheal. There are taenidia but no other cuticular components. The tracheolar cell envelops the tracheole and the outer and inner plasma membranes connect by folds in much the same way that a nonmyelinated axon is suspended within a Schwann cell (Figs. 29, 30, 31). For this reason, Edwards *et al.* (1958) called this connecting plasma membrane a "mestracheon" (see also Smith, 1960, 1963).

At each molt the connection between a tracheole and a trachea has to break and reform. Wigglesworth (1959a) has described how the tracheole

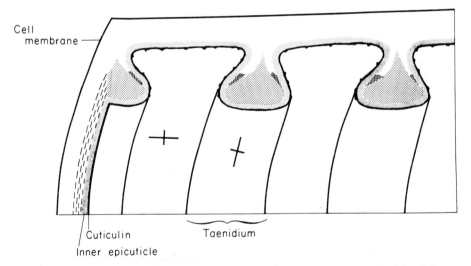

Fig. 28. Diagram of the cuticle structure in a trachea. The epicuticle lining the trachea is folded round the taenidia and contains chitin micelles oriented in the axis of the tube. The taenidia contain chitin micelles oriented in the length of the taenidium. The directions of greatest retardation in birefringence studies are shown by the arms of the crosses. In most tracheae there is no other cuticle between the taenidia and the plasma membrane of the tracheal cell. (After Locke, 1957.)

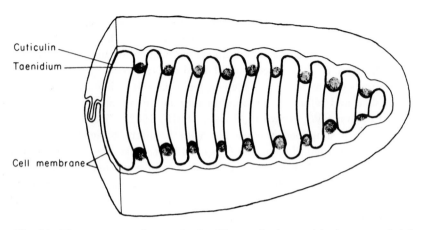

Fig. 29. The structure of a tracheole. The tracheolar cuticle is surrounded by a closely fitting sheath membrane connecting with the plasma membrane around the outside of the tracheolar cell by a "mestracheon" similar to the mesaxon suspending an axon inside a Schwann cell. There are taenidia in the folds of the cuticulin but no other cuticular structures. The frequency of the taenidia is always much greater than in tracheae even when they have similar diameters. (See Figs. 30 and 31.)

cell secretes a ring of adhesive which sticks the old tracheole to the new trachea. The absence of molting in tracheoles is clearly an adaptation which allows the uninterrupted diffusion of gases to the tissues in the immediate premolt period at a time when the new and old tracheal cuticles are widely separated and form a barrier to diffusion.

III. The Formation of the Cuticle

This account is written mainly from the viewpoint of a developmental biologist.

A. THE COORDINATION OF EVENTS

With the exception of the cement, which is secreted by the dermal glands, the remaining layers of the cuticle are secreted by the epithelial cells which undertake a series of syntheses in succession. They lay down the ecdysial membrane first and perhaps some components of the molting fluid, then the epicuticle and within it the endocuticle. The molting fluid becomes active and the products of digestion of the old endocuticle are resorbed and redeposited as the new endocuticle. The outer regions of the endocuticle are stabilized by quinone tanning and perhaps other processes, and finally a layer of wax is added to the outer surface of the epicuticle. The events leading to the formation of new cuticle take place for the most part from the outside inward. The components of the cuticle thus represent spatially the previous synthetic activities of the cells in a sort of fossilized series. Figure 32 is a diagram of cuticle structure showing the order in which the cells undertake the synthetic activities which result in its formation. Figure 33 shows the cycle of syntheses which the epidermal cells undergo in a number of larval molts. There is little overlap in the temporal succession of most of these events at molting, but two, the secretion of endocuticle and the secretion of wax, can occur concurrently and extend throughout the intermolt period. In addition to these gross synthetic events, the epidermal cells also control more subtle changes such as the change in plasticity of the cuticle allowing postecdysial expansion (Cottrell, 1962a,d,e; see Section III,E,F), the change in plasticity of the cuticle after feeding in *Rhodnius* (Bennet-Clark, 1962; see Section IV,B), and the movement of pigment granules (Raabe, 1959; L'Hélias, 1956; Nickerson, 1956; Ellis and Carlisle, 1961).

A fundamental problem concerns the control of the activities of the epidermal cells: are they continuously under the control of molting hormones which cause changes in synthetic activity by difference in structure or concentration, or does a molting hormone merely trigger the start of a series of events coordinated from within a target cell? These two alternatives represent extremes and perhaps some intermediate explanation is more probable. The two possibilities and an intermediate explanation are diagrammed in Fig. 34. In any of these schemes there is the further complication of modulation by the juvenile hormone.

Since many studies with ecdysone have been content to show merely the initiation of molting, a satisfying description of the complete control system must await further study, but recent work (see Section III,E,F,H) does suggest that more than one hormone may be needed for the complete succession of events described in Figs. 33 and 36.

The fact that an active prothoracic gland or ecdysone alone can cause molting favors the second possibility above. Karlson (1961) and Karlson and Sekeris (1962a), for example, favor a fairly direct relation between ecdysone and sclerotization of the cuticle according to the scheme in Fig. 35. H. W. Lewis (1962) has shown that at least four genes regulate phenolase activity in *Drosophila;* one is a structural gene concerned with the qualitative nature of the enzyme, its structure and properties, and three are control genes concerned with quantitative aspects. Since tyrosinase results from protyrosinase and an activating enzyme, many more genes are likely to be involved. Ecdysone could work upon one or more of these control genes.

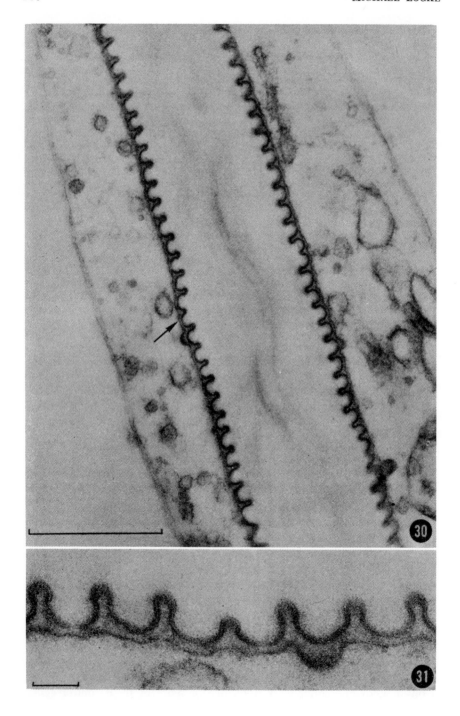

According to Clever (1962a,b, 1963), the puffing-pattern sequence in some loci of salivary gland chromosomes in last instar larvae of *Chironomus tentans* depends upon the dose of ecdysone injected. Each locus has its own threshold for activity and the duration of the puff depends upon the titer above this threshold. Puffs at loci *I-18-C* and *IV-2-B* can be induced experimentally within an hour of injection. During a normal pupal molt, puff *IV-2-B* appears 1 to 2 days later than puff *I-18-C*. This difference from the experimentally induced molting is explicable by the different reactivities of the two loci: during the normal molt, the titer of ecdysone rises gradually and the activity pattern of these genes is adjusted depending upon the altering hormonal titer. An objection to theories connecting ecdysone with a particular event in the molting sequence is that there is no evidence for a sequence of gene activity in an epidermal cell corresponding to the puffing pattern of polytene chromosomes in other tissues. Indeed all the evidence would tell us that the sequence is different, since we should define different types of differentiated cell in terms of the availability of particular loci for activation and synthesis. The puff pattern will certainly be related to syntheses in the salivary glands but this may tell us little about epidermal events. It must therefore remain a moot point whether all or only part of the molting sequence of activities can be accounted for by a varying titer of ecdysone and differential sensitivity of gene loci (see Sections III,E,F, and H).

It has been found that ecdysone is not a single substance but two compounds (Karlson, 1956) or even five compounds (Burdette and Bullock, 1963), so the argument that a single pure compound causes several events has lost some of its force. Nor does it follow that because one hormone, for example ecdysone, can elicit a complete sequence of activities under one set of circumstances, another hormone or agent may not be needed to stimulate one of the activities taken out of the context of its sequence, for example intermolt endocuticle deposition in *Triatoma* and wax synthesis in *Calpodes,* or the control of cuticle plasticity and hardening and darkening in *Calliphora* (Cottrell, 1962a–e; Fraenkel and Hsiao, 1962). Also, the epidermal cells can carry out at least one of the syntheses normally part of the molting sequence in response to wounding [the secretion of endocuticle (Wigglesworth, 1937)]. The most probable explanation is that the sequence is linked at the genetic level in the epidermal cells and that a stimulus like ecdysone

Fig. 30 and 31. Fig. 30, longitudinal section of a tracheole shortly after its formation in a tracheoblast in the testis membrane of *Rhodnius.* The cuticulin lining of the tube is closely invested by a plasma membrane (arrowed). The thickenings in this plasma membrane may be where small vacuoles are discharging their contents. × 37,000. Scale = 1 μm. Fig. 31, enlargement of Fig. 30 to show the cuticulin lining of the plasma membrane which invests it. The cuticulin is two layered at this stage. × 130,000. Scale = 1000 Å.

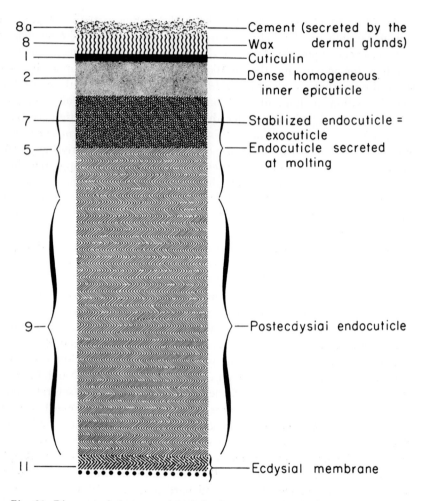

Fig. 32. Diagram of the structure of the integument showing the order in which the components are laid down. The cement is secreted by the dermal glands; all other layers are secreted by the epidermis. The ecdysial membrane has a double structure: it is partly a separate secretion and partly the product of an interaction between this secretion and the last-formed lamellae of the endocuticle. The layers present a sort of fossilized series of most of the previous synthetic events in an epidermal cell. Compare with Fig. 33.

acts initially in the sequence. Other stimuli—hormonal, nutritional, wounding—act at peripheral levels, and quantitatively control the expression of particular phases of the sequence. This system is potentially of great interest and value in development biology. The diagram given in Fig. 36 is an outline hypothesis to show what might be happening to control events in the epidermal cell. At the very least, it serves to point out our ignorance and suggest lines of approach.

In its simplest form the problem is one of intrinsic versus extrinsic control (Fig. 34). When we consider the probable origin of the control systems a rapprochement between the extreme views becomes possible.

The derivatives of the epidermis and the integument are shown diagrammatically in Fig. 37. Some of these are clearly related to the structural role of the cuticle, for example the normal integument, tracheae, tracheoles, and sense organs. The remainder of the differentiated structures fall into two groups: those synthesizing products in bulk, e.g., silk glands, salivary glands and the foregut, and those synthesizing molecules with a specific activity upon other cells, e.g., the corpus allatum and prothoracic glands, with the oenocytes perhaps falling somewhere between the two categories. The prime movers of cellular events in insects are environmental, acting through the nervous system (de Wilde, 1961). In origin, therefore, it is most probable that epidermal cells were influenced in their activities by nerves or neurohumors, a similar sequence of events being elicited in all cells. If we suppose that one or more of these events were hyperelevated in some epidermal cells which then supplied the products of their synthesis to other epidermal cells in which those events had virtually ceased, we should have the present situation, viz., the neurosecretory cells of the brain supply the thoracotropic hormone. This stimulates the prothoracic gland to produce ecdysone which influences other epidermal cells, even though such other epidermal cells have not lost the power of initiating these syntheses by agents (wounding, neurosecretion) other than ecdysone. In this way an intrinsic controlling factor may have become an extrinsic controlling agent working basally in the sequence. In this light, the intervention of neurosecretions and other agents in some of the events diagrammed in Fig. 36 seems very probable, and is supported by the observation of Ichikawa and Nishiitsutsuji-Uwo (1960) that isolated *Cynthia* abdomens in the presence of implanted brain will molt to the adult. Williams (1947), however, did not obtain this result in *Cecropia*. It is also supported by the findings of Kobayashi and Burdette (1961) who found that a brain extract potentiated the effect of ecdysone. They extracted brains and pupae of *Bombyx mori* and obtained active brain hormone and ecdysone. When various concentrations of these two hormones were introduced alone and in combination into isolated abdomens of *Calliphora erythrocephala* larvae, pupation occurred when brain hormone was added to concentrations of ecdysone which would not induce pupation alone. Brain hormone may therefore have a direct action on the tissues when acting synergistically with ecdysone in addition to its tropic action on the prothoracic gland. There is now also an accumulation of evidence that neuro-

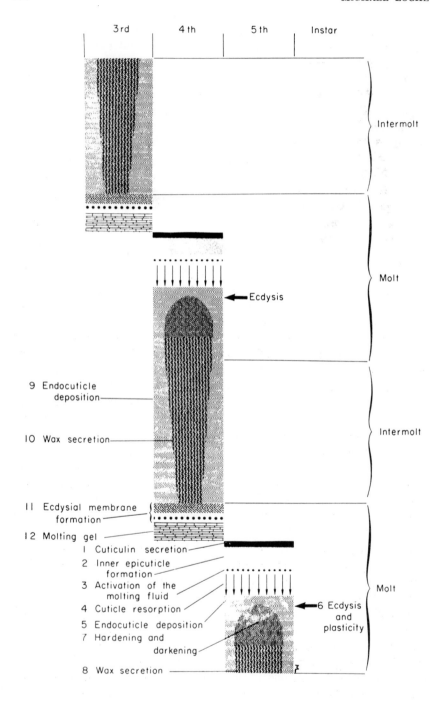

3rd   4th   5th   Instar

Intermolt

Molt

← Ecdysis

9 Endocuticle
   deposition—

Intermolt

10 Wax secretion—

11 Ecdysial membrane
   formation—
12 Molting gel—

1 Cuticulin secretion—
2 Inner epicuticle
   formation—
3 Activation of the
   molting fluid—
4 Cuticle resorption—   ← 6 Ecdysis
5 Endocuticle deposition—   and
7 Hardening and   plasticity
   darkening—

Molt

8 Wax secretion —

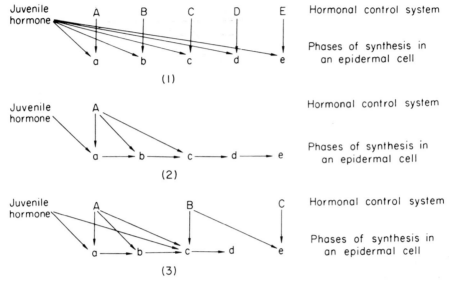

Fig. 34. The problem of the control of epidermal syntheses. A,B,C,D,E are possible controlling factors; abcde are phases of synthesis. In its simplest form the problem is one of extrinsic (1) versus intrinsic (2) control; how many events are triggered by one hormone or controlled by the titer of one hormone? The most probable type of answer is to be found in (3) with an interplay of several factors. Neglecting the juvenile hormone, a is caused by one hormone, b is the result of one hormone and an intrinsic factor, c results from the action of two hormones and an intrinsic factor, d from an intrinsic factor alone, and e from two hormones.

secretory cells are concerned with protein metabolism in general (Thomsen and Möller, 1961; Highnam, 1962; Clarke and Langley, 1962).

B. CHANGES IN THE EPITHELIUM

In the development of a cell, growth involves ribonucleic acid (RNA) and specific protein synthesis controlled by the nucleus. In insect epidermal cells these events have been followed histologically (Krishnakumaran, 1961a; Wigglesworth, 1963b), but there has not yet been any complete ul-

---

Fig. 33. The phases of activity in an epidermal cell. The diagram shows the main events in several stadia in a typical insect. Two of these events, endocuticle deposition (5 and 9) and wax secretion (8 and 10), occur in both the molt and intermolt periods; two more, plasticity (6) and hardening (7), may do so under certain conditions; 6, 7, 9, and 10 are the events for which there is most evidence for a control mechanism independent of or in addition to ecdysone. Compare with Figs. 32 and 36.

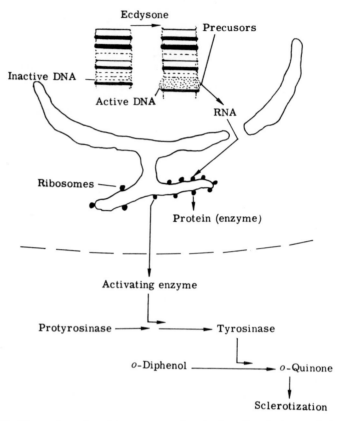

Fig. 35. The action of ecdysone upon sclerotization (hardening und darkening) according to Karlson. This model is of the form shown in Fig. 34(2). There is presumed to be a fairly direct relation between ecdysone and sclerotization. For more recent work which contradicts this hypothesis, see Section III,F. (After Karlson, 1961.)

trastructural study. Krishnakumaran also followed biochemically changes in total-body RNA in developing *Gryllus*. He found it to be highest in the midintermolt period, falling before both the first and second molt. Wigglesworth studied the effects of nutrition, distension, and molting hormone on *Rhodnius* by parabiosis experiments. He found that feeding and molting hormone resulted in activation of the nucleolus, the appearance of RNA in the cytoplasm, and a great increase in mitochondria. It would seem from these results that RNA and protein synthesis follow the activation of the system inducing molting. However, the activation of the epidermal cells by the molting hormone studied by Wigglesworth is only a part of the sequence shown in Fig. 33. If we think of molting as a superimposed event and a

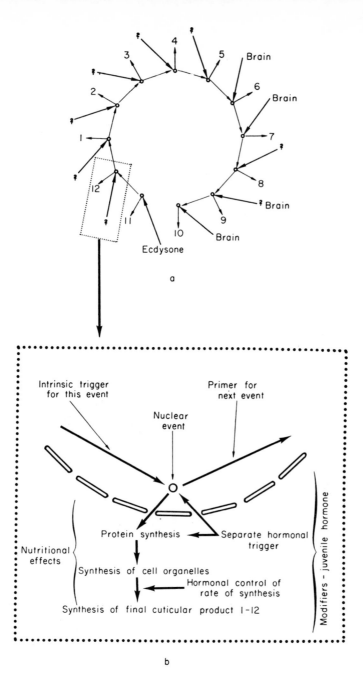

Fig. 36. The control of epidermal syntheses. (a) The numbers 1–12 represent the results of the syntheses shown in Figs. 32 and 33; each event (○) probably has some intrinsic factor relating it to the previous event and may be modulated or controlled by an external factor (arrows); (b) some of the factors which must participate in and control each event.

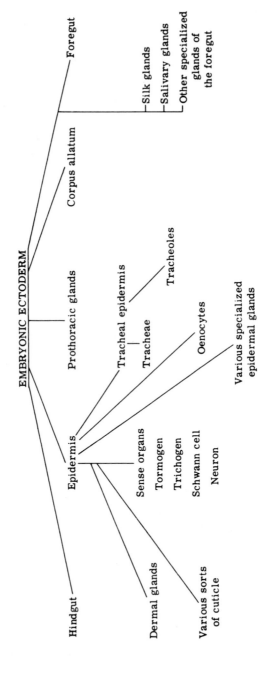

Fig. 37. The derivatives of the ectoderm. The derivatives of the ectoderm fall into three groups. There are those concerned in secreting cuticle in some form (the general epidermis, tracheal cells, sense organs), there are tissues or cells concerned in manufacturing molecules in bulk (silk glands, salivary glands, dermal glands, oenocytes), and there are tissues which synthesize molecules in smaller numbers but which influence the metabolism of other cells (prothoracic glands, corpus allatum).

cycle of alternating molt and intermolt periods (Fig. 33), then it is improbable that the preparation of the cell for molting by molting hormones is necessarily the preparation for intermolt events. We might, therefore, expect two phases of cellular activity, one preparing for the molt and one, postecdysial, preparing for the intermolt. Evidence for this hypothesis has come from autoradiographic studies upon fifth-instar *Calpodes* larvae. There are two peaks of RNA synthesis, one reaching a peak about 24 hours after ecdysis, and another with a slower buildup about 5 days later, three-fourths of the way through the stadium. The latter peak probably corresponds to activation by the molting hormone (Locke and Hurshman, 1963). These results are preliminary, but they favor the division of epithelial events into the two categories of molt and intermolt stressed in Fig. 33.

C. The New Epicuticle, Surface Patterns, and the Determination of Size

Immediately after the secretion of the ecdysial membrane and the molting gel, the cuticulin membrane of the epicuticle appears. Although it is of crucial interest, its formation has only been studied in trachea during the earlier days of electron microscopy (Locke, 1958a). In trachea it is at first smooth and not much thicker than a plasma membrane in thickness; later it thickens by gathering the dense homogeneous layer upon the face next to the cells and increases in diameter, the increase in the axis being taken up by the characteristic helical or annular shape of the tracheal cuticle. This increase in surface area of the newly formed cuticulin also takes place in the surface cuticle (Wigglesworth, 1933, 1954b; Locke, 1959a). In *Rhodnius* larval cuticle the increase results in a pattern of stellate pleats and in the adults in a pattern of transverse ripples. The increase in surface area is presumably responsible for the complex surface patterns of many insects.

The formation of the cuticulin is thus of interest for several reasons. First, its increase in area is coincident with the formation of the surface pattern and an understanding of the mechanism should therefore tell us how such patterns are formed. Second, the extent of the increase in area determines the limits of growth in the subsequent instar. The increase in area is, in turn, a direct measurement of growth which varies with feeding (Locke, 1958b) and an understanding of the links in the chain between nutrition and the area of cuticulin would provide an explanation for some of the problems outlined in Fig. 36b.

The only evidence on the nature of the increase in surface area and its relation to the subsequent pattern of the cuticulin comes from a study of tracheal formation (Locke, 1958a) in which it is supposed that the pattern is the result of simple physical forces generated during the expansion and subsequent buckling of the cuticulin. The newly formed cuticular tubes of the trachea can be treated as thin-walled, elastic cylinders with uniform properties varying only in radius. When they increase in surface area against the restraint of the sheathing epidermal cells, they can be treated as tubes increasing in diameter but buckling as if under the action of axial compression. Symmetrical buckling would be expected in biological material with a low Young's modulus. In symmetrical buckling, the buckling frequency n is related to the initial radius r, the wall thickness t, and Poisson's ratio (v) for the material, in the expression

$$\frac{1}{n} = \frac{\pi(rt)^{1/2}}{[12(1-v^2)]^{1/4}} \tag{1}$$

Poisson's ratio (v) must lie between $+\frac{1}{2}$ and -1, for steel $v = 0.3$, and for rubber $v = 0.46$–0.49 (Kaye and Laby, 1948); therefore, for biological materials, v should be close to 0.5.
For $v = 0.3$:

$$\frac{\pi}{[12(1-v^2)]^{1/4}} = 1.72$$

and $v = 0.5$:

$$\frac{\pi}{[12(1-v^2)]^{1/4}} = 1.81$$

Thus, there should be very little error in simplifying Eq. (1) above to

$$1/n = 1.8(rt)^{1/2} \tag{2}$$

Now if the formation of taenidial folds is comparable to buckling, it should be possible to describe it by this formula. The number n of half-waves of buckling per unit length is known and, by preparing whole mounts shortly before molting, the new and old trachea can be observed together and the initial radius r may also be measured. If log r is plotted against log n for tracheae of different size, the result should be a straight line of slope $-1/2$, the position of the line being fixed by the value of t, the initial thickness. Figure 38a shows such a graph for *Rhodnius* fifth-instar tracheae taken just before emergence from the fourth-instar exuvium. The slope is a fair approximation to $-1/2$. Data taken from newly buckled tracheae show an even closer approximation to the theoretical slope of $-1/2$ (Fig. 38b). The initial thickness of the cuticle calculated from Eq. (2) is 196 Å (S.D. 27). Measurements of cuticulin thickness with the electron microscope at the time of buckling lie in the range 160–200 Å, which is very close to the predicted value.

The hypothesis of expansion and buckling of the first-formed cuticulin

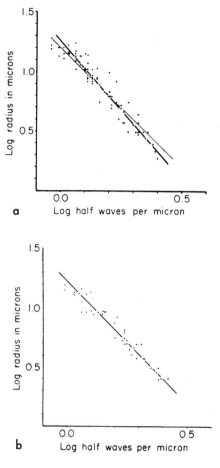

Fig. 38. The relation between taenidial frequency and the initial radius of tracheae before they increase in diameter. (a) Fourth-instar *Rhodnius* larvae just before molting to the fifth instar; (b) fourth-instar *Rhodnius* larvae, 10 days after feeding, before expansion of the new fifth-instar trachea is complete. In (a) a slope of —½ is shown by a thin line. In (b) the best-fitting straight line coincides with the line of slope —½, as predicted by the theory of expansion and buckling. The thickness of the cuticulin calculated from the data in (a) is 197 Å, and from the data in (b) it is 196 Å. See explanation in the text. Ordinate, log initial radius of trachea; abscissa in (a) log number of taenidia in 2 μm, in (b) log half waves of buckling in 1 μm.

cylinder thus accounts for the formation of annular taenidia. It can also account for taenidia of helical form. The previous argument has been simplified by treating the low-pitch helices which occur in large tracheae as if they were annuli. It seemed likely that a tangential shearing force added to the forces responsible for annular taenidial formation might cause a helix to form. The critical normal stress A_{cr} necessary to produce annular buckling from axial compression is given by

$$A_{cr} = \frac{Et}{r[3(1 - v^2)]^{1/2}} \tag{3}$$

where E = Young's modulus and v = Poisson's ratio for the material of the tube-wall of thickness t and radius r (Timoshenko, 1936, p. 441). Putting in reasonable values for v (= 0.48), t (= 0.02 μm), and r (= 20 μm), this becomes

$$A_{cr} = \frac{0.02E}{20[3(1 - 0.48^2)]^{1/2}} = (7 \times 10^{-4})E$$

The critical shear necessary to produce helical buckling is given by (Timoshenko, 1936, p. 486)

$$T_{cr} = \frac{E}{3\sqrt{2}(1 - v^2)^{3/4}} \left(\frac{t}{r}\right)^{3/2} \tag{4}$$

Putting in the same values as before, this becomes

$$T_{cr} = \frac{E}{3\sqrt{2}(1 - 0.48^2)^{3/4}} \left(\frac{0.02}{20}\right)^{3/2} = (9 \times 10^{-6})E$$

Thus, the critical shearing stress necessary to produce helical buckling is only a very small fraction of the axial stress necessary to produce annular buckling. Helical folds in tracheae are not comparable to those produced by torsion alone, for this induces helices with a high pitch with lobes in the circumference rather than the axis. The low-pitched spirals of tracheae probably result from an axial stress with only a slight shear component. This may be illustrated by a model. A thin rubber sleeve can be prepared by allowing rubber latex to dry over a glass rod. If this is lubricated with water, it will slide and fold freely on the rod. Axial compression on the rubber induces the formation of annuli, while twisting produces helices. Compression with a very slight twist produces helices with about the same pitch as are found in tracheae.

Thus the torsion necessary to change annuli into helices is very small. It seems probable that randomly occurring torsional stresses in the tissues could be responsible. The distribution of helices in tubes of different diameter supports this. Whereas the critical axial stress is proportional to the ratio of wall thickness to radius, the critical shear stress is proportional to this (ratio)$^{3/2}$. Thus, while the critical shear may be only just over 1% of the critical axial stress in tubes of radius 20 μm, it is over 8% in tubes of radius 1 μm. Small tubes are almost always annular and large tubes almost always helical.

The cuticular pattern in tracheoles can also be accounted for by the expansion and buckling hypothesis. The theory predicts a thickness for the tracheole cuticle of about 80–90 Å, which is similar to that observed with the electron microscope.

Simple models of expanding rubber latex have also been used to simulate the formation of patterns in the surface cuticle of *Rhodnius*. Expansion with uniform stress in the plane of the surface results in stellate patterns which can be converted to the side-to-side ripples of the adult by axial compression (Locke, 1959a).

We may conclude that many, if not all, surface patterns may be the result of the expansion of the newly formed cuticulin resulting in buckling of various sorts depending upon other superimposed stresses.

D. THE DEPOSITION AND RESORPTION OF THE ENDOCUTICLE

The endocuticle is a mixture of chitin with protein tightly bound to it, and other more labile proteins. Two outstanding problems of structure concern the mechanism by which the microfibers are laid down and oriented in lamellae, and whether the chitin or the protein are the prime movers in determining the fiber pattern.

The lamellae themselves may be the result of cycles of deposition on the part of the epidermal cells. In this connection, the recent work of Neville (1963a,b,c) is most relevant, although the lamellae he describes seem to be formed from several lamellae as defined in Section II,C, known from all endocuticles. He found evidence for daily growth layers in both rubber-like and solid cuticles in locusts and other insects. In lamellar rubberlike cuticle, that is, in chitinous lamellae glued together by thick layers of resilin, the daily growth layers occur in pairs and are due to zonal differences in the resilin, since they also occur in the pure resilin pad found in the wing hinge ligament. They are most conveniently seen in sections observed in UV light. Each pair represents 24 hours of growth. During night conditions, a wide, brightly fluorescent zone is laid down; during day conditions, a thin faintly fluorescent zone is deposited. Thus, most growth normally occurs during the night. The pattern of fluorescent growth zones in rubberlike cuticle could be modified by varying external conditions (temperature and light). Experimentally prolonged days and nights produced correspondingly wide, faintly or brightly fluorescent zones, in which the daily zonation pattern characteristic of control conditions had been suppressed. Neville thinks that daily growth layers are a general characteristic of the exopterygote skeleton. He found them in the solid cuticle of all of the exopterygote orders which he examined, and their occurrence was widespread throughout the skeleton. The alternating light and dark fluorescent bands are presumably due to variations in the synthesis of the two previously unknown amino acids reported by Andersen (1963) which link resilin chains together.

Although the lamellae may be the result of cyclical activity, the orientation of the component microfibers still presents a problem. In plants, the orientation of cellular fibers has been linked to the orientation of secretory components at the plasma membrane (Green, 1962; Preston, 1962), but endocuticle is secreted from a surface covered with microvilli which could scarcely transmit information upon direction of deposition (Fig. 2).

The adsorption of amino acids by pure chitin (Hackman, 1955) suggests that the chitin may be the molecule which is first ordered into microfibers. Chitin is also suggested as the prime mover by autoradiographic studies using amino acids as protein precursors and glucose as a chitin precursor. (Condoulis and Locke, 1966). The pattern of incorporation of amino acids is totally different from that of glucose. In *Calpodes* larvae after injections of tyrosine, incorporation is diffuse throughout the thickness of the endocuticle with some concentration next to the epidermal cells. This suggests either that there is continuous deposition of protein throughout the endocuticle or that some protein components are labile and in an equilibrium maintained with a high turnover, or both. Glucose, on the other hand, is incorporated in a sharp band corresponding to the number of lamellae laid down during the interval of incorporation. This would be expected if chitin were indeed the first endocuticular component deposited and Schmidt's subcuticular layer were to be concerned with its polymerization.

Resorption of the endocuticle in many insects can occur on two different occasions, during starvation and at molting, both consistent with its function as a food reserve.

In *Rhodnius* and other bugs the endocuticle is thick and crisp immediately after molting but becomes soft and paper-thin after some months of starvation. The function of the endocuticle as a food store during starvation merits more attention than it has received.

The endocuticle is also, by definition, the labile part of the cuticle which is resorbed at molting. There are several problems concerning its resorption which remain unsolved. At an early stage in molting the space between the old endocuticle and the new epicuticle (cuticulin and homogeneous layer) is occupied by the ecdysial membrane and the molting gel (Passoneau and Williams, 1953). Shortly afterwards the gel liquifies and becomes the molting fluid, and solution of the old cuticle begins. The new epicuticle is protected from solution but somehow the cells resorb the products of digestion through it while depositing the new cuticle. There are some reports that the molting fluid is resorbed through the mouth by swallowing. These are based upon experiments in which ligature of the head caused an accumulation of molting fluid between the old and new cuticles in the abdomen. They perhaps indicate that resorption is under hormonal control but not that absorption is through the mouth, for Jeuniaux (1958) has shown that dyes injected into the molting fluid appear rapidly in the blood and other internal organs. No mechanism has been proposed for the activation of the gel or the resorption of the products of digestion. It could be that the wax canals (the spaces in the epicuticle occupied by the lipid-water liquid crystals during wax secretion) make the epicuticle a membrane of very select permeability, allowing small activating molecules out and depolymerized chitin and protein back in. In this connection it is of interest that Jeuniaux (1957) records that the chitinase of the molting fluid is not resorbed but remains attached to undissolved endocuticle in the exuvium; presumably, it is too large a molecule. In *Schistocerca*, Phillips (1961) found the intima of the rectum to function in this way in exchange with the gut. The membrane acted as a molecular sieve, allowing the rapid exchange of water and salts and to a lesser extent amino acids and monosaccharides between the lumen and the epithelium. The hypothesis deserves further study.

Endocuticle resorption after solution by the molting fluid and the endocuticle deposition which immediately follows are molting events; but the great bulk of endocuticle deposition is an intermolt activity. In *Triatoma* (Hemiptera, Reduviidae) endocuticle is deposited after each blood meal and several meals may be needed before molting is induced, and in *Calpodes* endocuticle deposition is continuous while the animal feeds. The problem of control is an interesting one since endocuticle deposition follows both the peaks of RNA synthesis described by Locke and Hurshman (1963). Unless the prothoracic gland goes through two peaks of activity, intermolt endocuticle deposition must be the result of stimulation by something other than ecdysone. Nothing is known of the control of cuticle resorption during starvation, which is also an intermolt event.

The lability of the endocuticle, i.e., the possibility that some of it can be depolymerized and resorbed or redeposited, leads to an hypothesis to account for one of the oldest problems in cuticle biology, the active uptake of water vapor against an apparent concentration gradient. The paradox arises because the blood in *Tenebrio* larvae, for example, is isotonic with 2.12% NaCl equivalent to about 99% RH (Patton and Craig, 1939), and yet these larvae can take up water vapor from humidities as low as 90%. This phenomenon is widespread in insects (see Beament, 1961). It occurs in *Tenebrio* larvae above 90% RH (Buxton, 1930; Mellanby, 1932); in *Chortophaga* above 82% RH (Ludwig, 1937); in ticks above 92% RH (Lees, 1948); in *Xenopsylla* above 45–50% RH (Edney, 1947); and in *Thermobia* above 50% RH (A. J. Watson, personal communication). It is readily demonstrated in mealworms. Figure 39 shows the weight gained by

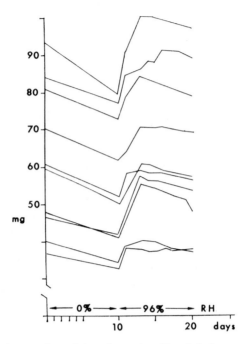

Fig. 39. The changes in weight of starving *Tenebrio* larvae suspended for 10 days of 0% RH and then exposed to 96% RH at 30°C. Ordinate, weight in milligrams; abscissa, time in days.

starving mealworms when placed in 96% RH after desiccation. The larvae were suspended so that they had no contact with any surface in their environment, and most of the weight gain must be due to water vapor uptake.

The paradox can be resolved if we accept that the fluid within the cuticle may not be in osmotic equilibrium with the blood. The firm links of the septate desmosomes between adjacent epidermal cells would suggest this possibility. Quintuple-layered junctions similar to the terminal bars between mouse capillary endothelial cells described by Muir and Peters (192) and Overton (1962) also occur in the epidermis of *Rhodnius*. If the hard cuticle of *Tenebrio* larvae contained 20% free water and only one-twentieth of the thickness of the cuticle were to be depolymerized, this would still be enough to give a solution saturated with respect to most of the amino acids and stronger than a molar solution of glucose. In this way the water in the cuticle might very well absorb water vapor in the environmental humidities quoted. The epidermal cells would then only be faced with the same problems of active transport in water vapor uptake as in cuticle resorption It is worth noting that water vapor uptake has always been observed on starving animals. This may account for the large difference in rate between water entering a living *Tenebrio* larva in a high humidity and leaving in a low humidity (Fig. 25). The rate of gain at high humidities is much greater than the rate of loss at low humidities. This would be expected if there were a constant pull from within (but see also Section II,F,4 on a hypothesis to explain this asymmetrical property of cuticle).

E. The Postecdysial Expansion of the New Cuticle

Immediately after release from the exuvium an insect expands by taking in water or air to increase in size prior to the hardening of the new cuticle. In cyclorraphous Diptera the splitting and shedding of the old cuticle has become independent of air swallowing, and expansion of the new cuticle can be delayed for considerable periods after emergence. This enabled Cottrell (1962a,d,e) to study the mechanism by which expansion is brought about, and the properties of the cuticle during expansion in the blow fly, *Calliphora erythrocephala*.

Newly emerged flies perform digging movements. Both the digging movements and the expansion process of newly emerged flies involve the production of positive internal pressures. Digging produces a characteristic cycle of pressure changes reaching a peak of 6 to 12 cm Hg. During expansion, two different pressure phenomena are detectable. First, there is a gradual rise and fall in the basic hemolymph pressure which reaches a maximum of 6 cm Hg in *Calliphora* and 9.5 cm in *Sarcophaga,* a few minutes after full wing extension, and then falls to atmospheric pressure in the next 20 minutes. Secondly, superimposed on the basic rise, there is a series of brief rhythmic pressure pulses which gradually decline and then cease about the time of full wing extension. Evidence obtained by blocking the proboscis or denervating the abdominal muscles of newly emerged flies indicates that the gradual rise in hemolymph pressure is attributable to air-swallowing, and the pressure pulses to the performance of "muscular efforts," i.e., simultaneous contractions of both the ptilinal and abdominal muscles. Experiments in which blow flies with proboscis and anus blocked at emergence were subjected to artificial pressures of 7 cm Hg show that expansion is possible only during the period when air-swallowing movements occur. At this time there is a change in the mechanical properties of the cuticle which is confined to the presumptive sclerites and enables them to expand. Before and after air-swallowing, application of identical artificial pressures merely results in the temporary distension of membranous areas and in the unfolding of the wing membrane.

These experiments show that the mechanical properties of the cuticle may be reversibly altered very rapidly and directly through the epidermis. The major part of the cuticle is not only within the body metabolic pool (Section I,C) but is also, at least at this time, subject to a more subtle control. There may be a parallel between the cuticle becoming plastic in this immediate postecdysial stage and in the plasticity induced in the cuticle of fed *Rhodnius;* see Section IV,C (Bennet-Clark, 1962).

F. THE HARDENING OF THE CUTICLE AND QUINONE TANNING

The chemistry of the hardening process has been reviewed by Dennell (1958), Hackman (1959), Karlson and Sekeris (1962b), Malek (1960, 1961), and Mason (1955), and will not be discussed here. The general hypothesis is that dihydric phenols diffuse outward through the cuticle to the epicuticle where they are oxidized to quinones which diffuse back, tanning and hardening the cuticle as they do so. Darkening is not necessarily coincident with hardening since albino insects are known which have equally hard cuticles. Using ^{14}C markers and autoradiographs of whole mounts, Fuzeau-Braesch (1959a,b) has also shown that hardening is distinct from the deposition of both yellow and black pigments in *Gryllus*. Melanization and sclerotization (= hardening) are two separate processes each with its own copper-protein phenolase (Malek, 1957; Jones and Sinclair, 1958). The route to the surface for the polyphenols is undoubtedly by way of the pore canals but the only route across the epicuticle is by way of the wax canals or perhaps within the lipid-water liquid crystals themselves. If the integrity of the wax canals or their contents is essential for the movement of polyphenols it would explain Kawase's (1961) observation that brief treatment of a silkworm pupa with ether prevents darkening. In this connection it is of interest that a number of epidermal defense glands show structures similar to the epicuticular lipids which could be concerned in the secretion of the quinone contents of the glands (T. Eisner and M. M. Salpeter, personal communication). Secretion of a polyphenol must present a problem of protection to any cell.

The control of hardening and darkening in the emerging blow fly *Calliphora erythrocephala* has recently been studied by Cottrell (1962a,b,c) and by Fraenkel and Hsiao (1962). The following is a brief account of their work.

Cottrell found that apart from certain specialized areas which are hardened and darkened before emergence, the cuticle of the newly emerged blow fly is soft and colorless. After emergence the body is inflated and expansion is completed in about 30 minutes at 22°C. Air pumping ceases some 10 minutes later. The first signs of darkening appear 20 minutes after air-pumping movements have stopped. Normal hardening and darkening is rapid so that within 40 minutes of its first appearance the cuticle of the intact insect appears quite black, while within 2 hours the elasticity of the pterothorax is established and the fly is capable of buzzing.

Blood transfusion experiments show that normal hardening and darkening is brought about by the release into the blood of an active factor.

The darkening factor can be detected in fly blood by injecting it into flies decapitated at the moment of emergence. Blood taken from flies at the moment of emergence or 24 hours after expansion produces little or no reaction, although blood taken between 3 minutes and 10 hours after emergence shows darkening activity. However, extracts of other tissues and many chemicals will also induce darkening because of nonspecific damaging effects. The assay can therefore only be used with certainty for detecting activity in fly blood.

Introduction of this factor into a newly emerged fly some 35 minutes prior to the time at which it would normally be released is sufficient to prevent expansion. It is normally released some 45 minutes before the appearance of the first signs of darkening and between 3 and 15 minutes after the fly has reached conditions suitable for expansion, that is, at about the time of initiation of air pumping. Decapitation at emergence will prevent the initiation of normal hardening and darkening but not of secondary darkening.

The critical period for the prevention of normal hardening and darkening by decapitation lies between 3 and 15 minutes after the fly has reached conditions suitable for expansion. Isolated abdomens behave in a manner similar to decapitated flies but their reactions are complicated by secondary darkening with damage. Allowing for the effects of secondary darkening it is possible to demonstrate the occurrence of the bloodborne darkening factor by means of ligatures placed at emergence between the thorax and the abdomen. Under these conditions only the head and thorax exhibit normal darkening. Evidently the head is concerned in the release or the control of the release of the bloodborne darkening factor.

The darkening factor will withstand boiling for 10 minutes and drying at 120°C for 20 minutes, but it does not retain its activity when kept in solution at room temperature for more than 24 hours. It is nondialyzable, relatively insoluble in organic solvents, and is inactivated by ethyl alcohol and the bacterial protease, subtilisin. It is probably proteinaceous and is certainly not tyrosine or any of the phenolic compounds at present thought to act as the precursors of the tanning agent.

Cottrell concludes that the general hardening and darkening of the newly emerged blow fly is controlled independently of the other processes involved in molting and is not one of the final events brought about by the increasing titer of ecdysone.

Fraenkel and Hsiao (1962) obtained similar results. They found that hardening is initiated by a signal form neurosecretory cells in the brain, transmitted to the thorax via both the central and stomatogastic nervous system. Blood from a 15-minute-old fly induces tanning when injected into the body of a head-ligated fly. The active factor in blood is probably neither a substrate nor an enzyme necessary for tanning. It could be a new hormone, because it is neither ecdysone, nor the juvenile hormone, nor a brain hormone. This factor is unspecific, being present in newly molted cockroaches (*Periplaneta*) and newly emerged adults of the beetle *Tenebrio*. They also found that a mechanism that inhibits tanning plays a role in the normal control of hardening and darkening.

There is little doubt from these two studies that, in some insects at least, ecdysone is not a trigger for all events in the molting sequence (see also Section III,H).

G. The Secretion of Wax

The secretion of wax, whether hydrocarbons or esters, presents a problem in transport. The epidermis, where at least the precursors are synthesized, is separated by a wide hydrophilic endocuticle from the surface of the epicuticle where the wax is deposited. Several hypotheses have been put forward to account for the movement of this inert material through the aqueous phase of the endocuticle.

1. The wax may be conveyed to the surface in a solvent which later evaporates, for in cockroaches there is a grease composed of a hard wax dissolved in long-chain alcohols and hydrocarbons (Beament, 1955; but see Gilby, 1962). In *Aphis fabae* the wax from the cornicles hardens immediately on a water surface, but remains liquid and translucent if kept below. Presumably it contains a volatile solvent (Edwards, personal communication). Although solvents may be present, their evaporation in quantity is not necessary for wax to appear at the surface, for in *Calpodes* wax continues to be secreted below glass or araldite plates (Locke, 1959b).

2. The wax may be secreted in a water-soluble form in combination with a protein as it is in the eggs of ticks (Lees and Beament, 1948). Although this is possible, such large molecules are not likely to be much easier to transport within the cuticle than the lipids themselves. There is as yet no evidence for such molecules in cuticle but they can exist in a liquid-crystalline form which would give an explanation similar to (5) below (see Cook and Martin, 1962a,b).

3. The wax may pass through the pore canals. The presence of lipid-water liquid crystals in all the pore canals examined makes this the most probable route, but it has been rejected as a general explanation since many insects secreting wax, as in *Calpodes,* lack pore canals. However, see (5) below.

4. The final stage of wax synthesis may take place close to the surface. In a number of insects an esterase has been detected in the pore canals and in the epicuticle where it could be concerned in wax synthesis (Locke, 1959b, 1961).

5. The wax may be transported in lipid-water liquid crystals continuous with the surface wax. These could be pulled out through the epicuticle by a slight evaporation of water or other small molecules in the micelle and/or by the surface tension and the spreading of the surface wax. It has recently been found that although there are no pore canals in the endocuticle through which wax is secreted in *Calpodes*, the texture is sufficiently coarse to permit the penetration of lipids which can be detected close to the plasma membrane of the epidermal cells (Fig. 3). This finding makes it less likely that the epicuticular esterase is concerned directly in wax synthesis. It could perhaps be concerned in maintaining the conditions for stability of the lipid-water system, in a wound response, or in some other epicuticular process.

H. The Control of Wax Secretion

Wax secretion is one of the last events at molting and in many insects carries on throughout the intermolt period (Fig. 33). The control of its synthesis is therefore of special interest (Locke, 1965). In most insects it is present in only trace amounts and is therefore difficult to study. However, it is readily observable in insects with wax blooms and particularly in the larva of *Calpodes ethlius* Stoll in which a small part of the integument is specialized for the secretion of wax in an amount and form which is readily measured (Figs. 16, 17, 18; Locke, 1959b, 1960a). Long straight filaments emerge at rates up to 5 μm per hour. Figure 40a shows how the rate of this specialized wax secretion varies during the last larval stadium. The effect of ligatures, excision, and implantation operations has been observed upon this wax secretion. Typical results are shown in Figs. 40b–e. Once secretion has begun, it is not greatly influenced by feeding (Fig. 40b). Starved larvae with the nerve cord cut continue to secrete. However, removal of the head stops secretion although pupation proceeds normally (Fig. 40c). Thus some factor in the head, with or without the prothoracic gland hormone, controls the rate of wax synthesis. After wax secretion has begun, removal of the brain has little effect (Fig. 40d) but removal of the corpus allatum–corpus cardiacum complex causes a sudden fall in rate of secretion which later increases before a delayed pupation (Fig. 40e). This experiment suggest that wax secretion is controlled by a neurosecretion stored in the corpus allatum–corpus cardiacum complex.

This is yet another example of an intermolt epidermal event which is controlled by the brain. It remains to be seen whether this specialized wax

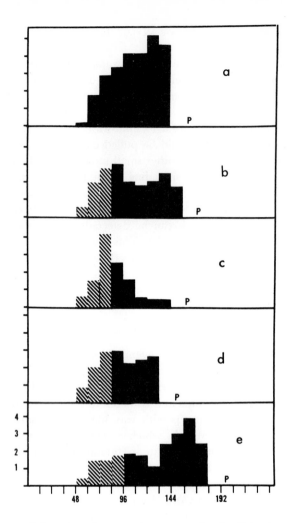

Fig. 40. Factors influencing the rate of wax secretion in *Calpodes ethlius* larvae. The wax takes the form of cylinders (see Figs. 16, 17, and 18) whose length is measured at 12 hourly intervals. P, time of pupation; ordinate, rate of wax secretion measured in microns of wax cylinder secreted per hour; abscissa, time in hours from the fourth-fifth molt. (a) The rate of wax secretion in a normal unoperated larva; (b) larva with the nerve cord cut between the thorax and abdomen at the time shown: secretion continues at about the same rate; (c) a larva with the head ligated; there is an immediate fall in the rate of wax secretion but the time of pupation is unaffected since the prothoracic gland has already been activated; (d) a larva with the brain removed: wax secretion continues; the brain is no longer necessary for the continuation of wax secretion at this stage; (e) a larva with the corpus allatum–corpus cardiacum removed. There is an immediate fall in the rate of wax secretion followed by a delayed rise to a peak and delayed pupation. Presumably, one or both of these organs stores and releases a neurosecretion which finds another outlet after their extirpation. (Selected data from Locke, 1965.)

secretion can be considered part of the sequence outlined in Fig. 33, or whether it is a special case, a secondary interpolation of a synthetic event no longer bearing any relation to the molting sequence in its control. Wax blooms also commonly occur in adults where they would be considered as intermolt events as, for example, the blooms on many Odonata, some cicadas, and many bugs.

IV. The Properties of the Epidermis

Many aspects of differentiation and the control of growth and form are dealt with in Wigglesworth (1959b), "The Control of Growth and Form," which is subtitled "A Study of the Epidermal Cell in an Insect." Some other aspects of insect epidermal morphogenesis are dealt with in Waddington (1962). The following account attempts to avoid overlapping with these discussions.

A. GROWTH AND THE CELLS

At each molt the integument may change in two ways. It may vary qualitatively (in form) and it may vary quantitatively in the area secreted (in size). The change of size and form from one stadium to the next and particularly the transformation to pupal and imaginal structure involves three sorts of change in the epidermal cells. In most holometabolic insects new structures may be derived at least in part from pockets of cells set aside in the late embryo. These changes in structure result from the recruitment of reserve cells. On the other hand, in hemimetabolous insects and some holometabola most of the cuticle of a later instar may be derived from cells which seem to function in earlier larval life. A changed pattern has resulted from the changed synthetic capabilities of the same cells as, for example, in *Galleria* (Marcus, 1962). There is, in addition, in all insects a third important change in most tissues, the loss of a large number of cells. Changed patterns may result from the selective survival of the original cell population. The form of the cuticle will depend upon the type of syntheses initiated, which will be the result of the interaction between the hormonal and cellular milieu and the potentiality of the cell. The size of the new integument, on the other hand, will probably bear some relation to the number of cells, the number formed and the number surviving. In spite of this, cell death at molting in insects has been surprisingly little studied.

In the tracheal epithelium of *Rhodnius* the cell area is constant in the intermolt period of different larval stadia. There is a burst of mitotic activity 4 days after the initiation of molting in a fourth larval instar (Fig. 41).

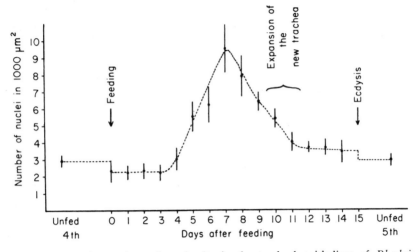

Fig. 41. The change in nuclear density in the tracheal epithelium of *Rhodnius* after the initiation of molting by feeding. The destruction of nuclei exceeds their production by mitoses on about the seventh day. The size of the new trachea is determined on about the tenth to eleventh days.

By the eighth day mitoses are almost absent. Cell destruction begins at about the same time as mitoses begin but continues until the ninth day or longer. The peak of destruction is at 7 to 8 days and is complete by the time the new trachea expands. Thus, in this example it is cell destruction which governs the number of cells and hence the area of new integument laid down. In searching for factors which control growth, therefore, we should do well to examine the causes of cell breakdown. It is indeed a remarkable process which has to create many times more units than it requires, and then must needs destroy the excess.

In the tergal epidermis of *Rhodnius* it has been shown that, whereas mitosis is dependent upon the separation of cells, cell destruction at metamorphosis is dependent upon a low level of juvenile hormone (Wigglesworth, 1963b). Juvenile hormone thus acts not only to direct the type of synthesis performed by the cell but also says whether it shall exist at all. Conversely, in the epidermis of the wing pads, a low level of juvenile hormone results in more cells surviving. However, the number of cells alone does not control the area of integument to be secreted, for in areas with a deficient tracheal supply many times the normal number of cells may survive without having any effect upon the size or form of the cuticle (M. Locke, unpublished observations, 1963). The control of size must therefore operate above the level of the cell although it is normally coupled with the processes governing cell number. This would also be expected from a consideration of the effects of the juvenile hormone. If growth were directly and constantly linked to cell numbers it would be difficult to visualize how the juvenile hormone could control both growth and form. On the other hand short and long wings in *Drosophila* brought about by selection are due to changes in cell number (Robertson, 1962). The sequence of events in the determination of size in an integumentary structure and the possible controlling factors are diagrammed in Fig. 42.

B. Nutrition and Cell Growth

Although the area of the new integument laid down and, hence, the size of an insect are related to feeding, there have been few quantitative investigations of the relation between food and cell growth. The problem is complicated because the new size can be a result of two factors, a direct nutritional one and an indirect one owing to the increased area of integument caused by the intake of food during a stadium. These two factors have been disentangled through some observations on tracheal growth in *Rhodnius* (Locke, 1958b). Stretching the trachea results in longer tracheae but the increase in diameter of the stretched tracheae is similar to unstretched ones. Thus the amount of growth manifested by the increase in diameter is not related to the initial cell density. Whatever their initial density, the cells increase in number and secrete a new cuticle with a diameter appropriate to the peripheral growth of tracheae. Increase in diameter varies with the formation of new tracheoles which is proportional to nutritional state by way of tissue growth. Some experiments also suggest that the nodes controlling the growth of the lateral tracheae also respond to a blood factor, perhaps nutritional. Lateral tracheae with nodes need no other connections in order to vary in growth and can only obtain the information upon how much to grow from the blood.

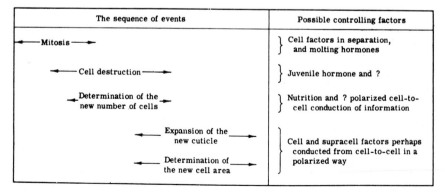

Fig. 42. The sequence of events in the determination of the size of the integument.

C. Intermolt Growth and the Restraint
of the Integument

In hard-bodied insects, intermolt growth is restricted to unfolding at the intersegmental membranes, but in soft-bodied insects the cuticle itself stretches. The form of the insect at the end of the instar is thus determined by the force available for stretching, the mechanical properties of the integument, and epidermal and muscular restraints.

In normal feeding, a *Rhodnius* larva imbibes blood until all the pleats are stretched out and the epicuticle is shiny. It becomes so turgid that it bounces when dropped. Bennet-Clark (1962) found that the endocuticle becomes less rigid during the early part of a meal and stretches more readily. Artificial inflation of bugs took many times longer if Ringer solution were used instead of blood. The loss of rigidity is under physiological control and depends upon some factor in the blood meal which diffuses into the hemolymph, perhaps then operating through the epidermis.

In caterpillars the problems are somewhat similar; they also grow within a stadium by the inflation of a plastic skeleton with food (M. Locke and R. Greenberg, unpublished observations, 1963). A caterpillar is to a first approximation a long tubular organism under uniform internal hydrostatic pressure from the contents of the gut and the blood. Now, in a cylinder under uniform internal hydrostatic pressure the hoop stress is twice the axial stress. This is readily verifiable in a caterpillar by blowing it up with Ringer solution through a hypodermic needle inserted in the head capsule. Above a critical pressure, the caterpillar explodes by tearing along its length. In spite of this, caterpillars when left to expand by themselves by feeding have approximately the same proportions at the end as at the beginning of a stadium. We are forced to conclude that the integument is mechanically anisotropic with a greater rigidity in the circumference than in the axis. The role of such anisotropic cuticles in the determination of form and their structural basis have yet to be explored.

The experiments show that the form of an insect at the end of a stadium may be markedly affected by the intrinsic properties of the integument, and some of these properties may be influenced by external factors working through the cells.

D. Epidermal Cells and Tracheoles

Like the tracheae, the concertina structure of the tracheoles is well adapted for allowing changes in length while resisting lateral deformation. Judging from the change in frequency of the taenidia they may increase in length by two to three times after their formation (Locke, 1958a). They may be likened to a very weak spring restrained by the cell membrane. In the integument, these tracheoles lie in an overdispersed fashion so that all cells are more or less equidistant from an oxygen source.

Wigglesworth (1959a, 1961b, 1962) has described how the epidermal cells, and presumably other tissues, send out processes which attach to any tracheole they come across, the even distribution of tracheoles being the result of the competing demands of all the epidermal cells. Under condition of lack of oxygen, more and more of these processes are put out and tracheoles may be fetched from distances as far as 100 μm. It is as if the epidermal cells had two states comparable to the alternate states of adhesive and pseudopodal versus nonadhesive and nonpseudopodal described for amebae (Bell, 1963), low oxygen tension and some of the stimuli present in wounds stimulating the formation of pseudopodia. Nothing is known of the ultrastructure of these tracheole-fetching processes or their contractile mechanism. Small patches of desmosomes are commonly found very close to the cuticle of the tracheole, which may perhaps be the termination of the epidermal processes.

E. The Coordination of Growth in Tracheae

Growth in a tracheal system is treelike, with new tracheoles and small tracheae being added for the most part terminally at each molt, while the main trunk increases in diameter (Wigglesworth, 1954a). In a number of insects in which the tracheae are parallel-sided tubes [although not in all, and not in large insects, for measurements on tracheae of noncircular cross section are of uncertain value: see the discussion in Buck (1962)] the branching pattern is such that the cross-sectional area of the diffusion path is constant before and after branching (Locke, 1958b). This implies that the increase in diameter at molting is related in main and peripheral tracheae. Even in tracheae in which this relation is not precise, the variabilty in terminal tracheation makes it probable that growth of large and small tracheae must be coordinated rather than independent. This poses the problem of the mechanism by which coordination of growth is achieved. The problem is not unlike the control of growth in hydroids (Berrill, 1961).

The simplest hypothesis to account for the coordination of tracheal growth, such that the increase in cross section of the main trachea is equal to the increase in cross section of the terminal tracheae, is that tracheal growth is determined by the tissues.

This has been confirmed by observing the growth of tracheae after they have been cut in various ways (Locke, 1958b). The tergal tracheae were cut in a segment upon one side of the abdomen in newly fed fourth-instar *Rhodnius* larvae. The two ends developed differently. At molting the new tracheae surrounding the end with tissue connections increased normally or supernormally in diameter with a rounded end (Fig. 43a). The end connected to the spiracle showed the smallest increase in diameter compatible with being formed round and old trachea, with a truncated end. The differences in growth cannot be attributed to wounding since they have had identical treatments. The two ends may have differed in the oxygen tension available to the tracheal cells and the tissues which they supplied, but when a branch of the isolated tergal trachea was also cut, its stump also failed to increase in diameter (Fig. 43c). The tergal trachea was also cut between the lateral trachea and the spiracle (Fig. 43b). Here the spiracle end tended to fail, while the length attached to the lateral trachea showed an increase in diameter, although both should have had an adequate oxygen supply. These results suggest that the diameter of a tergal trachea is determined from the tissue endings.

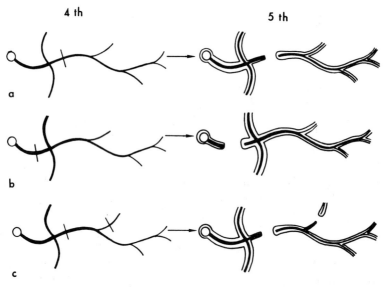

Fig. 43. The effect upon growth of cutting a trachea in *Rhodnius* larvae. Diagram of the spiracle and tergal trachea to one segment. The cut was made as shown on the fourth-instar larva, and the result was observed just before molting to the fifth instar. The new fifth-instar trachea is shown as an enveloping line around the fourth-instar trachea. (a) In a trachea cut between the lateral trachea and the branches to the tissues only the end with tissue connections grows normally; (b) in a trachea cut between the spiracle and the lateral trachea, only the trachea on the tissue side of the cut grows normally; (c) a cut tributary trachea fails to grow normally even when it is part of an isolated trunk with a low oxygen tension. (After Locke, 1958b.)

The lateral tracheae are exceptions to the normal branching rule in that they link up different spiracles instead of taking part in the segmental diffusion path from spiracle to tissue. It is therefore of particular interest to study the factors controlling their growth.

At each molt the remains of the old tracheae are drawn through the spiracles and shed with the exuvium. To allow this, the lateral tracheae break at the nodes, which are predetermined weak points having a characteristic structure. The effects of cutting the lateral tracheae varied with the position of the cuts relative to the node (Fig. 44). The new trachea on the cut end containing the node increased in diameter normally or almost normally in both directions from the node. The end without the node failed to increase in diameter by the amount expected from measurements upon the opposite side with the node. There can be no possibility of this growth being proportional to oxygen tension since both ends have a spiracular connection. When the cut was made through the node, damaging it, both ends tended to fail.

This interpretation of the part played by the node is also suggested by

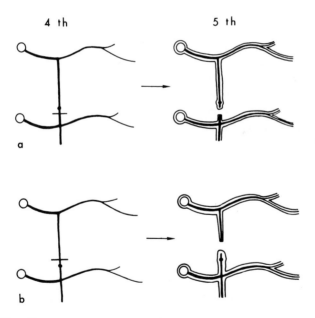

Fig. 44. The effect upon growth of cutting a lateral trachea in *Rhodnius* larvae. Diagram of two spiracles, a lateral trachea and the tergal trachea to two segments. The cut was made on the lateral trachea of the fourth instar as shown, and the result was observed just before molting to the fifth instar. The new fifth-instar trachea is shown as an enveloping line around the fourth-instar trachea. The node is represented by a thickening. (a) and (b) In a cut tergal trachea, only the length with the node increases normally in diameter. (After Locke, 1958b.)

the marked changes in tracheal diameter which may occur on each side
of it. The lateral tracheae pass forward from the abdomen to the thorax.
There is an abrupt change in diameter at a node between abdominal segments 1 and 2, the anterior trachea being excessively slender. This could
result from differential growth controlled through the node.

Large tracheae only increase normally in diameter when they are connected with an organizing center—a node, an epidermal wound, or tracheal
terminations in the tissues. This control must involve transport of the stimulus along the intervening tracheae.

When the lateral trachea between abdominal spiracles 2 and 3 is cut soon
after feeding, the long anterior length lacks a node and fails to increase
normally in diameter. The trachea itself is capable of growing without a
node since it does so when attached to a wound at its tip. It is not obvious
why it should fail, since it is attached at its other end to the tergal trachea
and tissue connections, which grow normally. If the stimulus for growth in
diameter moves from the tracheal endings to the spiracles by any process
similar to diffusion it should be possible to alter its course to pass along
the lateral trachea by severing its connections with the spiracle.

The anterior section of the lateral trachea between abdominal spiracles
2 and 3 was isolated without a node soon after feeding. At the same time
the connection with the spiracle was cut as close to the junction with the
tergal and lateral tracheae as possible. This did not cause complete isolation
from the rest of the tracheal system, for the tergal trachea connects by a
slender branch to a thoracic spiracle. The lateral trachea still failed to increase in diameter (Fig. 45).

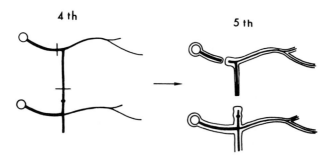

Fig. 45. The effect upon growth of cutting a trachea in *Rhodnius* larvae. Diagram
of two spiracles, a lateral trachea and the tergal tracheae to two segments. The
cuts were made on the fourth instar as shown, and the result was observed just
before molting to the fifth instar. The new fifth-instar trachea is shown as an
enveloping line around the fourth-instar trachea. The node is represented by a
thickening. The lateral trachea is uninfluenced by the stimulus from the tissue
endings that causes the tergal trachea to increase in diameter, even when the
connection to the spiracle is severed. (After Locke, 1958b.)

It might have been expected that the "strength" of the stimulus for increase in diameter from the tergal tracheal endings could induce a change similar to that brought about by a small wound, but there is a difference in the point of application of the stimulus. This leads to a concept of polarity of growth in the tracheal system. The experiment described above suggests that movement of the stimulus causing increase in diameter can only take place in one direction. This is not due to a simple concentration gradient but is an intrinsic property of the trachea.

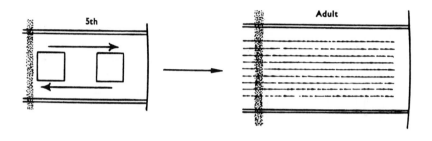

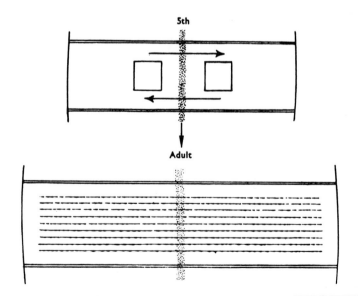

Fig. 46. Skin grafting in *Rhodnius*. Grafts were performed on the fifth instar, and the results were observed in the adult. Grafts can be exchanged from side to side, either mesial-to-lateral or lateral-to-lateral, without affecting the pattern in the adult. (After Locke, 1959a.)

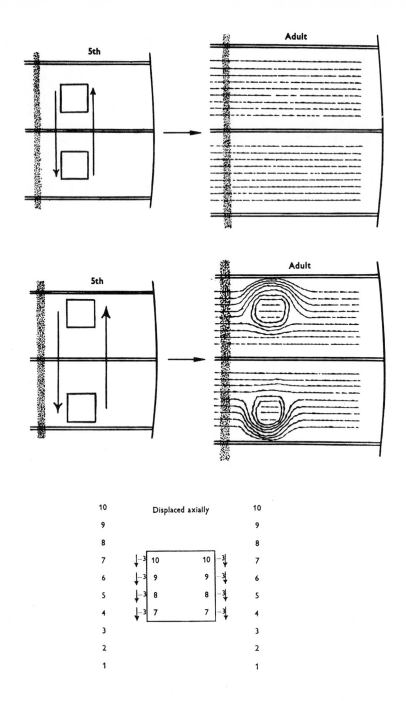

Fig. 47. Skin grafting in *Rhodnius*. Grafts were performed on the fifth instar and the results observed in the adult. Grafts can be exchanged from segment to segment without affecting the pattern in the adult as long as the relative position

The determination of growth and the secretion of the cuticle take place first at the spiracle and then proceed toward the terminal tracheae. It would be expected from the polarized conduction of information about how much to grow that growth should be completed first farthest from the source of control in this way.

If the tracheal cells are representative of other epidermal cells, the polarity of the control of growth, that is, the cell-to-cell transport of information about how much to grow, could be a most important property of the epidermis (see Section IV,F).

F. THE GRADIENT ORGANIZATION OF THE EPIDERMAL CELLS

The polarized behavior of the tracheal epidermis suggested that the body epidermal cells might have similar properties, and that their growth also might be controlled by way of polarized cell-to-cell transport of information upon how much to grow from localized areas. Some support for this hypothesis has come from skin-grating experiments on blood-sucking bugs (Locke, 1959a, 1960b).

As there is no individual or species-specific incompatibility to grafting in *Rhodnius* or *Triatoma*, it is possible to perform integumental grafts, sometimes using different species as markers. There is, however, another sort of incompatibility, a positional one.

The integument in *Rhodnius* and *Triatoma* has lent itself to these studies because of the surface pattern. In the larva, the pattern is one of stellate pleats interspersed between the bristle-bearing plaques. In the adult, the pattern changes to one of transverse ripples (see Section III,C). The nature of the ripple pattern makes it ideal for detecting displacements in the axis. It is relatively easy to interpret the way in which the ripple lines are displaced and link up in response to grafting operations.

Grafts will take readily with a marriage of host and graft patterns when displaced laterally, but any displacement axially within a segment results in a distorted pattern which is not incorporated into that of the host. The behavior of the cells in skin grafts deduced from changes in the cuticle pattern after a molt and an interpretation of the behavior are summarized in Figs. 46–53.

---

within the axis of each segment remains unchanged. When a graft from the front of a segment is transposed to the rear of a segment and vice versa, the pattern of both host and graft is distorted as shown. The levels from the front to the back of a segment have been assigned numerical values in the bottom diagram. The arrows indicate the direction in which the host pattern would be expected to be displaced, and the small figures show the difference in value between host and graft. (After Locke, 1959a.)

From these and other experiments, we can say that the epidermal cells
are arranged in a segmentally repeating gradient of some kind which we
can picture as in Fig. 54; the cells are alike from side to side but differ
axially. The characteristic which varies quantitatively is the recognition of
position in a linear axial order. When displaced from this order, the host
and graft patterns interact in a way predictable from their relative positions.

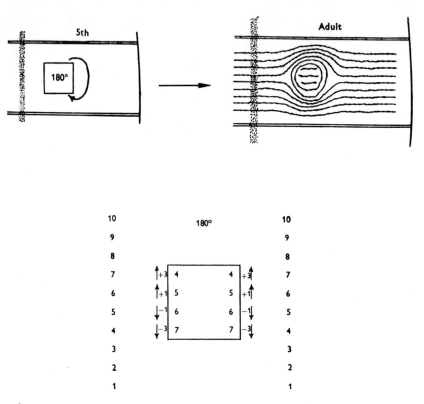

Fig. 48. Skin grating in *Rhodnius.* Grafts were performed on the fifth instar
and the results observed in the adult. When the relations in the axis are destroyed
by rotating a graft through 180°, the pattern in the adult is distorted. In the
lower diagram numerical values have been assigned to the levels within a segment.
The arrows predict the direction in which the host pattern should be displaced,
and the small figures give the difference in value between host and graft. (After
Locke, 1959a.)

G. The Gradient and the Control of Growth

It has been assumed that the ripple pattern of the cuticle is caused by
and reflects some similarly oriented mechanism within the cells of the epi-

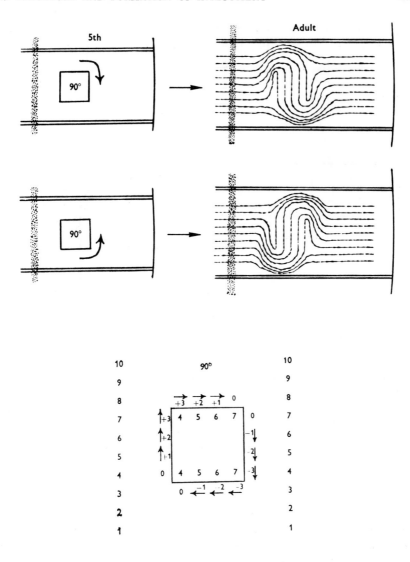

Fig. 49. Skin grafting in *Rhodnius*. Grafts were performed on the fifth instar and the results observed in the adult. When the relations in the axis are only partly destroyed by rotating through 90°, the pattern of distortion in the adult depends upon the direction of rotation. Numerical values have been assigned to the levels within a segment in the bottom diagram. The arrows show the direction in which the pattern should be displaced according to the difference in level between host and graft given by the small figures. (After Locke, 1959a.)

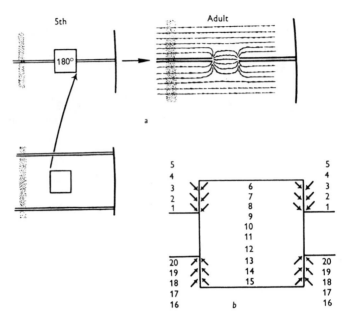

Fig. 50. Skin grafting in *Rhodnius.* Grafts were performed on the fifth instar and the results observed in the adult. When the grafts involve an intersegmental membrane, the distortions in the adult can still be predicted. (a) A graft from the center of a tergite rotated through 180° and implanted in a hole in the intersegmental membrane; (b) the levels within the segment have been assigned numerical values. The arrows in the graft and in the host show the direction predicted for the distortion of the ripple pattern. (After Locke, 1960b.)

thelium. The question arises whether any other property of the cells obeys similar rules. In the tracheal system it has been shown that epithelial continuity is necessary for the quantitative control of growth but not for the normal sequence of events at molting (Locke, 1958b). The tracheal epithelium is polarized with respect to this control: a trachea cut from its tissue connections molts but does not grow normally in diameter. If the gradient in the axis of the tergites is homologous with the polarity of growth control in the tracheal system, the cuticle isolated in a discontinuity pattern might be expected to behave like a trachea isolated from its tissue connections, molting normally but without the usual increase in size.

The effects of isolation in a discontinuity pattern upon growth in area have been followed, making use of the darkly pigmented sternal cuticle as a marker. The sternal cuticle also differs from the tergal in retaining bristles in the adult, but it has similar transverse ripple marks, and grafts to the tergites take satisfactorily. Squares were cut from the center of the sternites in third-instar larvae and implanted in similar positions in the tergites. Some

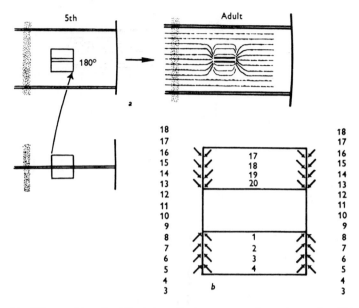

Fig. 51. Skin grafting in *Rhodnius*. Grafts were performed on the fifth instar and the results observed in the adult. (a) The converse of the experiment in Fig. 50: a graft with an intersegmental membrane has been rotated through 180° and implanted in the center of a tergite; (b) numerical values have been assigned to the levels within the segment; the arrows predict the direction of distortion expected in the host and the graft. (After Locke, 1960b.)

were rotated through 180° to induce the discontinuity pattern; others were implanted with normal orientation as controls. The changes in area could be followed by examining the exuviae. The differences between the isolated cuticle and the normally oriented transplant were most marked in the adult. In the rotated squares the transplant was reduced to a few bristles with no pigmented cuticle. In those with normal orientation, the transplant was marked by an oval of pigment with normally spaced and oriented bristles. These results are in agreement with the hypothesis derived from a study of growth in tracheae that epithelial continuity is necessary for the quantitative control of growth. The epithelial continuity must be of a particular kind, being part of an axial gradient in the normal tergite.

Isolated pieces of cuticle may have satisfied their capacity for uniting with cells of the same axial level by forming a concentric pattern, but they fail to grow harmoniously with the rest of the animal. Satisfaction of the capacity for completing the pattern is alone insufficient to induce normal growth; the integument must also be in the gradient of the whole segment. The quantitative control of growth is therefore not coincident with the behavior of the cells described as a gradient. If the amount of growth depended upon

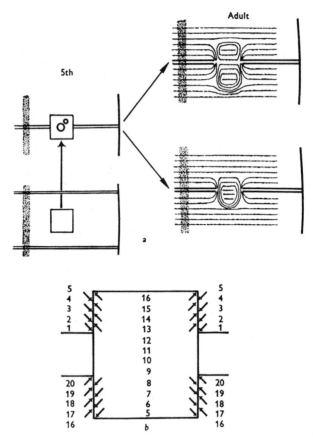

Fig. 52. Skin grafting in *Rhodnius*. Grafts were performed on the fifth instar and the effect observed in the adult. (a) The results of implanting a graft with normal orientation from the center of a tergite into a hole in the intersegmental membrane are variable with respect to the graft; the predicted result from a consideration of the numerical values in (b) is the one in the upper diagram in which the graft forms a double pattern. Frequently, however, only a single pattern forms. The predicted pattern depends upon an interaction between host and graft. If there is an anterior dominance and some grafts are isolated from the host influence, then the single pattern would be expected (see Fig. 53c). (b) Numerical values have been assigned to the axial levels in the host and graft. The arrows indicate the predicted direction of distortion. (After Locke, 1960b.)

a bloodborne factor alone, the converse might have been expected. That this is not so implies that in the normal animal the cells receive information about the growth to be made in some other way—presumably through the epithelium as in tracheae. The cuticle in a discontinuity pattern differs from normal cuticle in that the ridges do not go anywhere: they unite with them-

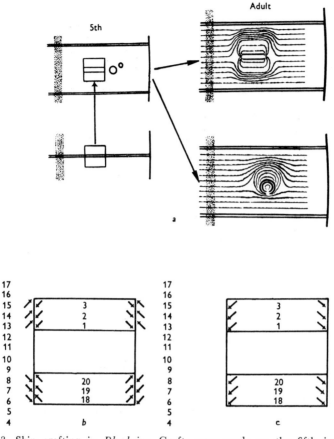

Fig. 53. Skin grafting in *Rhodnius.* Grafts were made on the fifth instar and the effect observed in the adult. (a) Implantations of grafts containing an intersegmental membrane with normal orientation give a variable result with respect to the graft; either a double or single pattern may result. The double pattern is explained as the result of an interaction between host and graft as predicted in (b). The single pattern would result if there is an anterior dominance in an isolated graft as predicted in (c). When there is no interaction with the host, the anterior dominance within the graft asserts itself. (b) Numerical values have been assigned to the levels within the host and graft. The arrows indicate the direction predicted for the distortion. (c) Numerical values have been assigned to the levels within an isolated graft. The arrows indicate the direction predicted for distortion of the ripple pattern if there is an intrinsic anterior dominance. (After Locke, 1960b.)

selves. If this is the cause of the failure of controlled growth it could mean that in the continuity of the normal segment the cells receive information from somewhere in the direction of the gradient. This may be the real significance of the cells maintaining contact with their own level. As revealed

Fig. 54. Skin grafting in *Rhodnius.* The results of the experiments described in Figs. 46–54 show that the epidermis is arranged in a segmentally repeating gradient with respect to an ability to recognize and react to the axial level of neighboring cells. Left to themselves, the anterior cells show a dominance in restoring the pattern. We can picture this behavior as in the diagram. Cells may be moved from side to side and from segment to segment in the same relative position in the axis without reacting. Any change in axial position brings incompatible cells together and creates a distortion of the pattern. (From Locke, 1960b.)

by the abnormal conditions of the experiments, the gradient appears as a curiosity. To the animal it may be a mechanism for maintaining the cells in a preferred order for the transport of growth stimuli.

In tracheae the nodes, which are the homologue of the intersegmental membrane, are the source of control of growth. It is therefore of interest that grafts containing the intersegmental membrane tend to survive whatever their orientation in the animal, unlike similar grafts without the intersegmental membrane. This may mean that the intersegmental membranes, like the nodes, are in some sense organizers for growth (Locke, 1959a, 1960b).

H. The Gradient and Segmental Organization

Whether or not the significance of the gradient behavior is the maintenance of a preferred order of cells for the transport of growth stimuli, it is probable that it has interest for segmental organization in general. The gradient behavior is not confined to areas with the simple ripple pattern.

On the second abdominal tergite, the ripple pattern is replaced in the midline by one of curved ridges. Transplants were made as in Fig. 55. The grafts survived, and the host responded by restoring the continuity of the ripples anteriorly or posteriorly according to the origin of the graft. Thus the gradient in the second abdominal segment resembles that in the fourth and fifth although the cuticular pattern is very different. The gradient, then, is a segmental phenomenon not restricted to the ripple pattern.

Similar results have also been obtained from grafting experiments upon the appendages of *Triatoma.* Integumental grafts moved around the femur take satisfactorily, whereas grafts changed in the axis cause considerable distortion. Grafts can also be exchanged between segments of an appendage (M. Locke, unpublished observations, 1963).

These experiments are a reminder that the cells which secrete the cuticle

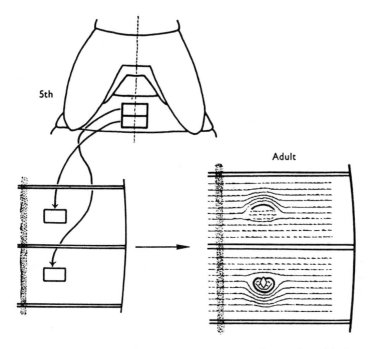

Fig. 55. Skin grafting in *Rhodnius.* Grafts were made on the fifth instar and the effect observed in the adult. The distortions of the host ripple pattern make it possible to recognize the axial level from which a graft has come, even though the graft itself may have a different pattern altogether. This verifies that the repeating gradient is a feature of segmentation and not a peculiarity of the ripple pattern. (After Locke, 1960b.)

have properties and an individuality which one would never suspect from the surface pattern alone.

V. Conclusion

In this chapter, I have tried to outline the general problem of qualitative and quantitative control of the epidermal syntheses which take place in the construction of one sort of integument. The polymorphic nature of the epidermal cell is a superimposed complication. Wigglesworth (1953c, 1959b, 1961a) stresses its polymorphism, either in the terminal differentiation which it may undergo to form structures other than integument *sensu strictu,* or in the sequential polymorphism of the type of integument modulated by the juvenile hormone. The elementary textbook picture of an epidermal cell with two arrows pointing to it, ecdysone and juvenile hormone, is no longer adequate or even correct. There are many factors, both extrinsic and intrinsic, which guide an epidermal cell in its choice, timing, and extent of syntheses.

As a metaphor, we may think of a two- or three-storied building containing dichotomously branched corridors. During embryology and later development, an epidermal cell follows a corridor toward differentiation of a particular sort. The factors here are cellular interactions, although the rate may be determined hormonally. The doors of specific development are progressively unlocked. Then, we have modulation by the juvenile hormone determining on which of the similar but not identical floors the corridor is located. Also, we have the contents of each room where a cell is at any moment, which tell it exactly how much, when, and what to do. We know the plan of the building, a little about the choice of floors and corridors, but nothing of the contents of the rooms.

For the moment, we may do little more than list some of the properties of an epidermal cell and the factors known to influence their behavior and syntheses.

a. Factors influencing the epidermal cells
   1. Intrinsic factors
   2. The internal environment
      Hormones
         Ecdysone
         Juvenile hormone
         Neurosecretions
      Factors dependent on feeding
         General nutrition
         Specific precursors
      Membrane permeability
      Oxygen tension
   3. The cellular environment
      The density of cells
      Cell to cell interactions
   4. The external environment
      Environmental features acting through the cuticle
      Wounding
b. The properties of the cells
      The regulation of numbers—mitoses and cell destruction
      The recognition of orientation
      The recognition of position in the segmental gradient
      The polarized conduction of information
      Tracheole-fetching behavior in response to lowered oxygen tension

Movement in response to wounding

The uptake of specific precursors

The uptake of water

Alterations in the form and number of cell organelles; e.g., the plasma membrane, which may be smooth, lobed, or microvillate

The synthesis of specific cuticular components

It is inevitable that a number of important points should be left out in a brief review. Fortunately, many of them have been well covered in recent literature. The general outline of cuticle structure is well known from the work of Richards (1951, 1953, 1958b) and Wigglesworth (1953a, 1957). The structure of chitin-protein complexes is excellently reviewed in Rudall (1963), and an account of the mechanical properties is given in Jensen and Weis-Fogh (1962). For information on the chemistry and permeability of cuticle see Chapters 3 and 4 by Hackman and Ebeling, respectively, in this volume. Comparisons between the structure of the cuticle in different groups of Arthropoda are given by Krishnakumaran (1961b, 1962). Lipids and the water relations of cuticle have been reviewed by Beament (1954, 1961). The diversity of chemical compounds synthesized by specialized glands is well brought out in Roth and Eisner (1962). The evolution of segmented structures is covered by Snodgrass (1958), and the ultrastructure of sense organs has been described by Slifer (1961).

Acknowledgments

I am grateful for the support of a grant from the United States Public Health Service. I am most grateful to Dr. John Edwards and Dr. J. A. L. Watson for helpful discussions on the manuscript.

References

Albrecht, F. O. (1953). "The Anatomy of the Migratory Locust." Oxford Univ. Press (Athlone), London and New York.

Andersen, S. O. (1963). *Biochim. Biophys. Acta* **69**, 249.

Ashurst, D. E. (1959). *Quart. J. Microsc. Sci.* **100**, 401.

Ashurst, D. E., and Chapman, J. A. (1961). *Quart. J. Microsc. Sci.* **102**, 463.

Baccetti, B. (1955). *"Redia"* **40**, 197.

Baccetti, B. (1956a). *"Redia"* **51**, 75.

Baccetti, B. (1956b). *"Redia"* **51**, 259.

Bailey, S. W. (1954). *Nature (London)* **173**, 503.

Baker, G. L., Pepper, J. H., Johnson, L. H., and Hastings, E. (1960). *J. Insect Physiol.* **5**, 47.

Beament, J. W. L. (1945). *J. Exp. Biol.* **21**, 115.

Beament, J. W. L. (1954). *Symp. Soc. Exp. Biol.* **8**, 94.

Beament, J. W. L. (1955). *J. Exp. Biol.* **32**, 514.

Beament, J. W. L. (1959). *J. Exp. Biol.* **36**, 391.

Beament, J. W. L. (1961). *Biol. Rev. Cambridge Phil. Soc.* **36**, 281.

Bell, L. G. E. (1963). *New Sci.* **18**, 103.

Bennet-Clark, H. C. (1962). *J. Insect Physiol.* **8**, 627.

Berrill, N. J. (1961). "Growth, Development, and Pattern." Freeman, San Francisco, California.

Bouligand, Y. (1965). *C. R. Acad. Sci.* **261**, 3665–3668.

Buck, J. (1962). *Annu. Rev. Entomol.* **7**, 27.

Burdette, W. J., and Bullock, M. W. (1963). *Science* **140**, 1311.

Buxton, P. A. (1930). *Proc. Roy. Soc., Ser. B* **106**, 560.

Chetverikov, S. S. (1915). *Bull. Soc. Entomol. Moscou* **1**, 14; translanted in *Smithson. Inst., Annu. Rep.* p. 441 (1918).

Clarke, K. U., and Langley, P. (1962). *Nature (London)* **194**, 160.

Clever, U. (1962a). *Chromosoma* **13**, 385.

Clever, U. (1962b). *J. Insect Physiol.* **8**, 357.

Clever, U. (1963). *Develop. Biol.* **6**, 73.

Condoulis, W. V., and Locke, M. 1966. *J. Insect Physiol.* **12**, 311.

Cook, W. H., and Martin, W. G. (1962a). *Can. J. Biochem. Physiol.* **40**, 1273.

Cook, W. H., and Martin, W. G. (1962b). *Biochim. Biophys. Acta* 56, 362.

Cottrell, C. B. (1962a). *Trans. Roy. Entomol. Soc. London* **114**, 317.

Cottrell, C. B. (1962b). *J. Exp. Biol.* **39**, 395.

Cottrell, C. B. (1962c). *J. Exp. Biol.* **39**, 413.

Cottrell C. B. (1962d). *J. Exp. Biol.* **39**, 431.

Cottrell, C. B. (1962e). *J. Exp. Biol.* **39**, 449.

Currey, J. D. (1962). *Nature (London)* **195**, 513.

Day, M. F. (1949). *Aust. J. Sci. Res. Ser. B* **2**, 421.

DeHaas, B. W., Johnson, L. H., Pepper, J. H., Hastings, E., and Baker, G. L. (1957). *Physiol. Zool.* **30**, 121.

de Mets, R. (1962). *Nature (London)* **196**, 77.

de Mets, R., and Jeuniaux, C. (1962). *Arch. Int. Physiol. Biochim.* **70**, 93.

Dennell, R. (1958). *Biol. Rev. Cambridge Phil. Soc.* **33**, 178.

Dennell, R., and Malek, S. R. A. (1955a). *Proc. Roy. Soc. Ser. B* **143**, 414.

Dennell, R., and Malek, S. R. A. (1955b). *Proc. Roy. Soc., Ser. B* **143**, 427.

de Wilde, J. (1961). *Bull. Res. Counc. Isr.* **10**, 36.

Edney, E. B. (1947). *Bull. Entomol. Res.* **38**, 263.

Edwards, G. A., Ruska, H., and de Harven, E. (1958). *Arch. Biol.* **69**, 353.

Eglinton, E. G., Hamilton, R. J., Raphael, R. A., and Gonzalez, A. G. (1962a). *Nature (London)* **193**, 739.

Eglinton, E. G., Gonzales, A. G., Hamilton, R. J., and Raphael, R. A. (1962b). *Phytochemistry* **1**, 89.

Ellis, P. E., and Carlisle, D. B. (1961). *Nature (London)* **190**, 368.

Fraenkel, G., and Hsiao, C. (1962). *Science* **138**, 27.

Fuzeau-Braesch, S. (1957). *C. R. Acad. Sci.* **244**, 1274.

Fuzeau-Braesch, S. (1959a). *C. R. Soc. Biol.* **153**, 57.

Fuzeau-Braesch, S. (1959b). *C. R. Acad. Sci.* **248**, 856.

Gilby, A. R. (1962). *Nature (London)* **195**, 729.

Gilmour, D. (1960). "Biochemistry of Insects." Academic Press, New York.

Green, P. B. (1962). *Science* **138**, 1404.

Hackman, R. H. (1955). *Aust. J. Biol. Sci.* **8**, 83.

Hackman, R. H. (1959). *In* "The Biochemistry of Insects" (L. Levenbook, ed.), pp. 48–57. Pergamon, Oxford.

Herzog, R. O. (1926). *Z. Angew. Chem.* **39**, 297.

Hess, A. (1958). *J. Biophys. Biochem. Cytol.* **4**, 731.

Highnam, K. C. (1962). *Quart. J. Microsc. Sci.* **103**, 57.

Hinton, H. E. (1961). *Sci. Progr. (London)* **49**, 329.

Holdgate, M. W., and Seal, M. (1956). *J. Exp. Biol.* **33**, 82.

Ichikawa, M., and Nishiitsutsuji-Uwo, J. (1960). *Mem. Coll. Sci., Univ. Kyoto, Ser. B* **27**, 9.

Jensen, M., and Weis-Fogh, T. (1962). *Phil. Trans. Roy. Soc. London, Ser. B* **245**, 137.

Jeuniaux, C. (1957). *Actes Soc. Linn. Bordeaux* **97**, 1.

Jeuniaux, C. (1958). *Arch. Int. Physiol. Biochim.* **66**, 121.

Jones, B. M. (1958). *Proc. Roy. Soc., Ser. B* **148**, 263.

Jones, B. M. (1959). *Proc. Roy. Phys. Soc. Edinburgh* **27**, 35.

Jones, B. M., and Sinclair, W. (1958). *Nature (London)* **181**, 927.

Juniper, B. E. (1960). *Proc. Int. Conf. Electron Microsc. 4th, 1958* Vol. 2, p. 489.

Karlson, P. (1956). *Ann. Sci. Natur., Zool. Biol. Anim.* [11] **18**, 125.

Karlson, P. (1961). *Symp. Deut. Ges. Endokrinol.* **8**, 90.

Karlson, P., and Sekeris, C. E. (1962a). *Biochim. Biophys. Acta* **63**, 489.

Karlson, P., and Sekeris, C. E. (1962b). *Nature (London)* **195**, 4837.

Kawase, S. (1961). *Nature (London)* **191**, 279.

Kaye, G. W. C., and Laby, T. H. (1948). "Physical and Chemical Constants." Longmans, Green, New York.

Keister, M. L. (1948). *J. Morphol.* **83**, 373.

Kobayashi, M., and Burdette, W. J. (1961). *Proc. Soc. Exp. Biol. Med.* **107**, 240.

Koch, J. C. (1917). *Amer. J. Anat.* **21**, 177.

Koidsumi, K. (1957). *J. Insect Physiol.* **1**, 40.

Kramer, S., and Wigglesworth, V. B. (1950). *Quart. J. Microsc. Sci.* **91**, 63.

Krishnakumaran, A. (1961a). *Nature (London)* **189**, 243.

Krishnakumaran, A. (1961b). *Presidency College Zool. Mag.* **8**, 23.

Krishnakumaran, A. (1962). *Zool. Jahrb., Abt. Anat. Ontog. Tiere* **80**, 49.

Lees, A. D. (1948). *Discuss. Faraday Soc.* **3**, 187.

Lees, A. D., and Beament, J. W. L. (1948). *Quart. J. Microsc. Sci.* **89**, 291.

Lewis, C. T. (1962). *Nature (London)* **193**, 904.

Lewis, H. W. (1962). *Biol. Bull.* **123**, 464.

L'Helias, C. (1956). *Année Biol.* **32**, 203.

Locke, M. (1957). *Quart. J. Microsc. Sci.* **98**, 487.

Locke, M. (1958a). *Quart. J. Microsc. Sci.* **99**, 29.

Locke, M. (1958b). *Quart. J. Microsc. Sci.* **99**, 373.

Locke, M. (1959a). *J. Exp. Biol.* **36**, 459.

Locke, M. (1959b). *Nature (London)* **184**, 1967.

Locke, M. (1960a). *Quart. J. Microsc. Sci.* **101**, 333.

Locke, M. (1960b). *J. Exp. Biol.* **37**, 398.

Locke, M. (1961). *J. Biophys. Biochem. Cytol.* **10**, 589.

Locke, M. (1965). *J. Insect Physiol.* **11**, 641.

Locke, M. (1967). *Advan. Morphog.* (M. Abercrombie, ed.), Vol. 6, 33.

Locke, M., and Hurshman, L. (1963). *Amer. Zool.* **3**, 250.

Lower, H. F. (1957a). *Zool. Jahrb., Abt. Anat. Ontog. Tiere* **76**, 165.

Lower, H. F. (1957b). *Biol. Bull.* **113**, 141.

Ludwig, D. (1937). *Physiol. Zool.* **10**, 342.

Luzzati, V., and Husson, F. (1962). *J. Cell Biol.* **12**, 207.

McBain, J. W. (1950). "Colloid Science." Heath, Boston, Massachusetts.

Malek, S. R. A. (1956). *Nature (London)* **178**, 1185.

Malek, S. R. A. (1957). *Nature (London)* **180**, 237.

Malek, S. R. A. (1958). *Proc. Roy. Soc., Ser. B* **149**, 557.

Malek, S. R. A. (1959). *J. Insect Physiol.* **2**, 298.

Malek, S. R. A. (1960). *Nature (London)* **185**, 56.

Malek, S. R. A. (1961). *Comp. Biochem. Physiol.* **2**, 35.

Marcus, W. (1962). *Arch. Entwicklungsmech. Organismen* **154**, 56.

Mason, H. S. (1955). *Advan. Enzymol.* **16**, 105.

Mellanby, K. (1932). *Proc. Roy. Soc., Ser. B* **111**, 376.

Mercer, E. H., and Brunet, D. P. (1959). *J. Biophys. Biochem. Cytol.* **5**, 257.

Millard, A., and Rudall, K. M. (1960). *J. Roy. Microsc. Soc.* **79**, 227.

Mitchison, J. M., and Swann, M. M. (1954). *J. Exp. Biol.* **31**, 461.

Monné, L. (1960). *Ark. Zool.* **2**, 287.

Muir, A. R., and Peters, A. (1962). *J. Cell Biol.* **12**, 443.

Neville, A. C. (1963a). *J. Insect Physiol.* **9**, 117.

Neville, A. C. (1963b). *Oikos* **14**, 1.

Neville, A. C. (1963c). *J. Insect Physiol.* **9**, 265.

Neville, A. C., and Luke, B. M., (1969). *Tissue & Cell* **1**, 689–707.

Nickerson, W. J. (1956). *Anti-Locust Bull.* p. 24.

Nilsson, O. (1957). *J. Ultrastruct. Res.* **1**, 170.

Overton, J. (1962). *Develop. Biol.* **4**, 532.

Passoneau, J. V., and Williams, C. M. (1953). *J. Exp. Biol.* **30**, 545.

Patton, R. L., and Craig, R. (1939). *J. Exp. Zool.* **81**, 437.

Phillips, J. E. (1961). Ph.D. Thesis, Cambridge University.

Pipa, R. L., and Cook, E. F. (1958). *J. Morphol.* **103**, 353.

Power, M. E. (1952). *J. Morphol.* **91**, 389.

Preston, R. D. (1962). *Electron Microsc. Proc. Int. Congr., 5th,* 1962 P2.

Raabe, M. (1959). *Bull. Soc. Zool. Fr.* **84**, 272.

Richards, A. G. (1951). "The Integument of Arthropods." North Central Publ.
Co., St. Paul, Minnesota.

Richards, A. G. (1953). *In* "Insect Physiology" (K. D. Roeder, ed.), pp. 1–54.
Wiley, New York.

Richards, A. G. (1955). *J. Morphol.* **96**, 537.

Richards, A. G. (1958a). *Z. Naturforsch.* b **13**, 812.

Richards, A. G. (1958b). *Ergeb. Biol.* **20**, 1.

Richards, A. G., and Anderson, T. F. (1942). *J. Morphol.* **71**, 135.

Richards, A. G., and Schneider, D. (1958). *Z. Naturforsch.* b **13**, 680.

Robertson, R. W. (1962). *Genet. Res.* **3**, 169.

Roff, W. J. (1956). "Fibres, Plastics and Rubbers." Butterworth, London.

Roth, L. M., and Eisner, T. (1962). *Annu. Rev. Entomol.* **7**, 107.

Rudall, K. M. (1963). *Advan. Insect Physiol.* **1**, 257–311.

Schmidt, E. L. (1956). *J. Morphol.* **99**, 211.

Selman, B. J. (1960). *J. Insect Physiol.* **4**, 235.

Slifer, E. H. (1961). *Int. Rev. Cytol.* **11**, 125.

Smith, D. S. (1960). *J. Biophys. Biochem. Cytol.* **10**, 123.

Smith, D. S. (1963). *J. Cell. Biol.* **16**, 323.

Smith, D. S., and Wigglesworth, V. B. (1959). *Nature (London)* **183**, 127.

Snodgrass, R. E. (1958). *Smithson. Misc. Collect 138.*

Stoeckenius, W. (1962). *J. Cell Biol.* **12**, 221.

Strich-Halbwachs, M. C. (1959). Thesis, L'Université de Strasbourg. Masson, Paris.

Thompson, T. (1964). *In* "Membranes in Development" (M. Locke, ed.), pp. 38–94. Academic Press, New York.

Thomsen, E., and Möller, L. (1961). *In* "The Ontogeny of Insects" (I. Hrdý, ed.) pp. 121–126. Academic Press, New York.

Thor, C. J. B., and Henderson, W. F. (1940). *Amer. Dyest. Rep.* **29**, 461.

Thorpe, W. H., and Crisp, D. J. (1947). *J. Exp. Biol.* **24**, 227.

Timoshenko, S. (1936). "Theory of Elastic Stability." McGraw-Hill, New York.

Waddington, C. H. (1962). "New Patterns in Genetics and Development." Columbia Univ. Press, New York.

Way, M. J. (1950). *Quart. J. Microsc. Sci.* **91**, 145.

Weaver, A. A. (1958). *Int. Congr. Entomol., 10th, 1956* p. 535.

Weis-Fogh, T. (1960). *J. Exptl. Biol.* **37**, 889.

Weis-Fogh, T. (1961a). *In* "The Cell and the Organism" (J. A. Ramsay and V. B. Wigglesworth, eds.), pp. 283–300. Cambridge Univ. Press, London and New York.

Weis-Fogh, T. (1961b). *J. Mol. Biol.* **3**, 520.

Weis-Fogh, T. (1961c). *J. Mol. Biol.* **3**, 648.

Whitten, J. M. (1959). *Syst. Zool.* **8**, 130.

Whitten, J. M. (1960). *J. Morphol.* **107**, 233.

Wigglesworth, V. B. (1933). *Quart. J. Microsc. Sci.* **76**, 269.

Wigglesworth, V. B. (1937). *J. Exp. Biol.* **14**, 369.

Wigglesworth, V. B. (1942). *Bull. Entomol. Res.* **33**, 205.

Wigglesworth, V. B. (1947). *Proc. Roy. Soc., Ser. B* **134**, 163.

Wigglesworth, V. B. (1948a). *Biol. Rev. Cambridge Phil. Soc.* **23**, 408.

Wigglesworth, V. B. (1948b). *Discuss. Faraday Soc.* **3**, 172.

Wigglesworth, V. B. (1948c). *Quart. J. Microsc. Sci.* **89**, 197.

Wigglesworth, V. B. (1953a). "The Principles of Insect Physiology." Methuen, London.

Wigglesworth, V. B. (1953b). *Quart. J. Microsc. Sci.* **94**, 507.

Wigglesworth, V. B. (1953c). *J. Embryol. Exp. Morphol.* **1**, 269.

Wigglesworth, V. B. (1954a). *Quart. J. Microsc. Sci.* **95**, 115.

Wigglesworth, V. B. (1954b). "The Physiology of Insect Metamorphosis." Cambridge Univ. Press, London and New York.

Wigglesworth, V. B. (1956). *Quart. J. Microsc. Sci.* **97**, 89.

Wigglesworth, V. B. (1957). *Annu. Rev. Entomol.* **2**.

Wigglesworth, V. B. (1959a). *J. Exp. Biol.* **36**, 632.

Wigglesworth, V. B. (1959b). "The Control of Growth and Form." Cornell Univ. Press, Ithaca, New York.

Wigglesworth, V. B. (1961a). *In* "Insect Polymorphism" (J. S. Kennedy, ed.), p. 103. Roy. Entomol. Soc., London.

Wigglesworth, V. B. (1961b). *Bull. Soc. Sci. Bretagne* **26**, 119.

Wigglesworth, V. B. (1962). *Proc. Int. Entomol. Congr. 11th, 1960* p. 626.

Wigglesworth, V. B. (1963a). *J. Insect Physiol.* **9**, 105.

Wigglesworth, V. B. (1963b). *J. Exp. Biol.* **40**, 231.

Williams, C. M. (1947). *Biol. Bull.* **93**, 89.

Wood, R. L. (1959). *J. Biophys. Biochem. Cytol.* **6**, 343.

Chapter 3

CHEMISTRY OF THE INSECT CUTICLE

R. H. Hackman

I. Introduction

The cuticle is that part of the integument external to the epidermal cell layer. It is a heterogeneous, noncellular membrane which is secreted by the epidermal cells and it is both the skin and skeleton of the insect. Publications concerning chemical aspects of the insect cuticle can be traced back over a period of 150 years or more. The greater part of these papers describe investigations of the cuticle on a histological or cytological level making use of staining reactions and histochemical tests. Few, if any, of these color reactions are specific and so deductions from them are inconclusive. However, this descriptive work is important and has provided the background information for chemists and biochemists attacking problems in this field by more rigorous methods. Methods now available enable useful information to be obtained from quite small amounts of material. The lack of sufficient material no doubt was in the past responsible, at least in part, for the shortage of information of an exact nature.

A detailed description of the structure and formation of the cuticle has been given in Chapter 2 of this volume. Briefly, the cuticle is divided into two principal layers—an inner relatively thick procuticle and a thin outer epicuticle. The procuticle is composed essentially of protein and chitin and the intimate association of the chitin and protein has suggested to people over the years that cuticle is a mucopolysaccharide. The epicuticle does not contain chitin but consists of lipoproteins, polyhydric phenols, and waxes. A cement layer or "tectocuticle" may also be present. Protein in the epicuticle, and sometimes also in the outer layers of the procuticle, is tanned, i.e., it becomes hard and insoluble in water—a process known as sclerotization. When the outer region of the procuticle becomes tanned the hard layer so formed is known as the exocuticle and the remainder of the procuticle, the inner soft portion, is known as the endocuticle. In some insects an additional zone, the mesocuticle, has been described. Mesocuticle is the name given to a lightly colored, stabilized but not hardened region between the exo- and endocuticles and one which stains differently from the regions on either side of it. The waterproofing properties of the cuticle play an important role in the regulation of water balance in the insect and the wax layer of the epicuticle is largely responsible for these waterproofing properties.

The cuticle, besides being the external covering of the insect, continues into the fore- and hindguts, lines the ducts of the dermal glands and the tracheal system and takes part in the formation of sense organs. Mouthparts are in part hardened cuticle. The cuticle is not uniform over the entire insect and its structure is modified for special purposes. Various parts differ from each other in appearance and other physical (and chemical) properties, e.g., infrared absorption studies on the worker ant *Dolichoderus quadripunctatus* (Torossian, 1968) show that neighboring areas of the abdominal cuticle differ in chemical composition. The cuticle varies from a thin, delicate, transparent membrane, such as that over a photoreceptor, to a thick, hard, horny armor such as occurs in a beetle. Most striking in its chemical resistance is the cuticle which lines the defense glands of bombardier beetles; this cuticle withstands violent chemical reactions taking place at 100°C (Aneshansley *et al.*, 1969). Rigid parts of the cuticle maintain body shape, give protection to the insect, and provide points of attachment and leverage for muscles. The softer parts, such as intersegmental membranes, are flexible and serve as hinges to permit the harder parts to move in relation to one another. The cuticle also serves as a food reserve because during starvation and at the time of molting its inner layers are absorbed.

A knowledge of the chemical and physical nature of the cuticle, as well as its mode of formation, is needed to understand the interactions that occur between components and to explain the results obtained with the electron microscope. In the description which follows some physical properties of the cuticle are discussed and then its chemical components are considered. Biochemical mechanisms associated with the cuticle have not yet been studied sufficiently to enable unifying principles to be found and so it is not possible to give a well-integrated account of the biochemistry of the cuticle. No attempt has been made to record the percentages of the various components present in the cuticle because such determination are of unknown accuracy and the errors are almost certainly high. A major difficulty is one of obtaining pieces of cuticle free from epidermis and other body tissues without removal of any cuticular components or alteration in their water content. Estimation of the total amount of any one component is also difficult. The information available in the earlier literature has been summarized by Richards (1951). Chitin and ash estimations are subject to the least errors and these values for a number of dried cuticles are given in Table I.

II. Physical Properties

A. INTRODUCTION

Cuticle is not a homogeneous structure; different parts of it may differ in composition and changes may occur in cuticular structure during the life of an insect. One of the most striking features of insect cuticle is its variability and this fact makes a study of its physical properties and their interpretation difficult. With the exception of those on wettability and per-

TABLE I

CHITIN AND ASH CONTENT OF INSECT CUTICLES
(PERCENT BY WEIGHT OF DRY CUTICLES)

Species	Cuticle	% chitin	% ash
Agrianome spinicollis	Larval	37.2[a]	0.40[a]
Anabrus simplex	Exuviae	14.25[b]	8.72[b]
Blatta orientalis adult	Abdominal tergite ♀, ♂	31.7, 35.0[c]	
Blatta orientalis adult	Abdominal sternite ♀, ♂	34.2, 36.0[c]	
Bombyx mori	Larval	45.4, 44.2[a,d]	0.35[a]
Calliphora augur	Puparia	37.1[a]	2.4[a]
Dytiscus sp.	Elytra	37.4[d]	
Galleria mellonella	Larval	38.3[c]	
Locusta migratoria	Elytra and wings	23.7[d]	
Locusta migratoria	Abdominal	31.8[d]	
Locusta migratoria	Femoral	36.9[d]	
Lucilia cuprina	Larval	52.5[a]	0.36[a]
Lucilia cuprina	Puparia	41.1, 39.6[a,e]	8.8, 8.1[a,e]
Lucilia cuprina	Pupal and molting membranes	10.1[e]	5.0[e]
Periplaneta americana	Abdominal and pronotal	33.8[a]	0.84[a]
Phormia regina	Larval	48.6[c]	
Phormia terranovae	Larval	59.5[d]	
Sarcophaga bullata	Larval		0.365[f]
Sarcophaga falculata	Larval	60[d]	0.56[g]
Sarcophaga falculata	Puparia	47[d]	1.9[g]
Sphinx ligustri	Larval	50.3[d]	1.5[g]
Tenebrio molitor	Larval	31.3[d]	
Tribolium confusum	Larval	22.8[c]	
Xylotrupes gideon	Pronotal	46.3[a]	0.49[a]

[a] Hackman and Goldberg (1971a).
[b] Johnson *et al.* (1952).
[c] Tsao and Richards (1952).
[d] Fraenkel and Rudall (1947).
[e] Gilby and McKellar (1970).
[f] Richards (1956).
[g] Trim (1941).

meability (see Chapter 4, this volume), there have been relatively few papers published on the physical properties of the cuticle. Information published up to 1950 has been collected by Richards (1951). The physical properties of the individual cuticular components are given in their appropriate sections.

The surface of most insects is covered by a discrete layer of lipid (for exceptions see Section III,F) which, in general, prevents the cuticle from being wetted by water In some insects the cuticle is covered with a layer of hydrofuge hairs which trap air, as in the plastron of aquatic insects. Insect cuticles shrink on drying and swell on rewetting, i.e., imbibition and swelling occur but little is known about the mechanism (see Richards, 1951). The rate of swelling is influenced by factors such as pH and the presence of salts. The isoelectric points of different layers of the cuticle, as determined by staining methods, fall in the range pH 3.5 to 5.5 and different isoelectric points is one factor in the differential staining of cuticles with dye mixtures. Scheie (1969) and Scheie and Smyth (1967, 1968) have studied the changes which occur in the electrical resistance of the integument of *Periplaneta americana* during its life cycle and they have attempted to relate the changes to physiological events. Dermal glands and their ducts are likely paths for ionic conduction across the integument (Scheie and Smyth, 1972).

Except for the epicuticular layer the cuticle is composed principally of chitin and protein, considered by some to be chitin microfibrils in a matrix of protein (see Section III,C). A question which still waits a final solution is whether the chains of chitin and protein are arranged parallel to each other or are mutually at right angles (for discussion, see Fraenkel and Rudall, 1940, 1947; Richards and Pipa, 1958; Rudall, 1950, 1963). Cuticle has a lamellate structure brought about by the deposition of planes of micro-fibrils which may either form preferred oriented layers or rotate progressively in the form of a helicoid. The ultrastructure of the cuticle has a profound effect on its physical properties and this aspect has been discussed in detail by Neville (1970) and Weis-Fogh (1970).

B. MECHANICAL PROPERTIES

1. Density and Hardness

Jensen and Weis-Fogh (1962) give 1.2 gm/ml as the density (i.e., specific gravity) of locust (*Schistocerca*) wing. Sclerotization causes cuticles to become hard and the degree of sclerotization determines the degree of hardness. Bailey (1954) measured the hardness of the mouthparts (heavily sclerotized structures) of a number of arthropods including six species of insects (from four orders). In each the hardness was approximately 3 on Moh's scale of hardness of minerals. Such insects, therefore, would have no difficulty in boring through many of the metal foils used to protect foodstuffs and other materials.

2. Tensile Strength and Elasticity

A series of papers by Weis-Fogh on the biology and physics of flight describe direct measurements of the strength and elasticity of cuticle and relate these to function. Parts of the insect cuticle are subject to bending and there-fore to compression as well as to tension. Insects which fly possess an elastic system which counteracts the effects of wing inertia and reduces the energy required to maintain flight. Three structures, wing muscle, the thoracic box and the wing hinges (i.e., muscle, cuticle and resilin; see Section III,A), are involved and to different extents in different species of insects (Weis-Fogh, 1965). The static coefficients of elasticity (i.e., the elastic modulus E or Young's modulus) of the hard cuticle of the hind tibia, the pleural wall, and the elastic ligament of the forewing of *Schistocerca* lay between 800 and 1000 kg/mm² and the tensile strengths between 8 and 10 kg/mm². At small loads the coefficients of elasticity for extension and compression were not significantly different. Thus, in spite of differences in structure and function these parts have similar properties. For the wing the dynamic and static modulus of elasticity were similar. Locust cuticle resembles a resin reinforced with cellulosic materials (Jensen and Weis-Fogh, 1962).

Hepburn (1972; Hepburn and Ball, 1973) has investigated the relation-ship between structure and mechanical properties for the cuticle of the beetle, *Pachynoda sinuata*. This cuticle contains crossed reticulate layers of chitin fibrils arranged in the preferred orientation together with proteins which are considered as glues. Unlike the locust tibia (see above) the whole cuticle and the chitin obey Hook's law in the elastic region. Although a two phase material, the cuticle does not meet the requirements of plywood mechanics; its design ensures reasonable functional isotropy from an inherently tough anisotropic fibrous structure.

Krzelj and Jeuniaux (1968) observed that one species of Coleoptera, *Malachius bipustulatus,* had elytra with a very low Young's modulus (186 kg/mm² instead of 500 to 1300 kg/mm², values similar to those given above for *Schistocerca*) which would indicate some peculiar construction of the elytra. As was to be expected the flexural rigidity of soft cuticles was found to be less than that of hard cuticles.

The prealar arm of *Schistocerca* contains a very tough thick ligament which consists of parallel lamellae of chitin (0.2 μm thick) and resilin (about 0.3 μm thick). When extended in the direction of the lamellae the structure behaves like a solid, but in a direction normal to the lamellae it behaves like a rubber with an elastic modulus of 0.2 kg/mm². Again the dynamic and static values are similar and the elastic efficiency is about 97% (Jensen and Weis-Fogh, 1962). The elastic tendon of the metathorax of dragon flies (*Aeshna* spp.) is composed of resilin without chitin and has an elastic modulus G of 0.064 kg/mm² (i.e., an elastic modulus E of about 0.14 kg/mm²), a tensile strength of 0.3 to 0.4 kg/mm² and a length at break-ing point of about three times the unstrained length (Weis-Fogh, 1961).

C. OPTICAL PROPERTIES

1. Absorption Spectra

Unsclerotized cuticles absorb UV light but are transparent to visible light; absorption increases with pigmentation, including sclerotization. Complex mixtures such as insect cuticles do not give useful infrared absorption spectra or attenuated total reflectance (ATR) spectra. The sampling depth of ATR is such as to include many of the chemical components of the cuticle (see, e.g., Barker, 1970).

2. Birefringence

Chitin shows strong positive form birefringence and the nonchitinous epi-cuticle is also birefringent. The amplitude of the birefringence varies considerably and depends both on the composition of the cuticle and the orientation of the molceules. Caveney (1971) and Neville and Caveney (1969) have shown that the outer region of the exocuticle of some scarabaeid beetles is optically active and the birefringence is anomalous in that it is positive in a direction perpendicular to the cuticle surface. Most insect exocuticles are optically inactive and the birefringence is positive parallel to the cuticle surface. The unusual optical properties of the scarabaeid cuticles are located in the outer 5 to 20 μm of the exocuticle.

3. Refractive Index

Refractive index has assumed some importance in studies on the arrangement of layers within the cuticle. Values lie between 1.5 and 1.7 but as discussed by Richards (1951) they can at best be no more than approximations. Weis-Fogh (1970) reports that, presumably for *Schistocerca,* the isotropic inner epicuticle has the highest refractive index (about 1.58) of any of the cuticular structures; the birefringent outer epicuticle has a lower refractive index, while for the dry isotropic protein the value is about 1.54. Caveney (1971) reports values of up to 1.70 for the refractive index of the reflecting cuticular layer in *Plusiotis* species, while that in a *Potosia* species was 1.595. The higher value in the *Plusiotis* species is due to the presence of uric acid; extraction of the uric acid reduces the value to that observed in *Potosia.*

4. Coloration

The colors of some insects have a physical and not a chemical basis. Such colors depend on the physical structure of the cuticle and many are interference colors produced by thin lamellae separated by material of a different refractive index. In some scarabaeid beetles the interference colors are determined by the helicoidal pitch of the lamellae in the outer exocuticle; the helicoidal architecture is counter-clockwise and uric acid occurs in the reflecting layers (Caveney, 1971; Neville and Caveney, 1969). Other colors may be due to the scattering, reflection, refraction, or diffraction of light. Examples of physical colors are the metallic, iridescent colors of the elytra of some beetles and the scales of some butterflies (e.g., see Schmidt and Paulus, 1970). A study of the patterns of microsculpture on the cuticles of beetles and some other insects has shown many to be in the form of regular ridges which act as diffraction gratings to give iridescent colors (Hinton, 1970; Hinton and Gibbs, 1971). Colors due to physical structure are readily distinguished from pigment colors because they are changed or destroyed by pressure, distortion, swelling, shrinking, or immersion in a medium of the same refractive index. The colors of some beetles of the genus *Dynastes* depend on the amount of water in the cuticles and thereby on the thickness of the thin films responsible for the interference colors. By varying the amounts of water the insect can change color (reversibly), e.g., from black to greenish yellow (Hinton and Jarman, 1973).

III. Cuticular Components

A. PROTEINS

Protein occurs in all layers of the cuticle and may account for more than half the total weight of the dry cuticle. Distribution of proteins within the cuticle has been investigated by a number of workers. Lipoprotein is considered to be present in the eipcuticle but the evidence for it, based on histochemical and staining reactions, is not conclusive. There appear to be only two papers which give information about the proteins of the epicuticle or procuticle (Dennell, 1958a; Hackman, 1971a). Dennell has stated that hydrolysis (HCl) and subsequent chromatographic examination showed that both the procuticle and epicuticle (which was contaminated with procuticle) of larvae of *Calliphora vomitoria* have the same qualitative amino acid composition—that of the intact cuticle. Hackman has shown disulfide groups (cystine) to be present in a zone immediately below the epicuticle of larvae of *Lucilia cuprina* and disulfide groups may stabilize the mesocuticle. The only protein from a specialized part of the cuticle which has been studied in detail is resilin (see below).

Trim (1941) pioneered investigations into the chemical properties and amino acid composition of cuticular proteins. Fraenkel and Rudall (1940, 1947), in an X-ray diffraction study of the cuticle, concluded that some cuticular proteins exist in the β-configuration. Early work indicated that the proteins did not contain the sulfur amino acids but it is now known that one or more of the amino acids cystine, cysteine, or methionine are present in low concentrations in many (and perhaps all) insect cuticles (Fukushi, 1967; Fukushi and Seki, 1965; Gilby and McKellar, 1970; Hackman, 1971a; Hackman and Goldberg, 1968, 1971a; Karlson et al., 1969; Krishnan, 1969; Krishnan and Sundara Rajulu, 1964; Srivastava, 1971a). Free amino acids occur in cuticles and these have been studied by Moorefield (1953).

The protein of the cuticle is heterogeneous as is shown by (1) the extraction of protein fractions of differing solubilities (soluble in water, salt solutions, aqueous urea, formic acid, or alkali) and amino acid composition (Andersen, 1971b; Hackman, 1953a,b; Hackman and Goldberg, 1958; Kalson et al., 1969; Mills et al., 1967a); (2) the electrophoretic behavior of extracted proteins (Fox and Mills, 1969; Hackman, 1953a, 1972; Hackman and Goldberg, 1958; Kinnear et al., 1971; Krishnan, 1969; Moorefield, 1953; Srivastava, 1970; Willis, 1970); and (3) serological techniques (Downe, 1962). Polyacrylamide gel electrophoresis has shown up to twenty protein bands from one protein fraction but the number of bands detected depends on the stage of the animal's development. When different protein fractions are compared it is often assumed that proteins with the same mobilities are identical. However, Hackman (1972), in a study of cuticular proteins from *Agrianome spinicollis,* has shown that proteins with the same mobility can have quite different affinities for the dyes used to reveal their positions. The isoelectric points of the major components of these *Agrianome* proteins lay between pH 3.35 and pH 5.7, which is in good agreement with work published on the isoelectric points of epicuticle, exocuticle, and endocuticle (for references, see Richards, 1951). A few components, present in very low concentrations, had isoelectric points up to pH 8.2 Willis found that the proteins in locust exo- and endocuticle synthesized before ecdysis had isoelectric points about pH 9.5, whereas protein in intersegmental membranes all had lower isoelectric points.

The KCl-soluble protein fraction from cuticles of the house cricket *Acheta domesticus* has been separated by gel filtration into five components all with molecular weights which are multiples of 7000 (Mills et al., 1967a). Earlier, Moorefield (1953) found that the water-soluble proteins from larval cuticles of *Sarcophaga crassipalpis* (which electrophoretically were heterogeneous) behaved in the ultracentrifuge as a monodisperse substance with an apparent molecular weight of 7000 to 8000. Each of the *Agrianome* cuticular protein fractions examined by Hackman (1972) contained many components, the major ones having molecular weights less than 18,000. The remaining components had molecular weights between 18,000 and approximately 100,000.

No simple relationship between the molecular weights of the components was apparent although some of the proteins readily form aggregates. Fraenkel and Rudall (1947) gave the name "arthropodin" to the water-soluble protein of insect cuticle, but in the light of what is now known about these proteins this name has little real meaning.

The solubility of a protein fraction gives information on the interaction of that fraction with other cuticular components (Andersen, 1971b; Hackman, 1953a,b, 1972; Hackman and Goldberg, 1958; Karlson et al., 1969). For example, Hackman and Goldberg have shown that the larval cuticle of the beetle *Agrianome spinicollis* contains 63% protein: 14% of this protein was not bound to other components (soluble in water); 2% was bound by weak bonds such as van der Waal's forces (soluble in 0.16 M aqueous sodium sulfate); 25% was bound by hydrogen bonds (soluble in 7 M aqueous urea); 3% was bound by electrovalent bonds or double covalent bonds or both such as would occur in Schiff's bases (soluble in 0.01 N aqueous sodium hydroxide); the remainder of the protein (56%) was bound covalently to chitin. Electrophoresis has shown each of these protein fractions to be heterogeneous. Those fractions adsorbed to other components were richer in tyrosine than those not adsorbed, confirming earlier observations of Hackman (1955a,b).

It is clear that the protein component of unhardened cuticles contains a large number of different proteins. The interactions which occur between cuticular components would, no doubt, account for some of the different proteins (see above and Section III,C) but the reason for the very high number is obscure. Hackman (1972) has suggested that, since the properties of the unhardened cuticle are such as would be given by a protein component which is polydisperse and consists of a number of similar but not identical proteins, the multiplicity of protein species may be the result of protein polymorphisms.

Since the protein component of the cuticle is a mixture of many different protein fractions it is difficult to interpret amino acid analyses of intact cuticle. Moreover, when puparia or exuviae have been used it has not always been stated that contaminating materials, e.g., molting membranes, enzymes from molting fluid, etc., have been removed. Qualitative and quantitative analyses have been reported for the amino acids in hydrolysates of intact cuticles, of different regions of the cuticle (including scales and hairs), and for protein fractions extracted from cuticles (Andersen, 1971b; Bodnaryk, 1971b; De Haas et al., 1957; Dennell, 1958a; Fukushi, 1967; Fukushi and Seki, 1965; Gilby and McKellar, 1970; Hackman, 1953a,b; Hackman and Goldberg, 1958, 1971a; Hunt, 1971; Johnson et al., 1952; Karlson et al., 1969; Kennaugh, 1958; Krishnan, 1969; Lipke et al., 1965a; Srivastava, 1971a; Trim, 1941). In contrast to fibrous proteins (e.g., silk fibroin) cuticular proteins have a high content of amino acids with bulky side chains. The following conclusions can be drawn (see also Section IV): (1) the protein components of the (untanned) larval cuticles of *Agrianome* (Coleoptera) and *Bombyx* (Lepidoptera) resemble each other but differ from that of *Lucilia* (Diptera); (2) the protein components of the puparia of the two blow flies *Calliphora* and *Lucilia* are similar but show some obvious differences from those of two *Drosophila* species (which are themselves similar) and *Musca;* (3) the protein components of the sclerotized cuticles of *Periplaneta, Anabrus,* and *Schistocerca* show similarities but differ from that of *Xylotrupes* (Coleoptera); (4) β-alanine occurs only in sclerotized cuticles; (5) there are significantly higher amounts of amino acids with polar residues (arginine, aspartic acid, glutamic acid, histidine, lysine, serine, threonine, and tyrosine) in soft cuticles than in sclerotized cuticles (in this sense puparia must be regarded as soft cuticles since they are derived from the larval cuticle and are not synthesized anew); (6) o-tyrosine occurs in the cuticles of late last-instar larvae of *Calliphora vomitoria* and of newly molted cockroaches (*Periplaneta americana*). In the hemimetabolous *Schistocerca* the cuticular proteins of the adult and immature stages are similar (Andersen, 1973), while in the holometabolous *Tenebrio* they differ markedly (Andersen et al., 1973).

Certain parts of the insect cuticle are rubberlike and this property is due to the presence of an insoluble protein, resilin, first described by Weis-Fogh (1960). Resilin may be associated with chitin (see Section II) and it occurs in those parts of the cuticle involved in springlike movements. As it occurs in the cuticle, resilin is colorless, hyaline, and swollen with water and when exposed to UV light it shows a strong blue fluorescence. It appears to be completely devoid of structure when viewed in the electron microscope. From a study of its physical and chemical properties Weis-Fogh concluded that resilin represents a three-dimensional protein network of high stability, the protein chains being linked to each other at regular intervals through di- and trityrosine residues, but the protein has little or no secondary structure of mechanical significance. The amino acid composition of resilin differs distinctly from that of other proteins. Andersen (1971b) has reviewed the occurrence, properties, structure, and deposition of resilin.

B. CHITIN

The name chitin was given by Odier (1823) to the material he isolated from the elytra of cockchafers and which was later shown to be identical with the "fungine" which had been isolated by Braconnot from fungi in 1811. Chitin occurs in the procuticle (endocuticle and exocuticle) of insects but is absent from the epicuticle. Some membranes with a structural similarity to cuticle contain no demonstrable chitin, e.g., the scales on the wings of some lepidopterous insects (for discussion, see Richards, 1947). Chitin may account for up to one-half or more of the weight of the dry cuticle and the chitin contents of some insect cuticles are given in Table I. Since chitin is usually considered to be the cuticular material insoluble in hot, dilute aqueous alkali, the chitins prepared from some cuticles contain a non-chitinous membrane of extreme chemical stability (Section III,I,3).

Three distinct crystalline forms of chitin are known, α, β, and γ (Rudall, 1962, 1963) which were distinguished from one another by X-ray diffraction. The forms differ from one another in the arrangement of the chains and the presence of bound molecules of water, properties which readily explain their different rates of hydrolytic breakdown (Hackman and Goldberg, 1965). Only α-chitin, the most stable form, is known to occur in the insect cuticle, although Rudall (1965) has suggested that the chitin in *Calliphora* larval cuticles may be related to γ-chitin. Some peritrophic membranes and cocoons contain β- or γ-chitin (Rudall and Kenchington, 1971).

a. Structure and Properties of Chitin. Chitin is a high molecular weight polysaccharide which consists predominantly, perhaps entirely, of unbranched chains of β-(1,4)-linked 2-acetamido-2-deoxy-D-glucose (i.e., *N*-acetyl-D-glucosamine) residues, the unit being a biose (Fig. 1). From X-ray diffraction and infrared spectroscopic studies it is concluded that in chitin the molecular chains are associated together in a highly ordered manner, the structure being stabilized by multiple hydrogen bonding (both intra- and intermolecular). In α-chitin equal numbers of chains run in opposite directions with screw axes lying along the chain direction. There is an alternation of the acetamido groups from one side of the molecule to the other between adjacent residues (see Fig. 1). A "bent" chain is the only arrangement which is sterically satisfactory and fits X-ray diffraction data. [For a discussion on the crystal structure of chitin, see Carlstrom (1957, 1962, 1966), Dweltz (1960), and Ramachandran and Ramakrishnan (1962)]. This arrangement is probably true only for highly crystalline regions which Hackman and Goldberg (1965) have shown may account for no more than one-third of the chitin in a cuticle. Some of the amino groups in chitin

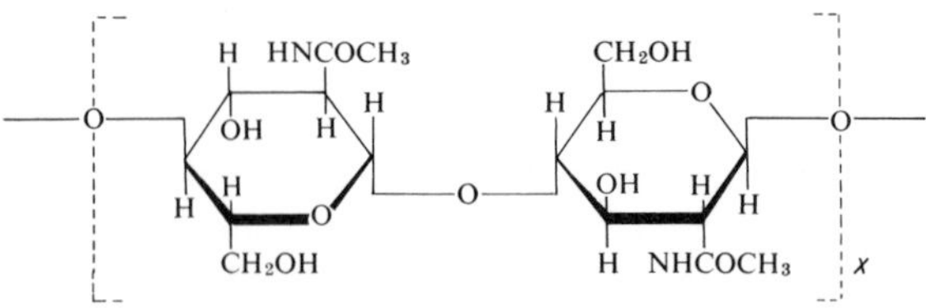

Fig. 1. Structure of chitin.

are not acetylated (Waterhouse *et al.*, 1961) and the acidic hydrolysis of chitin yields, in addition to glucosamine, small amounts of amino acids (Hackman, 1960). Structures based on analyses, such as those made by Giles *et al.* (1958), are of doubtful value since experience has shown that elementary analyses of chitin are very likely to be unreliable.

The intra- and intermolecular bonding between chitin chains mentioned above refers to chitins which have been freed as far as possible from protein and so may not be true of the chitin-protein associations which occur in cuticles. The β-1,4-linkages permit the backbone of the chitin chain to be oriented parallel to the microfibril and this gives the structure a high tensile strength. For a discussion on the arrangements of chitin chains within the cuticle and the properties which these confer on the cuticle, see Neville (1967).

Chitin is a colorless, amorphous solid, insoluble in water, dilute acids and alkalis, alcohol, and all organic solvents. It is soluble in concentrated mineral acids but undergoes rapid and extensive degradation (Hackman, 1962). Colloidal chitin prepared by precipitation from solution in concentrated acids is, therefore, a much degraded material. When chitin is treated with alkali hydroxides at high temperatures (about 160°C) acetyl groups are detached and the polysaccharide chitosan is formed. Hot dilute acids and alkalis also bring about some degradation of chitin (Hackman, 1954, 1962). Chitin can be dispersed in hot concentrated aqueous solutions of certain mineral salts, e.g., lithium thiocyanate, from which it can be precipitated by the addition of water, alcohol, or acetone. Chitin is polydisperse, i.e., it is not a single molecular species but consists of chains of varying lengths (and consequently molecular weights). The most recent reviews on the chemistry and properties of chitin are by Jeuniaux (1971) and Ward and Seib (1970).

For the physical properties of chitin see the reviews quoted above, the references to the work on the crystal structure of chitin, and Richards (1951). The data refer to chitin free from association with other cuticular components and so certainly most of it will not apply to chitin as it occurs in the cuticle (see Section III,C). Data on the optical rotation of chitin reported by Irvine (1909) are incorrect (see Hackman, 1962).

b. Detection and Estimation of Chitin. There are no histochemical tests which are specific for chitin. Tests such as the alkaline tetrazolium reaction (Pearse, 1961) or the periodic acid–Schiff reaction give positive results with a variety of different types of compounds. Richards (1951) has examined critically the color reactions which have been used to detect chitin. Runham's work (1961, 1962) and the comments on it by Salthouse (1962) show clearly that chitin cannot be identified by the periodic acid–Schiff (PAS) or Hale or Alcian blue reactions. Lack of solubility in hot dilute aqueous alkali is not a property peculiar to chitin (see Hackman and Goldberg, 1960, and Section III,I,3). Although lacking in specificity, histochemical reactions do have the advantage that the structure of the section being examined is preserved, whereas the more definitive methods given later destroy the structure of the section. Benjaminson (1969) has described the use of fluorescent conjugates of chitinase as stains for chitin, a method based on the affinity of an enzyme for its substrate. If great care is taken to purify the enzyme and the conjugate the method should be specific for chitin-type substrates. However, chitin in a structure is not necessarily accessible to chitinase (see Section III,C).

Chitin is usually identified by conversion to chitosan (heating with concentrated aqueous potassium hydroxide) which is identified by color reactions (iodine) or spherocrystal formation (with acids) (van Wisselingh, 1898; Campbell, 1929). It is a simple matter to demonstrate the presence of chitin in large structures by this method, but difficulties arise when minute structures or structures which will not withstand such drastic treatment are used. Failure to detect chitin in very small samples of material may mean that the sample has become transparent and has been lost, that it has disintegrated, that the percentage of chitin present is very low, or that chitin is absent.

A useful, and perhaps an essential, confirmative test for chitin, and one which requires very little material, is acidic hydrolysis and detection of glucosamine in the hydrolyzate (e.g., see Hackman, 1960; Hackman and Goldberg, 1965). This test is not specific for chitin because other polysaccharides yield glucosamine on hydrolysis. Partial hydrolysis and identification of oligosaccharides of glucosamine or N-acetyl-D-glucosamine (see Barker et al., 1958; Capon and Foster, 1970; Horowitz et al., 1957; Rupley, 1964) is more likely to be specific for chitin but it requires the use of lengthy and elaborate techniques and considerable amounts of the polysaccharide. X-ray diffraction is the only physical method by which chitin has been identified. The method has the disadvantages that the sample must be free from contaminating materials and that it is not applicable to extremely small samples. Enzymes will be of use for the identification of chitin only when pure preparations, specific for undegraded chitin, become available [however, see Hackman and Goldberg (1965), Jeuniaux (1963) and the comments made earlier on fluorescent conjugates of chitinase].

Most quantitative determinations of chitin are based on the weight of residue remaining after the sample has been repeatedly digested with hot dilute aqueous alkali (e.g., 1 N NaOH at 100°C for several days). This residue may contain nonchitinous membranes (see Section III,I,3). Some workers find it necessary to decolorize the product by oxidation with dilute aqueous potassium permanganate but usually this is unnecessary, adequate extraction with alkali being sufficient. An alternative method is prolonged treatment (up to 10 weeks) with Diaphanol (Fraenkel and Rudall, 1940). Treatment with 5% hydrochloric acid at 100°C as used by Fraenkel and Rudall (1940) degrades chitin (Hackman, 1954, 1962).

Quantitative estimations of chitin are subject to a number of sources of error, the most important of which are (1) degradation of the chitin, (2) unknown purity of the final product, and (3) actual loss of material on the walls of flasks, etc. during the preparation. The last is often a major source of error when handling small samples.

Ash is invariably present in chitin prepared by alkaline digestion but Richards (1956) has prepared an ash-free chitin by using polyethylene vessels instead of the usual glass flasks. When corrected for the presence of inorganic impurities, chitin would be expected to have, and is found to have, a nitrogen content of 6.9% (that of poly-N-acetyl-D-glucosamine). The many low values (when corrected for ash) reported in the literature appear inexplicable unless the nitrogen estimations are in error (see also discussion by Hackman and Goldberg, 1971a).

C. Glycoproteins

Hackman (1955b) has shown that there is a weak bonding between chitin and some of the cuticular proteins from larvae of the beetle *Diaphonia dorsalis* and that tyrosine-rich protein fractions are preferentially adsorbed. In larval cuticles of the beetle *Agrianome spinicollis* much of the protein is bound to chitin in the form of a stable complex (Hackman and Goldberg, 1958). In a more extensive study (Hackman, 1960) it has been established that protein is bound by covalent bonds to chitin to form glycoproteins. Glycoproteins from different species contained different amounts of chitin and protein and, like chitin, the glycoproteins are polydisperse. Protein would appear to be linked to chitin through aspartyl or histidyl residues or both. In view of the stability of the linkage to hot alkali and its instability to hot acid, the protein could be linked to an amino group of the chitin in the form of an N-acylglucosamine. Chitins from sources other than insects also occur as glycoproteins (Foster and Hackman, 1957; Hackman, 1960; Goffinet and Jeuniaux, 1969). Enzymic studies and acidic hydrolyses of cuticles from *Periplaneta americana* and *Sarcophaga bullata* have also led to the conclusion that a stable linkage exists between chitin and protein, proteins being linked to chitin through a nonaromatic amino acid (Lipke, 1971; Lipke and Geoghegan, 1971a,b; Lipke *et al.*, 1965a). Fragments of mucoprotein with glycosyl, N-acetylglucosaminyl, and peptidyl residues were isolated. Delachambre (1969) concludes that chitin in the form of a glycoprotein is PAS negative and that removal of the protein frees amino groups in the chitin which then gives a positive PAS reaction. Attwood and Zola (1967) failed to detect a chitin–protein complex in *Calliphora* larval cuticles but in view of the difficulties known to be inherent in the experimental techniques they used it is likely that their conclusions are in error.

In the work discussed above cuticles have been subjected to somewhat drastic chemical procedures which may have destroyed well-defined molecular species or macromolecular systems. Electron micrographs of cross sections of insect cuticles show a regular pattern of nonstaining centers surrounded by highly stained (PTA) material, which has been interpreted as rods of chitin surrounded by protein (Filshie, 1966; Neville and Luke, 1969; Rudall, 1969). The rods are more closely packed together in hard cuticles than in soft cuticles. X-ray diffraction patterns of chitinous structures show repeat periods corresponding to a spacing of six or eight glucosamine residues and Rudall (1963, 1969) suggests that this indicates association with protein at those intervals. As discussed above (Section III,B) these X-ray diffraction patterns are obtained from the more highly ordered regions of the structure and may not be representative of the bulk of the sample. The details of the X-ray diffraction patterns of soft cuticles are modified or destroyed by steaming or treatment with weak acids (Rudall, 1963). So, although covalent bonds are important in maintaining the integrity of chitin–protein complexes, weaker links are also involved. Sclerotization increases greatly the stability of these complexes. Rudall and Kenchington (1973), by X-ray diffraction studies, have shown that the chitin–protein complex is not affected by water or neutral salt solutions but is disrupted by treatment with 7 M urea, pronase, diaphonal, or alkali. Thus it appears that the integrity of the chitin–protein complex, as defined by X-ray diffraction, depends on proteins which are hydrogen bonded to the chitin.

Chitin, as it occurs naturally in association with protein, might be expected to have properties that differ somewhat from those of chitin when isolated as an amorphous powder. There is the apparently complete digestion (enzymic) of the chitin in the endocuticle at the time of molting. The enzymic digestion of chitin preparations is generally far from complete. Some chitinous membranes are resistant to attack by the enzyme chitinase and protein appears to protect the chitin. Much of the chitin in the peritrophic membrane is not degraded enzymically until the protein has been removed (De Mets and Jeuniaux, 1962). Earlier Jeuniaux (1959) described a chitinous membrane in a crustacean cuticle which was degraded by chitinase only after alkaline digestion to remove protein. Jeuniaux (1963) views chitin as being either free or bound; free chitin is readily hydrolyzed enzymically even in the presence of protein, bound chitin only after protein has been removed.

TABLE II

3,4-DIHYDRIC PHENOLS IN INSECT CUTICLES

Phenol 	Reference
Catechol, R=H	Dennell, 1958c; Malek, 1961
4-Methylcatechol, R=CH$_3$	Pryor *et al.*, 1947; Sapag-Hagar *et al.*, 1961a
3,4-Dihydroxybenzoic acid, R=COOH	Atkinson *et al.*, 1973; Hackman and Goldberg, 1963; Hackman *et al.*, 1948; Kawase, 1958b; Malek, 1961; Pryor *et al.*, 1946, 1947; Sapag-Hagar *et al.*, 1961a; Vercauteren and Aerts; 1959
3,4-Dihydroxyphenylacetic acid, R=CH$_2$—COOH	Hackman *et al.*, 1948; Malek, 1961; Pryor *et al.*, 1947; Schmalfusz, 1937; Schmalfusz and Müller, 1927; Schmalfusz *et al.*, 1933; Vercauteren and Aerts, 1959
3,4-Dihydroxyphenylpropionic acid, R=CH$_2$—CH$_2$—COOH	Malek, 1961
3,4-Dihydroxyphenyllactic acid, R=CH$_2$—CHOH—COOH	Pryor *et al.*, 1947; Schmalfusz *et al.*, 1933; Vercauteren and Aerts, 1959
3,4-Dihydroxyphenylalanine, R=CH$_2$—CHNH$_2$—COOH	Malek, 1961; Schmalfusz, 1937; Schmalfusz and Müller, 1927
3,4-Dihydroxybenzaldehyde, R=CHO 3,4-Dihydroxyphenylethanol, R=CH$_2$—CH$_2$OH 2-Hydroxy-3',4'-dihydroxy acetophenone, R=CO—CH$_2$OH	Atkinson *et al.*, 1973

D. PHENOLS

Phenols [especially 3,4- (i.e., *ortho-*) dihydric phenols] appear to occur in most, and perhaps all, insect cuticles as has been shown by histochemical tests and color reactions (e.g., Pryor, 1940; Locke and Krishnan, 1971). Those 3,4-dihydric phenols which have been identified are listed in Table II. Isolated cuticles were not always used but there is little doubt that the phenols are in the cuticles; however, dopa occurs in many tissues and so it is recorded as being present only if it has been identified in tegumental structures. Salicylic acid, *p*-hydroxybenzoic acid (and benzoic acid) have also been isolated from *P. americana* and *P. brunnae* cuticles (Atkinson *et al.*, 1973) and hydroquinone and pyrogallol from *C. vomitoria* (Dennell,

1958c). Additional dihydric phenols have been obtained on acidic hydrolysis of sclerotized cuticles and also from whole insects (e.g., N-acetyldopamine). These phenols and the role they and the other phenols play in sclerotization are discussed in Section IV.

E. Enzymes

Cuticular o-diphenoloxidase (EC 1.10.3.1, o-diphenol:oxygen oxidoreductase), which is sometimes referred to as a polyphenoloxidase, phenoloxidase, catechol oxidase, phenolase, tyrosinase, etc., takes part in sclerotization (Section IV). Bhagvat and Richter (1938) found that the "skin" of insects contained a catechol oxidase. Since the appearance of that paper there have been many reports of o-diphenoloxidases in insect cuticles, the enzyme being identified by histochemical tests (e.g., Dennell, 1947; Kawase, 1958a; Lai-Fook, 1966; Locke and Krishnan, 1971) and by extraction and, in the presence of selected substrates, measuring either oxygen uptake (e.g., Hackman and Goldberg, 1958, 1967, 1968; Ito, 1953; Kawase, 1960; Moorefield, 1953; Ohnishi, 1954; Pryor, 1955) or change in color development (e.g., Mills et al., 1968). Locke and Krishnan (1971) have studied the distribution of phenoloxidases during formation of the cuticle in Calpodes. The enzyme is very soluble in water and Kawase (1960) and Ito (1953) showed that in Drosophila larval cuticles its activity changed markedly during the life cycle of the larvae. In the last-instar larval cuticle of muscoid flies the activity of the water-soluble o-diphenoloxidase increases to a maximum just before puparium formation begins and then decreases rapidly (Hackman and Goldberg, 1967). The enzymes from six species (three genera) have similar properties and the preferred substrates are dopamine and 4-methylcatechol. 3,4-Dihydroxybenzoic acid and similar phenols are not oxidized. Mills et al. (1968) have isolated an enzyme with a similar specificity from cockroach cuticle. The enzyme appears to be confined to the exocuticle and it is most active immediately before the molt. Insoluble phenoloxidases of the laccase type, but different from that in the cockroach (Whitehead et al., 1960), occur in the puparial cuticle of Drosophila virilis and in the pupal cuticles of Bombyx mori and Papilio xuthus (Yamazaki, 1969), but the possibility that the enzymes may have been contaminants was not excluded. Insect phenolases are known to aggregate and become particulate (see Cottrell, 1964; Karlson et al., 1964).

Karlson and Liebau (1961) isolated from whole larvae of Calliphora erythrocephala a crystalline phenoloxidase which they assumed to be the cuticular enzyme. The enzyme is an o-dihydric phenolase and monohydric phenols are not oxidized. The preferred substrates are N-acetyldopamine, dopamine, and dopa although catechol and 4-methylcatechol are also oxidized; o-dihydric phenols with acidic sidechains are not oxidized. Heyneman (1965) has questioned Karlson's identification of his crystalline product with the enzyme and, from his own work on *Tenebrio* larval o-diphenoloxidase, has suggested that the enzyme is a much smaller molecule. Later, Munn and Greville (1969) isolated a crystalline protein ("calliphorin") from a number of insects (including larvae of *C. erythrocephala*) which had no enzymic activity but otherwise had properties similar to those of the crystalline enzyme isolated by Karlson and Liebau. Therefore, Karlson's group may have isolated not a crystalline enzyme but a protein to which enzyme has been adsorbed (or occluded) (Munn *et al.*, 1971; Munn and Bufton, 1973).

In a study of the larval integument of *Drosophila melanogaster*, Knowles and Fristrom (1967) reported the presence of ten enzymes which included a number of dehydrogenases, two phosphatases, an esterase, and an aminopeptidase, but the authors suggest that the enzymes may possibly have come from contaminating tissue. Without giving experimental evidence Hurst (1945) described a polyphenol peroxidase in the cuticles of several species of insects and he suggested the enzyme may be involved in sclerotization. Locke (1969b) has also suggested that a peroxidase takes part in the formation of hard cuticle in *Calpodes ethlius* and these observations on peroxidases are discussed in Section IV. Coles (1966) found a peroxidase in the epidermal cells of the locust.

F. LIPIDS

The wax layer of the epicuticle is responsible for the very low permeability of the cuticle to water, but since this topic is the subject matter of Chapter 4 of this volume, it will not be considered here. Cuticulin, as defined by Wigglesworth (1933, 1970), is a stabilized lipid which, when combined with protein, stiffens the cuticle prior to sclerotization. It is present in the epi-, exo-, and mesocuticle. Nothing is known about the chemical structure of cuticulin. Other workers, e.g., Locke (1966) and Filshie (1970b), have used cuticulin in a different sense, namely to describe one layer of the epicuticle. The insoluble membrane discussed in Section III,I,3 has some of the properties of long-chain paraffin hydrocarbons. Other structures such as the "oriented lipid layer" and "wax canal filaments" have been shown to be insoluble in lipid solvents and resistant to acids (Filshie, 1970a).

Cuticles of low permeability are thought to have a primary lipid layer beneath a shellac-like cement and a secondary layer of higher melting point on the outside of this material (Beament, 1955, 1959; Holdgate and Seal, 1956; Kramer and Wigglesworth, 1950). In some insects, e.g., *Calliphora* larvae, a surface layer of lipid is not present, the lipid being incorporated within the cuticle (Wolfe, 1954, 1955), while in some aquatic insects the lipid layer is probably absent (Beament, 1961b). The rectal cuticle of the desert locust does not appear to have a continuous lipid layer (Phillips and Dockrill, 1968). The thickness of the lipid layer has been calculated to be of the order of 0.25 μm (Beament, 1961a; Holdgate and Seal, 1956), but Lockey (1960) reports that the layers are much thinner because the surface areas, as measured by krypton adsorption, are much larger than formerly thought. Estimates of the thickness of the lipid layer can only be approximations because it is assumed that the lipid is uniformly distributed, that all the lipid can be extracted, and that the surface area of the cuticle is known.

Reviews on aspects of cuticular lipids have been given by Gilby (1965) and Jackson and Baker (1970). A summary of the composition of the lipids from six species of insect is given in Table III. Each lipid is a complex mixture but since each was a bulk sample obtained from a large number of insects, it is possible that the lipid from one insect may be less complex. The methods of extraction and separation into lipid classes also modify the results (see Gilby and McKellar, 1970). There were seasonal variations in the relative proportions of hydrocarbons in the lipids from a species of ant as well as minor variations in their relative proportions from head, thorax, and gaster (Cavill et al., 1970). Goodrich (1970) reports minor variations in the composition of the lipids from male and female blow flies as does Silhacek et al. (1972) for the cuticular hydrocarbons from male and female houseflies.

In addition to the results given in Table III, information is available on the hydrocarbons and fatty acids of other cuticular lipids. The major components of the hydrocarbons (alkanes) from the ant *Atta colombica* are two homologous series of trimethylalkanes (C_{34}–C_{39}) (Martin and Macconnell, 1970) while those from the ant *Myrmecia gulosa* are an homologous series of monomethylalkanes (C_{26}–C_{34}) (together with a little *n*-paraffin, C_{19}–C_{29}) (Cavill et al., 1970). In the lipids from the tobacco hookworm the hydrocarbons are both straight and branched chain alkanes (major components, heptacosane, and mono-, di-, and trimethylalkanes with equivalent chain lengths between C_{36} and C_{38}) (Nelson et al., 1971) and those in *Bombyx mori* wings are odd- and even-numbered alkanes, the major components being C_{34}–C_{36} (Suzuki et al., 1966). The hydrocarbons fractions from the cuticular lipids of the cockroaches *Leucophaea maderae* and *Blatta orientalis* are similar in composition and consist mainly of *n*-heptacosane and its 3-, 11-, and 13-methyl derivatives (Tartivita and Jackson, 1970). For the cockroaches *Periplaneta australasiae, P. brunnea,* and *P. fuliginosa,* the major hydrocarbons are 13-methylpentacosane and *n*-tricosane and its 3- and 11-methyl derivatives together with *cis*-9-tricosene in the males of *P.*

TABLE III

COMPOSITION OF INSECT CUTICULAR LIPIDS[a]

Content (% by weight)[b,c]

Lipid class	Anabrus simplex abdominal sclerites (1)	Bombyx mori exuviae (2,3,4)	Bombyx mori larva (4)	Lucilia cuprina Adults (5) ♂	Lucilia cuprina Adults (5) ♀	Lucilia cuprina Puparia (5)	Puparia (6) Free lipid	Puparia (6) Bound lipid	Pupal and molting membranes (6) Free lipid	Pupal and molting membranes (6) Bound lipid	Periplaneta americana nymphs (7,8)	Pteronarcys californica Adults (9)	Pteronarcys californica Naiads (9)	Tenebrio molitor larval exuviae (10)
Hydrocarbons	48–58	+	+	59.2	65.7	33.3	55.0	35.7	51.0	59.9	75–77	12	3	10
Fatty acids	15–18	+	–	8.8	5.4	15.6	14.9	21.5	3.4	8.4	7–11	49	12	5
Alcohols	–	–	–	–	–	–	–	–	–	–	–	–	–	55
Alkyl esters	9–11	+	+	15.8	16.1	25.6[d]	0.2	2.3	3.1	0.6	3–5	4	1	13
Aldehydes	–	–	–	–	–	–	–	0.5	–	–	8–9	–	–	–
Sterols	2–3	+	+	–	–	1.8	1.8	3.6	3.2	3.3	<1	18	1	–
Sterol esters	–	–	–	+	+	+	7.3	0.7	14.2	8.0	–	–	–	–
Glycerides	–	–	+	–	–	–	2.4	0.3	6.7	0.7	–	7	78	–
Acidic resins	12–14	–	–	–	–	–	–	–	–	–	–	–	–	–
Phospholipids	–	–	+	–	–	–	–	0.8	–	–	–	–	–	2

[a] After Hackman (1971b).

[b] +, present but amount not stated.

[c] Reference: (1) Baker et al. (1960); (2) Bergmann (1938); (3) Amin (1960); (4) Shikata (1960); (5) Goodrich, (1970); (6) Gilby and McKellar (1970); (7) Beatty and Gilby (1969); (8) Gilby and Cox (1963); (9) Armold et al. (1969); (10) Bursell and Clements (1967).

[d] Would include sterol esters and glycerides.

australasiae and *P. fuliginosa* (Jackson, 1970). The hydrocarbon fraction from these three *Periplaneta* species is quite different from that of *P. americana* (Table III) in which two-thirds of the hydrocarbon fraction (and half the total lipid) is *cis,cis*-6,9-heptacosadiene. This unsaturated compound is also the major hydrocarbon in *P. americana* oothecal lipid (Atkinson and Gilby, 1970). The fatty acids in the cuticular lipids from *Tenebrio molitor, Periplaneta americana, Schistocerca gregaria* and *Trogoderma granarium* are either saturated or unsaturated with chain lengths of C_{14}, C_{16}, C_{17}, or C_{18} (Cohen *et al.*, 1971; Thompson and Barlow, 1970). The cocoon lipid of *Bombyx mori* resembles closely that of the insect (Komatsu, 1969).

Blomquist *et al.* (1972) report the presence of monohydroxy secondary alcohols in the cuticular wax esters from grasshoppers and they suggest these alcohols are derived from the diet. The surface lipid from the pea aphid is a mixture of compounds similar to those in Table III (Stransky *et al.*, 1973). Jackson (1972) has extended his study on the hydrocarbons in cuticular lipids to include those from the cockroach *Periplaneta japonica* in which, unlike the other cockroaches, the major hydrocarbon is *cis*-9-nonacosene. The males, like those of *P. australasiae* and *P. fuliginosa*, have appreciable quantities of a hydrocarbon (*cis*-9-heptacosene) not found on the females.

As can be seen in Table III, hydrocarbons are the major component in the cuticular lipids from *Anabrus, Periplaneta*, and *Lucilia*. Less than 10% of the total lipid of both *Lucilia* puparia and pupal and molting membranes is bound lipid (extracted by solvents containing methanol). Free and bound lipids from both types of cuticle are similar in composition but the puparia have less lipid. Of the six species reported on, free alcohols were detected only in *Tenebrio* cuticles and 55% of the total lipid was a C_{27} diol. The adult stage of the stonefly *Pteronarcys* has nearly twice as much cuticular lipid as has the naiad stage and the major components in the two stages are free fatty acids and triglycerides, respectively. The lipids differ markedly in composition, the naiad cuticular lipid having the lower melting point, and the differences in composition are probably a result of the different needs for water conservation in the aquatic and terrestrial stages.

The cuticular lipid of *P. americana* is a soft grease, but when isolated from the cuticle it hardens on aging. Beament (1955) suggested that the hardening was due to loss of volatile paraffins and alcohols (C_8–C_{12}), but they have not been identified in the lipid and, moreover, such alcohols are toxic to the animal (Gilby, 1962; Gilby and Cox, 1963). The high proportion of nonvolatile, liquid components accounts for the mobility of the grease, the hardening with time is brought about by autoxidation, and it is suggested that polyhydric phenols (which are on the cuticle but not extracted with the lipid) act as antioxidants (Atkinson and Gilby, 1970).

There is evidence that the innermost layers of lipid are responsible for the low permeability of the cockroach cuticle to water (Beament, 1960), but there is no evidence to establish the presence of an oriented monolayer of lipid molecules. The composition of the cockroach cuticular lipid does not support a theory of impermeability to water based on a close-packed monolayer; an oriented monolayer, if formed, would be of the expanded interfacial type and unlikely to act as an efficient barrier to the passage of water. Further work is needed to explain the extraordinarily low permeability of the insect cuticle, a problem which has been discussed by Gilby and Cox (1963). The presence of aldehydes and acidic resins in the lipids suggests the possibility of impermeable, polymeric films, but the cuticles of many insects have a cement layer and resinlike components may be derived from this layer or may be precursors of it.

It has been reported that the rate of water loss through insect cuticles increases greatly at a certain temperature, known as the transition temperature for the insect; the increase was said to be due to disruption of the inner layer of lipid (Beament, 1958, 1959; Oloffs and Scudder, 1966; Ramsay, 1935; Wigglesworth, 1945). Other workers (Edney, 1957; Holdgate and Seal, 1956; Mead-Briggs, 1956) have doubted the existence of transition temperatures and consider that temperature and water loss are related exponentially, the supposed discontinuities being artifacts brought about by the choice of coordinates. More recent work on several species of insect confirms the fact that temperature and water loss are related exponentially (A. R. Gilby, unpublished).

There have been many reports on the composition of beeswax and the lipids secreted by some scale insects, but it is not known whether these lipids have any relation to cuticular lipids. The lipids consist principally of long-chain aliphatic hydrocarbons, acids, alcohols, and esters and, in a few instances, contain short-chain acids and alcohols. Sesterpene acids and alcohols may also be present. Reviews of this work have been given by Tamaki (1969) and Tulloch (1970, 1971) and additional references can be obtained from Hashimoto et al. (1971a,b), Rios and Quijano (1969), and Schildknecht and Vetter (1963). A fine, white, crystalline powder accumulates superficially on the cuticle of larvae of Samia cynthia ricini. The powder may be brushed off and it consists of alcohols (93%) together with hydrocarbons, esters, acids, and sterols. Of the alcohols, 99.4% is n-triacontanol and 0.6% n-octacosanol, but their biological function is unknown (Bowers and Thompson, 1965). Hydrocarbons and other compounds of the types discussed above have been identified in the lipids extracted from homogenates of whole insects. These lipids include those from the cuticle but none of the compounds can be unequivocally associated with the cuticle. Reviews on lipid metabolism in insects (e.g., Gilbert, 1967; Gilbert and O'Connor, 1970; Gilby, 1965) give information on this work.

Koidsumi (1957) found that larvae of *Bombyx mori* and of *Chilo simplex* became highly susceptible to attack by fungi when their cuticular lipids were removed. The lipids were found to have antifungal activity *in vitro* and the activity was associated with the acidic components.

G. Pigments

There is a considerable body of literature describing pigments in insects but often the location of the pigment is not given. Pigment granules occur in the hypodermal cells of many insects and in some insects the granules can be moved under hormonal influence to produce transient color changes. Colored hemolymphs may also be visible through the cuticle. The color of an insect may be due to the presence of more than one class of pigment and they may be located in different tissues, e.g., the green color of pupae of the cabbage white butterfly, *Pieris rapae crucivora,* is due to the combination of a yellow cuticle (carotenoid pigment) and an underlying bluish-green epidermis (tetrapyrrole pigment) (Ohtaki and Ohnishi, 1967). Several general reviews on pigments in insects have appeared recently (Fuzeau-Braesch, 1972; Goodwin, 1971; Vuillaume, 1969) which should be consulted for the references to and for further details on some of the work discussed below. Rowell (1971) has reviewed work on color variation in Acridoid grasshoppers. Optical effects, brought about by the structure of the cuticle, are responsible for the color of some insects and this topic is discussed in Section II,C,4.

a. Sclerotin and Melanin. The cuticles of some insects possess a black melanic pattern as well as the normal amber color of the hard cuticle (e.g., see Fogal and Fraenkel, 1969b; Hackman, 1967). The chemistry of sclerotin, the colored pigment generally formed when cuticles harden, is discussed in Section IV, but little is known about the structure of sclerotin(s) apart from their quinonoid and phenolic character. Melanin also has quinonoid and phenolic properties and an interesting problem in the field of cuticular pigments is that of the relation between melanin and sclerotin.

Melanin is a somewhat ill-defined term and many pigments which are dark in color and insoluble in most solvents have been referred to as melanins without regard to their chemistry. Sclerotin was originally described as a melanin, but spectroscopically they are different (Hackman, 1953c). Melanins are high molecular weight compounds formed by oxidation of polyhydric phenols, e.g., dopa, but a complete structural formula cannot as yet be written for a melanin. Nicolaus (1968) has reviewed the chemistry and formation of melanins and he classifies such pigments as eumelanins (black, indole-type), phaeomelanins (yellow, red, or brown), and allomelanins (black, catechol-type). Methods for the microdetection of melanins

are given by Hackman and Goldberg (1971b). Mason (1967) regards melanin as a homopolymer, a polymer built up from only one type of monomer,
but Nicolaus (1968; Nicolaus *et al.*, 1967) regards melanin as a polymer
built up from a number of different but related monomeric units, perhaps
linked in different ways. The latter type of structure is favored by Hempel
(1966; Nicolaus *et al.*, 1967) and Swan (Chapman *et al.*, 1970; King *et
al.*, 1970; Swan and Waggot, 1970). Most melanins of animal origin, including insects, are eumelanins (e.g., Hackman, 1967; Nicolaus, 1968), but melanins from *Calliphora augur* puparia and *Xylotrupes gideon* cuticle give
both indole and catechol degradation products and have nitrogen contents
intermediate between those of eumelanins and allomelanins. Thus these two
insect melanins are either mixtures of indole- and catechol-type melanins
or are polymers containing both indole and catechol units (Hackman and
Goldberg, 1971a,b).

Melanin is often bound to protein to form melanoproteins. In fact, it
is difficult to imagine that there could exist in a biological system quinones
and proteins without some reaction taking place between them. Methods
used to isolate melanins generally destroy the protein but cysteine survives
and remains attached to the melanin nucleus. The link is presumably covalent and of the thioether type. There is some evidence to suggest that
protein may also be linked to melanin by peptide bonds (see Nicolaus, 1968).

Although melanin formation is sometimes included in the general process
of sclerotization, the evidence suggests that sclerotization and melanization
are two independent processes. In a number of insects it has been shown
that the two processes are separated in time and, depending on the species,
either hardening of the cuticle or the appearance of black pigment may
occur first, e.g., *Calliphora* flies (Cottrell, 1962a), *Ephestia kühniella*
(Blaich, 1969), *Schistocerca gregaria* (Malek, 1957), and various Diptera
species (McLintock, 1964). In other insects melanin formation can be prevented while cuticular hardening continues normally, e.g., *Calliphora* larve
(Dennell, 1958c) and *Gryllus bimaculatus* (Fuzeau-Braesch, 1959a,b). A
study of enzyme activity and the utilization of radioactive phenolic compounds has shown the same process of cuticular tanning to take place in
both the melanic and nonmelanic strains of *Schistocerca gregaria* (Karlson
and Schlossberger-Raecke, 1962) and of *Lucilia cuprina* (Hackman and
Goldberg, 1968). In these two species melanization appears to proceed by
a discrete metabolic pathway.

Melanin would appear to have a protective function in insects, e.g., adaptive coloration and heat control. Deposition of melanin takes place where
a cuticle is damaged (mechanically or by a parasite) (e.g., Aoki and Yanase,
1970; Nappi and Stoffolano, 1970; Sannasi, 1969), implanted tissues frequently darken, and there are many references to dark-pigmented encapsula-

tions around parasites or other foreign bodies. Taylor (1969) has suggested that the polyhydric phenol–phenol oxidase system which produces melanin is directly involved in invertebrate immunity. Quinones are powerful oxidizing agents and they rapidly destroy enzymes, so there could be considerable advantage attached to the possession of a system which, in the event of damage, produces quinones able to inactivate rapidly enzymes and other proteins and to act as antibiotics. A melanin-producing system may therefore be one of the most primitive, nonspecific defense systems of all living forms.

b. Carotenoids (Fig. 2A). Carotenoids occur in many different species of insects, usually as conjugates with proteins .The colors of most carotenoids are yellow, orange, or red and these are the colors they give to insects. Together with a blue tetrapyrrole pigment (see below), carotenoids produce the green color of many insects. Carotenoids have been isolated from the "body wall" or integument of Orthoptera, Coleoptera, Lepidoptera, and Hemiptera and from the wings of Orthoptera; the compounds isolated were β-carotene (red), lutein (yellow), or astaxanthin (purple). Carotenoids are present in many other insect tissues (for references see the general reviews quoted above). Cheeseman *et al.* (1967) have reported on carotenoproteins in invertebrates.

c. Tetrapyrrole Pigments (Fig. 2B). Tetrapyrrole pigments (bile pigments) occur in the hemolymph of many insects, generally as protein conjugates and they are blue. Two pigments belonging to this class have been reported from integuments; biliverdin $1X_\alpha$ in Orthoptera and biliverdin $1X_\gamma$ in Lepidoptera. Pterobilin, previously thought to be mesobiliverdin, is now reported to be biliverdin $1X_\gamma$. The reported occurrence of biliverdin $1X_\alpha$ and $1X_\gamma$ if of considerable interest since they are generally associated with the degradation products of hemoglobin and so, *a priori,* would not have been expected to occur in these insects. Rüdiger (1970) has reviewed the occurrence of tetrapyrrole pigments in invertebrates. Some brown and yellow pigments of Orthoptera are said to be formed by oxidation of biliverdin (Passama-Vuillaume, 1964, 1966).

d. Pteridines (Fig. 2C). Pterins (i.e., pigments with a pteridine nucleus) are a class of fluorescent pigments which was first discovered in the wings of butterflies, but are now known to occur in other insect tissues, e.g., eyes. Xanthopterin (yellow), leucopterin (white), isoxanthopterin (white), erythropterin (red), chrysopterin (yellow), and sepiapterin (yellow) have been identified in the wings or integuments of Lepidoptera, Hemiptera, Hymenoptera, and Neuroptera. Of the pterins, xanthopterin is the most widespread in nature and its fluorescence and absorption spectrum make it readily identifiable in small amounts. Some of the white, yellow, orange, and red colors

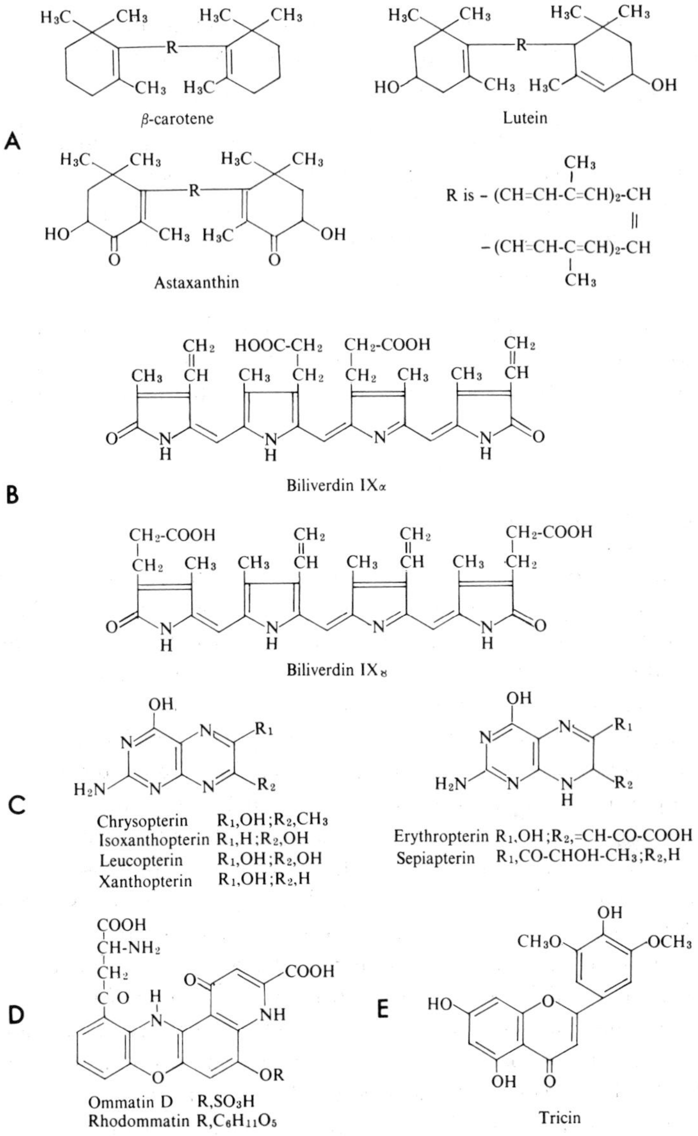

A

β-carotene

Lutein

Astaxanthin

$$R \text{ is } - (CH=CH-C=CH)_2-CH$$
$$\text{with } CH_3 \text{ substituents}$$
$$- (CH=CH-C=CH)_2-CH$$

B

Biliverdin IXα

Biliverdin IXᵦ

C

Chrysopterin R_1,OH;R_2,CH$_3$
Isoxanthopterin R_1,H;R_2,OH
Leucopterin R_1,OH;R_2,OH
Xanthopterin R_1,OH;R_2,H

Erythropterin R_1,OH;R_2,=CH-CO-COOH
Sepiapterin R_1,CO-CHOH-CH$_3$;R_2,H

D

Ommatin D   R,SO$_3$H
Rhodommatin R,C$_6$H$_{11}$O$_5$

E

Tricin

Fig. 2. Cuticular pigments.

in insects are due to differential concentrations of the above pterins. In contrast to melanins which are usually embedded in the cuticle, pterins are located in the epidermal cells or in the interlamellar spaces in wings, scales, and hairs. Ziegler and Harmsen (1969) have reviewed the biology of pteridines in insects and some additional references are Battaglini and Parisi (1968), Baust (1967), Berthold and Henze (1971), Collins and Kalnins (1970), Parisi and Battaglini (1970), and Pfeiler (1968).

e. Ommochromes (Fig. 2D). These pigments were first isolated from the ommatidia of insect eyes and red and brown ommochromes are present in the wings of Nymphalidae (Lepidoptera) and in the epidermis of some Orthoptera and Lepidoptera. Rhodommatin, xanthommatin, and ommatin D are the pigments most commonly found in these tissues.

f. Flavones (Fig. 2E). The wings of the marble white butterfly *Melanargia galathea* and the small heath butterfly *Coenonympha pamphilis* contain the flavone tricin both free and as a glycoside (Morris and Thomson, 1963, 1964).

H. Inorganic Constituents

The amount of ash formed when insect cuticles are incinerated is generally low (less than 1% of the dry weight, but in puparia and cast skins it may be higher) (Table I). Inorganic materials, therefore, form only a very small part of the cuticle. Richards (1956) has shown that in the larval cuticle of *Sarcophaga bullata* there is a higher concentration of ash components in the epicuticle than in the procuticle. So puparia and cast skins may be expected to give a higher percentage of ash than would the intact cuticle. Both puparia and cast skins contain the remains of secretions and care must be taken to remove these contaminants (see also comments below).

Richards (1956) has given quantitative values for the 25-elements in the ash of *S. bullata* larval cuticles. Na, K, Mg, Ca, Fe, Si, Cl, SO_4, CO_3, and PO_4 account for almost all the ash, Mg and K are the predominant cations, Fe is concentrated in the epicuticle, and the cations are balanced by the anions. Cuticles and cast skins from other species of insect also contain a large number of ions (Gilby and McKellar, 1970; Hackman and Goldberg, 1958, 1968; Sapag-Hagar *et al.*, 1961b; Stamm-Menendez and Dean-Guelbenzu, 1961; Stamm-Menendez and Fernandez, 1957).

Insect cuticles are not calcified as are crustacean cuticles but some specialized forms of calcification are known. Weaver (1958), in a survey of 147 species of immature insects (11 orders), has reported the presence of calcium oxalate crystals in the exo- or procuticle of thirteen species (twelve Lepidoptera and one Hymenoptera). Certain aquatic dipterous larvae (living in calcium-rich waters) belonging to the families Psychodidae (e.g., *Pericoma* spp.) and Stratiomyidae (e.g., *Sargus* spp.) have calcereous deposits in the form of warts or nodules on the outer surface of the cuticle (see Richards,

1951). The inner surfaces of the cast skins of some insects are covered with a deposit of calcium salts. These deposits are not part of the cuticle proper but are formed by materials from the malpighian tubules being secreted into the space between the old and new cuticles prior to ecdysis. Thus an inner shell is formed within the puparium of the fly *Acidia heraclei* (Keilin, 1921). In larvae of the fly *Rhagoletis cerasi* the calcium salts are excreted before separation of the old cuticle and they are deposited as granules in the procuticle of the old cuticle. With the transformation of the larval cuticle into the puparium the granules remain within the cuticular structure and may account for more than half of its weight (Wiesmann, 1938). Calcification appears to have replaced sclerotization in the puparium of *Musca autumnalis* (Fraenkel and Hsiao, 1967; Bodnaryk, 1972).

Calcium oxalate occurs in the oothecae of many, but not all, species of cockroach (Brunet, 1952; Clark, 1958; Ito, 1924; Roth, 1968; Stay *et al.*, 1960). Oothecae of *Periplaneta americana* contain 6.5% calcium (the only major cation present), but only 35% of the calcium was present as the oxalate (Hackman and Goldberg, 1960). The oothecae of *Blattella germanica* gave similar results. Parker and Rudall (1955) reported crystals of calcium citrate in oothecae of praying mantids, but Hackman and Goldberg (1960) failed to detect citrate in oothecae of the mantid *Orthodera ministralis,* the oothecae, in composition, resembling closely that of cockroaches. Kato and Kubomura (1956) have reported the identification (histochemically) of calcium carbonate in egg packets of Japanese mantids. Calcium oxalate, as a crystalline powder, covers the cocoon of the Lepidopteran *Malacosoma neustria testacea* (Ohnishi *et al.*, 1968).

The biological significance of inorganic ions in insect cuticles is not known, but Kikkawa *et al.* (1954, 1955) have suggested that certain metals (Fe, Cu, Co, Ni, Ti, Mo) are associated with melanoproteins and other natural pigments. Hackman and Goldberg (1960) and Kenchington and Flower (1969) have discussed the role calcium may play in the oothecae of cockroaches and mantids.

I. OTHER COMPONENTS

1. Sugars

Sugars other than glucosamine or N-acetyl-D-glucosamine (from chitin) have been identified in hydrolyzates of cuticles; e.g., in *Periplaneta americana* (arabinose, galactose, glucose, mannose, and xylose, together with ketoses and pentoses (Lipke *et al.*, 1965a; Lipke and Geoghegan, 1971a) and in *Lucilia cuprina* larvae (galactosamine and mannose (Gilby and McKellar, 1970). The role of these sugars is not known.

2. *Mucopolysaccharides*

On histochemical evidence acidic mucopolysaccharides have been reported to be present in some cuticles (e.g., see Martoja and Cantacuzene, 1969; Sannasi, 1968a) but their significance is obscure. In crustacea mucopolysaccharides appear to be involved in calcification of tissues.

3. *Insoluble Membranes*

The epicuticle of some insects contains a membrane which is very resistant to chemical action; in dipterous larvae and puparia this membrane has considerable mechanical strength. For example, acidic hydrolysis of blow fly larval cuticles (or puparial cases) leaves insoluble membranes which are ghosts of the original cuticle (Hackman, 1967; Hackman and Goldberg, 1968; Hackman and Saxena, 1964) and they are insoluble in boiling aqueous alkali and in all common organic solvents. The membranes are nitrogen free and, as judged by the infrared spectrum, appear to be built up from long chains of CH_2 groups (R. H. Hackman, unpublished). They resemble membranes isolated by less drastic means from blow fly larval cuticles and from cockroach cuticles by Dennell and Malek (1955a) and said to contain mostly "paraffin structures." A similarly resistant layer occurs in *Calpodes ethlius* (Filshie, 1970a). Filshie (1970a,b) suggests that the resistant membrane is composed of the outer epicuticle together with the cuticulin layer, the latter as defined by Locke (1966). Hot acid treatment removes the cuticulin leaving only the outer epicuticle.

IV. Hardening of the Cuticle

The cuticle of the newly ecdysed insect is soft and pale in color, but over a period of a few hours it darkens and becomes harder, a process known as sclerotization or tanning. Tanning occurs first in the outer layers of the cuticle and proceeds inwards. The water content of the cuticle decreases (i.e., it becomes less hydrophilic), the protein and chitin chains become more highly oriented and close packed (Fraenkel and Rudall, 1940, 1947; Section III,C) and inorganic components are extruded from the exocuticle (Richards, 1956). Similar changes occur during formation of the cockroach ootheca and fly puparium and they have been studied extensively as examples of sclerotization. However, amino acid analyses of their protein components differ markedly from those of cuticles which harden after a molt (Hackman and Goldberg, 1971a). Some cuticles are harder and darker than others, while some have in addition black melanic patterns (Section III,G). Hardness of a cuticle is not related to its chitin content (Campbell, 1929), but occasionally the term "heavily chitinized" is still used to describe a hard cuticle.

Sclerotization is considered to involve reaction between cuticular proteins and o-quinones, derived from o-dihydric phenols. o-Benzoquinones react with —NH$_2$ and —SH groups (Hackman and Todd, 1953; Mason and Peterson, 1965) and so in the cuticular protein they would react with a terminal amino group, a sulfhydryl group or the ϵ-amino group of lysine to form, in the first instance, an N-catechol protein which, in the presence of excess quinone, would be oxidized to the N-quinonoid protein. The quinonoid protein would then react with another —NH$_2$ or —SH group to form a disubstituted derivative, first as the colorless catechol form which would be oxidized by excess quinone to the colored quinonoid form. Cross-linked structures are formed in which the N of the —NH$_2$ groups and the S of the —SH groups in the protein chains become attached directly by covalent bonds to the aromatic nuclei of the quinones, the protein chains being linked end to end and also at intermediate points. Since reaction takes place with hydrophilic groups, the cuticle becomes less hydrophilic. Sclerotized cuticles are highly resistant to enzymolysis (Lipke and Geoghegan, 1971a) and this may indicate extensive hydrogen bonding between components. Chitin is bound to protein (Section III,C) and so would form part of the network. Dennell (1958c), Dennell and Malek (1955b), and Malek (1952), on histochemical evidence only, suggested that sclerotization involves a sterol–protein complex. Simple o-benzoquinones may not be efficient cross-linking agents. Polymerization and condensation would form larger molecules with many reactive positions: those able to bridge longer distances than can o-benzoquinones and those which would impart a deeper color to the cuticle. It is also possible that polymerized quinones or phenols are simply embedded in the cross-linked protein structure or are complexed to the protein in a manner analogous to the tanning of proteins with vegetable tannins. With time, covalent bonds would be formed and this could explain the slow continuous insolubilization of cuticular proteins observed by Andersen (1971b). The minimum number of cross-links needed to stabilize the cuticle is not known but, in view of what is known about other structural proteins, there is no reason to believe that it is great.

Formation of the cockroach ootheca is the best understood example of sclerotization, although it appears to be a special case. The ootheca is formed from the secretions of the left and right colleterial glands, the left gland secreting the structural protein, a phenoloxidase and a phenolic glucoside, while the right gland secretes a β-glucosidase. When the secretions mix, the glucosidase liberates the free phenol which is oxidized by the phenoloxidase to an o-quinone which cross-links the structural protein and so forming the hard, dark ootheca. Two phenolic glucosides have been identified in the left gland, the 4-o-β-glucosides of 3,4-dihydroxybenzoic acid and 3,4-dihy-droxybenzyl alcohol (Brunet and Kent, 1955; Kent and Brunet, 1959; Pau and Acheson, 1968; Pryor, 1940; Stay and Roth, 1962). The phenol oxidase is of the laccase type; it oxidizes a number of o- and p-dihydric phenols with the notable exceptions of dopa and dopamine (Whitehead et al., 1960, 1965). It has been suggested that calcium ions and calcium oxalate (which are in the ootheca) help to organize and stabilize the structural protein (Kenchington and Flower, 1969; Sundara Rajulu and Renganathan, 1966).

The puparium of flies is the sclerotized last larval cuticle (however, see Section, III,H). The tyrosine content of the larval hemolymph of blow flies reaches a maximum just before puparium formation and then decreases rapidly as the puparium hardens and darkens. If labeled tyrosine, dopa, dopamine, or N-acetyldopamine are injected into the larvae, most of the radioactivity becomes located in the puparia, thus showing that tyrosine or its derivatives (Section V,C) are incorporated into the sclerotized structure. Karlson has proposed that the quinone precursor is N-acetyldopamine, that the phenol oxidase originates from a proenzyme in the hemolymph with the activator in the cuticle, that the activator is a protease, and that the protein which reacts with the quinone may be the phenol oxidase itself (Karlson and Sekeris, 1964, 1966; see also Dennell, 1947; Fraenkel and Rudall, 1947; Hackman, 1956; Hackman and Goldberg, 1967, 1968; Hackman and Saxena, 1964; Pant and Lal, 1970). In *Calliphora erythrocephala* the activator of the proenzyme is in the outermost layers of the cuticle (Price and Hughes, 1971) and this would explain why sclerotization occurs first in the outer layers. A similar but not identical sclerotization pathway to that for blowflies (one involving tyrosine o-phosphate) has been suggested for *Drosophila* (Hodgetts and Konopka, 1973).

It has generally been assumed that the phenol oxidase needed to form quinones arises from a proenzyme in the hemolymph, the proenzyme moving into the cuticle shortly before sclerotization commences, but this view was questioned by Hackman and Goldberg (1967). Recent work (Locke and Krishnan, 1971; Mitchell et al., 1971) suggests that hemolymph enzymes may not be involved in sclerotization. A phenol oxidase appears to be secreted as part of the cuticle and only the phenolic substrates diffuse out-ward through the cuticle prior to sclerotization. If the phenol oxidase forms part of the cuticle as it is laid down then, from the work of Hackman and Goldberg (1967) on fly larvae, the enzyme must initially be in an inactive form.

o-Diphenoloxidases occur in all hemolymphs, in the plasma, in hemocytes, or in both and usually as proenzymes. Cuticular extracts contain an activator for the proenzyme which in some species is said to be a protein and in others a lipid, but the proenzyme can often be transformed into an active enzyme by heat, a change in pH, surface-active agents, or heavy metals (Ashida, 1971; Ashida and Ohnishi, 1967; Dohke, 1973; Hackman and Goldberg, 1967, 1968; Hoffmann *et al.*, 1970; Mills *et al.*, 1968; Ohnishi, 1953, 1954). The cuticular activator in *Calliphora* and *Drosophila* spp. is not species specific (Sin and Thomson, 1971). In *Drosophila melanogaster* (Paradi and Csukas-Szatloczky, 1969) and *Lucilia cuprina* (Hackman and Goldberg, 1968) the larval hemolymph contains either more than one prodiphenol oxidase or a proenzyme which with different activators gives enzymes with different specificities, while Mitchell (1966) has shown that the proenzymes in *Drosophila* larvae appear to be used for melanization in the adult. These observations may explain why hemolymph phenol oxidases are less specific than are cuticular phenol oxidases and, if used for sclerotization, why activity is still present in the hemolymph during and after sclerotin formation. Sin and Thomson (1971) have separated the hemolymph prophenol oxidases from *Calliphora* spp. into components similar to those isolated from homogenates of whole *Drosophila* (Mitchell and Weber, 1965; Mitchell *et al.*, 1967; Peeples *et al.*, 1969).

Cuticular sclerotization occurring after ecdysis, in the few species which have been investigated, appears to be similar to that outlined for the puparium of flies. Work has been confined principally to *Periplaneta* (for references, see Mills and Whitehead, 1970) but *N*-acetyldopamine is also considered to be the quinone precursor in *D. melanogaster, Schistocerca gregaria, Tenebrio molitor,* and adult *C. erythrocephela* (Karlson and Herrlich, 1965; Sekeris and Herrlich, 1966). When dewaxed abdomens of *S. gregaria* albino nymphs are dipped into solutions of catechol or tyrosine the entire cuticle takes on the normal black pattern; 3,4-dihydroxybenzoic acid and dopa only cause an intensification in the amber color of the cuticle (Jones and Sinclair, 1958). The absence of a differentiation along chemical lines (phenols containing or not containing nitrogen) indicates that sclerotization may be more specific than is generally thought.

Prior to cuticular hardening in adult *Calliphora* the cuticle becomes plastic, and Cottrell (1962b) suggests that plasticization may be a stage in the process of sclerotization. Many different labeled compounds derivable from tyrosine have been injected into insects and the percentage of the radioactivity incorporated into the cuticle measured. Tyrosine, dopa, dopamine, and *N*-acetyldopamine always give high rates of incorporation, very much higher than tyramine and its derivatives or *N*-acetylarterenol or *N*-acetylarterenone. Koeppe and Mills (1970) suggest that arterenol (noradrenaline) may be a sclerotizing substance in the cockroach cuticle, but Andersen (1971a) has shown in the locust cuticle that although this and related compounds are incorporated in low yields, it is unlikely that they are natural tanning agents since they do not give rise to the usual degradation products. Symbionts are essential for normal sclerotization to take place in some insects, e.g., *Reticulitermes assamensis* (Sannasi, 1968b), *Sitophilus granarius* (Schneider, 1954), *Blatella germanica,* and *Rhodnius prolixis* (Richards and Brooks, 1958) and they may supply essential aromatic compounds. Viles and Mills (1969)have speculated that in the cockroach dihydric phenols are stored as PAS-positive granules in the epidermal cells. Tomino (1963) has isolated 3,4-dihydroxyphenylacetic acid and its glucoside from female *Locusta migratoria* and suggested its participation in sclerotization.

Although radioactivity from labeled compounds is incorporated into the sclerotized cuticle nothing is known about the manner in which it is incorporated. The evidence for some of the suggested steps in the chemistry of cuticular sclerotization is still largely circumstantial and nowhere in the field of chemistry of the cuticle has speculation been more popular.

A theory of sclerotization based on *p*-quinones proposed by Dennell (1958a,b,c) in *Calliphora vomitoria* and extended to *Periplaneta americana* and *Popilius disjunctus* by Kennaugh (1958) has now little or no evidence to support it. Current views on sclerotization, as discussed in this section, have provided answers to the objections raised against *o*-quinone tanning. The formation of quinone precursors by a nonenzymic process, as suggested by Dennell, could not be controlled so as to synchronize with the biological differentiation of the cuticle. Other workers have been unable to demonstrate the presence of a nonspecific hydroxylation system in *T. molitor* or *C. erythrocephala* (Aerts *et al.,* 1962; Vercauteren *et al.,* 1962) or to isolate *p*-aminophenols (Hackman and Goldberg, 1963).

Pau *et al.* (1971) suggest, but without experimental evidence, that sclerotization in the ootheca of *P. americana* involves coupling of the protein tyrosyl residues with the oxidation products of *o*-dihydric phenols to give biphenyl links. The total amino acid composition they give for the proteins in the left colleterial gland is very similar to that given by Hackman and Goldberg (1971a) for the hardened ootheca except for lysine, the concentration of which is much higher in the gland proteins. In an earlier paper Hackman and Goldberg (1963) had shown that on hydrolysis the white ootheca yielded 2.7 times as much lysine as did the fully hardened ootheca. Thus lysine may take part in sclerotization. Pau *et al.* also suggest, again without evidence, that one protein fraction, amounting to 15% of the total, is the phenol oxidase. This concentration is very high for an enzyme. The hardened ootheca must contain this protein fraction to give it the required proline content and so it forms part of the structural protein. By polyacrylamide gel electrophoresis they separated the water-soluble proteins from the gland into three fractions. Previously Adiyodi and Nayar (1966) showed four protein fractions to be present while Adiyodi (1968) separated the water-soluble proteins from the left colleterial gland of another cockroach

(*Nauphoeta cinerea*) into a maximum of eight fractions, the number varying according to the reproductive state of the female.

In 1945 Hurst suggested that cuticular hardening may involve peroxidases and Pennell and Tsuyuki (1965) have described an oxidase–peroxidase system in a number of insects. Recently Locke (1969b) has proposed that a peroxidase takes part in the hardening of the proleg spines of *Calpodes ethlius;* however, no attempt was made to distinguish between true peroxidases, atypical peroxidases, artificial peroxidases, and model peroxidases (see Saunders *et al.,* 1964). At neutral pH and at low concentrations of peroxide, catalase acts as a peroxidase and some quinones, e.g., as produced from tyrosine by phenol oxidases, act as very efficient catalases. Peroxidases catalyze the oxidation of polyhydric phenols and in some organisms may mediate the oxidation of phenols to melanin (see Bayse and Morrison, 1971). Peroxidases may take part in the formation of resilin (Section V,A).

Andersen (1970, 1971a) and Andersen and Barrett (1971) have isolated from acidic hydrolyzates of cuticles a number of catechol derivatives and they propose that N-acetyldopamine can be involved in sclerotization via its sidechain and that quinone tanning and sidechain tanning can go on simultaneously in the same cuticle. Protein chains are said to be cross-linked through the first (or β) carbon atom in the sidechain, but on steric considerations this type of structure is unlikely to be formed readily and so would not be expected to make a worthwhile contribution to cross-linking. It has not been proposed how such a compound could be formed and the suggested stability of the compound to hydrolysis is not that to be expected from such a structure. Again Atkinson *et al.* (1973) have identified these catechol derivatives in water extracts of cuticles as well as in acidic hydrolysates. The catechol derivatives could arise from tyrosine and, if associated with protein, they may be complexed to it and not covalently bound in the manner assumed by Andersen. Thus acids would give a more rapid extraction than does water. Their function in the cuticle is unknown but they may act as antioxidants. Andersen (1972a) has extended his ideas on sclerotization to include reaction of N-acetyldopamine with both lysine and tyrosine residues in the cuticular proteins and with N-acetyldopamine already bound to the protein; the crosslinks would be of a mixed nature.

β-Alanine occurs only in hydrolyzates of sclerotized cuticles (Section III,A), the amount present is directly related to the degree of sclerotization and it is in an N-terminal position. Its absence in unhardened cuticles suggests involvement in sclerotization, but how it participates is not known. For the metabolism of β-alanine in *Musca* see Ross and Monroe (1972a,b). In some species much of the β-alanine is derived from a peptide β-alanyl-L-tyrosine in the hemolymph (Bodnaryk, 1971a,b). Apart from this work few attempts have been made to relate amino acid composition with cuticular hardness and in some only protein fractions or extracted cuticles have been used (see Hackman and Goldberg, 1971a). Hard cuticles generally contain more of the polar amino acids than do soft cuticles (Section III,A). The puparial protein of *L. cuprina* contains, besides β-alanine, very much more arginine, serine, and cystine and substantially more histidine but less alanine than the larval cuticle from which it is formed, but the significance of these differences is not known (Hackman and Goldberg, 1971a). In *Anabrus simplex* (De Haas *et al.*, 1957) and *S. gregaria* (Andersen, 1971b) cuticular hardness may be associated with a high alanine content. Halogenated tyrosines have been identified in scleroproteins, and Andersen (1972b) has identified 3-chlorotyrosine in enzymic hydrolyzates of insect cuticular proteins.

Cuticular proteins contain low concentrations of one or more of the sulfur-containing amino acids, cysteine, cystine, or methionine (Section III,A). In *L. cuprina* larvae cystine occurs in the outer layers of the cuticle and during puparium formation its concentration is more than doubled, which suggests that it plays a part in the hardening process, perhaps that of protein stabilization before quinone tanning occurs (Hackman, 1971a). The sulfur content of cuticles is not accounted for fully as amino acids or as melanin-bound sulfur and so sulfur may be combined with quinones. Sclerotization may include reaction of quinones with both —NH₂ and —SH groups. Disulfide bonds are present in the cuticles of the collembolan *Sminthurus punctatus* and the thysanuran *Machilis variabilis* and hardness is said to be due to disulfide bonding (Krishnan, 1969; Krishnan and Sundara Rajulu, 1964). Tanning involving sulfur has been reported in other groups of arthropods (see Hackman, 1971b).

Quinones are intensely colored and so an explanation is needed for hard cuticles with little or no color. Catechol–protein complexes are colorless (see above), so hard, colorless cuticles could be formed if a large excess of quinone is avoided. Should a process of autotanning occur in which tyrosyl groups in the protein are the quinone precursors (see Brown, 1949, 1950; Hackman, 1953c) then colorless structures cross-linked with catechols would result because an excess of quinone would not be available. Because of steric hindrance the formation of many cross-links by such a process would appear to be unlikely but there could possibly be enough to stabilize the protein. Tyrosyl residues in proteins are oxidized by phenol oxidases, the nature of the reaction depending on whether the tyrosyl residue is N-terminal, C-terminal or located within the chain (e.g., see Cory *et al.*, 1962; Lissitzky and Rolland, 1962; Lissitzky *et al.*, 1962; Rolland and Lissitzky, 1962; Sizer, 1953; Yasunobu *et al.*, 1959). Cross-linking through the β-carbon atom of N-acetyldopamine has been suggested as an explanation for nearly colorless cuticles; this scheme has been commented on above.

The hormones β-ecdysone and bursicon take part, directly or indirectly, in sclerotization, and additional factors, as yet unidentified, also appear to be involved (see Delachambre, 1971; Edwards, 1966; Mills *et al.*, 1965; Price, 1970; Weir, 1970; Zdarek and Fraenkel, 1969). β-Ecdysone (20-hydroxyecdysone) is the prothoracic gland hormone; it is a steroid and is secreted after the gland has been activated by the brain (see Berkoff, 1969; Horn, 1971). Bursicon is a protein and is secreted by the brain.

Ecdysone appears to influence many of the steps in the process of sclerotization and this makes it difficult to know which are the primary and which the secondary targets. Ecdysone initiates sclerotization by producing *N*-acetyldopamine indirectly thrugh the stimulation of mRNA and by inducing formation of the enzyme dopa decarboxylase (Section V,C). The enzyme is considered to occur in epidermal cells (Sekeris and Karlson, 1966), but Whitehead (1969, 1970a) suggests it is in hemocytes adhering to the cells. Hemocytes synthesize dopa and dopamine from tyrosine, especially after bursicon is released, and ecdysone appears to control the development of some hemocytes (Crossley, 1964). The ecdysone titer in *Calliphora* larvae never reaches the level of exogenous hormone needed to induce puparium formation and Zdarek and Fraenkel (1970) have concluded that ecdysone acts by the accumulation of covert effects which finally lead to the overt response (puparium formation). Besides inducing molting, ecdysone induces formation of the phenol oxidase activator, and contraction of the larval cuticle, prior to puparium formation in flies, brings about apolysis (Beck and Shane, 1969; Madhavan and Schneiderman, 1968) and controls calcification in the purparium of *Musca autumnalis* (Berreur and Fraenkel, 1969; Fraenkel and Hsiao, 1965, 1967).

Bursicon controls hardening and darkening in insects by determining the time at which ecdysis occurs (Fraenkel and Hsiao, 1965; Mills, 1967; Vincent, 1971). It controls water loss and so may be the antidiuretic hormone in cockroaches (Mills and Whitehead, 1970). Bursicon affects deposition of endocuticle and metabolism of tyrosine and tyrosine *o*-phosphate (Fogal and Fraenkel, 1969a; Seligman *et al.*, 1969). The synthesis of dopamine from tyrosine via tyramine in cockroach hemocytes is initiated by bursicon (Whitehead, 1969).

The process of sclerotization is complex and the need for definitive and unequivocal biochemical and chemical evidence is patently obvious. At best there is only circumstantial evidence to support crucial steps in the reactions which are thought to take place. It would appear from the mass of data now available that a good description of the chemistry of cuticular hardening is well within our reach. However, the solution remains elusive.

V. Biosynthesis of Cuticular Components

Many attempts have been made to relate biochemical changes and morphological events in insects, e.g., changes which occur in the concentrations and types of amino acids, peptides, and proteins in the hemolymph of an insect during its life, or gain in chitin and loss in glycogen (see Wyatt, 1968). Unless radioactive markers are used, such observations neither establish a relationship nor give information on biosynthetic mechanisms. At the time of molting in an insect the inner layers of the old cuticle are degraded by enzymes in the molting fluid, the products of digestion are absorbed, and at least some are stored for use in the synthesis of the new cuticle. Fat-body sequesters protein and Locke and Collins (1966, 1968) suggest that some of the protein may come from the old cuticle. Information on biosynthetic pathways that lead to cuticular components is often fragmentary, but much more work has been published on metabolic processes in the insect body. In the latter case, even though only a small number of species have been studied, the results suggest that, in general, insects synthesize and degrade compounds by methods which are basically similar to those used by other organisms. Little is known about the action of hormones on cuticle secretion by the epidermal cells; the literature has been reviewed by Marks (1970).

A. Proteins

Autoradiographic studies show that hemolymph proteins and amino acids are utilized in the synthesis of new cuticle in *Calliphora, Galleria, Locusta,* and *Periplaneta,* but it is not known if the proteins are used directly or as their constituent amino acids (Fox and Mills, 1971; Kinnear *et al.,* 1971; Srivastava, 1971b; Tobe and Loughton, 1969). Resilin (Section III,A) appears to be secreted by specialized epidermal cells as a soluble protein which quickly becomes insoluble. The chains are cross-linked through di- and tri-tyrosine residues; cross-linking takes place outside the cell membrane and a peroxidase may be involved. Radiographic studies show that the cross-links are formerly from tyrosine. The rate of incorporation of tyrosine into resilin is very rapid and there is little or no turnover of resilin in the adult locust (Andersen, 1971b).

In *Periplaneta americana* and *Manduca sexta* radiotracer studies, electrophoretic analyses, and immunological techniques have led to the suggestion that hemolymph proteins, to which phenols (dopamine or its metabolites) are bonded, are transported through the epidermal cells, and are incorporated in the cuticle. The phenols would take part in sclerotization (Fox *et al.,* 1972; Koeppe and Gilbert, 1973; Koeppe and Mills, 1972).

B. Chitin

During the pupal molt in *Hyalophora cecropia* larvae, in which the cuticle has been rendered radioactive, a little isotopic carbon is transferred from the cuticle to glycogen, sugars, and lipids. The specific activity of the pupal cuticular chitin is as high as that in the previous larval cuticle, so material from the larval cuticle is used for synthesis of that of the pupa. Since most of the pupal chitin is added after the remains of the larval cuticle are shed, degradation products, presumably from the old cuticle by action of the molting fluid, must be stored temporarily in the animal (Bade and Wyatt, 1962). In the cockroach *Periplaneta americana* trehalose in the hemolymph appears to be a precursor of chitin and there is a rapid conversion of proline to trehalose (Lipke *et al.*, 1965a,b,c). Bade and Wyatt did not find high radioactivity in hemolymph sugars but the results of these two groups can be reconciled. *Hyalophora cecropia* larval cuticle contains approximately twice as much chitin as does the pupal cuticle and so could provide all the precursor material required for the new cuticle; after it has molted the cockroach has increased in size, a significant amount of chitin is lost when the exuviae is shed, and so the synthesis of additional chitin would be necessary. Lipke *et al.* (1965b) have also shown that both the synthesis and degradation of chitin occur as continuous processes throughout the nymphal life of the cockroach. In *Dixippus morosus,* Janda and Socha (1970) concluded that degradation products of chitin from the old cuticle are used in the newly secreted cuticle.

In the desert locust (*Schistocerca gregaria*) Candy and Kilby (1962) have shown that there is a high rate of chitin formation during and shortly after molting and that this is paralleled by a high rate of incorporation into chitin of radioactivity from [^{14}C]glucose injected 24 hours earlier. Since the insects do not feed during the first 24 hours after molting, the chitin must be synthesized from reserves within the insect. The enzymic formation of UDP-N-acetylglucosamine was established but there was no indication of the synthesis of chitin from this uridine complex. The same observation was made by Fristrom (1968) when using imaginal disks of *Drosophila melanogaster* and by Surholt and Zebe (1972) in *Locusta migratoria*. Chitin synthesis requires concurrent protein synthesis (Clever and Bultmann, 1972), and ecdysome appears to control the changing sequence of chitin-synthesizing and -catabolizing systems through the action of proenzymes or the synthesis of enzymes (Kimura, 1973).

Jaworski and his group (Jaworski *et al.,* 1963; Krueger and Jaworski, 1966; Porter and Jaworski, 1965) have identified a chitin synthetase in high- and low-speed particulate fractions from late last-instar larvae and prepupae of *Prodenia eridania*. When the enzyme is incubated with UDP-[^{14}C]acetyl-glucosamine, N-acetylglucosamine, and chitodextrin, an insoluble radioactive material results which is degraded by chitinase. The enzyme would be a chitin-UDP-acetylglucosaminyltransferase (EC 2.4.1.16). A glucosamine-phosphate isomerase (EC 5.3.1.10) isolated from young adult *Musca domes-*

tica required fructose 6-phosphate and inorganic ammonia for the synthesis of glucose 6-phosphate (Benson and Friedman, 1970) and this isomerase may be a link in the pathway leading to chitin. The above results are similar to the earlier work of Glaser and Brown (1957a,b) who showed that a cell-free preparation from the mold *Neurospora crassa* catalyzed the synthesis of an insoluble chitinlike polysaccharide from UDP-*N*-acetyl-D-glucosamine and chitodextrin. The *N*-acetyl-D-glucosamine was incorporated as an intact unit and the reaction proceeded by glycosyl transfer from the nucleotide-linked sugar to performed chitodextrin chains. Other workers have since obtained similar results using cell-free and particulate preparations from molds. The antibiotic Polyoxin D is a specific inhibitor for mold chitin synthetase and brings about the accumulation of UDP-*N*-acetylglucosamine (Endo *et al.*, 1970; Keller and Cabib, 1971). Griseofulvin causes anomalous development of the cuticle in *Aedes atropalpus* larvae (Anderson, 1966) but it is not known if the antibiotic interferes with chitin synthesis or with the organization of chitinous structures (see Bent and Moore, 1966).

C. Phenols

A number of dihydric and monohydric phenols have been isolated from insect cuticles (Section III,D), some of which are required for sclerotization of the cuticle (Section IV) and some to prevent autoxidation of lipids (Section III,E), but the biological role of the others is not known. Phenols have been identified in many species of insects, but a study of their biosynthesis has been made only in the blow fly larva and the cockroach. The biosynthesis of catecholamines in insects has been discussed by Sekeris and Karlson (1966) and a summary of this and other information is given below and in Fig. 3. See also the discussion in Section IV on the role of hormones in sclerotization.

In early third instar larvae of *Calliphora erythrocephala* tyrosine is mainly catabolized (transaminated) to 4-hydroxyphenylpyruvic acid which is converted to the corresponding lactic acid and propionic acid derivatives. There is also some decarboxylation of tyrosine to give tyramine which is *N*-acetylated. In older larvae (just before pupation) these metabolites become minor ones, the principal reaction being hydroxylation of tyrosine to give dopa, followed by decarboxylation to dopamine, *N*-acetylation of the dopamine, and formation of the 4-*O*-β-glucoside. The hormone ecdysone is probably responsible for the change in tyrosine metabolism and ecdysone induces the formation of the dopa decarboxylase necessary to produce dopamine. Some dopamine is oxidatively deaminated to 3,4-dihydroxyphenylacetic acid, and tryramine and *N*-acetyltyramine are hydroxylated to dopamine and *N*-acetyldopamine. Tyrosine probably arises by hydroxylation of phenyl-

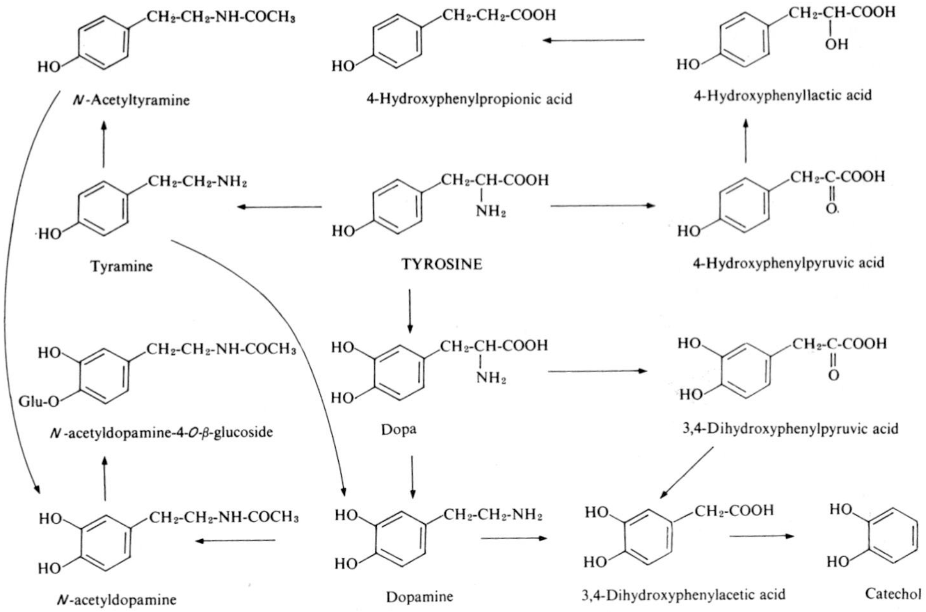

Fig. 3. Metabolism of tyrosine in insects.

alanine and tyrosine may be accumulated temporarily as tyrosine-O-phosphate (cf. Lunan and Mitchell, 1969; Mitchell and Lunan, 1964; Seligman et al., 1969). In Drosophila Henderson and Glassman (1969) have suggested that tyrosinase metabolites are stored as macromolecules. A phenol oxidase catalyzes the hydroxylation of tyrosine to dopa and a transacetylase involving acetyl-CoA brings about the N-acetylations. Mills and Lake (1971) and Mills et al. (1967b) have shown a similar series of reactions to take place in the cockroach, including the oxidation of 4-hydroxyphenylpyruvic acid to 4-hydroxyphenylacetic acid. In this insect tyramine is converted to dopamine (Whitehead, 1969). On completion of sclerotization of the cockroach cuticle the metabolism of tyrosine shifts from one of decarboxylation to a more extensive catabolism of the sidechain (Murdock et al., 1970a,b; Hopkins et al., 1971). When dopa is incubated with blow fly larval homogenates (in the absence of phenol oxidases) 3,4-dihydroxyphenylpyruvic acid is formed which, when kept at pH 8, undergoes oxidation to 3,4-dihydroxyphenylacetic acid and catechol (Pryor, 1962).

Very little is known about the biosynthesis of the other phenols discussed in Section III,D, but their structures, and also those of the catechol derivatives isolated from sclerotized cuticles by Andersen (1970, 1971a), Andersen and Barrett (1971), and Atkinson et al. (1973) are such as to suggest their synthesis from tyrosine or dopa. Dennell (1958c) has suggested that hydroquinone and pyrogallol arise from the nonspecific hydroxylation of phenyl-alanine. The 4-O-β-glucoside of N-acetyldopamine is presumably a storage product for later use by the insect (Sekeris, 1964). Fat-body, the left col-leterial gland and the integument of the cockroach have been implicated in the synthesis of the 4-O-β-glucoside of 3,4-dihydroxybenzoic acid, this dihydric phenol, as well as 3,4-dihydroxybenzyl alcohol, being synthesized from tyrosine by reactions which involve hemocytes and juvenile hormone (Lake *et al.*, 1970; Shaaya and Sekeris, 1970; Takahashi and Ohnishi, 1966; Whitehead, 1970b). As is described in Section V,E, tyrosine and catechol metabolites may be channeled into melanin formation.

D. LIPIDS

The metabolism of lipids in insects has been the subject of reviews (e.g., Gilbert, 1967; Gilbert and O'Connor, 1970; Gilby, 1965), but little is said about cuticular lipids. Jackson and Baker (1970) have discussed the biosyn-thesis of cuticular lipids with particular reference to their own work. Oeno-cytes are considered to take part in the synthesis of cuticular lipids (Locke, 1969a; Philogene and McFarlane, 1967) and it has been shown that the integument (presumably the epidermal cells) can synthesize hydrocarbons from acetate or other fatty acids (Jackson and Baker, 1970; Nelson, 1969). There is little correlation between the fatty acid composition of cuticle and fat-body tissue (Thompson and Barlow, 1970).

Incubation of insect integument with labeled acetate, or its injection into insects, leads to the formation of labeled long-chain hydrocarbons and satu-rated and unsaturated acids. So, in insects it appears that these compounds are synthesized from acetate in a way similar to that which takes place in other organisms. Other fatty acids, including long-chain fatty acids, are also incorporated into hydrocarbons and, in the cockroach, *cis,cis*-6,9-hepta-cosadiene can be derived from linoleate (see references given above). The lipids from adult *Lucilia cuprina* contain more branched-chain alkanes and very much less nonocosane than do the lipids from the puparium, which sug-gests that different biosynthetic pathways operate in the adult and larval stages (Goodrich, 1970). Also the absence of branched-chain alkanes from the diet of the tobacco hookworm and their presence in the integument indicates that this insect is able to synthesize them (Nelson *et al.*, 1971). Biosynthesis of cuticular hydrocarbons in *Periplaneta* appears to proceed along an elongation–decarobxylation pathway (Conrad and Jackson, 1971). Dietary n-alkanes (C_{27}, C_{29}, C_{31}) are incorporated unchanged into the cuti-cular lipids of the grasshopper *Melanoplus sanguinipes,* while secondary alco-hols are metabolic products of the oxidation of n-alkanes (Blomquist and Jackson, 1973).

Additional references are available for the synthesis of lipids in whole insects, but it is not known if the results have any relevance to cuticular lipids. For example, in *Drosophila melanogaster* (Kiyomoto and Keith, 1970), *Galleria mellonella* (Thompson and Barlow, 1971), and *Lucilia sericata* (Lindsay and Barlow, 1971) fatty acids can be synthesized by more than one pathway; in *Bombyx mori* dietary fatty acids inhibit, but sucrose accelerates fatty acid synthesis (Horie and Nakasone, 1971), while in the same insect the synthesis of long-chain saturated and unsaturated fatty acids from glucose by way of pyruvate has been confirmed (Horie *et al.*, 1968), and dietary amino acids, especially leucine and glutamate, are used by the house fly for lipid synthesis (Kon and Monroe, 1971).

E. PIGMENTS

Little is known about the regulation of the processes which lead to pigmentation. In larvae of the hawk moth, *Dicranura vinula*, low concentrations of ecdysone induce the formation of ommochromes (Bückmann, 1959). Hormonal control of pigmentation in *Locusta* has been discussed by Girardie (1967) and in the larval integument of *Leucania* by Ogura *et al.* (1971). For a general discussion on the subject see Fuzeau-Braesch (1971, 1972). Some cuticular pigments, e.g., carotenoids and flavones are of plant origin and are obtained from the diet.

a. Sclerotin and Melanin. The synthesis of sclerotin has been discussed in Section IV. In vertebrates melanin is produced by specialized organelles known as melanosomes which give the uniformly dense particles of melanin. Melanin in insects is a much more finely divided pigment, but information about its development and deposition is very limited. In the puparium of a *Lucilia cuprina* mutant, melanin is present as a layer beneath the epicuticle (Hackman, 1967) and in this mutant, as in insects with melanic patterns, there must be localized melanin synthesis and deposition. An enzyme with a high activity on precursors of melanin occurs as a proenzyme in the hemolymph of many insects (e.g., Hackman and Goldberg, 1968; Ohnishi, 1954) and it is generally assumed that such a proenzyme is the precursor of the phenol oxidase needed for melanin formation. If this is so, then the phenolic precursor of melanin, and perhaps also the activator of the proenzyme, must be at the site where melanin deposition is to take place.

Many different phenolic compounds can be used as precursors of melanin, e.g., dopa and catechol derivatives (see Nicolaus, 1968). In biological systems the former would give rise to eumelanins and the latter to allomelanins by reactions such as are given in Fig. 4. Melanins can be formed from dopamine and adrenaline by analogous reactions. A phenol oxidase is necessary for the initial oxidation but the remainder of the synthesis proceeds even

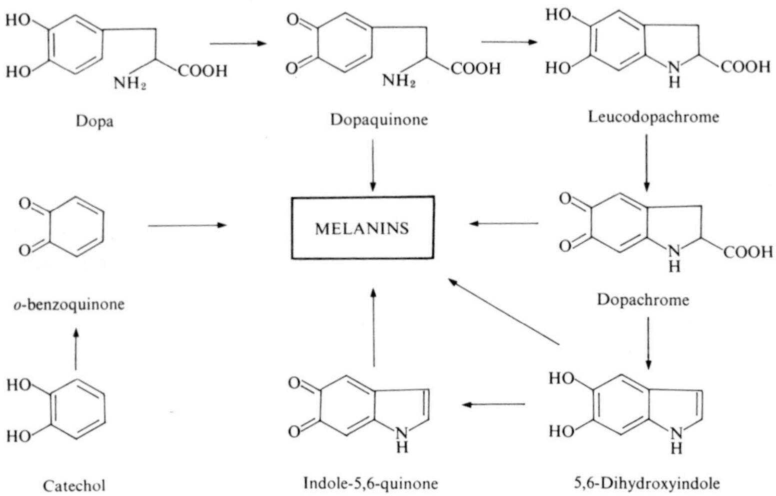

Fig. 4. Biosynthetic pathways leading to melanins.

in the absence of enzymes and the intermediates may exist in the form of zwitterions or as the isomeric 1,4-quinones.

b. Carotenoids. Insects, like other animals, derive carotenoids from their food and they are able to metabolize these to related pigments, e.g., β-carotene is oxidized to keto and hydroxy derivatives in such insects as *Locusta migratoria* (Goodwin and Srisukh, 1951), *Carausius morosus* (Willig, 1969), and *Leptinotarsa decemlineata* (Czeczuga, 1971; Leuenberger and Thommen, 1970). Of these insects only in *Locusta* is the derived pigment located in the integument and the keto carotenoid is astaxanthin. Many insects are known to be unable to synthesize carotenoids when maintained on a carotenoid-free diet but Przibram and Lederer (1933) report such a synthesis by the mantid *Sphodomantis bioculata;* reviewers have questioned the work.

c. Tetrapyrrole Pigments. These pigments appear to be formed from small precursors and not from pyrrole compounds ingested with the food. Injection of [1,2-^{14}C]glycine into *Pieris brassicae* larvae, *Mantis religiosa,* or *Locusta migratoria* results in the activity being incorporated into biliverdin IX_γ (*Pieris*) and biliverdin IX_α (*Mantis* and *Locusta*) in a specific manner corresponding to the usual way bile pigments are synthesized by animals (see Rüdiger, 1970).

d. Pteridines. Purines are readily converted to pteridines by first opening the imidazole ring to give a diaminopyrimidine which is then reacted with a suitable dicarbonyl compound such as glyoxylic acid. Such conversions take place in insects and steps in the reaction pathway have been followed by the use of labeled compounds, e.g., Weygand and Waldschmidt (1955) and Weygand *et al.* (1961). Synthesis of pteridines in the wings of *Colias eurytheme* follows the expected pathways (Watt, 1967). It is clear that in insects pteridines are formed from purine precursors (e.g., guanosine) and that they initially have a 3-carbon side chain in the 6 position and are hydrogenated in the 5, 6, 7, and 8 positions. Oxidations, catalyzed by xanthine dehydrogenase, subsequently convert pterin to isoxanthopterin, and xanthopterin to leucopterin. Erythropterin requires substitution in position 7. It appears that all oxidized, nonsubstituted pterins are metabolic end products and folic acid from the diet appears to be a precursor of pteridines in some insects (for references, see Ziegler and Harmsen, 1969). In *Mylothris chloris aganthina* the C-6 substituted pterines are converted to either xanthopterin and leucopterin or isoxanthopterin depending on the oxygen pressure (Harmsen, 1969). The integument of a *Bombyx mori* mutant contains a deaminase which catalyzes the hydrolysis of sepiapterin to xanthopterin-B_2 and ammonia (Tsusue, 1971).

e. Ommochromes. Ommochromes were first studied as eye pigments in mutants of *Drosophila* and *Ephestia* and are of historic interest because successive steps in their biosynthesis were shown to be associated with specific genes. As mentioned above ecdysone induces the formation of ommochromes. Ommochromes are phenoxazone compounds and are formed from tryptophan via kynurenine and 3-hydroxykynurenine by well-established biosynthetic reactions. Butenandt (1957) has reviewed his group's work in this field. There is evidence to suggest that pterins are involved in the hydroxylation of kynurenine and so may be involved in the biosynthesis of ommochromes (see Ziegler and Harmsen, 1969). Ommochromes are divided into two groups, ommatins which are of low molecular weight (e.g., rhodommatin) and ommins which are of high molecular weight and contain sulfur in the ring system. Ommins occur in the integument of *Gryllus bimaculatus* and Linzen (1970) has shown that sulfur from both [^{35}S]methionine and [^{35}S]cysteine is incorporated into the pigments but that labeled sulfate and sulfite are not.

VI. Conclusion

The insect integument, i.e., the cuticle and the epidermal cell layer, is a versatile experimental material and has been used for studying fundamental problems of biology. The activities of the single layer of epidermal cells vary according to the stage of development of the insect and at accurately predetermined times they synthesize the quite different chemical components of the cuticle. Developmental biologists thus find much of interest in this field as do endocrinologists when trying to understand the hormonal controls which bring about these changes in activities, while the biosynthesis and orientation of the macromolecular components provides an intriguing challenge. The cuticle protects the insect from death by desiccation while many toxicants, e.g., contact insecticides, have to penetrate the cuticle to be effective. The consequences of any malformation of the cuticle may be fatal for the insect, e.g., inability to fly, desiccation or lack of protection from mechanical injury and from predators. A thorough understanding of the structure, function, composition, and synthesis of the cuticle should lead in time to the development of methods for the control of insect pests.

From the chemistry of the cuticle as outlined in this chapter, it will have become obvious to the critical reader that many important gaps exist in our knowledge and that there are many crucial points which need elucidation. Not the least of these is the biosynthesis of the cuticular components. Although some information has appeared, we are still largely ignorant as to how these materials are elaborated and transported at precisely determined periods in the formation of the cuticle. Only detailed and systematic studies will establish the mechanisms and actual reaction pathways which take place in the living insect.

Significant advances have taken place when chemists and biochemists have interested themselves in the nature of the cuticle. This is not to deny the importance of the many outstanding contributions made by biologists. It is to be hoped that problems associated with the cuticle will continue to be attractive to both chemists and biochemists and that there will be collaboration between workers in different disciplines. Steady progress in this field, as in any other, demands an uncompromising methodology and with the results obtained by the different techniques and methods now in use care must be taken to avoid speculation leading to conclusions which have no foundation.

Cuticles from a variety of insect species have been studied by various workers but detailed information is available for only a few species. Although it is tempting to draw sweeping generalizations from this information, it is rarely justified unless they are stated in the most general terms. There is great diversity of form among insects and it is likely that in fine detail any one cuticle may be different from another. Electron micrographs have already shown that differences in fine structure occur between insects of different species, e.g., Filshie (1970b) and Locke (1966, and Chapter 2, this volume). Clearly work must be done on a much greater number of species from all orders before conclusions can be applied to insects in general.

References

Adiyodi, K. G. (1968). *J. Insect Physiol.* **14**, 309.
Adiyodi, K. G., and Nayar, K. K. (1966). *Curr. Sci.* **35**, 587.

Aerts, F., Vercauteren, R. E., and Decleir, W. (1962). *Proc. Int. Congr. Entomol., 11th, 1960* Vol. 3, p. 171.

Amin, E. S. (1960). *J. Chem. Soc., London* p. 1410.

Andersen, S. O. (1970). *J. Insect Physiol.* 16, 1951.

Andersen, S. O. (1971a). *Insect Biochem.* 1, 157.

Andersen, S. O. (1971b). *Comp. Biochem.* 26C, 633–657.

Andersen, S. O. (1972a). *J. Insect Physiol.* 18, 527.

Andersen, S. O. (1972b). *Acta Chem. Scand.* 26, 3097.

Andersen, S. O. (1973). *J. Insect Physiol.* 19, 1603.

Andersen, S. O., and Barrett, F. M. (1971). *J. Insect Physiol.* 17, 69.

Andersen, S. O., Chase, A. M., and Willis, J. H. (1973). *Insect Biochem.* 3, 171.

Anderson, J. F. (1966). *J. Econ. Entomol.* 59, 1476.

Aneshansley, D. J., Eisner, T., Widom, J. M., and Widom, B. (1969). *Science* 165, 61.

Aoki, J., and Yanase, K. (1970). *J. Invertebr. Pathol.* 16, 459.

Armold, M. T., Blomquist, G. J., and Jackson, L. L. (1969). *Comp. Biochem. Physiol.* 31, 685.

Ashida, M. (1971). *Arch. Biochem. Biophys.* 144, 749.

Ashida, M., and Ohnishi, E. (1967). *Arch. Biochem. Biophys.* 122, 411.

Atkinson, P., and Gilby, A. R. (1970). *Science* 168, 992.

Atkinson, P., Brown, W. V., and Gilby, A. R. (1973). *Insect Biochem.* 3, 309.

Attwood, M. M., and Zola, H. (1967). *Comp. Biochem. Physiol.* 20, 993.

Bade, M. L., and Wyatt, G. R. (1962). *Biochem. J.* 83, 470.

Bailey, S. W. (1954). *Nature (London)* 173, 503.

Baker, G., Pepper, J. H., Johnson, L. H., and Hastings, E. (1960). *J. Insect Physiol.* 5, 47.

Barker, R. J. (1970). *J. Insect Physiol.* 16, 1921.

Barker, S. A., Foster, A. B., Stacey, M., and Webber, J. M. (1958). *J. Chem. Soc., London* p. 2218.

Battaglini, P., and Parisi, G. (1968). *Boll. Soc. Ital. Biol. Sper.* 44, 1975.

Baust, J. G. (1967). *Zoologica (New York)* 52, 15.

Bayse, G. S., and Morrison, M. (1971). *Biochim. Biophys. Acta* 244, 77.

Beament, J. W. L. (1955). *J. Exp. Biol.* 32, 514.

Beament, J. W. L. (1958). *J. Exp. Biol.* 35, 494.

Beament, J. W. L. (1959). *J. Exp. Biol.* 36, 391.

Beament, J. W. L. (1960). *Nature (London)* 186, 408.

Beament, J. W. L. (1961a). *Biol. Rev. Cambridge Phil. Soc.* 36, 281.

Beament, J. W. L. (1961b). *J. Exp. Biol.* 38, 277.

Beatty, I. M., and Gilby, A. R. (1969). *Naturwissenschaften* 56, 373.

Beck, S. D., and Shane, J. L. (1969). *J. Insect Physiol.* 15, 721.

Benjaminson, M. A. (1969). *Stain Technol.* 44, 27.

Benson, R. L., and Friedman, S. (1970). *J. Biol. Chem.* 245, 2219.

Bent, K. J., and Moore, R. H. (1966). *Symp. Soc. Gen. Microbiol.* 16, 82–110.

Bergmann, W. (1938). *Ann. Entomol. Soc. Amer.* 31, 315.

Berkoff, C. E. (1969). *Quart. Rev., Chem. Soc.* 23, 372.

Berreur, P., and Fraenkel, G. (1969). *Science* 164, 1132.

Berthold, G., and Henze, M. (1971). *J. Insect Physiol.* 17, 2375.

Bhagvat, K., and Richter, D. (1938). *Biochem. J.* 32, 1397.

Blaich, R. (1969). *Zool. Jahrb., Abt. Anat. Ontog. Tiere* 86, 576.

Blomquist, G. J., and Jackson, L. L. (1973). *J. Insect Physiol.* 19, 1639.

Blomquist, G. J., Soliday, C. L., Byers, B. A., Brakke, J. W., and Jackson, L. L. (1972). *Lipids* **7**, 356.

Bodnaryk, R. P. (1971a). *J. Insect Physiol.* **17**, 1201.

Bodnaryk, R. P. (1971b). *Insect Biochem.* **1**, 228.

Bodnaryk, R. P. (1972). *Insect Biochem.* **2**, 119.

Bowers, W. S., and Thompson, M. J. (1965). *J. Insect Physiol.* **11**, 1003.

Braconnot, H. (1811). *Ann. Chim. (Paris)* [1] **79**, 265.

Brown, C. H. (1949). *Exp. Cell Res., Suppl.* **1**, 351.

Brown, C. H. (1950). *Nature (London)* **165**, 275.

Brunet, P. C. J. (1952). *Quart. J. Microsc. Sci.* **93**, 47.

Brunet, P. C. J., and Kent, P. W. (1955). *Proc. Roy. Soc., Ser. B* **144**, 259.

Bückmann, D. (1959). *J. Insect Physiol.* **3**, 159.

Bursell, E., and Clements, A. N. (1967). *J. Insect Physiol.* **13**, 1671.

Butenandt, A. (1957). *Angew. Chem.* **69**, 16.

Campbell, F. L. (1929). *Ann. Entomol. Soc. Amer.* **22**, 401.

Candy, D. J., and Kilby, B. A. (1962). *J. Exp. Biol.* **39**, 129.

Capon, B., and Foster, R. L. (1970). *J. Chem. Soc., C* p. 1654.

Carlström, D. (1957). *J. Biophys. Biochem. Cytol.* **3**, 669.

Carlström, D. (1962). *Biochim. Biophys. Acta* **59**, 361.

Carlström, D. (1966). *Funct. Organ. Compound Eye, Proc. Int. Symp., 1965* pp. 15–19.

Caveney, S. (1971). *Proc. Roy. Soc., Ser. B* **178**, 205.

Cavill, G. W. K., Clark, D. V., Howden, M. E. H., and Wyllie, S. G. (1970). *J. Insect Physiol.* **16**, 1721.

Chapman, R. F., Percival, A., and Swan, G. A. (1970). *J. Chem. Soc., C* p. 1664.

Cheeseman, D. F., Lee, W. L., and Zagalsky, P. F. (1967). *Biol. Rev. Cambridge Phil. Soc.* **42**, 131.

Clark, E. W. (1958). *Ann. Entomol. Soc. Amer.* **51**, 142.

Clever, U., and Bultmann, H. (1972). *Cell Differentiation* **1**, 37.

Cohen, E., Ikan, R., and Shulov, A. (1971). *Entomol. Exp. Appl.* **14**, 315.

Coles, G. C. (1966). *J. Insect Physiol.* **12**, 679.

Collins, R. P., and Kalnins, K. (1970). *J. Insect Physiol.,* **16**, 1587.

Conrad, C. W., and Jackson, L. L. (1971). *J. Insect Physiol.* **17**, 1907.

Cory, J. G., Bigelow, C. C., and Frieden, E. (1962). *Biochemistry* **1**, 419.

Cottrell, C. B. (1962a). *J. Exp. Biol.* **39**, 395.

Cottrell, C. B. (1962b). *J. Exp. Biol.* **39**, 449.

Cottrell, C. B. (1964). *Advan. Insect Physiol.* **2**, 175.

Crossley, A. C. S. (1964). *J. Exp. Zool.* **157**, 375.

Czeczuga, B. (1971). *J. Insect Physiol.* **17**, 2017.

De Haas, B. W., Johnson, L. H., Pepper, J. H., Hastings, E., and Baker, G. L. (1957). *Physiol. Zool.* **30**, 121.

Delachambre, J. (1969). *Histochemie* **20**, 58.

Delachambre, J. (1971). *J. Insect Physiol.* **17**, 2481.

de Mets, R., and Jeuniaux, C. (1962). *Arch. Int. Physiol. Biochim.* **70**, 93.

Dennell, R. (1947). *Proc. Roy. Soc., Ser. B* **134**, 79.

Dennell, R. (1958a). *Proc. Roy. Soc., Ser. B* **148**, 270.

Dennell, R. (1958b). *Proc. Roy. Soc., Ser. B* **148**, 280.

Dennell, R. (1958c). *Proc. Roy. Soc., Ser. B* **149**, 176.

Dennell, R., and Malek, S. R. A. (1955a). *Proc. Roy. Soc., Ser. B* **143**, 239.

Dennell, R., and Malek, S. R. A. (1955b). *Proc. Roy. Soc., Ser. B* **143**, 414.
Dohke, K. (1973). *Arch. Biochem. Biophys.* **157**, 203 and 210.
Downe, A. E. R. (1962). *Can. J. Zool.* **40**, 957.
Dweltz, N. E. (1960). *Biochim. Biophys. Acta* **44**, 416.
Edney, E. B. (1957). "The Water Relations of Terrestrial Arthropods." Cambridge Univ. Press, London and New York.
Edwards, J. S. (1966). *J. Insect Physiol.* **12**, 1423.
Endo, A., Kakiki, K., and Misato, T. (1970). *J. Bacteriol.* **104**, 189.
Filshie, B. K. (1966). Ph.D. Thesis, Australian National University, Canberra, Australia.
Filshie, B. K. (1970a). *Tissue & Cell* **2**, 181.
Filshie, B. K. (1970b). *Tissue & Cell* **2**, 479.
Fogal, W., and Fraenkel, G. (1969a). *J. Insect Physiol.* **15**, 1235.
Fogal, W., and Fraenkel, G. (1969b). *J. Insect Physiol.* **15**, 1437.
Foster, A. B., and Hackman, R. H. (1957). *Nature (London)* **180**, 40.
Fox, F. R., and Mills, R. R. (1969). *Comp. Biochem. Physiol.* **29**, 1187.
Fox, F. R., and Mills, R. R. (1971). *J. Insect Physiol.* **17**, 2363.
Fox, F. R., Seed, J. R., and Mills, R. R. (1972). *J. Insect Physiol.* **18**, 2065.
Fraenkel, G., and Hsiao, C. (1965). *J. Insect Physiol.* **11**, 513.
Fraenkel, G., and Hsiao, C. (1967). *J. Insect Physiol.* **13**, 1387.
Fraenkel, G., and Rudall, K. M. (1940). *Proc. Roy. Soc., Ser B* **129**, 1.
Fraenkel, G., and Rudall, K. M. (1947). *Proc. Roy. Soc., Ser. B* **134**, 111.
Fristrom, J. W. (1968). *J. Insect Physiol.* **14**, 729.
Fukushi, Y. (1967). *Jap. J. Genet.* **42**, 11.
Fukushi, Y., and Seki, T. (1965). *Jap. J. Genet.* **40**, 203.
Fuzeau-Braesch, S. (1959a). *C.R. Acad. Sci.* **248**, 856.
Fuzeau-Braesch, S. (1959b). *C.R. Soc. Biol.* **153**, 57.
Fuzeau-Braesch, S. (1971). *Arch. Zool. Exp. Gen.* **112**, 625.
Fuzeau-Braesch, S. (1972). *Annu. Rev. Entomol.* **17**, 403.
Gilbert, L. I .(1967). *Advan. Insect Physiol.* **4**, 70.
Gilbert, L. I., and O'Connor, J. D. (1970). *In* "Chemical Zoology" (M. Florkin and B. T. Scheer, eds.), Vol. 5, Part A, pp. 229–253. Academic Press, New York.
Gilby, A. R. (1962). *Nature (London)* **195**, 729.
Gilby, A. R. (1965). *Annu. Rev. Entomol.* **10**, 141.
Gilby, A. R., and Cox, M. (1963). *J. Insect Physiol.* **9**, 671.
Gilby, A. R., and McKellar, J. W. (1970). *J. Insect Physiol.* **16**, 1517.
Giles, C. H., Hassan, A. S. A., Laidlaw, M., and Subramanian, R. V. R. (1958). *J. Soc. Dyers Colour.* **74**, 647.
Girardie, A. (1967). *Bull. Biol. Fr. Belg.* **101**, 79.
Glaser, L., and Brown, D. H. (1957a). *Biochim. Biophys. Acta* **23**, 449.
Glaser, L., and Brown, D. H. (1957b). *J. Biol. Chem.* **228**, 729.
Goffinet, G., and Jeuniaux, C. (1969). *Comp. Biochem. Physiol.* **29**, 277.
Goodrich, B. S. (1970). *J. Lipid Res.* **11**, 1.
Goodwin, T. W. (1971). *In* "Chemical Zoology" (M. Florkin and B. T. Scheer, eds.), Vol. 6, Part B, pp 279–306. Academic Press, New York.
Goodwin, T. W., and Srisukh, S. (1951). *Biochem. J.* **48**, 199.
Hackman, R. H. (1953a). *Biochem. J.* **54**, 362.
Hackman, R. H. (1953b). *Biochem. J.* **54**, 367.
Hackman, R. H. (1953c). *Biochem. J.* **54**, 371.

Hackman, R. H. (1954). *Aust. J. Biol. Sci.* **7**, 168.

Hackman, R. H. (1955a). *Aust. J. Biol. Sci.* **8**, 83.

Hackman, R. H. (1955b). *Aust. J. Biol. Sci.* **8**, 530.

Hackman, R. H. (1956). *Aust. J. Biol. Sci.* **9**, 400.

Hackman, R. H. (1960). *Aust. J. Biol. Sci.* **13**, 568.

Hackman, R. H. (1962). *Aust. J. Biol. Sci.* **15**, 526.

Hackman, R. H. (1967). *Nature (London)* **216**, 163.

Hackman, R. H. (1971a). *J. Insect Physiol.* **17**, 1065.

Hackman, R. H. (1971b). *In* "Chemical Zoology" (M. Florkin and B. T. Scheer, eds.), Vol. 6, Part B, pp. 1–62. Academic Press, New York.

Hackman, R. H. (1972). *Insect Biochem.* **2**, 235.

Hackman, R. H., and Goldberg, M. (1958). *J. Insect Physiol.* **2**, 221.

Hackman, R. H., and Goldberg, M. (1960). *J. Insect Physiol.* **5**, 73.

Hackman, R. H., and Goldberg, M. (1963). *Biochim. Biophys. Acta* **71**, 738.

Hackman, R. H., and Goldberg, M. (1965). *Aust. J. Biol. Sci.* **18**, 935.

Hackman, R. H., and Goldberg, M. (1967). *J. Insect Physiol.* **13**, 531.

Hackman, R. H., and Goldberg, M. (1968). *J. Insect Physiol.* **14**, 765.

Hackman, R. H., and Goldberg, M. (1971a). *J. Insect Physiol.* **17**, 335.

Hackman, R. H., and Goldberg, M. (1971b). *Anal. Biochem.* **41**, 279.

Hackman, R. H., and Saxena, K. N. (1964). *Aust. J. Biol. Sci.* **17**, 803.

Hackman, R. H., and Todd, A. R. (1953). *Biochem. J.* **55**, 631.

Hackman, R. H., Pryor, M. G. M., and Todd, A. R. (1948). *Biochem. J.* **43**, 474.

Harmsen, R. (1969). *J. Insect Physiol.* **15**, 2239.

Hashimoto, A., Yoshida, H., Mukai, K., and Kitaoka, S. (1971a). *Nippon Nogei Kagaku Kaishi* **45**, 96.

Hashimoto, A., Hirotani, A., Mukai, K., and Kitaoka, S. (1971b). *Nippon Nogei Kagaku Kaishi* **45**, 100.

Hempel, K. (1966). *Struct. Contr. Melanocyte, Proc. Int. Pigment Cell Conf., 6th, 1965* pp. 158–175.

Henderson, A. S., and Glassman, E. (1969). *J. Insect Physiol.* **15**, 2345.

Hepburn, H. R. (1972). *J. Insect Physiol.* **18**, 815.

Hepburn, H. R., and Ball, A. (1973). *J. Mater. Sci.* **8**, 618.

Heyneman, R. A. (1965). *Biochem. Biophys. Res. Commun.* **21**, 162.

Hinton, H. E. (1970). *In* "Insect Ultrastructure" (A. C. Neville, ed.), pp. 41–58. Blackwell, Oxford.

Hinton, H. E., and Gibbs, D. F. (1971). *J. Insect Physiol.* **17**, 1023.

Hinton, H. E., and Jarman, G. M. (1973). *J. Insect Physiol.* **19**, 533.

Hoffmann, J. A., Porte, A., and Joly, P. (1970). *C.R. Acad. Sci., Ser. D* **270**, 629.

Holdgate, M. W., and Seal, M. (1956). *J. Exp. Biol.* **33**, 82.

Hodgetts, R. B. and Konopka, R. J. (1973). *J. Insect Physiol.* **19**, 1211.

Hopkins, T. L., Murdock, L. L., and Wirtz, R. A. (1971). *Insect Biochem.* **1**, 97.

Horie, Y., and Nakasone, S. (1971). *J. Insect Physiol.* **17**, 1441.

Horie, Y., Nakasone, S., and Ito, T. (1968). *J. Insect Physiol.* **14**, 971.

Horn, D. H. S. (1971). *In* "Naturally Occurring Insecticides" (M. Jacobson and D. G. Crosby, eds.), pp. 333–459. Dekker, New York.

Horowitz, S. T., Roseman, S., and Blumenthal, H. J. (1957). *J. Amer. Chem. Soc.* **79**, 5046.

Hunt, S. (1971). *Experientia* **27**, 1030.

Hurst, H. (1945). *Nature (London)* **156**, 194.

Irvine, J. C. (1909). *J. Chem. Soc., London* **95**, 564.

Ito, H. (1924). *Arch. Anat. Microsc. Morphol. Exp.* **20**, 343.

Ito, H. (1953). *Bull. Sericult. Exp. Sta. Tokyo* **14**, 115.

Jackson, L. L. (1970). *Lipids* **5**, 38.

Jackson, L. L. (1972). *Comp. Biochem. Physiol.* **41B**, 331.

Jackson, L. L., and Baker, G. L. (1970). *Lipids* **5**, 239.

Janda, V., and Socha, R. (1970). *J. Insect Physiol.* **16**, 2051.

Jaworski, E., Wang, L., and Marco, G. (1963). *Nature (London)* **198**, 790.

Jensen, M., and Weis-Fogh, T. (1962). *Phil. Trans. Roy. Soc. London, Ser. B* **245**, 137.

Jeuniaux, C. (1959). *Arch. Int. Physiol. Biochim.* **67**, 516.

Jeuniaux, C. (1963). "Chitine et Chitinolyse." Masson, Paris.

Jeuniaux, C. (1971). *Compr. Biochem.* **26C**, 595–632.

Johnson, L. H., Pepper, J. H., Banning, M. N. B., Hastings, E., and Clark, R. S. (1952). *Physiol. Zool.* **25**, 250.

Jones, B. M., and Sinclair, W. (1958). *Nature (London)* **181**, 926.

Karlson, P., and Herrlich, P. (1965). *J. Insect Physiol.* **11**, 79.

Karlson, P., and Liebau, H. (1961). *Hoppe-Seyler's Z. Physiol. Chem.* **326**, 135.

Karlson, P., and Schlossberger-Raecke, I. (1962). *J. Insect Physiol.* **8**, 441.

Karlson, P., and Sekeris, C. E. (1964). *Comp. Biochem.* **6**, 221–243.

Karlson, P., and Sekeris, C. E. (1966). *Recent Progr. Horm. Res.* **22**, 473.

Karlson, P., Mergenhagen, D., and Sekeris, C. E. (1964). *Hoppe-Selyer's Z. Physiol. Chem.* **338**, 42.

Karlson, P., Sekeri, K. E., and Marmaras, V. I. (1969). *J. Insect Physiol.* **15**, 319.

Kato, K., and Kubomura, K. (1956). *Sci. Rep. Saitama Univ., Ser. B* **2**, 165.

Kawase, S. (1958a). *Nippon Sanshigaku Zasshi* **27**, 321.

Kawase, S. (1958b). *Nippon Sanshigaku Zasshi* **27**, 327.

Kawase, S. (1960). *J. Insect Physiol.* **5**, 335.

Keilin, D. (1921). *Quart. J. Microsc. Sci.* **65**, 611.

Keller, F. A., and Cabib, E. (1971). *J. Biol. Chem.* **246**, 160 and 4376.

Kenchington, W., and Flower, N. E. (1969). *J. Microsc. (Oxford)* **89**, 263.

Kennaugh, J. H. (1958). *J. Insect Physiol.* **2**, 97.

Kent, P. W., and Brunet, P. C. J. (1959). *Tetrahedron* **7**, 252.

Kikkawa, H., Ogita, Z., and Fujito, S. (1954). *Kagaku (Tokyo)* **24**, 528.

Kikkawa, H., Ogita, Z., and Fujito, S. (1955). *Science* **121**, 43.

Kimura, S. (1973). *J. Insect Physiol.* **19**, 2177.

King, J. A. G., Percival, A., Robson, N. C., and Swan, G. A. (1970). *J. Chem. Soc., C* p. 1418.

Kinnear, J. F., Martin, M-D., and Thomson, J. A. (1971). *Aust. J. Biol. Sci.* **24**, 275.

Kiyomoto, R. K., and Keith, A. D. (1970). *Lipids* **5**, 617.

Knowles, B. B., and Fristrom, J. W. (1967). *J. Insect Physiol.* **13**, 731.

Koeppe, J. K., and Gilbert, L. I. (1973). *J. Insect Physiol.* **19**, 615.

Koeppe, J. K., and Mills, R. R. (1970). *Biochem. J.* **119**, 66p.

Koeppe, J. K., and Mills, R. R. (1972). *J. Insect Physiol.* **18**, 465.

Koidsumi, K. (1957). *J. Insect Physiol.* **1**, 40.

Komatsu, K. (1969). *Sanshi Shikenjo Hokoku* **23**, 499.

Kon, R. T., and Monroe, R. E. (1971). *Ann. Entomol. Soc. Amer.* **64**, 247.

Kramer, S., and Wigglesworth, V. B. (1950). *Quart. J. Microsc. Sci.* **91**, 63.

Krishnan, G. (1969). *Acta Histochem.* **34**, 212.

Krishnan, G., and Sundara Rajulu, G. (1964). *Curr. Sci.* **33**, 639.

Krueger, H. R., and Jaworski, E. G. (1966). *J. Econ. Entomol.* **59**, 229.

Krzelj, S., and Jeuniaux, C. (1968). *Ann. Soc. Roy. Zool. Belg.* **98**, 87.

Lai-Fook, J. (1966). *J. Insect Physiol.* **12**, 195.

Lake, C. R., Mills, R. R., and Brunet, P. C. J. (1970). *Biochim. Biophys. Acta* **215**, 226.

Leuenberger, F., and Thommen, H. (1970). *J. Insect Physiol.* **16**, 1855.

Lindsay, O. B., and Barlow, J. S. (1971). *Comp. Biochem. Physiol. B* **39**, 823.

Linzen, B. (1970). *Hoppe-Seyler's Z. Physiol. Chem.* **351**, 622.

Lipke, H. (1971). *Insect Biochem.* **1**, 189.

Lipke, H., and Geoghegan, T. (1971a). *J. Insect Physiol.* **17**, 415.

Lipke, H., and Geoghegan, T. (1971b). *Biochem. J.* **125**, 703.

Lipke, H., Grainger, M. M., and Siakotos, A. N. (1965a). *J. Biol. Chem.* **240**, 594.

Lipke, H., Graves, B., and Leto, S. (1965b). *J. Biol. Chem.* **240**, 601.

Lipke, H., Leto, S., and Graves, B. (1965c). *J. Insect Physiol.* **11**, 1225.

Lissitzky, S., and Rolland, M. (1962). *Biochim. Biophys. Acta* **56**, 95.

Lissitzky, S., Rolland, M., Reynaud, J., and Lasry, S. (1962). *Biochim. Biophys. Acta* **65**, 481.

Locke, M. (1966). *J. Morphol.* **118**, 461.

Locke, M. (1969a). *Tissue & Cell* **1**, 103.

Locke, M. (1969b). *Tissue & Cell* **1**, 555.

Locke, M., and Collins, J. V. (1966). *Nature (London)* **210**, 552.

Locke, M., and Collins, J. V. (1968). *J. Cell Biol.* **36**, 453.

Locke, M., and Krishnan, N. (1971). *Tissue & Cell* **3**, 103.

Lockey, K. H. (1960). *J. Exp. Biol.* **37**, 316.

Lunan, K. D., and Mitchell, H. K. (1969). *Arch. Biochem. Biophys.* **132**, 450.

McLintock, J. (1964). *Nature (London)* **201**, 1245.

Madhavan, K., and Schneiderman, H. A. (1968). *J. Insect Physiol.* **14**, 777.

Malek, S. R. A. (1952). *Nature (London)* **170**, 850.

Malek, S. R. A. (1957). *Nature (London)* **180**, 237.

Malek, S. R. A. (1961). *Comp. Biochem. Physiol.* **2**, 35.

Marks, E. P. (1970). *Gen. Comp. Endocrinol.* **15**, 289.

Martin, M. M., and Macconnell, J. G. (1970). *Tetrahedron* **26**, 307.

Martoja, R., and Cantacuzene, A. M. (1969). *C.R. Acad. Sci., Ser. D* **26B**, 697.

Mason, H. S. (1967). *Advan. Biol. Skin* **8**, 293–312.

Mason, H. S., and Peterson, E. W. (1965). *Biochim. Biophys. Acta* **111**, 134.

Meade-Briggs, A. R. (1956). *J. Exp. Biol.* **33**, 737.

Mills, R. R. (1967). *J. Insect Physiol.* **13**, 815.

Mills, R. R., and Lake, C. R. (1971). *Insect Biochem.* **1**, 264.

Mills, R. R., and Whitehead, D. L. (1970). *J. Insect Physiol.* **16**, 331.

Mills, R. R., Mathur, R. B., and Guerra, A. A. (1965). *J. Insect Physiol.* **11**, 1047.

Mills, R. R., Greenslade, F. C., Fox, F. R., and Nielsen, D. J. (1967a). *Comp. Biochem. Physiol.* **22**, 327.

Mills, R. R., Lake, C. R., and Alworth, W. L. (1967b). *J. Insect Physiol.* **13**, 1539.

Mills, R. R., Androuny, S., and Fox, F. R. (1968). *J. Insect Physiol.* **14**, 603.

Mitchell, H. K. (1966). *J. Insect Physiol.* **12**, 755.

Mitchell, H. K., and Lunan, K. D. (1964). *Arch. Biochem. Biophys.* **106**, 219.
Mitchell, H. K., and Weber, U. M. (1965). *Science* **148**, 964.
Mitchell, H. K., Weber, U. M., and Scharr, G. (1967). *Genetics* **57**, 357.
Mitchell, H. K., Weber-Tracy, U. M., and Scharr, G. (1971). *J. Exp. Zool.* **176**, 429.
Moorefield, H. H. (1953). Ph.D. Thesis, University of Illinois, Urbana.
Morris, S. J., and Thomson, R. H. (1963). *J. Insect Physiol.* **9**, 391.
Morris, S. J., and Thomson, R. H. (1964). *J. Insect Physiol.* **10**, 377.
Munn, E. A. and Bufton, S. F. (1973). *Eur. J. Biochem.* **35**, 3.
Munn, E. A., Feinstein, A., and Greville, G. D. (1971). *Biochem. J.* **124**, 367.
Munn, E. A., and Greville, G. D. (1969). *J. Insect Physiol.* **15**, 1935.
Murdock, L. L., Hopkins, T. L., and Wirtz, R. A. (1970a). *J. Insect Physiol.* **16**, 555.
Murdock, L. L., Hopkins, T. L., and Wirtz, R. A. (1970b). *Comp. Biochem. Physiol.* **36**, 535.
Nappi, A. J., and Stoffolano, J. G. (1970). *J. Parasitol.* **56**, 246.
Nelson, D. R. (1969). *Nature (London)* **221**, 854.
Nelson, D. R., Sukkestad, D. R., and Terranova, A. C. (1971). *Life Sci., Part II* **10**, 411.
Neville, A. C. (1967). *Advan. Insect Physiol.* **4**, 213.
Neville, A. C. (1970). *In* "Insect Ultrastructure" (A. C. Neville, ed.), pp. 17–39. Blackwell, Oxford.
Neville, A. C., and Caveney, S. (1969). *Biol. Rev. Cambridge Phil. Soc.* **44**, 531.
Neville, A. C., and Luke, B. M. (1969). *Tissue & Cell* **1**, 689.
Nicolaus, R. A. (1968). "Melanins." Hermann, Paris.
Nicolaus, R. A., Hempel, K., and Mason, H. S. (1967). *Advan. Biol. Skin.* **8**, 313–317.
Odier, A. (1823). *Mem. Soc. Hist. Natur. Paris* **1**, 29.
Ogura, N., Yagi, S., and Fukaya, M. (1971). *Appl. Entomol. Zool.* **6**, 93.
Ohnishi, E. (1953). *Jap. J. Zool.* **11**, 69.
Ohnishi, E. (1954). *Annot. Zool. Jap.* **27**, 33.
Ohnishi, E., Takahashi, S. Y., Sonobe, H., and Hayashi, T. (1968). *Science* **160**, 783.
Ohtaki, T., and Ohnishi, E. (1967). *J. Insect Physiol.* **13**, 1569.
Oloffs, P. C., and Scudder, G. G. C. (1966). *Can. J. Zool.* **44**, 621.
Pant, R., and Lal, D. M. (1970). *Indian J. Biochem.* **7**, 57.
Paradi, E., and Csukas-Szatloczky, I. (1969). *Acta Biol. (Budapest)* **20**, 373.
Parisi, G., and Battaglini, P. (1970). *Boll. Soc. Ital. Biol. Sper.* **46**, 207.
Parker, K. D., and Rudall, K. M. (1955). *Biochim. Biophys. Acta* **17**, 287.
Passama-Vuillaume, M. (1964). *C.R. Acad. Sci.* **258**, 6549.
Passama-Vuillaume, M. (1966). *C.R. Acad. Sci.* **262**, 1597.
Pau, R. N., and Acheson, R. M. (1968). *Biochim. Biophys. Acta* **158**, 206.
Pau, R. N., Brunet, P. C. J., and Williams, M. J. (1971). *Proc. Roy. Soc., Ser. B* **177**, 565.
Pearse, A. G. E. (1961). "Histochemistry, Theoretical and Applied," 2nd ed. Churchill, London.
Peeples, E. E., Geisler, A., Whitcraft, C. J., and Oliver, C. P. (1969). *Biochem. Genet.* **3**, 563.
Pennell, J. T., and Tsuyuki, H. (1965). *Can. J. Zool.* **43**, 587.
Pfeiler, E. J. (1968). *J. Res. Lepidoptera* **7**, 183.
Phillips, J. E., and Dockrill, A. A. (1968). *J. Exp. Biol.* **48**, 521.

Philogene, B. J. R., and McFarlane, J. E. (1967). *Can. J. Zool.* **45**, 181.

Porter, C. A., and Jaworski, E. G. (1965). *J. Insect Physiol.* **11**, 1151.

Price, G. M. (1970). *Nature (London)* **228**, 876.

Price, G. M., and Hughes, L. (1971). *Biochem. J.* **123**, 21p.

Pryor, M. G. M. (1940). *Proc. Roy. Soc., Ser. B* **128**, 378 and 393.

Pryor, M. G. M. (1955). *J. Exp. Biol.* **32**, 468.

Pryor, M. G. M. (1962). *Comp. Biochem.* **4B**, 371–396.

Pryor, M. G. M., Russell, P. B., and Todd, A. R. (1946). *Biochem. J.* **40**, 627.

Pryor, M. G. M., Russell, P. B., and Todd, A. R. (1947). *Nature (London)* **159**, 399.

Przibram, H., and Lederer, E. (1933). *Anz. Oesterr. Akad. Wiss., Math.-Naturwiss. Kl. No.* 17.

Ramachandran, G. N., and Ramakrishnan, C. (1962). *Biochim. Biophys. Acta* **63**, 307.

Ramsay, J. A. (1935). *J. Exp. Biol.* **12**, 373.

Richards, A. G. (1947). *Ann. Entomol. Soc. Amer.* **40**, 227.

Richards, A. G. (1951). "The Integument of Arthropods." Univ. of Minnesota Press, Minneapolis.

Richards, A. G. (1956). *J. Histochem. Cytochem.* **4**, 140.

Richards, A. G., and Brooks, M. A. (1958). *Annu. Rev. Entomol.* **3**, 37.

Richards, A. G., and Pipa, R. L. (1958). *Smithson. Misc. Collect.* **137**, 247.

Rios, T., and Quijano, L. (1969). *Tetrahedron Lett.* p. 1317.

Rolland, M., and Lissitzky, S. (1962). *Biochim. Biophys. Acta* **56**, 83.

Ross, R. H., and Monroe, R. E. (1972a). *J. Insect Physiol.* **18**, 791 and 1593.

Ross, R. H., and Monroe, R. E. (1972b). *Insect Biochem.* **2**, 460.

Roth, L. M. (1968). *Psyche* **75**, 99.

Rowell, C. H. F. (1971). *Advan. Insect Physiol.* **8**, 145.

Rudall, K. M. (1950). *Progr. Biophys. Biophys. Chem.* **1**, 39.

Rudall, K. M. (1962). *Sci. Basis Med.* p. 203.

Rudall, K. M. (1963). *Advan. Insect Physiol.* **1**, 257.

Rudall, K. M. (1965). *Struct. Funct. Connect. Skeletal Tissue, Proc. Advan. Study Inst., 1964* pp. 191–196.

Rudall K. M. (1969). *J. Polym. Sci., Part C* **28**, 83.

Rudall, K. M., and Kenchington, W. (1971). *Annu. Rev. Entomol.* **16**, 73.

Rudall, K. M., and Kenchington, W. (1973). *Biol. Rev. Cambridge Phil. Soc.* **48**, 597.

Rüdiger, W. (1970). *Naturwissenschaften,* **57**, 331.

Runham, N. W. (1961). *J. Histochem. Cytochem.* **9**, 87.

Runham, N. W. (1962). *J. Histochem. Cytochem.* **10**, 504.

Rupley, J. A. (1964). *Biochim. Biophys. Acta* **83**, 245.

Salthouse, T. N. (1962). *J. Histochem. Cytochem.* **10**, 109.

Sannasi, A. (1968a). *Zool. Jahrb., Abt. Allg. Zool. Physiol. Tiere* **74**, 319.

Sannasi, A. (1968b). *Experientia* **24**, 1238.

Sannasi, A. (1969). *J. Invertebr. Pathol.* **13**, 4.

Sapag-Hagar, M., Gonzalez-Gonzalez, M. P., and Stamm-Menendez, M. D. (1961a). *Rev. Espan. Fisiol.* **17**, 89.

Sapag-Hagar, M., Stamm-Menendez, M. D., and Dean-Guelbenzu, M. (1961b). *Rev. Espan. Fisiol.* **17**, 107.

Saunders, B. C., Holmes-Siedle, A. G., and Stark, B. P. (1964). "Peroxidase. The Properties and Uses of a Versatile Enzyme and of Some Related Catalysts." Butterworth, London.

Scheie, P. O. (1969). *Comp. Biochem. Physiol.* **29**, 479.

Scheie, P. O., and Smyth, T., Jr. (1967). *Comp. Biochem. Physiol.* **21**, 547.

Scheie, P. O., and Smyth, T., Jr. (1968). *Comp. Biochem. Physiol.* **26**, 399.

Scheie, P., and Smyth, T., Jr. (1972). *Comp. Biochem. Physiol.* **43A**, 469.

Schildknecht, H., and Vetter, H. (1963). *Fette, Seifen, Anstrichm.* **65**, 551.

Schmalfusz, H. (1937). *Biochem. Z.* **294**, 112.

Schmalfusz, H., and Müller, H. P. (1927). *Biochem. Z.* **183**, 362.

Schmalfusz, H., Heider, A., and Winkelmann, K. (1933). *Biochem. Z.* **257**, 188.

Schmidt, K., and Paulus, H. (1970). *Z. Morphol. Tiere* **66**, 224.

Schneider, H. (1954). *Naturwissenschaften* **41**, 147.

Sekeris, C. E. (1964). *Science* **144**, 419.

Sekeris, C. E., and Herrlich, P. (1966). *Hoppe-Seyler's Z. Physiol. Chem.* **344**, 267.

Sekeris, C. E., and Karlson, P. (1966). *Pharmacol. Rev.* **18**, 89.

Seligman, M., Friedman, S., and Fraenkel, G. (1969). *J. Insect Physiol.* **15**, 553 and 1085.

Shaaya, E., and Sekeris, C. E. (1970). *J. Insect Physiol.* **16**, 323.

Shikata, M. (1960). *Nippon Sanshigaku Zasshi* **29**, 391.

Silhacek, D. L., Carlson, D. A., Mayer, M. S., and James, J. D. (1972). *J. Insect Physiol.* **18**, 347.

Sin, Y. T., and Thomson, J. A. (1971). *Insect Biochem.* **1**, 56.

Sizer, I. W. (1953). *Advan. Enzymol.* **14**, 129.

Srivastava, R. P. (1970). *J. Insect Physiol.* **16**, 2345.

Srivastava, R. P. (1971a). *J. Insect Physiol.* **17**, 189.

Srivastava, R. P. (1971b). *J. Insect Physiol.* **17**, 261.

Stamm-Menendez, M. D., and Dean-Guelbenzu, M. (1961). *Rev. Espan. Fisiol.* **17**, 163.

Stamm-Menendez, M. D., and Fernandez, F. (1957). *Rev. Espan. Fisiol.* **13**, 225.

Stay, B., and Roth, L. M. (1962). *Ann. Entomol. Soc. Amer.* **55**, 124.

Stay, B., King, A., and Roth, L. M. (1960). *Ann. Entomol. Soc. Amer.* **53**, 79.

Stransky, K., Ubik, K., Holman, J., and Streibl, M. (1973). *Collect. Czech. Chem. Commun.* **38**, 770.

Sundara Rajulu, G., and Renganathan, K. (1966). *Naturwissenschaften* **53**, 136.

Surholt, B., and Zebe, E. (1972). *J. Comp. Physiol.* **78**, 75.

Suzuki, M., Ikekawa, N., Akaike, E., and Tsuda, K. (1966). *Chem. Pharm. Bull.* **14**, 837.

Swan, G. A., and Waggott, A. (1970). *J. Chem. Soc., C* p. 1409.

Takahashi, S. Y., and Ohnishi, E. (1966). *J. Biochem. (Tokyo)* **60**, 473.

Tamaki, Y. (1969). *Bochu-Kagaku* **34**, 86.

Tartivita, K., and Jackson, L. L. (1970). *Lipids* **5**, 35.

Taylor, R. L. (1969). *J. Invertebr. Pathol.* **14**, 427.

Thompson, S. N., and Barlow, J. S. (1970). *Comp. Biochem. Physiol.* **36**, 103.

Thompson, S. N., and Barlow, J. S. (1971). *Comp. Biochem. Physiol. B* **38**, 333.

Tobe, S. S., and Loughton, B. G. (1969). *J. Insect Physiol.* **15**, 1331.

Tomino, S. (1963). *Experientia* **19**, 231.

Torossian, C. (1968). *Insectes Soc.* **15**, 73.

Trim, A. R. (1941). *Biochem. J.* **35**, 1088.

Tsao, C-H., and Richards, A. G. (1952). *Ann. Entomol. Soc. Amer.* **45**, 585.

Tsusue, M. (1971). *J. Biochem. (Tokyo)* **69**, 781.

Tulloch, A. P. (1970). *Lipids* **5**, 247.

Tulloch, A. P. (1971). *Chem. Phys. Lipids* **6**, 235.

Van Wisselingh, C. (1898). *Jahrb. Wiss. Bot.* **31**, 619.

Vercauteren, R. E., and Aerts, F. (1959). *Naturwissenschaften* **46**, 449.

Vercauteren, R. E., Aerts, F., and Decleir, W. (1962). *Proc. Int. Congr. Entomol., 11th, 1960* Vol. 3, p. 167.

Viles, J. M., and Mills, R. R. (1969). *J. Insect Physiol.* **15**, 1079.

Vincent, J. F. V. (1971). *J. Insect Physiol.* **17**, 625.

Vuillaume, M. (1969). "Les Pigments des Invertébrés. Biochimie et Biologie des Colorations," Les Grandes Problèmes de la Biologie, Monographie 10. Masson, Paris.

Ward, K., and Seib, P. A. (1970). *In* "The Carbohydrates" (W. Pigman and D. Horton, eds.), 2nd ed., Vol. 2, pp. 435–438. Academic Press, New York.

Waterhouse, D. F., Hackman, R. H., and McKellar, J. W. (1961). *J. Insect Physiol.* **6**, 96.

Watt, W. B. (1967). *J. Biol. Chem.* **242**, 565.

Weaver, A. A. (1958). *Proc. Int. Congr. Entomol., 10th, 1956* Vol. 1, p. 535.

Weir, S. B. (1970). *Nature (London)* **228**, 580.

Weis-Fogh, T. (1960). *J. Exp. Biol.* **37**, 889.

Weis-Fogh, T. (1961). *J. Mol. Biol.* **3**, 520, 648.

Weis-Fogh, T. (1965). *Proc. Int. Congr. Entomol., 12th, 1964* p. 186.

Weis-Fogh, T. (1970). *In* "Insect Ultrastructure" (A. C. Neville, ed.), pp. 165–185. Blackwell, Oxford.

Weygand, F., and Waldschmidt, M. (1955). *Angew. Chem.* **67**, 328.

Weygand, F., Simon, H., Dahms, G., Waldschmidt, M., Schliep, H. J., and Wacker, H. (1961). *Angew. Chem.* **73**, 402.

Whitehead, D. L. (1969). *Nature (London)* **224**, 721.

Whitehead, D. L. (1970a). *FEBS Lett.* **7**, 263.

Whitehead, D. L. (1970b). *Biochem. J.* **119**, 65p.

Whitehead, D. L., Brunet, P. C. J., and Kent, P. W. (1960). *Nature (London)* **185**, 610.

Whitehead, D. L., Brunet, P. C. J., and Kent, P. W. (1965). *Proc. Cent. Afr. Sci. Med. Congr., 1963* pp. 351 and 365.

Wiesmann, R. (1938). *Vierteljahresschr. Naturforsch. Ges. Zuerich, Beiblatt* **83**, 127.

Wigglesworth, V. B. (1933). *Quart. J. Microsc. Sci.* **76**, 269.

Wigglesworth, V. B. (1945). *J. Exp. Biol.* **21**, 97.

Wigglesworth, V. B. (1970). *Tissue & Cell* **2**, 155.

Willig, A. (1969). *J. Insect Physiol.* **15**, 1907.

Willis, J. H. (1970). *Amer. Zool.* **10**, 320.

Wolfe, L. S. (1954). *Quart. J. Microsc. Sci.* **95**, 49.

Wolfe, L. S. (1955). *Quart. J. Microsc. Sci.* **96**, 181.

Wyatt, G. R. (1968). *In* "Metamorposis: A Problem in Developmental Biology" (W. Etkin and L. I. Gilbert, eds.), pp. 143–184. Meredith Corp., New York.

Yamazaki, H. I. (1969). *J. Insect Physiol.* **15**, 2203.

Yasunobu, K. T., Peterson, E. W., and Mason, H. S. (1959). *J. Biol. Chem.* **234**, 3291.

Zdarek, J., and Fraenkel, G. (1969). *Proc. Nat. Acad. Sci. U.S.* **64**, 565.

Zdarek, J., and Fraenkel, G. (1970). *Proc. Nat. Acad. Sci. U.S.* **67**, 331.

Ziegler, I., and Harmsen, R. (1969). *Advan. Insect Physiol.* **6**, 139.

Chapter 4

PERMEABILITY OF INSECT CUTICLE

Walter Ebeling

I. Introduction

The insect physiologist is apt to derive interest in cuticle permeability primarily from its importance in the transpiration and absorption of water, as well as the broad physiological and ecological implications of these factors. The economic entomologist has, likewise, come to realize the potentialities in insect control, from the interesting fact that insects and other arthropods are protected from a lethal rate of desiccation by an extremely thin and fragile epicuticular lipid layer, and from environmental and other toxicants by one or more of the cuticular layers. With many species, disruption of these cuticular barriers has proved to be an effective means of or a valuable adjunct to control measures. The toxicologist, in turn, is concerned with the routes of penetration through the cuticle for polar and apolar insecticides and with the role of the cuticle as a barrier, a storage area, or even in some cases as an activator, for toxicants.

From the following pages it will be apparent that investigators with purely academic interests have contributed their full share toward the elucidation of the intricate problems connected with cuticle permeability, particularly those relating to the penetration of water. Likewise, substantial fundamental contributions concerning the waterproofing mechanism of insect cuticle and the nature of the lipoidal and aqueous pathways for the penetration of insecticides have been made by researchers whose investigations have been directed primarily toward the objectives of pest control. If this review reflects an unusual degree of rapport in the efforts of the two groups of investigators, it may be because the integument is, possibly more than any other organ of the insect's body, the object of such mutual interests. Both groups of investigators have made substantial contributions to our understanding of cuticle permeability since the publication of the first edition of this treatise. Accordingly, it is a welcome opportunity to review these contributions in this revision.

II. Basic Features of Insect Integument

Progress in our understanding of the permeability of insect cuticle depends to a large extent on the progress made in basic investigations on cuticular structure and chemistry—subjects reviewed in this volume by Locke in Chapter 2 and Hackman in Chapter 3, respectively. A good concise review of cuticular characteristics that influence permeability was presented by Noble-Nesbitt (1970b). However, a brief discussion of elementary features, along with a sketch of an idealized section of insect integument (Fig. 1), will enable the reader to visualize the pathways of penetration.

The insect cuticle, as well as the arthropod cuticle in general, is secreted or deposited on the outer surface of the epidermis, composed of cells (Fig. 1, h–n) which control or supply such external structures as setae, hairs, bristles, and ducts, which are of great interest in relation to permeability. Some investigators believe they form, along with the articular and intersegmental membranes, the only pathways of entry for insecticides. The epidermal cells are separated from the body cavity by the *basement membrane,* which appears to be continous and homogeneous, having no internal structures or pores discernible with the light microscope (Wigglesworth, 1933; Woods, 1929).

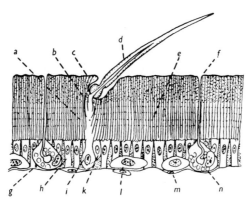

Fig. 1. Section of insect integument: a, laminated endocuticle; b, exocuticle; c, epicuticle; d, sensilla; e, pore canals; f, duct of dermal gland; g, basement membrane; h, epidermal cell; i, trichogen cell; k, tormogen cell; l, oenocyte; m, hemocyte adherent to basement membrane; n, dermal cell. (From Wigglesworth, 1948b.)

When the cuticle is first secreted, it is soft and pliable and may remain that way on many insects, particularly larvae and some aquatic species. With most adults, however, the outer portion of the cuticle becomes hard and dark (sclerotized). Therefore, it is essential that it be divided into a series of plates or *sclerites* between which are folds of soft and flexible membrane that allow for stretching and distension, as well as for articulation. This membranous material is a continuation of the sclerotized cuticle and is homologous in structure. Correlated with the fact that they are not sclerotized and probably because they are usually thinner, the membranous portions of the cuticle are more permeable.

The *procuticle* or *chitinous cuticle* (Fig. 1,a and b) comprises the bulk of the cuticle. Although different layers can be distinguished (Richards, 1951, 1958), the procuticle is essentially a hydrophilic chitin—protein complex containing a considerable quantity of water. Dehydration occurring during hardening and darkening decreases its permeability.

Setae (bristles, hairs, etc.) are sensilla that arise from pits in the cuticle surface in a kind of ball-and-socket arrangement (Fig. 1,d). Setae are formed by two cells: a sensillum-forming cell (*trichogen*) and a socket-forming cell (*tormogen*) (Fig. 1,i and k) and, of course, whatever cells that give rise to the sensory nerve(s). In the pit, the cuticle (articular membrane) is relatively thin and so is the cuticle of the seta, etc., particularly in the case of some of the chemoreceptors. Cuticles of olfactory sense hairs are perforated by many minute pores (Slifer *et al.,* 1959; Steinbrecht and Müller, 1971). Cytoplasmic filaments, arising in the trichogen and tormogen cells, pass up through the cuticle to enter the seta, forming a cytoplasm continuum in close contact with chemicals that may be deposited on the surface of the cuticle. The significance of setae in relation to permeability will be discussed further in Section IV,B,3.

On the surface of an insect may also be found mere outgrowths of the cuticle called spines, if large, and spinules, if they are small. These do not arise from pits and have no direct connection with epidermal cells, and are not sensilla.

Arising from the epidermal cells and extending through the procuticle of most insects, and sometimes part way into the overlying epicuticle, are usually minute ducts, called *pore canals* (Fig. 1,e). They are so narrow that they can be adequately distinguished only in species in which they are exceptionally large, as in the blow fly larvae, or with the aid of an electron microscope. Much larger pores, the *dermal gland ducts* (Fig. 1,f), arise from dermal glands (Fig. 1,n), and reach to the surface of the epicuticle (Richards, 1951; Wigglesworth, 1948b; Scheie *et al.,* 1968).

Arthropods possess an outer nonchitinous layer, the *epicuticle,* generally about 1 μm in thickness (only a small fraction of the thickness of the cuticle). Wigglesworth (1947) recognized four distinct sublayers termed, from inside to outside, the cuticulin, polyphenol layer, wax layer, and the "cement" layer.

The cuticulin layer is considered to be composed of conjugated protein of the lipoprotein group and which Wigglesworth (1933, 1947) believes is secreted by the large oenocytes (Fig. 1,l) which may be found below the epidermal cells. Wigglesworth (1947, 1948a) believes that the pore canals deposit a fluid possessing dihydroxyphenols which serve to tan the cuticulin layer. Over the surface of the hydrophilic tanned protein, a wax layer appears. The wax is conducted to the surface of the cuticulin by *wax canals* arising from the ends of the pore canals. They are discernible only with the aid of an electron microscope (Locke, 1961, 1964, 1965; Gluud, 1968). They will be discussed in greater detail in Section IV,A,6. The wax layer appears shortly before molting, whereas *after* molting, the dermal glands secrete a fluid that spreads over the surface of the wax layer of some insects to form the "cement," which Beament (1955) believes is similar to shellac. After the cement is secreted, the dermal ducts probably usually become occluded as Way (1950) observed in the larvae of *Diataraxia oleracea.*

The epicuticular lipid or wax layer is concerned in such important processes as adsorption and wetting, and has been shown to be the principal barrier to the passage of water into or out of the arthropod cuticle (Ramsay, 1935a,b; Wigglesworth, 1945; Beament, 1945, 1965; Lees, 1947; Winston, 1967; Winston and Beament, 1969).

The insect wax layers generally range in thickness from 0.1 to 0.4 μm, with a probable average of about 0.25 μm, although the soft "grease" of the cockroach is about 0.6 μm in thickness (Beament, 1945). The waxes of the various insect species show a wide range of melting points, up to over 100°C, with a corresponding range of hardness and crystallinity. They consist of alcohols, hydrocarbons, wax acids, esters, diols, and sometimes aldehydes, phospholipids, and other minor constituents. Bursell and Clements (1967) point out that the long-chain alcohols appear to predominate in the hard-cuticle waxes and the proportion of hydrocarbon is high in the soft waxes and greases. Beeswax is particularly high in esters (Warth, 1956; Tulloch, 1970). The epidermis of the American cockroach can synthesize hydrocarbons from esters (Nelson, 1969) and fatty acids (Jackson and Baker, 1970). The chemistry of insect cuticular lipids has been reviewed by Jackson and Baker (1970) and is treated by Hackman in the preceding chapter in this volume.

III. Permeability—Artificial Membranes and Lipid Films

When reduced to its barest essentials for an understanding of permeability, the cuticle can be thought of as a very thin lipid film supported by a much thicker hydrophilic, predominantly chitin–protein–mucopolysaccharide membrane, which is in turn deposited on the outer surface of living cells that exert a regulatory effect on permeability phenomena. The "hydrophilic membrane" is pierced by myriads of pores.

It may be useful to consider to what extent research with artificial membranes and lipid films has (1) served as a stimulus and a guide to experimentation with living insects and in the interpretation of results, and (2) to what extent the generalizations inferred from work with artificial membranes and films must be modified as the result of the structural and biochemical aspects of the insect integument which are not artificially reproducible.

A. PENETRATION OF WATER THROUGH SOLID LIPID LAYERS

In view of the apparent lack of information on the penetration of water through insect lipid, the experiments of Lovegren and Feuge (1954) are of interest. These investigators worked primarily with films of acetostearin products, which are di- and triglycerides containing one and two acetyl groups, respectively, and solidify into unique waxlike solids. Permeability constants were greatly affected by the vapor-pressure differences across the films and the vapor pressures at the surfaces of the films. With one product, maintaining one side of the film at a vapor pressure of nearly 0 mm Hg, while increasing the vapor pressure on the other side from 6.17 to 18.24, increased the permeability constant from 44.7 to 133. When the original vapor pressure on both sides of the film was raised to a high value (18.73), the permeability constant from an equivalent vapor-pressure difference increased still further to 410.

Lovegren and Feuge (1954) believe that the diffusing molecule of water passes into the film by being absorbed on an active spot on the external surface of the film. In this position it vibrates until it acquires sufficient energy to evaporate, then passes on to the next active spot. The process is repeated until the molecule emerges on the other side of the film. If a sufficient quantity of water diffuses through the film, there is a weakening or even a disruption of the intra- and intermolecular bonds originally in the film and the resistance to diffusion is reduced.

At a constant relative humidity, vapor pressure increases with temperature and in nearly all instances the rate of diffusion of water through a lipid film likewise increases with temperature. However, the effect of temperature on the physical composition of the film may lead to variations from theoretical expectations.

Taylor et al. (1936) have considered the common observation that water passes through organic materials and not through metals. They believe the difference may be due to the greater intermolecular spacings in the organic materials. For metals, these spaces can be calculated to be too small for the passage of water, but the center-to-center distance for wax molecules is about three times the diameter of the water molecule.

The water permeability of solid lipids is known to be greatly increased by the addition of liquid hydrocarbons. Thus, increasing the percentage of liquid cottonseed oil in highly hydrogenated cottonseed oil (solid) from 0 to 40% increased the water-permeability constant from 1.3×10^{-12} to about 420×10^{-12}, a 325-fold increase (Landmann et al., 1960).

1. Effect of Hydrogen-Ion Concentration

The influence of pH on the penetration of lipid layers was investigated (Ebeling and Clark, 1962) by using 15 μm layers of beeswax and paraffin wax that were prepared by means of a microtome. Sections of such wax layers were placed on a black surface. As soon as the layers were penetrated the black color of the substrate became visible from above. While an increase in pH did not affect the rate of penetration of water through paraffin wax, it greatly increased the rate of penetration through beeswax. Beeswax had a relatively high proportion of esters when compared with other insect waxes. In a recent analysis, utilizing the latest analytic procedures, beeswax was found to consist of 65% esters, 23% hydrocarbons, 12% free acids, and 1% free alcohols (Tulloch, 1970). It is likely that the ester linkages in beeswax are hydrolyzed in the presence of a base, resulting in saponification of the ester fraction as well as the acids, and causing the layer to be more permeable. With KOH as the base, no penetration of water into beeswax occurred in 24 hours, at pH 8. The period required for penetration then decreased exponentially from 460 minutes at pH 9 to less than 15 seconds at pH 14 (1 N KOH) (see tabulation).

pH	Penetration time (minutes)
7	No penetration
8	No penetration
9	460
10	45
11	4
13	0.25
14	<0.25

The effect of pH on the permeability of beeswax to 2% solutions of nicotine sulfate and sodium arsenite was about the same as its effect on the permeability of beeswax to water. The penetration of arsenite ions was confirmed by precipitating the sodium arsenite from the penetrated solution with acid, both for the 15 μm films and the wax-impregnated paper described later.

An aqueous solution of a strong base (5 N KOH) penetrated 50 μm beeswax in 1 minute, and 10^{-3} N KOH penetrated in 2 minutes, while a weak base (5 N NH$_4$OH) penetrated in 9 minutes. Solutions of either strong or weak acids, as well as distilled water, however, did not penetrate to any appreciable extent.

A grade of newspaper stock that had no resistance to the penetration of water was dipped in a 10% solution of white beeswax in carbon tetrachloride. After the solvent had evaporated, pieces of waxed paper were placed on a strip of universal indicator paper on a plate of glass supported above a mirror. A change in color of the indicator could be detected at the instant of complete penetration of water from droplets placed on the waxed paper. In five trials, the average period for penetration of a normal KOH solution (pH 14) was 14 minutes. Distilled water alone evaporated before there was any evidence of penetration.

When the same waxed paper was heated to temperatures above the melting point of wax and then cooled, it never returned to its original white color but retained the more translucent appearance of melted wax. Penetration of the KOH solution then required only about 1 minute, indicating some irreversible change in the orientation of the wax molecules on the cellulose fibers, accompanied by an increased permeability. The matter of "transition temperatures" will be discussed more fully in Section IV,A,2.

Glynne Jones and Edwards (1952) also worked with 15 μm layers of beeswax prepared with a microtome and found that a sodium salt of DNOC (4,6-dinitro-o-cresol) in aqueous solution would penetrate these layers.

Their original idea was that a slight hydrolysis might take place in the salt solution, resulting in free sodium hydroxide, which might have an "emulsifying action" on the wax. However, they believed this possibility was precluded by their finding that the ammonium salt of DNOC was also capable of penetrating the wax membrane. Our experiments showed that an ammonium hydroxide solution can also penetrate beeswax. It is well known that NH_4OH has the same type of saponifying action on an ester (ammonolysis) as NaOH, but at a slower rate.

2. Effect of Surfactants

Although water at a normal pH does not appreciably penetrate white beeswax or paraffin wax, penetration can be increased by the addition of surfactants to such an extent that a 15 μm layer can be penetrated almost instantly (Ebeling and Clark, 1962). Newspaper stock dipped in 10% beeswax was usually penetrated in less than one minute by 1% solutions of anionic and nonionic surfactants, compared with an average of 14 minutes, in five trials, for a normal solution of KOH (pH 14). Oil emulsions can be made to penetrate rapidly by using appropriate emulsifiers. The best emulsifiers for the penetration of beeswax are those of high pH, such as triethanolamine or monethanolamine, but these were of no advantage in the penetration of paraffin wax layers.

3. Effect of Molecular Organization

Films of waxes or fatty substances are most effective barriers to the passage of water in liquid or vapor form if their molecules are oriented and packed in an appropriate manner. The striking effect of orientation and packing on monolayers resting on a free water surface was demonstrated by Sebba and Briscoe (1940), who showed that the optimum lateral pressure on a monolayer of alcohol (C_{20}) decreased the evaporation of water by 99%.

The molecular species involved is also important. Langmuir and Schaefer (1943) state that a film of cetyl alcohol (C_{16}) on water is 100 times more effective in decreasing evaporation than a comparable film of oleic acid (C_{18}). This emphasizes the importance of working with pure samples and avoiding contamination in experiments with synthetic water barriers, for in this instance 1% of oleic acid in a cetyl alcohol film reduces resistance to evaporation by 50%.

B. Penetration of Water through Lipids on Artificial Membranes

Lipid films on artificial membranes have contributed much to our understanding of cuticle permeability. Their use in this kind of investigation appears to be justified, provided that the supporting membrane has the ability to organize the adjacent molecules of lipid, and provided also that the lipid material applied to the membranes is of a type that is amenable to organization. The important consideration in the present context is that the epicuticular lipids are composed principally of "polar" [possessing polar (hydroxyl) and nonpolar (alkyl) ends] molecules which can orient themselves in a monolayer on the hydrophilic and water-bearing tanned-protein surface of the cuticulin. This monolayer of organized lipid has substantial waterproofing qualities and moreover may order the arrangement of still other layers of epicuticular lipids for an unknown distance above it, at least in insect species with particularly water-impermeable epicuticles. In lipid layers on artificial hydrophilic membranes, or in biophysical models, the lipid must likewise consist of polar molecules to have any meaningful relevance to the insect epicuticle; hydrocarbons (e.g., paraffin wax) are worthless except as a control to emphasize the physicochemical mechanism involved in cuticular permeability.

Hurst (1950) reported that the electron diffraction pattern of lipid layers on collodion membranes is similar to those on the epicuticle of *Calliphora* pupae. These lipid layers were found to consist of a three-dimensional arrangement of orthorhombic microcrystals, oriented perpendicular to the surface of the membrane. However, a perpendicular arrangement is not in accord with the theory of molecular orientation of lipids at the lipid/cuticulin interface favored by Beament (1964), as will be discussed later.

1. Experiments with Celluloid Substrates

Much basic information on the permeability of insect wax layers was obtained by Alexander *et al.* (1944a) in experiments with artificial membranes resembling insect cuticle in performance as closely as possible. The most suitable membranes available were sheets of transparent celluloid, 0.003 inch (0.070 mm) thick, which remained rigid when wet. They were attached to cells designed to facilitate the gravimetric determination of water movement through the membranes. The cells were placed in a desiccator at $23° \pm 0.5°C$. The rate of diffusion of water through the 11 cm² of the celluloid membranes was about 5 mg per hour (0.45 mg/cm²/hour), comparable to the rate of diffusion through the average insect cuticle from which the protective wax layer has been removed.

The above investigators also investigated the effect of wax films in retarding the diffusion of water through the membranes. The substances used as water barriers, besides the natural waxes, were various hydrocarbons, carboxylic acids, alcohols, esters, and other miscellaneous waxlike materials. The thickness of the films was generally about 0.2 μm, or approximately 50 to 100 molecules in depth, and they were practically invisible. Beeswax was found to be the most effective of all twenty-nine substances tested. No liquid film offered any appreciable resistance to water loss under the con-

ditions of the experiment. Beeswax and paraffin wax reduced evaporation
by 83% and 57%, respectively. In a separate experiment these two waxes
were applied in films ranging up to 4×10^{-5} gm/cm². The curves A and
B in Fig. 2 show the striking difference in the properties of the two waxes.
Beeswax attained its maximum efficiency in retarding evaporation in a film
that was only 0.02 μm in thickness. From the data of Müller (1930), who
showed that a C_{30} hydrocarbon chain was about 40 Å long, Alexander *et
al.* (1944a) calculated that the beeswax film consisted of five monolayers,
presuming the molecules to be normal to the surface.

No additional amount of beeswax contributed to the impermeability of
the film. On the other hand, increasing the quantity of paraffin wax on
a film composed of that material resulted in decreasing water loss, in an
exponential relationship, up to the maximum quantity used in the experiment. Alexander *et al.* (1944a) concluded that the polar groups in the
celluloid substrate probably cause orientation of the first layer or layers of
molecules of an acid, alcohol, ester, etc., whereas paraffin wax, having no
polar groups, would crystallize at random, molecules resting on the surface of
the hydrophilic membrane providing no greater barrier to water loss than
those farther removed from that interface.

2. *Experiments with a Wing-Membrane Substrate*

Beament (1945) refined the technique of Alexander *et al.* (1944a) by
using membranes believed to simulate more closely the insect epicuticle in

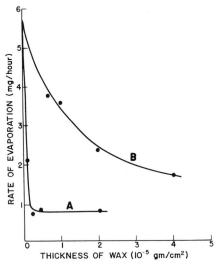

Fig. 2. Effect of thickness of wax on passage of water through films of beeswax
(A) and paraffin wax (B). (Redrawn from Alexander *et al.,* 1944a).

its ability to orient wax molecules. When cockroach grease was placed on a *Pieris* wing membrane, at 22°C, permeability dropped rapidly during the first 6 hours from 15 to 3.8 mg/cm²/hour, then was gradually reduced to a minimum of 0.35 in 100 hours. Beament believed that in 6 hours an oriented and perhaps compressed monolayer of molecules had been formed. He assumed that such a film could not be removed by alumina dust for the dust was unable to increase water permeability significantly. He concluded that adsorption by inert dusts cannot overcome the orientational bonds of a wax layer deposited on a membrane. However, Wigglesworth (1942) found that when sprinkled on cockroaches that had been killed to eliminate the possibility of abrasion through the insects' movements, the powder caused a rapid water loss. Water loss was 46.8% in 48 hours, compared with 32.0% when insects were rubbed with alumina to facilitate the passage of water by disruption of the impermeable wax barrier. This indicates that the constituents of the water barrier in the cuticle of the cockroach are much more amenable to adsorption than an oriented and compressed monolayer of cockroach grease on butterfly wing. The implications of this difference in results on a dead cockroach and a biophysical model will be discussed later.

Olson and O'Brien (1963) considered the grease removed from the epicuticle of the American cockroach by 1 ml of water rinse to be the "A" layer and the grease subsequently removed with a 1 ml acetone rinse as the "B" layer. Neither contained proteinaceous epicuticle. The relative solubilities of lipophilic and lipophobic compounds in the two layers suggested that the "B" layer, with the alkyl ends of the molecules directed outwards, is particularly hydrophobic.

C. Adsorption and Migration of Wax Molecules

In the experiment the results of which are shown in Fig. 2, Alexander *et al.* (1944a) allowed dusts to settle on the wax films to determine whether they affected the waterproofing properties of the films. The manner of application of the dusts would appear to rule out the possibility of removal or disruption of the lipid film by abrasive action. This leaves adsorption and surface migration of the lipid as the probable cause for increased water permeability, a supposition that is supported by the finding that the highly sorptive activated gas-mask charcoal was superior to carborundum in increasing the permeability of the wax films despite its relative softness and larger particle size. As with activated carbon, acid activation of a clay by increasing its sorptivity for wax, greatly increases its insecticidal efficacy, even though its abrasiveness remains unaltered (Majumder *et al.*, 1959; Ebeling, 1971). This is not meant to imply that insect cuticle cannot be abraded and thereby made to lose its barrier properties. In fact, abrasion of the epicuticle as by rubbing with sand paper or abrasive dusts has played an important role in permeability experiments.

Adsorption in Relation to the Organized Wax Layer

Beeswax possesses both amorphous and crystalline fractions (Ebeling, 1961). The following mechanism was proposed by Ebeling (1961) for the removal of the amorphous constituents of a beeswax film: Molecules in a solid are in constant motion within their own loci, the dimensions of which are circumscribed by the surrounding molecules. However, if one of these surrounding molecules is removed, its place will be taken by its neighboring molecule. It is proposed that, when the molecules at the surface of a wax film which are amenable to adsorption are removed, those below will take their place and that a progressive displacement of molecules from the amorphous wax fractions will occur in this manner down through the film. Eventually minute pathways are formed in an otherwise crystalline matrix. A sufficient number of molecules is removed from the wax film greatly to increase its permeability to light and to water, even though the depth of the film is unchanged. The same process was presumed to take place when sorptive dust rests on the epicuticle of an insect. Removal of molecules from the organized lipid water barrier at the cuticulin/lipid interface may be presumed to take place in the same way, but with possibly greater difficulty. However, a breakthrough in our understanding of how wax reaches the surface of the cuticulin layer, via "wax canals," to be discussed later, has obviated the necessity for presuming that disruption of the organized lipid barrier is required to bring about a lethal rate of water loss.

Ebeling (1961) found that powders that were said by the manufacturers to have a pore diameter of at least 20 Å and which also had a large specific surface (e.g., 300 m^2/gm) were the only dehydrated powders he investigated that had great insecticidal efficacy. For prolonged effectiveness, however, a powder should have a low sorptivity for atmospheric moisture while retaining its ability to absorb wax. Certain fluorinated silica aerogels proved to be particularly effective.

D. Penetration of Organic Molecules

Organic molecules, particularly in liquid form, penetrate relatively rapidly into membranes and waxy films. Petroleum oils are particularly effective in penetrating lipid layers, increasing in rate of penetration with decreasing distillation range. If they contact a water-bearing membrane underlying the lipid film, their penetrativity can be further increased by the addition of a surface-active solute to decrease the interfacial tension oil/water. As an example, Wigglesworth (1942) placed a drop of dyed oil on a piece of blotting paper saturated with water. The oil appeared at the lower surface in about $1\frac{1}{2}$ hours. The same oil containing 5% oleic acid penetrated in $1\frac{1}{2}$ minutes. This phenomenon may also be demonstrated with strips of blotter paper serving as wicks to draw the oil upward (Ebeling and Clark, 1962). The blotter must contain much water if the addition of oleic acid (or surfactant) is to increase capillary rise. No difference in capillary rise can be demonstrated if the blotter is merely damp.

Even with organic solids, penetration is sometimes remarkably rapid. When a particle of powdered oil-soluble red dye was placed on a layer of beeswax, it dissolved in the wax and spread from the initial point of contact. At 26.7°C the progress of the dye into the wax was found to be at the rate of about 125 μm per day, a distance that is about 500 times the thickness of the average insect epicuticular lipid layer (Ebeling, 1961). Evidence of lateral migration of organic insecticides in the cuticular lipid layer is presented in Section IV,B,3,a.

IV. Permeability—Insect Cuticle

Because the previous section dealt primarily with the penetration of water into artificial membrane systems, it would seem to be appropriate to follow with a discussion of the transpiration and uptake of water by the insect integument per se. More particularly, however, a discussion of water permeability is appropriate at this point because of the great contributions that have been made toward our knowledge of the insect cuticle, particularly the epicuticle, through the investigations on such permeability of the cuticle. This should also aid in the understanding and interpretation of other aspects of permeability, to be discussed later.

A. WATER

Because most terrestrial animals live under conditions in which the humidity is at least appreciably below saturation, and in many instances extremely low, their means of conserving water is obviously an important aspect of their structure and physiology. Whereas water content is a function of the volume of an animal, evaporation is a function of the surface area (Kennedy, 1927). Smaller animals, including the arthropods, are thus placed at a disadvantage with respect to water conservation and one would expect to find among them the most highly specialized structural and physiological mechanisms for this purpose. Nevertheless, there is great variation among

species, dependent upon the extent to which ecological adaptations can share
the burden of water conservation (Edney, 1956; Lees, 1948).

1. Arthropods Lacking an Epicuticular Lipid Layer

Not all arthropods are protected against a lethal rate of water loss by
an epicuticular lipid layer. Amphipods, sowbugs, pillbugs, millipedes, centipedes, and probably at least some whip scorpions, have neither an epicuticular lipid protective layer nor the tracheal spiracles of insects. Therefore, they
lose water rapidly in dry air, are generally nocturnal in habits, and must
confine themselves to damp environments (Edney, 1949, 1951, 1968; Cloudsley-Thompson, 1968). Besides seeking damp environments to avoid excessive cuticular transpiration, such arthropods may also obtain liquids from
their prey or drink free water from a wet substrate. They lose water at remarkably high humidities. Edney (1951) found that, after desiccation in
dry air for 45 minutes or at 50% RH for 23 hours, at room temperature,
none out of four species of sowbugs, one species of pillbug, and one species
of millipede could live or recover weight if subsequently kept at 95% RH.
The pillbug *Armadillidium* sp. could recover weight if kept at 95% RH,
but all others required a saturated atmosphere. The whip scorpion which,
like the other arthropods mentioned, loses water in direct proportion to the
saturation deficit of the atmosphere, suggesting the absence of an epicuticular lipid layer, lost water readily to the air at 26°C at relative humidities
up to 95% (Crawford and Cloudsley-Thompson, 1971).

The aquatic larvae of some insects also lack a lipid water barrier. A. G.
Richards (personal correspondence) finds that the larvae of chironomids and
some mosquitoes dry up at about the same rate as a drop of water of the
same weight.

A decreased transpiration rate of living isopods after considerable desiccation was noted by Bursell (1955). He suggested that the depletion of water
in the bodies of these arthropods may have resulted in a concentration of
body fluids and a consequent shrinkage of the endocuticle which he considered to be the principal water barrier of these isopods. According to this
theory, the shrinkage of endocuticle brings about a closer packing of the im-
pregnating lipid molecules and decreased permeability to water. (In insects
the endocuticle is considered to be the most permeable layer of the procuticle.) Dehydration of cuticle also decreases permeability (page 287).

With arachnids and insects, which are covered with a protective lipid
layer, there is little increase in water loss with increasing temperature until
the "transition temperature" (p. 288) is reached. However, at temperatures
above transition, the rate of water loss of such arthropods is then for some
time even greater than that of the forms lacking a protective lipid layer,
until the two have reached the same degree of water depletion. With three

isopods (*Armadillidium, Porcellio,* and *Philoscia*) and a millipede (*Glomeris*), Edney (1951) showed a straight line relationship between saturation deficit (mm Hg) and rate of evaporation (mg/cm²/hour) as temperatures were increased from 10° to 60°C (Fig. 3). This is characteristic of arthropods lacking a protective epicuticular wax layer and spiracles. With the cockroach *Blattella germanica,* there was no appreciable water loss by evaporation until the "transition temperature" (30°C) was reached. Above 40°C the rate of water loss with *Blattella* was much more rapid than with the other arthropods. Cloudsley-Thompson (1955) found that by the time temperature had reached 60°C (and the saturation deficiency of the atmosphere 150 mm Hg), sowbugs (*Porcellio*), millipedes (*Oxidus*), spiders (*Lycosa,* and insects (*Pieris* larvae) averaged about the same degree of water loss.

Arthropods without a lipid protective layer are subject to a double hazard. Not only are they subject to a lethal rate of water loss at ordinary temperatures, when their environment becomes too dry, but they are also subject to a lethal rate of water uptake by osmosis in environments with an excess of free water (Cloudsley-Thompson, 1968). For example, in California, where rain is infrequent, the amphipod *Talitroides sylvaticus* Haswell sometimes inhabits damp locations in gardens, such as under patches of ivy. Rain

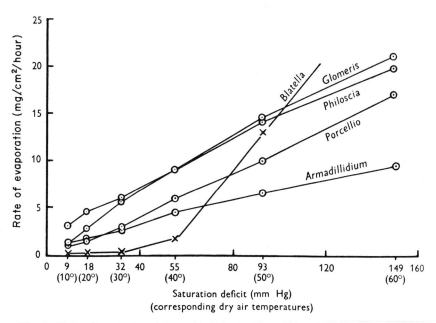

Fig. 3. Rate of water loss from two sowbugs (*Porcellio* and *Philoscia*), a pillbug (*Armadillidium*), a millipede (*Glomeris*), and a cockroach (*Blattella*), in a 1-hour exposure in dry air at temperatures from 10° to 60°C. (From Edney, 1951.)

may create a hazardous situation and the amphipods seek dry locations, often invading houses. There the atmosphere is too dry and they soon perish. *Porcellio scaber* migrates to the upper parts of trees at night in search of lower humidities at which it can lose excess water absorbed from its daytime saturated or near saturated shelters in woodland litter (Den Boer, 1961). However, some insects with an epicuticular lipid layer face a similar problem. The termite *Cryptotermes brevis* (Kalotermitidae), normally inhabiting dry areas, becomes bloated and finally dies at 86% RH (Collins, 1969). Apparently in this case cuticular transpiration plus excretion cannot remove metabolic water sufficiently rapidly (Bursell, 1964). Another kalotermitid, *Kalotermes approximatus,* can survive similar conditions of high humidity. They void fecal pellets as discrete dry pellets when the humidity is low, but as amorphous liquid masses when humidity is high (Hetrick, 1961). Collins (1969) has observed the same phenomenon in various genera of other termites.

2. The Epicuticular Lipid

The only site of water loss in insects that was recognized by Mellanby (1932) was the tracheal system. Consequently it appeared reasonable to Gunn (1933) that the great increase in water loss from the cockroach that he found to occur at temperatures above 30°C should be attributable to increased ventilation at the higher temperatures. Ramsay (1935a,b) found as Gunn (1933) had reported previously, that evaporation of water from the bodies of dead or live cockroaches increased abruptly above 30°C. He also noted that this was true of dead cockroaches whose spiracles had been blocked with wax. The explanation suggested was that the sudden break in the evaporation-rate curve was the result of a phase change in some cuticular constituent, possibly the melting of a fat. He found that with fat-impregnated artificial models no such abrupt break in the evaporation curve occurred. Ramsay then made the important observation that small drops of water placed on the bodies (but not the wings) of cockroaches persisted for indefinite periods at room temperature, but when the temperature was raised to above 30°C, the drops disappeared rapidly. Microscopic observation revealed that the drops were covered by a film of "fatty substance." This was subjected to a "change of state" at 30°C, allowing for rapid evaporation of the water covered by the film.

Even before Ramsay's highly significant experiment, interest in the epicuticular lipid as a barrier to the passage of water from the insect cuticle was initiated by Kühnelt (1928, 1939), who not only demonstrated the lipoid nature of this barrier, but showed that its disruption caused abnormal water loss. There followed an active interest in the control of insects by the application of finely divided powders to decrease the water impermeability of the epicuticular lipid, beginning with the experiments of Zacher and Künicke (1931) and Wigglesworth (1933) and continuing to this day. Experiments with finely divided powders played an important part in demonstrating the function of the protective lipid, especially after the proof by Alexander *et al.* (1944b), Wigglesworth (1944, 1945), Kalmus (1944), Parkin (1944), and Beament (1945) that the powders caused desiccation by removing this material.

Lees (1948) found that water loss through the cuticles of normal ticks in dry air at room temperature amounts to less than 0.2% of that lost from a free water surface, yet the removal of the lipid barrier as by abrasion or by extraction with solvents or detergents can result in complete desiccation of these arthropods within a few hours.

In *Sarcophaga,* the lipid appears to be distributed in depth throughout the cuticle, which is impermeable to water until the entire epicuticle has been abraded (Dennell, 1946; Richards *et al.,* 1953). Richards (1957) found that abrasion to such depth also eliminated the barriers to ions (measured by electrical impedance) and oxygen (measured in a polarographic cell), although the major barrier to ions was found to be the outer epicuticle. Light surface abrasion increased ion penetration and probably oxygen penetration. Richards found that treatment of the isolated cuticle with chloroform greatly increased the rate of penetration of water but did not necessarily affect the rate of penetration of ions or oxygen. He therefore suggested that a series of different barriers are present, and that extrapolation from data on one type of substance (e.g., water) to others (e.g., ions) must be done only with due caution.

Waterproofing in depth has also been shown for the adult stag beetle, *Lucanus* (Lafon, 1943; Beament, 1960). The impermeability of the *Lucanus* cuticle may be the result of its dryness. Richards (1958) pointed out that if one air-dries cuticle it becomes impermeable to solvents. Only hydrated cuticle is freely permeable. He states that the impermeability of dry cuticle is due to arthropodin and that dry arthropodin is impermeable both by itself and when associated with chitin in cuticle.

Wigglesworth (1945) exposed an *Agriotes* larva to dry air for successive 15-minute periods at 30°C. The rates of water loss expressed in mg/cm²/hour were as follows: 17.2, 11.6, 9.6, 7.6, 4.4, 2.8. He believed that the low final value may have been influenced by the depletion of water in the body but that the rapid initial decrease must have resulted from a progressive decrease in the permeability of the cuticle. This he attributed to the tendency of water to pass through layers of organic material at a decreasing rate with decreasing water content. Desiccation decreases the water permeability of all membranes, whether they be artificial or derived from animals or plants, dead or alive (King, 1944; Mitchell, 1956).

The cuticular wax is probably unevenly distributed over the insect body and some limited areas, such as humidity receptors and chemoreceptors, have no wax on them (Slifer, 1954; Beament, 1957).

3. Transition Temperatures and Their Significance

As in the experiments with artificial membranes (Section III,A,1), Wigglesworth (1945) and Beament (1945) demonstrated with both live and dead insects that an increase in temperature to a point termed the "critical temperature" resulted in what was believed to be a permanent change in phase or a "transition" in the packed and oriented layer of lipid on the tanned-protein cuticulin substrate, resulting in a sudden and rapid increase in water loss. The critical temperatures were somewhat lower than the melting point of the lipid and varied with the insect species, rising with increases in the carbon-chain length of the lipids, which for most insects are hard waxes. The epicuticular lipid of the cockroach, *Blattella germanica,* which is a soft grease, was found to have a transition point at about 30°C, as Ramsay (1935a,b) had observed for cockroach grease on water droplets; epicuticular lipids of the leaf-eating larvae of *Pieris brassicae* and the sawfly *Nematus ribesii* had similar transition points. *Tenebrio molitor* and *Rhodnius prolixus,* insects from dry environments which can survive long periods without feeding, were found to have transition points 20° to 30°C higher than *Blattella.* Hafez *et al.* (1970) found two ticks, both with very low cuticular permeability to water, to have high transition temperatures: 52°C for *Hyalomona dromedarii* Koch and 63°C for *Ornithodorus savignyi* (Audouin). They observed that the permeability/temperature curves for both species were L-shaped, suggesting the presence of a single epicuticular wax layer.

Except for *Blattella,* the impermeabilities of the insect lipids investigated by Beament (1945) did not return to their original value after they were heated to above the critical temperature.

Beament (1958a, 1959) called attention to the fact that investigations on the evaporation of water from the cuticle at *air* temperatures are useless in physicochemical interpretations of cuticle permeability. Data on *cuticle* temperatures are required, and these are somewhat less than ambient temperatures since the insect is cooled by its evaporating water. Using specially constructed instruments of a high degree of sophistication, he obtained data for the water loss from a large nymph of a cockroach (*Periplaneta americana*) in relation to air temperature and the corresponding cuticle temperature, with the relative humidity not exceeding 10%. The relation of evaporation to *air* temperature was indeed an exponential function, showing no discontinuity, as had been observed by Edney (1951) and Holdgate and

Seal (1956), but the relation of evaporation to *cuticle* temperature was expressed by a curve showing a sudden increase in water loss at 29.7°C. However, a further refinement in the interpretation of the data was effected by recognizing, as Mead-Briggs (1956) and Edney (1956) had pointed out, that the actual amount of water passing through an insect's cuticle at any temperature will be proportional to the permeability of the membrane at that temperature and to the saturation deficiency.* Dividing each datum by the corresponding mean saturation deficiency of the air, Beament obtained the evaporation (or cuticle permeability) per unit of saturation deficiency. He considered that the resulting curve showed the true relationship between permeability and temperature and firmly established the validity of the transition-point concept.

Beament (1958a, 1959) found that the cuticle of the cockroach goes through a transition at a sharply defined temperature, when its permeability increases abruptly, but that does not necessarily implicate the epicuticular lipid layer as the source of the transition. This was demonstrated to be the case, however, in a subsequent experiment in which Beament (1959) re-examined the evaporation from a drop of water covered with the cuticular grease of the cockroach (see Ramsay, 1935b). He used artificial surfaces on which to place the drop, so as to avoid the complication of passage of water from the body of the cockroach into the drop. The sudden breakdown of waterproofing occurred at a temperature of 32.7°C. The similarity of the temperature/permeability function between cockroach cuticle and cockroach grease on water was proof of what had been suspected by many investigators from circumstantial evidence, namely, that the epicuticular grease of the cockroach is the significant factor limiting water loss.

The permeability/temperature relationships for other insects differ from that of the cockroach according to whether they are covered with (1) a hard wax layer, or (2) two layers of hard wax separated by a cement layer. The permeability temperature curve for the latter depicts transition points for the two wax layers, the second at a much higher temperature than the first (Beament, 1959, 1961a).

A change is the epicuticular lipids at the "critical temperature" on a basis that is independent of the tests on rate of evaporation is indicated by a sudden change in the dielectric constant of cockroach cuticle at about 30°C (Beament, 1961a).

Many investigations with insects and other arthropods have suggested that the greater rates of water loss are associated with those arthropods inhabiting moist environments (Ahearn, 1970). Therefore it is to be expected

---

* At a given temperature, saturation deficiency is the difference between the pressure of water vapor at saturation and that actually obtaining.

that some studies would be made of water loss in desert insects, in which the mechanisms for prevention of water loss would be expected to be most efficient. In experiments with Sahara Desert species, Delye (1969) found a critical temperature of about 60°C for the solpugid *Othoes saharae* and 63°–65°C for the cockroach *Eremiaphila saharae,* whereas the beetle *Priomotheca coronata* apparently had no critical temperature. The three species are all native to the Sahara Desert. Ahearn (1970) made similar studies in Arizona with three tenebrionid beetles: *Eleodes armata, Centrioptera muricata,* and *Cryptoglossa verrucosa.* He concluded that the total water loss in these beetles was from transpiratory losses from the cuticle and spiracles, from defecation, and from the release of defensive quinone droplets or oral fluids. Starvation of the beetles before experimentation eliminated water loss by defecation and sealing mouthparts prevented oral discharges.

Freshly killed specimens of the three beetles had higher transpiration rates over long and short exposures than did living insects of the same species, a possible indication of the cessation of active water retention by cuticular and spiracular regulation in dead animals. Much evidence for cuticular regulation of water loss and uptake has been accumulated. Cuticular regulation is also implied by the effect of toxicants on water loss, to be discussed later. Active retention of water by spider mites has been proposed (Winston and Nelson, 1965). On the other hand, the difference in water loss between live and dead spiders has been attributed to an active secretion of water inward by epidermal cells (Davies and Edney, 1952).

In tenebrionid beetles the spiracles empty into spaces between the thorax and abdomen and the dorsal cuticle (Cloudsley-Thompson, 1956). The subelytral cavity of the abdomen empties to the outside through only one small orifice, above the anus, which can be blocked. In Ahern's (1970) investigation spiracular losses of water from the thorax occurred only at the intersegmental areas. Cuticular transpiration was measured by killing the beetles in cyanide gas and sealing with nail polish the anal, oral, and intersegmental regions.

Although low in absolute rate, cuticular transpiration was a greater source of water loss than respiratory transpiration in the three beetle species investigated by Ahearn at temperatures from 25° to 42.5°C at 0% RH, suggesting the importance of spiracular control in maintaining water balance. At 40°C for *E. armata* and 42.5°C for *C. muricata* and *C. verrucosa,* there was a marked increase in respiratory transpiration over previously low and practically uniform rates, indicating a temperature-induced breakdown in spiracular water regulation due to increased respiratory activity. Cuticular transpiration maintained a steady linear rate of increase over the same temperature range. (Cuticular transpiration was also abruptly increased, but at a higher temperature, as will be discussed later.) Ahearn considered the success of tenebrionid beetles in desert habitats to be the result of highly impermeable cuticles and well-regulated spiracular control mechanisms for minimizing water loss.

In an investigation of another desert arthropod, the scorpion *Hadrurus arizonensis,* Hadley (1970) determined cuticular water loss by killing the scorpions with cyanide gas and sealing their book lungs with nail polish. (Comparison of cyanide with pure nitrogen as killing agents had shown no significant difference between the effect of the two gases on cuticular properties.) Cuticular and total water loss was not significantly different between 25° and 35°C, suggesting that virtually all water is lost through the cuticle at these temperatures. Above 38°C cuticular water loss continued to increase with temperature, but at a rate well below that for total water loss, indicating, as in the previously discussed investigation with tenebrionid beetles, that respiratory transpiration becomes the major pathway of water loss at high temperatures.

Cuticular water loss in dead scorpions (*Centruroides sculpturatus*) was much greater than total water loss in living scorpions, indicating an active, energy-requiring, cuticular water-retaining mechanism in the live animals. Regardless of their degree of dehydration, scorpions were unable to absorb significant quantities of water from near-saturated atmospheres or moist substrates, as some arthropods are able to do. [A list of such species was prepared by Noble-Nesbitt (1969).] Desert-inhabiting scorpions avoid a severe negative water balance by possessing an extremely impermeable cuticle (lowest of all rates of water loss recorded for arthropods), and very low metabolic rates. They may drink when water is present, but like most desert predatory arthropods, they obtain body fluids primarily from captured prey.

a. Transition Temperatures Inversely Related to Transpiration Rates at Normal Temperatures. Edney (1967) concluded that arthropods with high transition temperatures have low cuticular transpiration rates. This was found to be the case in recent investigations on scorpions (Hadley, 1970) and the tenebrionid beetles (Ahearn, 1970) inhabiting the Arizona deserts. Figure 4 shows the results obtained by Ahearn with three species of tenebrionids. Results were expressed as "weight loss (%/mm Hg/hour × 1000)" to give values which were not affected by the changing saturation deficits at different temperatures. Plotting such data against temperature, curves were obtained showing abrupt change at a sharply defined temperature, such as the curves depicting Beament's (1958a) experiments with the American cockroach, although critical temperatures for the beetles were found to be much higher. With the three species of beetles investigated by Ahearn, transition temperatures were inversely related to the rate of cuticular transpiration at lower temperatures.

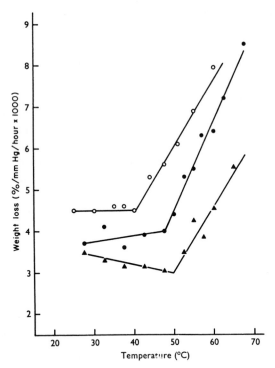

Fig. 4. Abrupt rise in rate of water loss for three species of tenebrionid beetles at the "critical temperature" points. ○, *Eleodes armata;* ●, *Centrioptera muricata;* ▲, *Cryptoglossa verrucosa.* (From Ahearn, 1970).

b. The Physicochemical Interpretation of the Permeability/Temperature Function. Beament (1961a) concludes from his long series of permeability studies that, on an insect that has a thick grease layer such as possessed by the cockroach, one monolayer is specially arranged, but that on insects even more waterproof than the cockroach, this specially impermeable state must extend through a much greater thickness of wax. Cockroaches can be washed with water to remove all grease not actually attached to the surface of the cuticle, thus removing 95% of it and leaving what is calculated to be a monolayer, yet this results in only a threefold decrease in permeability to water instead of the expected twentyfold decrease (Beament, 1960).

For some years after the development of the concept of the water-impermeable lipid monolayer in insect epicuticle, the molecules comprising this monolayer were considered to be vertically oriented (Beament, 1961a). They were believed to be oriented so that their hydroxyl ends comprise the hydrophilic outer surface before the cuticulin substrate becomes tanned by quinones. These quinones were believed to originate at least in part from the oxidation of dihydroxyphenols issuing from pore canals arising from epidermal cells (Wigglesworth, 1947). According to this theory the tanning process causes the cuticulin to become hydrophilic and the lipid monolayer is then reversed, with hydrophobic alkyl ends outward. This concept of the monolayer persists to this day except that Beament now believes that the organized lipid consists of "tilt-packed" long-chain molecules with alkyl ends outward and with the molecules inclined at about 24.5°. Judging from the zig-zag arrangement of the carbon atoms in the molecules, and with the polar groups larger in cross-section than the chains, the molecules would provide the closest contact of chains and greatest impermeability to water when inclined at 24.5°. Beament believes that the sudden and marked increase in water loss of insects at a definite "critical temperature" for each species (30°C for the American cockroach) is not caused by a reorientation of the organized lipid water barrier, as formerly believed, but by a change in the position of the molecules so that they occupy a *mean* vertical position, resulting in a much greater space for the escape of water from the water-bearing cuticulin substrate (Beament, 1964).

Another interpretation of transition temperatures is offered by Locke (1965). He suggests that the lipid monolayer on the cuticulin might be composed of "liquid crystals" reaching the surface of the cuticulin via "wax canals" to be discussed in Section III,A,6. He believes that phase changes in these liquid-crystalline systems might cause the abrupt change in water loss at "critical temperatures."

4. The Overturning of Lipid Monolayers and Its Possible Significance

Beament (1960, 1961b) believes that not only may the facultative reversal of a monomolecular layer explain the way in which the aquatic insect may change the wettability of its surface, but also a number of other biological phenomena that can be explained only on the basis of a reversal in wetting properties to change capillary pressures. Examples are the emptying of the tracheal system at eclosion (Sikes and Wigglesworth, 1931) and at ecdysis (Wigglesworth, 1938), the production of the air film of the plastron while submerged in water, and the appearance of air in the aeroscopic sponges of eggshell (Wigglesworth and Beament, 1950).

The overturning of the molecules of a monolayer of lipid was substantiated in the experiments of Langmuir (1938) and Rideal and Tadayon (1954a,b), who worked with monolayers of organic acids resting on various metals. The motivating factor, however, was the attraction of water for the hydroxyl ends of molecules of oleic acid and stearic acid, respectively, resting on various metals, rather than a change in the polarity of the substrate.

5. Permeability of Cuticle Below the Wax Layer

The predominant influence of the epicuticular wax layer on most insects has diverted attention from the fact that the remainder of the cuticle likewise has considerable resistance to the passage of water. Treherne (1957) calculated that the total area for the diffusion of aqueous solutions through the portion of the cuticle below the wax layer in the grasshopper *Schisto-cerca gregaria* was very limited. Diffusion was only 0.24% of that which would take place through an equivalent area of unstirred water. Beament (1961a) states that the permeability of relatively water-saturated cuticle below the lipid layer is only two- or threefold greater than that of wax and is thus about equivalent to that of a good leather. Among the three layers of the procuticle, the endocuticle is considerably more permeable than the mesocuticle and exocuticle (Noble-Nesbitt, 1970b). As stated previously, Lafon (1943) and Richards (1958a) have presented evidence that the chitin–protein portion of the cuticle can become highly impermeable in areas where it becomes dry. The probable role of derm pores and wax canals in the permeability of the portion of the cuticle below the wax layer will be discussed in Section III,A,9.

It will be shown in Section IV,C,4,a that the chitin–protein portion of the cuticle differs from the epicuticular lipid in the important respect that it is hydrophilic and selectively permeable (see Alexandrov, 1935). Therefore the penetration of water-soluble compounds is greatly enhanced if they are present as undissociated molecules rather than as ions. In this respect they perform as they would in artificial semipermeable membranes such as collodion (Hoskins, 1932). Treherne (1957) has demonstrated that the cuticular wax is the limiting factor in the cuticular penetration of a slowly penetrating molecule such as urea, whereas with a small, lipoid-soluble molecule such as ethanol, it accounts for only about half the resistance offered by the entire cuticle.

6. Wax Canals Break the Continuity of the Organized Lipid Layer

In the experiments of Alexander *et al.* (1944a), the compounds most effective in retarding water loss were those which contained polar groups that could become oriented toward the water-bearing celluloid membrane. The question remains as to how a selective adsorption, removing only amorphous fractions, could ultimately affect the interfacial layer or layers of oriented and presumably tightly packed molecules so as to increase their permeability to water. The experiment of Alexander *et al.* (1944a) indicated that sufficient lipid can be removed by a sorptive dust (gas-mask charcoal) to increase the water permeability of a beeswax film on a celluloid layer, but, as stated previously, Beament (1945) was not able to obtain such results with alumina dust allowed to settle on a monolayer of cockroach grease on butterfly wing. Presuming that the organized lipid layer on the cuticulin of the intact insect generally cannot be removed by sorption, as implied by Beament's experiment, the problem still remains as to how sorptive (and generally nonabrasive) dusts cause abnormal and even lethal water loss. A possible solution to the problem has been indicated by the discovery of Locke (1964, 1965) and Gluud (1968) of filamentous "wax canals" penetrating the cuticulin of the insects they investigated (see Chapter 2). Wax canals issue from pore canals, which arise from the epidermal cells, penetrate the procuticle and sometimes into the epicuticle, but not to the surface of the cuticulin layer. Wax canals issue from the terminal ends of the pore canals and generally branch out and reach the cuticulin/lipid interface of the epicuticle in enormous numbers. Even pore canals are commonly so small that they cannot be resolved as ducts by a light microscope (Richards, 1951) and wax canals were revealed, of course, only by means of electron microscopes. Wax canals provide a route to the surface for liquid waxes similar in appearance to the liquid crystalline phases of lipid-water systems. Locke (1965) believes that upon reaching the surface of the cuticulin, lipid-water liquid crystals spread out to cover it with a monolayer that is not static, but is being continually "moved, lost, and replaced."

The controversy as to whether the organized lipid barrier bound to the cuticulin can be removed by sorption has never been of more than academic interest, for the fact remains that at least some species of insects can be killed by means of sorptive dusts more rapidly than by the most potent of modern insecticides (Ebeling and Wagner, 1959; Wagner and Ebeling, 1959). A light, fluffy, amorphous, nonabrasive powder can desiccate insects with hard epicuticular wax when the insects are restrained from movement or are killed to eliminate the slightest chance for abrasive action. Now with the revelation of wax canals reaching the protein/lipid interface of the epicuticle, it is reasonable to suppose that the fate of the organized lipid may have little or no relevance to the action of sorptive powders. We do not know whether wax canals exist in most insect species, but it is at least reasonable to assume that wax canals may provide the means of transport for almost[*] all species with a wax water barrier. Wax can move through homogeneous cuticle, as indicated by its presence on tracheal walls despite the absence of any discernible canals in the tracheal cuticle. However, a canal system would appear to be an important advantage in the case of the much thicker cuticle of the exoskeleton. In any case, in species where wax canals exist, it is evident that a considerable portion of the lipid barrier rests over the mouths of the myriads of canals. These portions of lipid, unorganized and probably in continuous movement, may be expected to be much more readily affected by the adsorption process, initiated by dust resting on the surface

---

[*] It is of interest in this connection that cross sections of the cuticle of the twospotted spider mite, *Tetranychus urticae* Koch, failed to reveal any sign of pore canals, yet these mites possess an epicuticular lipid layer (Gibbs and Morrison, 1959).

of the epicuticle, than the organized lipid which is bound to tanned protein (Ebeling, 1971).

7. Wax Canals as Pathways for Artificial Desiccation

Locke (1965), who was the first investigator to reveal and to determine the significance of wax canals, was also the first to be led by this important discovery to speculate on their probable role in artificially induced desiccation of insects. When insects are immersed in oil, droplets of water gather at the surface (Wigglesworth, 1942). Locke pointed out that the dissolution of surface wax initiates an outward flow of all cuticular lipids and the water droplets which appear on the surface of the insect could have passed through the spaces vacated by the lipid-water liquid crystals. Even the crystallization of wax into the "blooms" appearing on the surfaces of the cuticles of some insect species "drags" the middle-phase filaments through the wax canals and leaves the latter filled with water. Locke concluded that "a similar effect would be expected for adsorbent dusts at the surface. With the wax in the canals replaced by water the subsequent desiccation is readily explained." Myriads of tiny pathways for the passage of water would be involved in this phenomenon. In accordance with the "pinhole effect" (Brown and Escombe, 1900) a very rapid movement of water through the cuticle may be expected.

The elytron of the carabid beetle, *Pristonychus complanatus,* lends itself particularly well for the photography of wax sorption by silica aerogels (Ebeling, 1971). The elytron is black and a thin film of silica aerogel, being translucent, can hardly be distinguished against the black substrate. Figure 5 shows a portion of elytron on which the silica aerogel Dri-die 67 was allowed to settle from a cloud confined in an inverted beaker. When the aerogel absorbs wax it becomes white. Portions of the elytron on which sorption (adsorption and absorption) has not yet taken place remain black for a while. Sorption generally begins at the anterior end of the elytron and spreads posteriorly, sometimes forming a sharp line of demarcation as it advances over the elytron; this tendency is shown in Fig. 5.

The delicate lacy structure of the silica aggregates was preserved, even though the lipid film migrated over the silica filaments for as far as 0.16 mm. [Fluorinated silica aerogels have a positive electrostatic charge. Films settling out of a cloud to rest on an insect's cuticle consist of filaments of silica particles oriented at right angles to the cuticle (Fig. 5) much as iron filings on a magnet.] It appears that the large quantity of lipid absorbed by the silica film could not have been supplied solely by the insect epicuticle and that apparently large quantities were drawn from deep within the cuticle as postulated by Locke (1965). Carabid beetles dusted with silica aerogel died of desiccation in an average of 6 hours at 25.5°C and 58% RH. Under

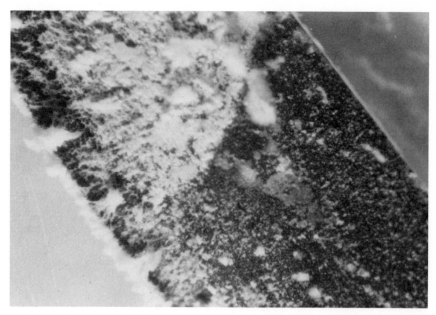

Fig. 5. Absorption of lipid from the dorsal surface of an elytron of a carabid beetle by a film of silica aerogel. The aggregates of silica are translucent and invisible against a black background until they become saturated with lipid. (From Ebeling, 1971.)

the same conditions it required an average of 17 hours to kill *Tribolium confusum*. Both species were placed in petri dishes containing 2 cm³ of Dri-die 67 per dish. *Tribolium confusum* weighs only about 3% as much as *P. complanatus* and, other things being equal, would become desiccated more rapidly because of the greater ratio of body surface to volume. This was interpreted to imply that it is more difficult to remove the cuticular lipids of *T. confusum* by sorption (Ebeling, 1971).

Rideal and Tadayon (1954a,b) found that ¹⁴C-labeled stearic acid molecules, in a mixture of stearic acid and paraffin wax, could be "overturned" and anchored to new surfaces (various metals) and that the molecules could also migrate along the metal or mineral surfaces for considerable distances. This is not surprising in view of the ability of molecules to migrate away from crystals on mineral or metal surfaces (Volmer, 1932) and the penetration of solid insecticides deeply into blocks of dried mud (Hadaway and Barlow, 1951).

The great distance the wax molecules migrated up into the layer of silica aerogel in the experiment depicted in Fig. 5, as well as in similar experiments with beeswax (Ebeling, 1961), is a striking demonstration of the mobility of adsorbed films, too often regarded by biologists as static systems.

8. Water Loss Caused by Toxicants

Increased cuticle water permeability resulting from the action of petroleum, petroleum plus oil-soluble surfactants, and surfactants alone (Wigglesworth, 1945; Ebeling and Wagner, 1959) is easily visualized. Surface lipid may be removed or rendered more water permeable. The mechanism for increased cuticle water permeability resulting from toxic action is more obscure. Normally transpiration takes place in insects via the cuticle, the tracheal system, or both (Mellanby, 1935). Wigglesworth (1941) attributed the loss of weight in pyrethrum-treated insects to spasmodic contractions. Ingram obtained a twofold increase in weight loss in flies within a 2-hour period and a fivefold increase in cockroaches, by abdominal injection or topical application of pyrethrins. He believed that the loss was caused by a neurotoxic action of the pyrethrins on the secretory activity of epidermal cells. This belief is supported by the observation that the great difference in water content of blood and cuticle of insects could not be maintained passively, indicating that a "water pump" is located in the epidermis (Winston, 1967; Winston and Beamont, 1969). Buck and Keister (1949) found that DDT-poisoned blow flies, *Phormia regina,* lost most of their water by increased transpiration through the spiracles. Mansingh (1965) found that malathion-poisoned cockroaches, *Blattella germanica,* lost water because of increased transpiration through the spiracles and intersegmental membranes of the body wall. These membranes were observed to be exposed in the paralyzed cockroaches. When the abdominal spiracles were sealed with nail polish, cockroaches lost about 25% less water than those with open spiracles. Mansingh concluded that water loss in insects is positively correlated with degree of intoxication, whereas the manner in which the water is lost depends on the insect species. The type of toxicant can also be shown to be an important factor. Ebeling and Pence (1957) found that arthropods killed by some toxicants remain turgid for several days and retain a lifelike appearance while the same species killed by certain other toxicants become completely shriveled (Fig. 6). The influence of the type of toxicant was again strikingly demonstrated by Chattoraj and Sharma (1964). They applied insecticides to cockroaches (*Periplaneta americana*) in doses that resulted in the insects becoming moribund within an hour. Free water was observed on the bodies of all insecticide-treated cockroaches, a phenomenon reported by Ingram (1955) for pyrethrum-treated cockroaches. This was not seen on the control insects. The insects were weighed 2 hours after treatment. Air was passed through calcium chloride and then over the cockroaches to remove this surface water; the air was then passed through another calcium

Fig. 6. Twospotted spider mite, *Tetranychus urticae* Koch, after treatment with chlorobenzilate (left) and parathion, showing great difference in effects of the two acaricides on rate of desiccation. (From Ebeling and Pence, 1957.)

chloride tube to collect the water. The results of the experiment are shown in Fig. 7. Of the 12 insecticides, allethrin caused the greatest water loss—over 5 times more than aldrin, which caused the least water loss. With both allethrin and aldrin, water loss was measured after anesthetizing the insects with CO_2 and cooling them to $1°$ to $2°C$. With allethrin this resulted in a 30.0% and with aldrin a 21.6% reduction in net increase in transpiration, when compared with the control, but water loss was still much higher than in the control kept at room temperature ($20°C$). Untreated cockroaches lost no measurable amount of water at $1°$ to $2°C$.

The insecticides were applied on the pronotum, but water was found in a film or in droplets, depending on the insecticide, over the entire body surface. Chattoraj and Sharma concluded that insecticide-induced water loss takes place through the general body surface as well as through openings such as mouth, anus, and spiracles.

Many investigators have reported various histological changes in the tissues of insects, principally vacuolation, contraction, and many other types of de-

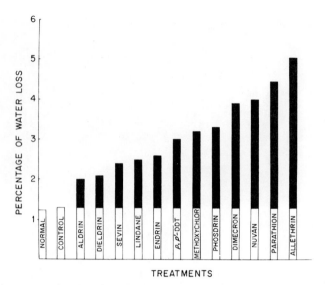

TREATMENTS

Fig. 7. Percent water loss from American cockroach, 2 hours after treatment with various insecticides. (From Chattoraj and Sharma, 1964.)

generative aberrations. Sharma and Chattoraj (1964), working with a number of insecticides, showed photomicrographs indicating a shrinkage and vacuolation of all affected tissues of treated insects and an expulsion of water.

9. Active Uptake of Water

As explained in Section IV,A,1, terrestrial anthropods without an epicuticular lipid layer lack a regulatory mechanism and may absorb water at a lethal rate when excess moisture is present in their environment. The following discussion deals only with terrestrial arthropods possessing an epicuticular lipid layer.

Subterranean termites (Rhinotermitidae) provide a classic example of the dependence of an insect on cuticular absorption of water. After they have lost a certain amount of water in a drywood structure they must descend to their moist subterranean galleries to absorb water before they can again continue feeding in the wood above ground. They cannot be reared in a continuously dry environment, even in the presence of abundant food and water (Collins, 1969). Even most drywood termites (Kalotermitidae), which confine their activities to the above-ground wood structure, thrive best at 90 to 97% RH (Pence, 1956; Collins, 1969).

Govaerts and Leclercq (1946) called attention to certain species of insects that gain weight while fasting in an atmosphere saturated with water vapor.

Whether the insects were "hygroscopic" or not, their body water came into equilibrium with the atmospheric moisture. Placing them in an atmosphere saturated with water, of which 8% was heavy water, resulted in the various species having 8% of heavy water in their bodies 5 to 13 days later. The above investigators concluded that there was a continuous exchange between the atmospheric water vapor and the body water of both "hygroscopic" and "nonhygroscopic" insects and that the body water is completely replaced by water molecules from the atmosphere within a few days.

Arthropods may take up water in moist food, metabolize it from dry food, drink it or obtain it via their cuticle or tracheae (Edney, 1957). They can take up water via the cuticle when spiracles are blocked (Browning, 1954) and, in any case, the tracheae are homologous with integument and, except for the tracheoles, a surface wax layer is generally assumed on the basis of nonwettability (Richards, 1951). Thus penetration through essentially similar membranes is involved.

Water uptake when mouth and anus were blocked was demonstrated on some insects (Beament, 1945). On the other hand, blocking of the anus in desiccated *Thermobia* and *Tenebrio* prevents absorption of water from the atmosphere by these insects (Noble-Nesbitt, 1970a). However, blockage of the anus might also result in a nervous inhibition which would in turn inhibit water uptake (Okasha, 1971).

Various kinds of stress (centrifugation, exposure to 45°C, burning of the integument, submergence in water for 1 hour) did not inhibit water uptake in *Thermobia domestica*. Five weeks of starvation resulted in an extreme hydration of the insects when compared with feeding insects kept under the same temperature and humidity conditions. Thus the body weight and therefore the volume of the starving insects is maintained by the retention of a higher proportion of water (Okasha, 1971).

Damage to the waterproofing lipid layer of the cuticle of ticks stops active intake of water until the damaged area is repaired (Lees, 1947). A droplet of water (0.5 mm³) on the thorax of a cockroach will disappear in 8 to 10 minutes, whereas it requires several hours for it to disappear by evaporation from a surface such as polytetrafluorethylene, when the droplet is covered with a film of cockroach grease, just as it is covered when resting anywhere on the body of a cockroach. Tritiated water has been traced through the cuticle to the epidermal cells and hemolymph. The surface lipid of the cockroach was found to act as a semipermeable membrane. Dissolved salts in penetrating aqueous solutions remained on the lipid surface (Beament, 1965).

An interesting aspect of Beament's experiments is that tests with many cockroaches revealed that water was most rapidly absorbed through dark cuticle, at an intermediate rate through the cuticle of the abdomen and tegmen

(forewing), and so slowly through intersegmental membranes and other un-
pigmented regions that the disappearance of the water could be attributed
to evaporation. The latter regions are those most permeable to apolar liquids.
In insects in which the epidermis had separated from the old cuticle prior
to molting, or which were more advanced in the molting cycle and had two
cuticles, or when the insects were paralyzed by hyperactivity,* water uptake,
if any, was slow. Apparently a layer of normal epidermal cells in intimate
association with the cuticle was required for normal water uptake. As might
be expected, water uptake was also very slow or nil if insects were anesthetized or decapitated.

Some relative humidities from which certain arthropods can absorb water
are as follows: partially hydrated larvae of the cigarette beetle, *Lasioderma
serricorne*, at humidities down to 43%, the lowest so far reported for
arthropods; the firebrat, *Thermobia domestica*, 45% (Beament *et al.*, 1964;
Noble-Nesbitt, 1969); the prepupae of fleas 50% (Edney, 1947); booklice,
Liposcelis, 60% (Knülle and Spadafora, 1969); larvae of the oriental rat
flea, *Xenopsylla cheopis*, 65% (Knülle, 1967); the grain mite, *Acarus siro*,
71% (Knülle, 1965); the grasshopper, *Chortophaga*, 82% (Ludwig, 1937);
nymphs of the cockroach, *Arenivaga* sp., 82.5% (Edney, 1966); the meso-
stigmatid mite, *Laelops eschidna*, 90% (Wharton and Kanungo, 1962); the
larva of the yellow mealworm, *Tenebrio molitor*, 90% (Mellanby, 1932);
the tick *Ixodes*, 92% (Lees, 1946); and the tick *Ornithodorus*, 92%
(Browning, 1954).

With both *Tenebrio* larvae (Mellanby, 1932) and ticks (Lees, 1946) near
equilibrium humidity, exchanges of water are limited by the relative humidity. Thus ticks took up water at 25°C and 95% RH [saturation deficiency
(SD) 1.2 mm Hg] yet lost water at 15°C and 90% RH (SD 1.3 mm Hg).
At lower humidities, however, they lost water at a rate related to saturation
deficiency, as is usual for arthropods (see Buxton, 1932; Johnson, 1942).
The lowest limit at which the desert cockroach *Arenivaga* would take up
water (82.5%) was found by Edney (1966) to be unaffected by temperature
within the range of 10° to 30°C. Thus saturation deficiency (SD) was not
the limiting factor. For example, at 15°C and 75% RH, SD is 3.2 mm Hg
and all insects lost weight, whereas at 30°C and 87.5% RH the SD is higher
(3.9 mm Hg) and all insects gained weight.

Previously desiccated and starved firebrats regain lost water in humidities
above 95% RH at room temperatures, but lose water at lower humidities.
Normal water content is reached most rapidly at the higher temperatures
and humidities (Noble-Nesbitt, 1969). Water uptake ceased at the later

---

* Hyperactivity of the nervous system "which precedes paralysis whether brought
about by DDT, self-promoted struggling or direct stimulation of the insect" (Beament,
1958b).

stages of the molting cycle of *Thermobia domestica* (Okasha, 1971). In *Chortophaga,* uptake has been reported to take place above 82% RH, but it is only at 96% RH or higher that the total weight increases by as much as 15%. However, the uptake continues for no longer than 24 hours, after which weight is lost (Ludwig, 1937). The bed bug *(Cimex lectularius)*, which normally inhabits damp cracks in walls for most of its life, loses water in nearly saturated atmosphere, where some other species would be taking up water continuously (Mellanby, 1932).

Knülle and Spadafora (1969) found that specimens of *Liposcelis rufus* lose water at relative humidities of 53% or less. At 25°C and 33% RH they lost 50% of their water in 11 days, but most of them recovered the lost water within 6 to 7 hours after transfer to an RH of about 58%. Water was regained in humid air 38 times faster than it was lost at 33% RH. The flat and contracted abdomen of the dehydrated insect became inflated, returning to its original size and shape. To return a psocid to its normal turgidity, 25 μg of water must be transported, requiring a minimum energy of 4.3×10^{-4} cal.

Finlayson (1933) observed that if psocids (*L. divinatorius*) were deprived of adequate moisture in culture dishes, they became flattened and lethargic, but when moisture, absorbed on paper, was added to the culture dish, the insects became filled out and active within 2 or 3 hours and began laying eggs.

In the experiments made by Knülle and Spadafora, when food was available, females of *L. rufus* died within 2 to 3 weeks at all humidities below 58%, the "critical equilibrium humidity" for all species of *Liposcelis* so far investigated. Above this level, they survive for more than 6 months. Females of *L. knullei* lost water twice as fast as *L. rufus* and died within 1 week at all humidities below their critical level. *Liposcelis bostrychophilus* survived for only 10 days. Egg laying ceased below the critical equilibrium humidity (Knülle and Spadafora, 1969).

Transfer of water from a lower to a higher concentration, as from an atmosphere of 60% RH to a hemolymph in equilibrium with approximately 99% RH, requires work. Evidence for the epidermis being the source of this energy was first produced by Lees (1946), who pointed out that ticks cannot take up water while asphyxiated, indicating involvement of secretory activity by the epidermal cells. The evidence for a "water pump" located in the epidermis, as developed by Winston and Beament (1969), has already been presented. These investigators concluded that the anatomy of insects is such that the source of the energy for the operation of this "pump" must be within the epidermal cells. Noble-Nesbitt (1969) found that, even during brief periods of locomotor activity, the water uptake of firebrats is often arrested, indicating that these insects can utilize their metabolic resources to suit the exigencies of the moment. However, there is not invariably an inverse relationship between uptake and locomotor activity, indicating that water uptake is a process under the control of the insect. Anesthesia inhibits water uptake by firebrats, but uptake is resumed when the insects recover. As might be expected from the energy output required for water uptake, after death insects can only lose water and at a more rapid rate than when they are alive. The mechanism for the translocation of the water remains an intriguing problem. Obviously the normal metabolism of an arthropod supplies sufficient energy for the sorption of the amounts of water that have been recorded.

In an investigation of the mite *Echinolaelops echidorinus* (Berlese), Kanungo (1965) found that dehydrated mites gained water at the rate of 1.0 ± 0.5 μg/hour/mite and hydrated mites lost water at the rate of 1.0 ± 0.59 μg/hour/mite. Both hydrating and desiccating females consumed approximately 0.17 μl of O_2/mite/hour at 80°F. Assuming that mites produce the same amount of energy per unit of oxygen consumed as do other aerobic forms (approximately 5 cal per 1 ml of O_2 consumed), the minimum amount of oxygen required for the sorption of water from the air at 90% RH is 0.0006 μl of O_2/μg of water. Thus Kanungo concluded that for the active uptake of water from unsaturated air, a mite utilized only an undetectable fraction of a percent of the total energy it produced.

Edney (1966) pointed out that uptake of water by the cockroach *Arenivaga* involves movement of water from 82.5% to 99% RH, the latter being the equilibrium humidity for the insect's hemolymph; thus it involves movement against an osmotic gradient of more than 100 atmospheres. Yet the energy required is small and easily provided by the insect's metabolism. The greatest rate of uptake observed in *Arenivaga*, which weighs about 100 mg, was 6 mg of water per day. The insect's oxygen uptake (about 22 μl/hour) produces 2.6 cal/day. Concentration of water through the observed gradient takes about 6 cal/gm, so the energy expended in absorbing 6 mg of water is 0.036 cal/day or about 0.01 mg of carbohydrate fuel, or even less of fat, well within the resources of the insect.

The net uptake of water depends upon metabolism being in progress. Here one is justified in distinguishing between the physical permeability properties, as possessed by a dead membrane, and physiological permeability which depends on the continuous expenditure of energy; physiologically active substances such as Cu^{2+} ions, adrenaline, and hormones may also alter the balance (Harris, 1956).

Edney (1947) proposed that prepupae of fleas are able to adsorb water that has gathered on the cuticle by condensation of vapor and that this water is then transferred into the body. Ticks can absorb water from either humid air or water, however; condensation of water on the cuticle is not required. After excessive desiccation, they appear to lose the ability to control the amount of water absorbed from free water and may die from absorbing excessive quantities. The unfed tick attains equilibrium between 90 and 95% RH. Above this point it gains water and at lower humidities it loses water. The equilibrium humidity increases as the tick increases in age, since active absorption depends on vital processes. Likewise rate of desiccation increases with age, for ticks have an ability, apparently rare among arthropods, of retarding to a limited extent the rate of transpiration. To the extent that transpiration is affected by vital activity, it too approaches the passive condition, that is, an accelerated rate, with advancing age.

Beament (1954) has calculated that it may require a reduction of only about 1% in the water content of the cuticulin to initiate an active absorption of water at 90% RH. He concludes that living cells should be able to regulate the water content of cuticulin to this extent and thus take up water from humidities lower than that in equilibrium with the body fluids. It is understandable, therefore, how a tick can take up water from the atmosphere at a relative humidity of 90% or even lower. However, Lees (1946) has pointed out that ticks will also *lose* water under the same humidity conditions if the wax layer is abraded. Even if a very small percentage of the wax were removed at small isolated points, the "pin-hole" accelerating effect would result in losses far greater than could be compensated by the relatively feeble absorptive forces of the remaining unabraded cuticle. Beament (1954) found that, while repair activity may proceed for several days after abrasion, water uptake may recommence in 1 day, since only a monolayer of lipid was required for active absorption to recommence.

Because water is absorbed into the bodies of insects against various degrees of gradient, requiring active control by the epidermal cells, the latter must in some way change the gradients in the cuticle (Beament, 1954). Wigglesworth (1933) proposed that this might be accomplished by epidermal cells exerting their effect by the passage of substances they synthesize via the cytoplasmic processes they extend up through the pore canals to the epidermis. Lees (1946) believed that, in ticks, water penetrates the cuticle via pore canals filled with cytoplasm. He believed their tufted external ends, which expose a large surface of cytoplasm, might actively absorb water from the epicuticle even though they do not penetrate this layer. Water uptake ceased when the pores were blocked. Active adsorption of water likewise takes place in *Tenebrio*, which is also known to have cytoplasm-filled pores (Mellanby, 1932). Active uptake of water is not known to occur in insects that do not have pore canals (Beament, 1965).

Pore canals do not reach the outer surface of the cuticulin (the cuticulin/lipid interface), so the discovery of the wax canals leading from the ends of the pore canals to the cuticulin surface (Locke, 1964, 1965) contributed to a possible understanding of the valve mechanism implied by the passage of water through insect cuticle against considerable gradients and both inward and outward. These pores must be at least about 100 Å in diameter to accommodate the spherical micelles in which form lipids (30 to 50 Å in length) are believed to be transported via these canals to the cuticulin/lipid interface (Locke, 1964, 1965; Beament, 1965). Beament (1965) suggests that through natural selection the size of the pores may have been determined by the length of the lipid molecules they conduct to the cuticulin surface in micelle form. Both water and lipid occur in the pores. Beament postulates that if the epidermal cells can produce, in the cuticle underlying the pores, a water activity which will result in a spherically concave meniscus at the top of the column of water in the pore, the vapor pressure at the meniscus will be reduced. To take *Tenebrio* as an example, its equilibrium humidity is 88% RH. Such a humidity would be given by a pore diameter of 200 Å at a capillary pressure of about 150 atmospheres. If the insect's cuticle can produce tensions of this magnitude, it can establish such a relative humidity in the air surrounding it and take up water from higher humidities. The size of the pore determines the curvature of the meniscus, and, provided the necessary tensions are produced in the cuticle below the pore, it also determines the equilibrium humidity. Thus the equilibrium humidities previously given for the various arthropods, ranging from 45% to 92% RH, are believed to be related to the size of the pores in the outer cuticulin. Reduced to simplest terms, the pores are "valves"—capillaries filled with water. Suction pulls the column down, tension is released, and the water returns to a near-flat meniscus sealed by a monolayer of lipid. How the arthropod develops sufficient "suction" remains to date an unanswered question.

B. Insecticides

The economic motive has provided a stimulus to investigations on the permeability of insect cuticle to chemical compounds, in a wide variety of forms and formulations, which is far in excess of anything that could be expected from normal academic interest. The fact that the bulk of the work has been done with poisonous compounds should not detract from the value of the data as a contribution toward the elucidation of the general principles of cuticle permeability.

The most obvious influence of the cuticle on the effectiveness of insecticides is its effect as a protective barrier to decrease the speed of action of a toxicant picked up by contact or applied topically. However, even within a group of related (e.g. organophosphorus) compounds, the penetration of some may be hindred by the cuticle in reaching sites of toxic action while that of others is not (Lovell, 1963).

1. Pathways for Penetration as Affected by Cuticle Structure

The epicuticular lipid layer is for most insects the primary barrier to the penetration of water or aqueous solutions. However, it can also be a formidable barrier to lipophilic substances if they are not properly formulated. The "cement" or "tectocuticle" layer appears to provide a surprisingly important barrier in view of its exceedingly thin and fragile nature. The cuticle below the lipid layer presents both lipid and aqueous pathways for penetration, greatly influenced by structural peculiarities such as thickness, sclerotization, conjunctivae, articular membranes, and the dermal gland ducts and pore canals.

The two types of channels traversing most of the cuticle in cockroaches, the dermal gland ducts and pore canals, have long been known from conventional microscopy and have been implicated in the transport of water, wax, and insecticides by many investigators since their role in the diffusion of oily solutions was recognized by Wigglesworth (1942).

The hard, tough cuticle of insects is remarkably permeable to the organic insecticides now in general favor, although the soft, tender, human skin is relatively impermeable thereto. This results in varying degrees of "absorptive selectivity," which, as O'Brien (1961) points out, is generally greater for chlorinated hydrocarbons than for organophosphorous compounds.

a. Areas of Least Resistance in the Cuticle. The density of tactile and chemosensory* setae is a complicating factor in the study of cuticle permeability, for if they are sufficiently dense they may prevent contact between the insecticide and the epicuticle (Klinger, 1936; Ossowski, 1944; David and Gardiner, 1950). Some of the cuticles so protected are thin, as in the case of *Euproctis* (Klinger, 1936), and would be expected to be readily penetrated if contacted.

On the other hand, if the setae are not sufficiently abundant to prevent contact of the insecticide, they, as well as spines and sensilla, facilitate its entry since the cuticle covering these processes is much thinner than that of the remaining areas (Richards, 1951). As stated previously, the cuticle of olfactory sense hairs is perforated by many minute pores (Slifer *et al.*, 1959; Steinbrecht and Müller, 1971). The articular membranes at their bases are especially permeable (Wilcoxon and Hartzell, 1933), and it should be borne in mind that the protoplasm of the epidermal cells extends upwards to that area (Richards, 1951). Wilcoxon and Hartzell (1933), Witt (1947),

---

* This statement applies to chemosensory setae although, as stated previously, they have uncovered areas, usually near the tips, through which the insecticide presumably can readily penetrate.

Wiesmann (1949), and Pfaff (1952) believed that these processes, their articular membranes, and the various conjunctivae, might be the only points of entry for various insecticides, in the different insect species which they investigated. Wigglesworth (1942) found them to be the only points of entry, for pyrethrum extract in oil, into adults and the later periods of nymphal instars of *Rhodnius,* although general penetration was observed for the thinner cuticles present immediately after each molt.

Although a compound could be expected to penetrate at the points of least resistance, this does not necessarily imply that the remainder of the cuticle is impermeable. The absence of dermal gland pores and pore canals, for example, does not preclude the passage of insecticide into cuticular tissue. Lewis (1965) found that the quantity of DDT absorbed per unit area by the lens of the compound eye of *Phormia* was one-third the quantity absorbed by the coxae and sternal-coxal membranes and less than one-sixth the quantity absorbed by the antennae. None of these regions has hinged setae and thin articulating membranes but in addition the lens of the compound eye has no dermal gland pores and pore canals. Likewise the ability of the cuticle to retain a large percentage of topically applied insecticide (see Section IV,2,d) appears to imply considerable penetration of cuticular tissue per se. Although Pfaff (1952) had concluded that parathion does not penetrate the homogeneous portions of the cuticle of *Periplaneta,* the experiments of Olson and O'Brien (1963) with paraoxon led them to doubt this conclusion. They found that considerable dimethoate *is* retained in isolated cuticle and presumed that this may also occur with parathion.

O'Kane *et al.* (1933) found that the weakly sclerotized areas on heads of cockroaches were more permeable to nicotine than areas of thicker and harder cuticle even when sensory structures were present on the latter. When droplets of insecticide were applied to many groups of hairs and spines and to all articular and intersegmental membranes, the membranous regions were found to be the areas of least resistance to penetration.

Cuticular sensilla on the plantar surface of their tarsi may account for the high susceptibility of certain adult flies and bees to toxicant residues in comparison with the beetle *Epilachna* and the bug *Oncopeltus,* which lack tarsal chemoreceptors. Additionally, *Oncopeltus* contacts the surface only with its claws (Hayes and Liu, 1947; Sarkaria and Patton, 1949). However, the influence of wear or abrasion must also be considered in such areas (Slifer, 1950).

Wigglesworth (1941) concluded that dermal glands and pore canals can figure prominently in penetration. However, pore canals must not be assumed to be essential for penetration in all species, for mosquito larvae are highly susceptible to oils and contact insecticides despite the absence of pore canals in their cuticles (Richards, 1941; Brown, 1951). The two-spotted mite, *Tetranychus urticae* Koch, shows no signs of pore canals, yet it is susceptible to many contact acaricides (Gibbs and Morrison, 1959). Richards and Anderson (1942a) believed that the pore canals probably were important in accelerating the penetration of aqueous solutions, but appeared not to be penetrated by oils and oil-soluble substances unless such substances were emulsified. They believed that oil penetrates through the general matrix of the cuticle.

Although Klinger (1936) reported structures like hair sockets or dermal glands to be the routes of least resistance, he believed that, where such structures were absent, the insecticide would pass through the epicuticle to the pore canals and then along their cytoplasmic strands. It is now known that pore canals are connected with the surface of the cuticulin via "wax canals." Locke (1964, 1965) sees no obstacle to the passage of apolar compounds through the wax canals.

In the grasshopper *Oedipoda* the areas of the cuticle with different degrees of permeability have been delineated in great detail by Nicoli (1958).

b. Cuticle Thickness. Many insects, particularly the adults, are protected against the entry of contact insecticides by thick and sclerotized cuticles. Likewise, larvae and nymphs, as they develop within a given instar, become progressively less permeable as their cuticles thicken (Brown, 1951). Klinger (1936), working with four species of lepidopterous larvae, found an increase in thickness of cuticle from the fourth to the fifth instar that ranged from 24 to 437%. With all four species the susceptibility of the larvae to pyrethrins was greatly decreased with increasing thickness of cuticle. However, the possibility of a concomitant decrease in the lipid content of the cuticle and its effect on permeability should be taken into consideration (see Pepper and Hastings, 1943). Likewise, the usual increase in micellar packing, generally with partial dehydration, would tend to decrease permeability. Finally, in experiments in which the permeability is presumed to be related to the period required to kill the insect, it should be borne in mind that the intrinsic susceptibility of the insect, as well as the permeability, may change with age. Another factor that must be taken into consideration is the proximity of the loci of application of an insecticide to the site of toxic action, generally the central nervous system (CNS). In comparison with this factor, thickness of cuticle and sclerotization are believed by some investigators to be of little importance (Fisher, 1952; Lewis, 1965; Gerolt, 1969). Gerolt (1969) particularly believes these factors to be of little importance because, he says, toxicants do not reach the CNS via the hemolymph. If this be the case, data based on speed of toxic action as a criterion of cuticle permeability are invalid. Gerolt believes topically applied toxicants spread laterally within the integument and then reach the site of toxic action via the integuments of the trachea, a subject discussed in greater detail in Section IV,B,3,a, below.

Thin cuticles can be highly impermeable. Using various aqueous solutions, it was shown that, among dipterous larvae, the impermeable cuticle of *Corethrus plumicornis* was only 2 μm thick, while the highly permeable cuticle of *Chironomus plumosus* was as much as 7 μm thick (Alexandrov, 1935). Experiments with *C. plumosus,* however, showed the usual trend of decreasing permeability with increasing age of larvae. *Tenebrio* larvae have thin cuticles of low permeability. The cuticle of the clothes moth (*Tineola*) is thin while that of the wax moth (*Galleria*) is thick, but the clothes moth can endure a dry environment while the wax moth cannot.

Beament (1957) found that, with dead cockroaches, water, organic solutions, or solids penetrated more rapidly in the thin and soft parts of the cuticle, whereas with live cockroaches water penetrated most rapidly through the thick, hard, and dark parts of the cuticle.

c. Tracheal Penetration. The spiracles and tracheae of insects are lined with cuticular layers homologous with those of the exterior surface. They are not wetted by aqueous solutions or suspensions applied as sprays unless suitable surfactants are added (McIndoo, 1916; Moore, 1921; O'Kane *et al.,* 1930; Wilcoxon and Hartzell, 1931; Gabler, 1939; Ebeling, 1939). There is good reason to believe that the lining of the tracheae is waxy and is even more hydrophobic than the surface of the cuticle (Beament, 1954).

If the cement or tectocuticle that overlies the waxy layer of the typical cuticle is not present in tracheae, this may account for its more hydrophobic nature (Beament, 1954). This possibility is enhanced by Locke's (1956) belief that there are no cement-producing glands in the tracheae. In fact, Richards and Korda (1950) in their extensive study of tracheae of all arthropod groups never saw any indication of gland ducts of any kind. Neither did they find pore canals or wax canals.

Water with appropriate surfactants, petroleum oil, or oil emulsions can enter the tracheae (Hoskins, 1940). Although continuous,* the walls of the tracheae are probably generally very thin; those of *Periplaneta* and *Phormia* larvae were found to be only 0.01 to 0.02 μm in thickness (Richards and Anderson, 1942b; Richards and Korda, 1950). Roy *et al.* (1943) found that when a solution of pyrethrins in kerosene stained red with Sudan III was injected into a large thoracic trunk through the spiracular opening, there was a wide and diffuse staining of the legs and thoracic muscles surrounding the tracheal trunk because of the diffusion of the injected fluid through the tracheal wall. Likewise, when finely powdered pyrethrum was gently blown into trachea via spiracular openings, paralysis followed the same order

---

* No holes or pores have been detected in the walls of tracheae (Weber, 1933) or tracheoles (Richards and Anderson, 1942a; Richards and Korda, 1950).

as when kerosene-pyrethrins solutions were injected. The investigators believed the tracheal trunks contain a certain amount of moisture which can dissolve pyrethrins as effectively as the hemocoele fluid. They ruled out any direct pathway for pyrethrins to the ganglia via the tracheal system and believed that they diffused in the fluid state, through the tracheal walls to the hemocoele, and were carried with the body fluid to the ganglia.

Richards and Weygandt (1945), investigating the penetration of over 100 dyed fat solvents into the tracheae of mosquito larvae, found the best dye penetration with the benzine-cymene series and some of the cyclic (terpene) hydrocarbons. The least dye penetration was obtained with the ketones, amines, chlorinated ethers, and nitroparaffins. The permeability of the tracheal membranes appeared to be of the same general order as that of dialysis membranes. The selective accumulation of Sudan dyes in the central nervous system indicated that lipid-soluble substances with low water solubility tend to accumulate in the insect nervous system, regardless of their mode of entry into the body.

Gerolt (1969) was led to believe, from extensive experimentation with several insecticides and several insect species, that the tracheal system provides the principal pathways by which topically applied toxicants reach the CNS, but as discussed in Section IV,B,3,b, this was disputed by other investigators.

In assessing the permeability of the surface cuticle, spiracular penetration must be avoided if liquids with a contact angle on wax of less than 90° are to be used.

d. Penetration Pathways Revealed by Natural Selection. An insight into what are apparently at least potential transcuticular pathways for the entry of nonelectrolytes into the body cavity of an insect should be provided by whatever morphological mechanisms insects may have developed by natural selection to resist the penetration of such compounds. No better example could be provided than the interesting adaptations developed by certain pentatomid bugs for protection against the penetration of lipoid-soluble poisons secreted as liquid sprays by glands in their cuticles. These bugs are themselves immune to the poison, which consists of short chain, mostly unsaturated aldehydes such as 2-hexenal, *trans*-3-heptenal, 2-octenal, and 2-decenal, each mixed with 50 to 60% of tridecane, the latter providing excellent solubility and penetrativity. In these pentatomids the wax filaments reach the surface of the cuticular layer in close juxtaposition (Fig. 8), which greatly decreases their ability to transport apolar substances to the pore canals. To further reduce the efficacy of this transport system, exposure of the insect to the poison solution results in a blockage of the canal containing the juxtaposed filaments, with a plug of presumably hardened wax. Gluud

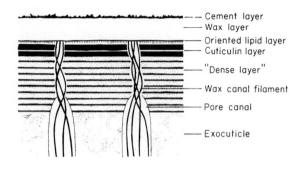

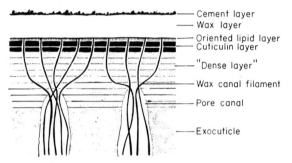

Fig. 8. Wax canals as they leave the pore canal and penetrate the epicuticle. Above, close juxtaposition of the wax filaments in pentatomids that secrete lipid-soluble poisons from cuticular glands; below, branched-out wax filaments of other insects. (From Gluud, 1968.)

(1968) was able to demonstrate this interesting phenomenon with *Pyrrhocoris apterus,* using a hexenal/tridecane solution and obtaining electronmicrographs with a magnification of 100,000. Figure 8 shows the difference in the juxtaposed wax filaments of poisonous pentatomids and the branched-out filaments of the other insects investigated. In most insects each wax filament, after leaving the pore canal, proceeds individually through the cuticulin of the epicuticle as explained in Section IV,6 and Chapter 2 in this volume.

In addition to the clustering of wax canals to decrease penetration of poison, the poison-secreting pentatomids have in their cuticles prominent wax-storage areas (*Wachseinlagerungen*) which could absorb poison that may have penetrated through the epicuticle despite the reduced capacity of the pore canals to conduct apolar compounds (Gluud, 1968). These apparent morphological adaptations to reduce the rate of penetration of natural organic poisons would appear to suggest that penetration of organic insecticides directly through the cuticle—even homogeneous portions of it—must ordinarily be at least a potential route of entry into the insect's body cavity. This should be borne in mind when considering the conflicting conclusions of various investigators concerning the cuticular pathways for nonelectrolytes, as presented below in Sections IV,B,3,a and b.

2. Physicochemical Considerations

The factors relating to the gross structure of the cuticle that could affect the locale and the rate of penetration have heretofore been considered. However, regardless of the extent to which a physical peculiarity might aid penetration, a certain amount of cuticle remains to be penetrated except for the very limited percentage of the body surface covered by the tips of chemoreceptors which, as discussed previously, are not covered with cuticle. Derm pores, pore canals, and tracheae offer passages through cuticle that may largely obviate the necessity for penetration of homogeneous cuticle tissue, but it is nevertheless evident that such tissue is permeable. Gerolt (1969, 1970) found that insecticides, including dieldrin applied in crystalline form, migrated laterally in insect endocuticles, indicating penetration of homogeneous tissue. Insecticides also penetrate the cuticle over eyes (Lewis, 1965) and through tracheal walls (Roy et al., 1943), which are homologous with the general integument and, in addition, have no pore canals or dermal ducts. The physicochemical factors that pertain to permeability of homogeneous cuticle tissue may be separated into three classes: liposolubility, surface migration, and phase partition. With regard to insecticides, the factors pertaining to their fate in the cuticle, such as storage and metabolism, will be discussed.

a. *Liposolubility.* Contact insecticides as a class are predominantly lipid soluble. The fact that such insecticides can penetrate the cuticle when applied as pure crystals shows that they are able to follow lipid pathways either by simple dissolution in the lipid or by migration on the surfaces of intermicellar spaces. Olson and O'Brien (1963), applying their solutes in very small volumes of highly volatile solvents, concluded that lipids in the cuticles are important in determining the facility of penetration of a given solute. Their data suggested that the epicuticular grease of an adult male cockroach (*Periplaneta americana*) does not represent the bulk of these lipids in the cuticle.

The first stage in the uptake of an insecticide by an insect appears to be a simple dissolution of the substance in the epicuticular wax (Armstrong et al., 1952; McIntosh, 1957). Its solubility in cuticular waxes may be influenced by the consistency of the wax, as was indicated by Pradhan et al. (1952) in the case of DDT. They found that DDT crystals dissolve far

more readily and with greater velocity in the softer waxes and that the insect
with softer cuticular wax is usually more susceptible to the action of DDT
than the insect with harder wax. The great difference in the consistency
of insect waxes is illustrated by the difference in melting point of wax extracted from the exuviae of the larvae of *Euproctis lunata* (47° to 54°C)
and of the larvae of *Trogoderma granaria* (77° to 83°C). However, it
should be pointed out that the German cockroach, *Blattella germanica,* has
a considerable natural tolerance to DDT, despite its soft "grease."

Nair (1959) found that three species of beetles showed decreasing susceptibility to five contact insecticides in the order *Bruchus* < *Tribolium* < *Calandra*. Differences in susceptibility were attributed to the soft epicuticular
wax layer of *Bruchus,* compared with the very hard wax of *Tribolium*. The
fact that *Calandra* showed the least susceptibility to the insecticides was
attributed to a protective cement layer that delayed the passage of insecticide
to the underlying wax layer. Nair (1957) had similarly previously found
that the three species showed the same ratio of resistance to desiccation resulting from exposure to finely divided powders.

The relationship of the liposolubility of certain types of insecticides to
cuticular penetration has been demonstrated (Dresden and Krijgsman,
1948; Bohm, 1951; Richards, 1951). In some species of insects the rate
of penetration of liposoluble insecticides appears to have a simple relationship to the total amount of lipid in the insect integument, indicating that
these insecticides progress through the cuticle entirely via lipid pathways
and by simple dissolution. For example, Pepper and Hastings (1943) found
that the susceptibility of the larvae of the sugar beet webworm, *Loxostege
sticticalis* L., to pyrethrum dusts and sprays and to petroleum oil decreased
with each successive instar. Likewise, the amount of extractable lipid decreased sharply, amounting to 11.7, 3.9, and 0.2% of the entire weight of
the exoskeleton in the third, fourth, and fifth instars, respectively. On the
other hand, a number of investigators have found an inverse relationship
between lipid content and toxicity for a number of insect species and for
both organophosphorus compounds and chlorinated hydrocarbons, particularly the latter (see Bennett and Thomas, 1963). However, in such cases
it may be that cuticular lipids occur in quantities greater than those required
for optimum penetration of liposoluble toxicants and the latter may be held
back from the subcuticular sites of toxic action by being absorbed in the
excess lipid. This subject will be discussed more fully in Section IV,C,2,d.

The epicuticular lipid layer, or this layer along with the "cement" at the
surface, can be an effective barrier even for lipid-soluble toxicants if they
are not properly formulated. For example, Wigglesworth (1942) found that
normally fifth-instar nymphs of *Rhodnius* were unaffected by a dry rotenone
powder (90% rotenone) after 3 weeks of contact with the powder. Those

rubbed with alumina powder to remove the epicuticular lipid showed toxic symptoms in 8 hours and were dead in less than 24 hours. Likewise, nymphs normally unaffected by an aqueous solution containing 2% nicotine were moribund within 20 minutes, after they had been rubbed with alumina dust before treatment. On the other hand, Winteringham *et al.* (1955), using acetone solutions, found no significant difference in the percent of pyrethroids absorbed by live adult house flies and those which were killed in hot water, a treatment which Olson and O'Brien (1963) found removes the epicuticular wax layer. It is reasonable, of course, that an organic insecticide applied as a dust or in an aqueous solution should penetrate the cuticle more rapidly when an insect's protective lipid layer is removed and that, therefore, when applied in an acetone solution, the presence or absence of a lipid barrier would be of no consequence with reference to rate of penetration.

Buerger and O'Brien (1965) investigated the penetration rates of sublethal quantities of radioactive DDT, famphur, and dimethoate, nonelectrolytes with a wide range of polarity (olive oil–water partition coefficients 199, 19.2, and 0.593, respectively). (Famphur and dimethoate can be used as systemic insecticides.) These compounds were applied in a 1 μl drop of acetone to the surface of integuments of six animals: three insects (*Tenebrio molitor, Acheta domesticus,* and *Periplaneta americana*) and three vertebrates (garden toad, chameleon, horned toad). Penetration of insecticide occurred in two phases: a brief period of extremely rapid penetration was followed by a long period of much slower penetration. In the second phase, which was the only phase for which actual rate could be measured, rate of penetration increased with increasing polarity of the compound except with the cricket, for which famphur penetrated faster than dimethoate. In this experiment the use of acetone presumably eliminated the outer (apolar) barrier as a factor of practical consequence, so that the major barrier became the relatively polar cuticulin and procuticle into which the rate of penetration of compounds might be expected to increase with an increase in their polarity. Buerger and O'Brien pointed out that, by contrast, when compounds are applied in an external aqueous phase, as had been done by Brown and Scott (1934) with human skin and by Treherne (1956, 1957) with rabbit skin and locust integument, the rate of penetration decreases with increasing polarity because the lipid barrier of the integument is the rate-limiting factor.

b. Surface Migration. Richards and Cutkomp (1946) concluded, from the similarity of the toxicity of DDT to certain animals when emulsions are injected as contrasted to the dissimilarity of effect from external applications, that chitinous cuticles facilitate the entry of DDT into the animal body by selectively concentrating the compound by adsorption phenomena. Lord (1948) found DDT and its analogues to be readily adsorbed from colloidal suspension by chitin, but not by wool, silica powder, or ground cellulose, the latter quite similar to chitin in chemical structure except for absence of amino groups.

Especially noteworthy in the paper by Richards and Cutkomp was their speculation that a kind of adsorption is involved "that apparently utilizes the surfaces of known intermicellar spaces." Fan *et al.* (1948) assumed that DDT can spread on surfaces in the "cuticle matrix" by surface diffusion of the adsorbed layer in a manner more or less analogous to that analyzed earlier by Volmer (1932, 1938).

Volmer had investigated the migration of benzophenone over the surfaces of various solids. Possibly a more appropriate example in the present instance would be the rapid and deep penetration of organic insecticides, such as DDT, dieldrin, and lindane, from large crystals into porous dried mud blocks (Hadaway and Barlow, 1951; Barlow and Hadaway, 1958); even more pertinent is the finding by Rideal and Tadayon (1954b) that molecules of stearic acid migrated over a monolayer of stearic acid more rapidly than over various metals. Likewise the migration of beeswax over aggregates of silica aerogel (Ebeling, 1961, 1971) resulted in layers that were visible, and therefore obviously many molecules in thickness. The migration of organic molecules on both metal and organic substrates is therefore experimentally established. Organic molecules can also move over wet surfaces, utilizing their ability to reduce surface tension (Marangoni effect) (see Bikerman, 1958).

Apparently lacking in the field of insect toxicology are precise calculations of the relationship of the molecular dimensions of the toxicant to the dimensions of the available spaces in the cuticle. Such calculations would be comparable to those of the size relationships of classes of herbicide compounds to the spaces between the microfibrillar strands of the epidermal wall of the leaf (see Mitchell *et al.*, 1960).

c. Phase Partition. Phase partition may improve the penetration of some types of toxicants or may even be absolutely essential. The role that phase partition can play is strikingly demonstrated by the combined action of two solvents of contrasting polarity, described previously (Section II,D) in improving penetration. Wigglesworth (1941) was impressed by this phenomenon as an example of the importance of the partition coefficient of a substance between oil and water in determining the rate at which it will leave its oily base not only for more rapid penetration through the aqueous phase of the procuticle but also for entering the tissues of the insect.

A substance may itself possess considerable ability to pass from the wax to the underlying water-bearing protein or it may be aided in this respect by a suitable solvent. Webb and Green (1945) used diphenylamine as a toxicant and determined the efficiency of a number of solvents of widely differing physical properties in increasing its effectiveness against the sheep ked, *Melophagus ovinus*. They recognized that penetration of cuticle involves passage through a thin wax layer as well as a much thicker layer composed of both protein and wax, that is, polar and nonpolar elements. Although only those insecticide solvents penetrating beeswax readily were effective, not *all* the good wax solvents were satisfactory. It was therefore concluded that wax solubility was not the only requirement for effective action. Obviously a solvent not able to leave the wax phase (i.e., with partition coefficient nil) could not be an effective carrier below the superficial wax layer.

Nevertheless, although both high wax solubility and adequate partition coefficient were found among the efficient solvents, others having both these properties were ineffective. It was found necessary that the "carrier" must also act as a mutual solvent for toxicant and water.

Webb and Green (1945) concluded that the "carrier efficiency" of a solvent depends on its ability to (1) carry the toxicant through the lipid layer to the interface lipid/water-bearing protein; (2) concentrate it there in sufficient quantity to form an adequate diffusion gradient, a task that requires a sufficiently high phase-partition coefficient; and (3) act as a mutual solvent for the toxicant and the water that permeates the cuticulin and the procuticle. A fourth requirement is that the solvent should not volatilize to a degree that would allow the toxicant to crystallize out of solution before it has a chance to penetrate.

Webb and Green presumed that portions of the insecticides also penetrated the cuticle via lipid routes and may or may not have been slightly aided in this respect by the solvents. It was also recognized that the pore canals may have facilitated diffusion of the insecticides through the cuticle. Finally, it must be borne in mind that, as in all investigations in which penetration phenomena are inferred from mortality data, physiological interactions are a distinct possibility when mixtures of chemical are used.

Extensive laboratory and field experimentation with mosquitoes led Hadaway *et al.* (1970) to conclude that both lipid solubility and partitioning properties of insecticides are important factors in insecticide uptake by insect cuticle and its movement into an aqueous phase. A number of carbamates were rated numerically by dividing their *n*-hexane solubility by their partition coefficient for *n*-hexane and water. The fastest-acting compounds had a value of more than 0.1, the intermediate group from 0.01 to 0.05, and the slowest acting had a value of 0.005. Lipid solubility and partition coefficient were considered to be complementary in that low lipid solubility could be compensated by a favorable water-lipid partition and an unfavorable partition by high solubility. Lipid solubility was more important for a dry insecticide residue than for an insecticide applied topically in a solvent. Two insecticides might have similar insecticidal efficacy when applied topically in solution but the one with lower lipid solubility was slower acting as a crystalline deposit.

Olson and O'Brien (1963) found that the half-time for penetration of the highly polar phosphoric acid *from the surface grease* through the cuticle for the American cockroach was only 16 minutes compared with 1584 minutes for DDT, an insecticide well known for its high liposolubility.

Burt and Lord (1968) and Burt *et al.* (1971) compared the penetration of two compounds that differ about as widely in phase partition as any that are available among organic insecticides. When they were applied to the metathoracic sterna of adult male American cockroaches in 1 μl of acetone solution to give an LD_{95}, five times more toxicant was required with diazinon than with pyrethrin I, yet 75% of the diazoxon but only 30% of the pyrethrin had penetrated the cuticle in 2 hours. Diazoxon possesses considerable polarity and is distributed between the solid and fluid fractions of cockroaches in the ratio of 3.7:1 compared with 30,000:1 for pyrethrin I. Despite its excellent ability to penetrate lipids, the relative inability of pyrethrin I to partition to aqueous phases in the portion of the cuticle below the lipid layer greatly retarded its rate of penetration when compared with a compound with adequate lipid solubility combined with an apparently ideal partition coefficient between the lipid and polar elements of the cuticle.

The above investigations presented an apparently convincing argument for the important role of phase partition in facilitating the penetration or, at least, increasing the insecticidal efficiency of solvent–toxicant combinations. However, Ebeling and Clark (1962), using adult male German cockroaches as test insects, were not able to improve the effectiveness of solutions of pyrethrins in kerosene by manipulations that might be expected to enhance phase partition, such as the addition of oil-soluble surface-active agents or solvents of considerable polarity, such as 10% n-propyl alcohol or 5% oleic acid. In fact, insecticidal effectiveness was significantly decreased. It is possible that solvents with maximum ability to penetrate insect cuticle, such as the light petroleum fractions, are most effective when allowed to follow, along with their dissolved toxicants, the purely lipid pathways.

d. Retention of Absorbed Toxicant by the Cuticle. Although lipid solubility is required of an insecticide in order that it might penetrate the cuticle, it is also desirable that it not be excessively soluble in the cuticular lipid. This might result in too much toxicant being retained by the cuticle and an insufficient quantity reaching sites of toxic action in the interior. DDT and chemically allied compounds, which are both freely liposoluble and slow in toxic action, are particularly susceptible. Lewis (1965) suggested that DDT accumulates in the integument of the blow fly, *Phormia terranovae,* but does not diffuse freely into the hemolymph. Weevils surviving the toxic effect of insecticides are usually those with the higher lipid content (Reiser *et al.,* 1953; Moore *et al.,* 1967). Vinson and Law (1971) found protein and lipid content of the cuticle to be higher and penetration of DDT to be slower, in larvae of resistant tobacco budworms, *Heliothis virescens* (F.), than in the larvae of a nonresistant strain. However, resistance is not correlated with lipid content in all insects (Matsubara, 1960; Fast and Brown, 1962; Fast, 1964; Perry, 1964). The "holding capacity" of lipids increases with temperature, the steep rise in the curve depicting this relationship occurring between 20° and 30°C, precisely the range in which negative temperature-mortality coefficients have been found to occur (Munson *et al.,* 1954).

Armstrong *et al.* (1951) determined the amount of each isomer of benzene hexachloride that was retained on or in the lipid layer of the epicuticle of granary weevils (*Sitophilus granarius*), and the amount penetrating beyond this layer. They showed that although the γ-isomer was only about 5% as soluble in the insect lipid as the δ-isomer, 5.6 times more penetrated into the bodies of the weevils. On the other hand, the solubilities of the α- and β-isomers appear to have been too low, for they did not penetrate into the insect bodies in appreciable quantities. The γ-isomer thus appears to have had the optimum degree of lipid solubility. The great difference in the rate of penetration of the δ- and β-isomers of BHC through the cuticle of the American cockroach was noted by Nakajima *et al.* (1969).

Similarly, Afifi and Roan (1961) found that, although ^{32}P-labeled diazinon penetrated into the cuticle of the American cockroach rapidly, much of it was retained in the cuticle. Indeed, it did not pass from the cuticle to the circulatory system to become generally distributed as rapidly as did other organophosphorus compounds they had studied previously.

Matsumura (1963) found that, within one hour after application, approximately 83% of the malathion entering the cuticle of *Periplaneta americana* could be accounted for by absorption, rather than penetration. When combined absorption and penetration was asymptotically reaching a maximum, this percentage was still approximately 70% (Fig. 9). He showed that the deviation from the theoretical rate in the diffusion of malathion through the cuticle was caused by absorption of this insecticide by the proteinaceous portions of the cuticle, for the absorbed toxicant could be extracted from the cuticle by treating it with hot water, but not with a wax solvent.

In apparent contradiction to the above finding, Bennett and Thomas

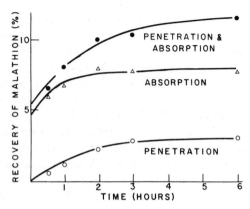

Fig. 9. Absorption and penetration of malathion in the cuticle of the American cockroach. (From Matsumura, 1963.)

(1963) showed that percent mortality of alfalfa weevils, *Hypera postica,* treated with malathion decreased with increasing lipid content of these insects. The association of organophosphorus compounds with the cuticular protein may be the most important consideration when penetration is the factor limiting the rate of toxic action and lipids may be the most important consideration when the initial pick-up of the insecticide by the insect is the rate-limiting factor.

Although a high lipid content might offer some protection against insecticides, it is not sufficient to account for the high tolerance of most species that have become genetically resistant through selection. However, high lipid content might coexist with other defense mechanisms to contribute to these high tolerances (Spiller, 1958; Winteringham and Hewlett, 1964; Kalra, 1970).

e. Metabolism of Toxicants in the Cuticle. The cuticle may be more than a mere physical barrier. Sternburg and Kearns (1952) concluded from their experiments with grasshoppers that topically applied DDT apparently is degraded into DDE in the cuticle and that the DDE is then transported to adjacent regions. Quraishi (1970) found that the external wash of DDT-treated house fly puparia yielded DDT and DDE in the approximate ratio of 19:1, indicating partial metabolism of DDT to DDE even on the surface of the cuticle. Therefore it may be assumed that a much greater degree of metabolism took place during the passage of the insecticide through the entire cuticle. The extracts of adult flies which emerged from the DDT-treated puparia contained only DDE. It is not known what proportion of the metabolism took place within the body of the adult fly.

Ahmed and Gardiner (1970) observed that locusts (*Schistocerca gregaria*) showed more typical symptoms of organophosphorus poisoning after topical applications of [³H]malathion than after injections. However, there was little or no difference when malaoxon, the hydrolytic degradation product of malathion, was used instead of malathion. Of the two compounds, malaoxon is the more potent cholinesterase inhibitor. The experiment indicated that topically applied malathion was metabolized to malaoxon by its passage through the insect cuticle, and that this accounts for the greater insecticidal efficacy of the topically applied dose. The cuticle was said to "activate" malathion. The more dilute the topically applied kerosene solutions were, the farther they spread and the more rapidly the toxicant penetrated, increasing the rate of activation into the more active malaoxon (Ahmed and Gardiner, 1970).

f. Penetration in Relation to Insecticide Resistance. Penetration of insecticides can be affected by genetic factors related to insecticide resistance. Matsumura and Brown (1961, 1963) found that an increased tolerance to toxicants in the larvae of a strain of *Aedes aegypti* was accompanied by as much as a twelvefold decrease in absorption and retention of toxicant and not by enzymatic detoxification. Retarded penetration accounts for at least part of the resistance shown by house flies of the Rutgers strain (Forgash *et al.*, 1962) and the SKA strain (El Basheir, 1967) against diazinon, by another strain against pyrethrin I (Fine *et al.*, 1967), and by the resistance shown by the tobacco budworm, *Heliothis virescens,* against DDT, endrin, and toxaphene-DDT (2:1) mixtures (Vinson and Brazzel, 1966).

The "penetration factor" in the SKA house fly strain was isolated by crossing experiments (Sawicki and Lord, 1970). Delayed penetration was responsible for slower knockdown and happened only with strain SKA and strain 348, which carried SKA's third chromosome. The penetration factor is selected for and increases resistance to many unrelated (nonspecific) insecticides. Sawicki and Lord consider it to be potentially very dangerous, deserving much more study.

Precaution is desirable in evaluating the influence of cuticle permeability on insect resistance to toxicants. As Oppenoorth (1958) and Oppenoorth and Dresden (1958) have demonstrated, if the resistance of a strain of insects is evident from topical application of the toxicant, but not from injection, this does not necessarily mean that the difference between the resistant and nonresistant strains is in the cuticle. It is possible in such a situation that resistance factors are present which protect the insect from toxication only if the toxicant is entering the body slowly, e.g., when it is entering through the cuticle.

It is evident from the above that the cuticle is involved in some of the most pressing problems in pest control, including the resistance problem,

and that a better understanding is needed of this organ, particularly in relation to the pathways for the penetration of toxicants.

3. The Cuticle as a Transport System for Nonelectrolytes

As the foregoing portion of the section implies, for many years it has been
assumed that contact insecticides enter the insect by penetrating through
the integument and being carried by the hemolymph to the next barrier,
the medullary sheath which encloses the central nervous system, the most
common site of toxic action. The literature from which such an assumption
might be drawn is too voluminous to be reviewed here, but the classic experiment is well illustrated by some research done by LeRoux and Morrison
(1954) on house flies with [^{14}C]DDT in benzol. They concluded that DDT
penetrates cuticle at the point of application and is transported to other
parts of the body by way of the hemolymph. For example, activity from
the labeled DDT was found in the hemolymph of the cervical region as
soon as 30 seconds after application to the tibiofemoral membrane of one
metathoracic leg. Ligation by means of cotton thread at the pedicel or cervix
greatly retarded movement of hemolymph and resulted in little or no activity
on the side of the ligature opposite to the locus of application. Hemolymph
did not accumulate DDT or its metabolites, but other tissues and organs
absorbed it and accumulated it. This was considered to be especially true
of the integument, which is not surprising in view of the great affinity of
the integument for DDT, as had previously been determined by Richards
and Cutkomp (1946), Sternberg et al. (1950), Richards (1951), and Hoffman et al. (1952).

To the present day, most investigators would concur with Burt and Lord
(1968) that, after penetration of the cuticle, a toxicant spreads rapidly
throughout the body of the insect via the hemolymph, and that close agreement between chemical and biological estimates suggest that it invades the
nerve cord from the hemolymph.

a. Evidence That an Insecticide Must Spread Laterally on or in the Cuticle
to Reach Portals of Entry through the Integument. Gerolt (1969, 1970) made
a series of experiments which led him to believe that insecticides can pass
directly from the cuticle to the hemolymph in significant quantities only
when dissolved in certain solvents. His experiments were made with 3- to
4-day-old female house flies (Musca domestica), mature adult locusts
(Schistocerca gregaria), fly larvae (Sarcophaga barbata), German cockroaches (Blattella germanica), and a few other species to a lesser extent.
His conclusions were based on seven types of experimental data.

1. Decreasing insecticidal effectiveness as the toxicant was applied at increasing distances from the central nervous system (CNS). Distance would

not be expected to be a significant factor if the toxicant were carried to the CNS by the hemolymph.

2. Olive oil injected into the body of the insect did not block the movement of toxicant, but did so when present as a film on the surface.

3. Crystalline masses of 0.2 μg of dieldrin introduced into the abdominal cavity produced no effect in 7 hours compared with a complete knockdown in 65 minutes with a surface deposit of the same quantity of dieldrin per house fly. Similar results were obtained with chlorophenvinphos, dichlorvos, and methyl parathion.

4. [^{14}C]dieldrin in benzene applied to the air-exposed outer side of isolated insect integument was not recovered in saline solution on the other side of the integument.

5. Beeswax acts as an absorbent for apolar insecticides. Beeswax collars around the "waists" of house flies and cockroaches prevented migration of insecticide, from crystals, via the integumental route and greatly retarded toxication, in striking contrast to rapid paralysis when the collars were not present. Wax collars on the coxa-trochanter area of German cockroaches protected the insects from toxic effects of dieldrin applied to the tarsus and tibia. This implies the absence of penetration through one or two articular membranes.

6. Lateral migration of [^{14}C]dieldrin in the integument could be followed by means of autoradiography. The label was situated in the endocuticle-hypodermis area. [Endocuticle is considered to be much more permeable than exocuticle and mesocuticle (Noble-Nesbitt, 1970b).]

7. [^{14}C]dieldrin from dry deposits was detected in tracheal trunks. Distribution of label was rather uneven, but increased with time. The gut, Malpighian tubes, fat-bodies, and hemolymph gave no detectable image. In evaluating this observation, it should be borne in mind that localization of toxicant, as indicated by radioautographs, does not necessarily imply that other surrounding areas contain no radioactivity, particularly when brief exposure times are employed with low-energy isotopes. Longer film exposure, e.g., 3 or 4 weeks or longer, might reveal a spread of label not indicated by an exposure time of a period of only 24 hours (M. S. Quraishi, personal communication). No localization of dieldrin in tracheae in thoracic ganglia was observed in radioautographs prepared by Nakajima et al. (1969) or Sellers and Guthrie (1971), indicating that the tracheae were not the avenue of transport all the way to the CNS, and implying a role for the hemolymph.

Gerolt (1969, 1970) concluded from the above experiments that an organic insecticide spreads laterally within the integument, probably primarily in the endocuticle, and reaches the site of toxic action via the integument of the tracheal system. To some extent, his conclusions appeared to be substantiated (and to some extent contradicted) by the results of other investigators. Lewis (1962) found that oil films spread over the entire body of the blow fly *Phormia terranovae* within a few minutes, a prerequisite for the entry of insecticides via the tracheal system as proposed by Gerolt. The film ceased to spread after an insect died. This led Lewis to the hypothesis that "a thin film of oil progresses as a result of very small elastic movements of the cuticle surface of the active insect." However, Locke (1965) believed that the rapid spread of the oil "is most easily explained if the surface monolayer of wax is liquid, allowing rapid diffusion." This might explain the cessation of spread when an insect dies and the monolayer of wax at the interface of the cuticulin and the remainder of the wax layer is no longer continuously supplied with wax in liquid form from the wax canals.

The rapid spread of toxic oil solutions was also reported for cockroaches, on which [^{14}C]lindane and [^{14}C]DDT in oil applied to the prothorax spread all over the insects, *including the wings,* in a few minutes (Quraishi and Poonawalla, 1969). The extensive lateral peripheral spread of [^{3}H]nicotine from the point of topical application was strikingly shown in a sagittal section of an American cockroach by means of frozen whole-body radioautography (Nakajima *et al.,* 1969; Kurihara *et al.,* 1970). Although the nicotine (41 μg in 10 μl of ethanol per insect) was applied to the apex of the third to fifth abdominal tergites, the radioautograph showed that within 5 minutes it had covered both dorsal and ventral cuticles of the abdomen (Fig. 10A) before entering the abdomen. The extensive distribution of this very penetrating compound in 3 hours is shown in Fig. 10C. Unike lindane, which accumulated on the peripheral regions of the CNS, nicotine penetrated the CNS rapidly. The above investigators believe that this is consistent with the supposition that nicotine affects synapses by mimicking acetylcholine. In some of the radioautographs shown in the papers by Nakajima *et al.* (1969) and Kurihara *et al.* (1970) concentrations of label may be seen that appear to indicate points of entry at intersegmental membranes as was so clearly demonstrated by Quraishi and Poonawalla (1969).

In view of the use of ethanol as a solvent by Nakajima *et al.* (1969), the writer observed the behavior of 2.5 μl droplets of ethanol on different surfaces at 23°C and 45% RH. A droplet spread on clean glass to a diameter of 8 mm and evaporated in 2 minutes. On beeswax a droplet had a diameter of 3.5 mm and evaporated in 3.25 minutes. On the pronotum of an American cockroach nymph a droplet had a diameter of 4 mm and disappeared, presumably by evaporation, in 4 minutes. (On abdominal tergites the behavior of droplets of ethanol was approximately the same.) Droplets did not spread from their initial area of contact. If ethanol aids the rapid lateral migration of nicotine, such as radioautographically monitored by Nakajima *et al.*

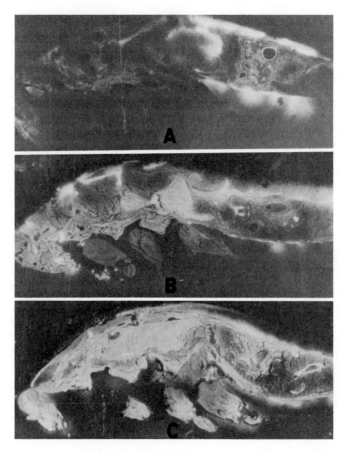

Fig. 10. Radioautographs of sagittal sections of American cockroaches receiving topical applications of [³H]nicotine in ethanol on dorsal tergites. (A) 5 minutes, (B) 1 hour, and (C) 3 hours after application of toxicant. (From Nakajima *et al.*, 1969; and Kurihara *et al.*, 1970.)

(1969), it is a type of migration that is not observable with the aid of a stereomicroscope.

Lewis (1965) applied DDT to structurally different parts of the cuticle of the blow fly *Phormia terranovae*. The DDT was applied in supersaturated lanolin solutions which rested on the cuticle in the form of tiny spheres that spread over a relatively small and easily measured area. Because the solution was supersaturated, a reservoir of crystals allowed for the constant concentration of DDT in the drop as the toxicant was taken up by the cuticle (at a faster rate than lanolin). Lewis found that, after an initial rapid rate of sorption of toxicant, the rate diminished exponentially. He assumed that if the insecticide had diffused readily through the local hypodermal cells and into the hemolymph and other tissues, no change in rate of DDT uptake would occur until the exhaustion of the external supply of the toxicant was approached. He concluded that, after a brief initial period of rapid penetration, the rate of diffusion of insecticide molecules through the cuticle is governed by the epidermal cells and that the insecticide does not diffuse very freely into the hemolymph. Nevertheless, in living flies the absorption rate did not approach zero asymptotically, as it did in dead flies, but reached a steady value. Lewis concluded that the rather steady state of absorption by live flies probably results from the removal of DDT from the cuticle, adjacent to the spheres of lanolin-DDT solution, and its passage at a slow but steady rate to the hemolymph via the epidermal cells. The great decrease in the entry rate of insecticides after the death of an insect has also been noted by Sternburg et al. (1950), Hoffman et al. (1952), Tahori and Hoskins (1953), and LeRoux and Morrison (1954). It indicates that hemolymph has at least some* function in the transport of insecticide to absorption sites in living insects, thus allowing for continued cuticular penetration, a role that ceases after death, when the hemolymph becomes a barrier.

Using freeze-drying and radioautographic techniques, Quraishi and Poonawalla (1969) studied the diffusion of topically applied [^{14}C]DDT into the house fly, *Musca domestica,* and its distribution in the internal organs. Benzene solutions of [^{14}C]DDT were applied by microsyringe on the center of the mesonotum and only the flies on which the solution had spread over a very small area in the mesonotum were selected for further study. Radioactivity was not detected from the surface of untreated cuticle. Nevertheless, insecticide was found entering the body of the insect through certain portals of entry situated within the untreated cuticle, all of them areas associated with thin membranes, such as setal pits and conjunctivae. These portals of entry were clearly revealed in radioautographs. The cuticle around these areas was not radioactive, and it was concluded that the insecticide did not diffuse through the homogeneous cuticle. In this respect the conclusions reached by Quraishi and Poonawalla with their modern procedures agreed with the older investigations of Wiesmann (1949), Witt (1947), and Pfaff (1952).

Unlike Gerolt (1969, 1970), Quraishi and Poonawalla did not locate

---

* An obvious factor influencing to what extent the hemolymph acts as a transport medium is the polarity of the insecticide. Burt (1970) found the moderately water-soluble diazoxon at just-lethal dose to have a concentration in the hemolymph of American cockroaches of about 2×10^{-6} M, whereas the extremely hydrophobic pyrethrin I at a just-lethal dose had a concentration in the hemolymph of less than 2×10^{-8} M.

radioactivity in the deeper layers of the cuticle. They believed that the DDT spread laterally via the wax monolayer over the entire surface of the insect. They were unable to trace the [^{14}C]DDT in the wax monolayer through radioautographs because the entire wax layer was lost during the processes of washing and embedding. As stated previously, it is Locke's (1965) belief that rapid diffusion of insecticide over the entire body of an insect, such as has been discussed in this section, is most easily explained if the surface monolayer of wax is liquid.

Another way in which the investigation of Quraishi and Poonawalla differed from that of Gerolt is that their radioautographic study led them to conclude that, after its lateral spread in the epicuticular monolayer of wax, DDT entered the body of the insect via "sutures, membranes, bases of the setae and possibly other portals of entry" whereas Gerolt believed the tracheal system provided the principal route of entry. The finding of LeRoux and Morrison (1954) that activity of [^{14}C]DDT applied to the tibio-femoral membrane of one metathoracic leg could be detected 30 seconds later in the hemolymph of the cervical region, would appear to rule out the tracheae as a necessary portal of entry and to favor the conclusion reached by Quraishi and Poonawalla concerning the portals of entry. Roy *et al.* (1943) concluded that the tracheal system did not provide a direct avenue of transport of pyrethrins to the CNS, but that they diffused through the tracheal walls and were carried to the ganglia by the haemolymph (Section, IV,B,1,c).

Poonawalla and Quraishi (1970) applied [^{14}C]DDT in benzene to the mesonota of American cockroaches; again, only those insects on which the insecticide spread over a very small area were selected for further study. Using doses of the toxicant varying between 35 and 2900 nanograms and "counting" the flies with a gas flow counter immediately after treatment and at 2-, 6-, and 24-hour intervals, they found that the rate of penetration was dose dependent; the higher the dose applied the lower the percentage absorbed, even at the lowest of the nanogram doses. Radioautographs made within 15 minutes showed uniform radioactivity over the entire alimentary canal, indicating the role of the hemolymph in the distribution of the toxicant.

b. Evidence Against the Necessity for Lateral Spread of Insecticide on or in the Cuticle to Reach Portals of Entry. In a recent investigation, Moriarty and French (1971) challenged the conclusions of Gerolt (1969, 1970), Quraishi and Poonawalla (1969) and, by inference, the conclusions reached by other investigators who believe nonelectrolytes must migrate laterally on or in the cuticle of an insect to find sutures, bases of setae, or tracheae in order to enter the body of the insect in significant quantity and to reach sites of toxic action. Using the cockroach *Periplaneta americana* as the test insect, Moriarty and French applied pure recrystallized dieldrin as a 2% (w/v) solution in 1,4-dioxan in 1 μl quantities onto the (1) center of the pronotum, (2) middle of the basal half of the upper forewing, (3) middle of the distal half of the upper forewing, and (4) both the basal and distal halves of the upper forewing.

Each cockroach was prepared for analysis at five periods ranging from 2 to 76 hours after application of the toxicant. For analysis, the basal and distal halves of the upper forewing were cut off and these parts and the rest of the body were rinsed with methanol to remove the dieldrin on the surface or in superficial lipids. Each part was then ground with sand and anhydrous sodium sulfate, covered with hexane, and stored at −20°C until analyzed by gas chromatography.

In the methanol rinse, an average of 95.4% of the dieldrin recovered was from that part of the body on which it was applied; the other two parts of the body contained less than 0.1 μg. When either the basal or the apical half of the wing was dosed, some dieldrin was usually found in the body rinse 76 hours later, but never in the rinse from the other half of the wing. Wing tissue never contained dieldrin unless the insecticide had been applied to that part of the wing. Whether only the wing base or tip was dosed, the amount of dieldrin in either half was similar. These results would appear to be the opposite of the aforementioned results of Quraishi and Poonawalla (1969), but in the latter investigation the insecticides may have been transported by the oil in which they were dissolved.

Moriarty and French concluded from the above findings that lateral spread of dieldrin in epicuticular lipids must be an unimportant factor. They found that there was little difference in the three sites studied in the rate of dieldrin uptake from the cuticular surface, for the amount of insecticide in the pronotum was not distinguished from that in the rest of the body.

c. Further Evidence against Tracheal Transport of Toxicant to the CNS.
Olson (1970) also obtained results that appear at variance with those of Gerolt (1969, 1970). He recovered 3% of topically applied DDT from the hemolymph of the American cockroach and concluded that the hemolymph is strongly implicated in the translocation of DDT. LeRoux and Morrison (1954), in their experiments with [14C]DDT topically applied to house flies, withdrew hemolymph samples through the coxal or cervical (neck) membranes. Radioactivity in the hemolymph of the cervical region was noted in as little as 30 seconds after application to the leg. The hemolymph appeared to transport but not to accumulate DDT, a claim that had also been made by Tahori and Hoskins (1953).

Possibly the most striking evidence against the necessity for the tracheal system for the transport of toxicant to the CNS and in favor of the hemolymph as a transport vehicle was obtained by P. E. Burt and his co-workers from three experiments designed specifically to test the aforementioned conclusions drawn by Gerolt (1969) from his experiments. The American cockroach was used as the test insect (Burt and Goodchild, 1969; Burt et al., 1971).

1. Hemolymph was collected from cockroaches poisoned with LD_{95}'s (0.45 μg) of pyrethrin I. They discharged the hemolymph continuously at the rate of 0.85 μl per minute onto exposed abdominal nerve cords in untreated cockroaches while the cords were monitored electrophysiologically for abnormal symptoms. Spontaneous nervous activity in sixth abdominal ganglia increased tenfold.

2. Pyrethrin I was applied, in doses sufficient for an LD_{95}, in three ways: (a) to the metathoracic sterna dissolved in 1 μl of acetone; (b) to the metathoracic sterna dissolved in 1 μl of a solvent-free emulsion of saline; (c) introduced into the tracheal system via a metathoracic spiracle in 0.1 μl of a solvent-free emulsion. (The emulsion contained 0.1% of the emulsifier Ethylan-TU.) Little difference was found between the efficacy of topical and tracheal treatments and the acetone did not increase speed of action or ultimate toxicity of the pyrethrin I applied topically.

3. After exposing the sixth abdominal ganglion and associated tracheal system in an isolated abdomen, solvent-free emulsion of pyrethrin I was introduced in 0.15 μl quantities into one of the longitudinal ventral tracheae close to the ganglion. Large doses (0.5 and 7.5 μg) of pyrethrin I in the tracheae decreased spontaneous activity, but without the initial burst of greatly increased activity associated with the external irrigation of ganglia with pyrethrin I in saline.* [In autoradiographs prepared by Sellers and Guthrie (1971) no localization of [^{3}H]dieldrin in tracheae was revealed in thoracic ganglion preparations.]

The above experiments suggest that the hemolymph from cockroaches poisoned with pyrethrin I contains a substance toxic to nerve tissue, presumably pyrethrin I. Tracheal transport of pyrethrin I via the tracheal system was no more efficient than the same dose applied externally; in fact, it appeared to be a little less efficient, for all topically treated insects were prostrate and severely affected in 48 hours, compared with only 80% of those receiving pyrethrin I via the tracheal system. Acetone increased neither the speed of action nor the ultimate toxicity of topically applied pyrethrin I.

Benezet and Forgash (1972), applying [^{14}C]malathion in 1 μl quantities

---

* In a discussion at a symposium on "Penetration through Insect Cuticle," Gerolt (1970) pointed out that the solubility of insecticides is much less in saline solution than in hemolymph and therefore partition to other tissues would be greater in saline.

of acetone to the thoraces of house flies, found that penetration was highest at the site of application and that the malathion moved through the cuticle and into the hemolymph within 15 seconds. Some concentration of malathion occurred at sutures, but radioautographs indicated sclerotized and membranous areas in adjacent locations had the same activity. They observed no concentration of activity in membranous areas, as reported by Quraishi and Poonawalla (1969) when using [^{14}C]DDT. There was no indication that the tracheal system was involved in insecticide transport, as reported by Gerolt (1969). It was concluded that the insecticide was transported to sites of toxic action via the hemolymph. This is not necessarily contrary to the findings of Gerolt (1969), who also found that the insecticide could be transported via the hemolymph when topically applied in a solvent.

How different investigators, each doing apparently sound and carefully conducted research, can reach diametrically opposite conclusions is the question that remains unanswered by the experiments described in the foregoing pages.

4. Aqueous Solutions

Recognizing that most of the information on cuticle permeability to insecticides had in the past been primarily qualitative, Treherne (1957) made an investigation on the diffusion of a series of nonelectrolytes through isolated cuticle of the grasshopper *Schistocerca gregaria* for the purpose of obtaining quantitative data. Cuticle from the third abdominal tergum was carefully removed and freed from underlying organs and the epidermal surface was washed in a gentle stream of saline solution. Care was taken to avoid damage to either the epidermis or the epicuticle. The cuticle was held in position in a "diffusion apparatus" placed so that the lower (epicuticular) surface of the cuticle was in contact with the test substance in an oxygenated and stirred saline solution. Above the cuticle was the saline solution only. The amount of this substance diffusing through the cuticle was measured by determining its concentration in the liquid above the inner surface of the cuticle.

The permeability constant ($P \times 10^7$ cm sec^{-1}) for seven nonelectrolytes varied from 12.23 for ethanol to 0.19 for urea. Cuticles were then used from which the epicuticular wax had been removed with chloroform, after which the removal of the wax was confirmed by the ammoniacal silver nitrate treatment. The removal of the epicuticular wax, which was shown not to affect the remainder of the cuticle, greatly increased the rate of diffusion of the above substances. However, of particular interest is the fact that the effect of wax removal was least for ethanol, the substance that penetrated most rapidly through the whole cuticle, and greatest for urea, the substance that penetrated least rapidly. Thus the resistance to diffusion of the cuticle remaining after chloroform treatment, expressed as a percentage of that through the whole cuticle, is as follows: ethanol, 49.5%; phenylthiourea, 44.4%; *m*-tolylurea, 37.2%; diethylthiourea, 12.6%; thiourea, 3.4%; dimethylurea, 2.6%; and urea, 1.1%. These figures show that the wax layer was almost exclusively the limiting factor in the diffusion of urea, but accounted for only about half of the resistance to the diffusion of ethanol.

Olson and O'Brien (1963) investigated the *in vivo* penetration of a series of six radioactive insecticidal solutes through the pronotum of *Periplaneta*. When the solutes gained entry through the epicuticular lipid via a small volume of a volatile organic solvent, penetration of the remainder of the cuticle was proportional to their water solubility.

Treherne (1957) was able to calculate that the diffusion of molecules through the layers of the cuticle below the wax layer is only about 0.24% of that through an equivalent quantity of unstirred water, which indicates that the area for free diffusion is very small in the insect cuticle.

If there is any tendency whatever for water to penetrate under normal conditions, one would expect that the quantity penetrating would at least be in proportion to the period of exposure (see McIndoo, 1916; Richards and Fan, 1949). However, Lovegren and Feuge (1954) showed, additionally, that the presence of water in a lipid film weakens intermolecular bonds and results in more rapid penetration of water subsequently passing through.

It must not be assumed that prolonged contact of an aqueous solution will favor the penetration of any solute. For example, Olson and O'Brien (1963) found that when phosphoric acid was applied to a cockroach in a droplet of water, no penetration of the solute occurred until the water had evaporated. However, when dissolved in acetone, the same solute penetrated rapidly.

a. Influence of pH on the Lipid vs. the Remainder of the Cuticle. The penetration of both plant cuticle (see Orgell, 1957) and insect cuticle (see Richardson and Shepard, 1930; Hoskins, 1932, 1940; Alexandrov, 1935; Ricks and Hoskins, 1948) by ionizable compounds is increased by adjusting pH in such a way as to repress dissociation. The undissociated molecule penetrates more readily than its ions. Penetration of plant or insect cuticle by acid substances increases with decreasing pH and penetration by basic substances increases with increasing pH.

Hoskins (1932) stated that the experiments he made on the penetration of sodium arsenite through the cuticles of mosquito pupae can be duplicated with collodion and other membranes. The relationship of pH to the penetration of ionizable compounds as described above would pertain, of course, only to selectively permeable membranes. The cuticle below the epicuticular lipid layer is such a membrane. However, this relationship of pH does not pertain to the lipid itself. As shown in Section III,A,1, the rate of penetration of aqueous solutions through beeswax is proportional to the pH value, regardless of the compound dissolved.

The lipid layer, being apolar, is a barrier to water and its dissolved ions. Its resistance to both water and ions is destroyed by saponification, whereas the portion of the cuticle below the epicuticular lipid, being both hydrophilic and selectively permeable, retains its ability to select between molecules and their ions. Thus, even when it was dissolved in a normal KOH solution, which rapidly saponified the lipids, sodium arsenite was found in our laboratory (Ebeling and Clark, 1962) to be less effective against cockroaches than it was in distilled water. The ability of the selectively permeable portions of the cuticle to repel arsenite ions more than offset any advantage resulting from the destruction of the lipid as a barrier to aqueous solutions and ions. On the other hand, the penetration of a nicotine sulfate solution is favored by a high pH both with respect to increased penetration of lipid and a repression of ionization. The striking effect of the addition of a base to a nicotine sulfate solution on its insecticidal effectiveness is well known to entomologists.

Richards (1951) pointed out that the rate of penetration of ions through membranes is generally in a definite order—for salts with a common anion: $K > Na > Ca > Mg > Sr > Li$; for salts with a common cation: $No_3 > Cl > HCO_3 > HPO_4$.

Alexandrov (1935) showed that weak acids and bases penetrated the cuticles of certain dipterous larvae readily. He considered the cuticle to be a "selective permeable membrane" penetrated easily by acetic acid and NH_4OH and with difficulty by strong acids and alkalis. This is of special interest in view of the fact that in our laboratory (Ebeling and Clark, 1962) we found no penetration of beeswax by acetic acid and found sodium hydroxide (strong base) solution to penetrate nine times more readily than an ammonia (weak base) solution of the same normality (Section III,A,1). In Alexandrov's experiments, *Chironomus* larvae with their contents pressed out, including the epidermis, were filled, saclike, with water containing either *Paramaecium* or chemical indicators. The possibility of checking biological against chemical indicators would appear to increase the reliability and significance of experiments of this nature. Also the absence of a surface lipid on muscoid larvae would obviate errors that could result from a disruption of the epicuticular lipid layer of most insects. It is worthy of note that Alexandrov's results agree with those of toxicologists who have presumed permeability phenomena from the insecticidal action of electrolytes.

The exact location of the ion barrier is not known for most insects. In *Sarcophaga* larvae, the epicuticles of which possess no distinct lipid layer, Richards (1957) found the outer epicuticle to be the major barrier to ions,

the chloroform-soluble fraction of the inner epicuticle to be the water barrier, while the inner epicuticle as a whole proved to be a lesser barrier to
ions. The procuticle was no barrier to water or water solutes, but was perhaps a barrier to hydrophobic compounds. Concerning insects in general,
Richards (1958a) suggested, on the basis of more limited data on *Periplaneta, Galleria,* and *Tipula,* that distinct sublayers will be found to serve
the function of barriers to water, dissolved bases, electrolytes, and hydrophobic nonelectrolytes, respectively.

In considering the penetration of ions, it is important to recognize, as
Richards (1957) pointed out, that permeability values are not constant.
With isolated *Sarcophaga* cuticles he could shift the permeability values over
a range of three- to tenfold, reversibly, by treatments that affect the degree
of hydration of the cuticle.

b. Effect of Pretreatment with Oil. As stated in Section III,A, the presence
of liquid hydrocarbons greatly increases the permeability of solid lipids to
water. In our laboratory (Ebeling and Clark, 1962) we found that it required
more than 100 hours for a 50% mortality of *Tribolium confusum* adults
treated with an aqueous mist of 4% Dylox (2,2,2-trichloro/1-hydroxyethyl
phosphonate). However, when the beetles were treated 2 hours previously
with a "light medium" petroleum spray-oil mist the LT[50] was decreased
to 27 hours and pretreatment with oil plus 1% glyceryl monooleate decreased
the period still further to 3 hours. The oil mist itself caused no mortality.

Despite the demonstrated value of petroleum as a pretreatment, no way
was found of effectively utilizing this advantage by combining an aqueous
solution and an oil emulsion and applying them simultaneously. This is in
accord with the experience of Turner (see Olson and O'Brien, 1963) who
found that the simultaneous application of oleyl alcohol with aqueous solutions of nicotine or Dylox was of no benefit in killing grain beetles, but
that a previous *dip* of the beetles in the alcohol increased the mortalities.

c. Effect of Surfactants. It is difficult to evaluate the influence of a surfactant in increasing the penetration of an insecticide through the general
cuticular surface when insecticides are applied as sprays. Even if the spiracles
are blocked, a surfactant may affect the efficacy of an insecticide by its influence on wetting, spreading, and run-off rather than by its influence on cuticular penetration. In an experiment made by Ebeling and Reierson (1972)
with adult males of the American cockroach, *Periplaneta americana,* as test
insects, the abdominal spiracles of half of the insects were blocked with fingernail polish, a material that commonly has been used for blocking insect
spiracles. The cockroach was anesthetized, its wings were clipped, its spiracles were blocked, and it was taped to a rod resting on a 100 ml jelly jar
so that the insect's abdomen hung down into the jar. After the fingernail

polish was dry, an aqueous solution was poured into the jar until the water level reached the top of the abdomen. Examination of tracheae, when the abdomens of cockroaches were immersed for 20 minutes in aqueous solutions of crystal violet dye, showed that water penetrates the tracheae when surfactant is added, but that the spiracles can be effectively blocked with fingernail polish. The anal openings of all cockroaches were also blocked, whether the spiracles were left unsealed or not.

The abdomen of each insect was submerged in an aqueous solution for periods of 20, 30, or 40 minutes. Then the insecticide on the surface was rinsed off and the insect was placed in a gallon jar with layers of paper towelling on the bottom. Submersion of the abdomens of the cockroaches eliminated the usual variability in wetting and spreading as a factor influencing insecticidal efficacy. The insecticide used was 2 w/v % NaF and the surfactant was 0.1% Triton X-100, an alkyl aryl polyether alcohol (nonionic). One hundred and fifty-five cockroaches were used in the experiment. The results are shown in Table I.

By comparing the percent knockdown or mortality between tests where surfactant was added and where it was not added, it is evident that surfactant facilitated the penetration of toxic solutions through the superficial

TABLE I

EFFECT OF SURFACTANT ON INSECTICIDAL EFFICACY OF SODIUM FLUORIDE
SOLUTIONS AGAINST MALE AMERICAN COCKROACHES WITH OR WITHOUT
SPIRACLES BLOCKED[a]

Insecticide	Period abdomen immersed (minutes)	Surfactant	Condition of spiracles	Moribund or dead in 24 hours (%)
NaF 2%	20	None	Open	0.0
NaF 2%	20	Triton X-100, 0.1%	Open	86.7
NaF 2%	20	None	Blocked	0.0
NaF 2%	20	Triton X-100, 0.1%	Blocked	46.7
None	20	None	Blocked	0.0
None	20	Triton X-100, 0.1%	Blocked	20.0
None	20	Triton X-100, 0.1%	Open	20.0
NaF 2%	30	Triton X-100, 0.1%	Blocked	40.0
NaF 2%	40	Triton X-100, 0.1%	Blocked	60.0
None	30	Triton X-100, 0.1%	Blocked	30.0
None	40	Triton X-100, 0.1%	Blocked	80.0
None	40	None	Blocked	10.0

[a] In the first 7 tests, 15 cockroaches, and in the remaining 5 tests, 10 cockroaches, were used per test. Average temperature 22.8°C, RH 41%.

protective layer of grease. Blockage of spiracles did not prevent toxication. This would be expected on the basis of the penetration of much thicker films of beeswax and paraffin wax that was facilitated by surfactants, as described in Section III,A,2. Presumably the rate of penetration through aqueous pathways below the lipid layer would also be increased by a surfactant.

Further evidence of cuticular penetration was provided by weighing cockroaches before and after 40 minutes of abdominal immersion, comparing water alone with water plus 0.1% Triton X-100. Whereas the average increase in weight of five cockroaches with unsealed spiracles, and when no surfactant was added to the water, was 3.4%, it was 9.5% when spiracles were sealed and surfactant was added. A 9.5% increase in body weight represents about 15.8% increase in the body content of water, a remarkable quantity, particularly in view of the fact that the area of penetration was limited to the abdominal cuticle. Among the five cockroaches with unsealed spiracles, immersed in water plus surfactant, the average increase in weight was 11.4% (19.0% of total water content). At odds of 19:1, among the cockroaches immersed in surfactant, those with open spiracles did not absorb significantly more water than those with sealed spiracles, but both groups absorbed significantly more water than the cockroaches with sealed spiracles immersed in water to which no surfactant was added.

Table I shows that the surfactant itself was insecticidal and in fact at the longer periods of exposure no significant effect could be found from the addition of NaF. In both 40-minute tests in which surfactant was added, all insects knocked down were also dead in 24 hours. It would be interesting to know to what extent, if any, death may have resulted from excessive hydration or from autointoxication, rather than from direct toxication by surfactant or insecticide.

V. Conclusions

During the past decade two important developments have greatly influenced our concepts regarding cuticle permeability. It is now known that "wax canals" branch out from the ends of pore canals to transport wax to the cuticulin/wax interface of the epicuticle (Locke, 1964, 1965; Gluud, 1968). It is not known whether all terrestrial insect species possess wax canals, but it is reasonable to believe they do. The packed and oriented lipid monolayer that covers most of the cuticulin surface has been shown to be an efficient barrier to transpiration in biophysical models (Beament, 1945), but it is now evident that *in situ* it is interrupted by the wax filaments issuing from the openings of myriads of wax canals. The wax filaments are believed to consist of "lipid-water liquid crystals," which, when withdrawn at an abnormal rate, are replaced by water (Locke, 1965). This apparently provides the means for bringing about lethal rates of water loss in insects without the necessity for removing the organized lipid barrier which covers the cuticulin. Likewise the latest findings regarding the pore structure of the cuticulin may be a basis for an understanding of the mechanism for water uptake (Beament, 1965).

Previous concepts of the influence of "transition temperatures" on the organized lipid water barier may also undergo change. If the water barrier at the cuticulin/wax interface consists of "lipid crystals," it follows that phase changes in these liquid-crystalline systems might cause an abrupt change in water loss at "critical temperatures" (Locke, 1965).

The epidermis has received attention in recent years with regard to the regulatory role of epidermal cells as a "cuticular water pump" (Winston, 1967; Winston and Beament, 1969) and this may not only increase our understanding of normal transpiration and uptake of water, but also abnormal cuticular water loss such as appears to be caused by toxification of epidermal cells by insecticides (Chattoraj and Sharma, 1964).

Wax canals may not only clarify our concepts regarding transpiration and active water uptake, but also our concepts regarding the possible pathways for the penetration of insecticides through the otherwise homogeneous tissue of the cuticulin. On the other hand, considerable evidence developed in recent years indicates that insecticides do not ordinarily penetrate directly through the cuticle at the point of contact, but must first migrate laterally in the endocuticle (Gerolt, 1969) or the organized lipid water barrier of the epicuticle (Quraishi and Poonawalla, 1970) to reach portals of entry such as tracheae or articular membranes. If the lipid-water liquid crystals issuing from wax canals spread out over the surface of the cuticulin to provide it with a liquid and moving monolayer, rather than one that is solid and static, it is easier to understand the lateral movement of insecticides in that layer as reported by Quraishi and Poonawalla (1969).

Gerolt (1969) does not believe the hemolymph to be the means of insecticide transport unless the insecticide is dissolved in a suitable solvent. If this proves to be universally true, it will leave many toxicologists with the task of reappraising their concepts regarding the pathways by which insecticides applied as dusts or wettable powders reach the central nervous system. However, not all toxicologists accept Gerolt's conclusion as to the subservient role of the hemolymph in insect toxication. Evidence favoring the traditional concepts has been obtained from recent research directed specifically toward a more thorough examination of this problem by such investigators as Burt *et al.* (1971) and Moriarty and French (1971). The epicenter of the controversy is currently in Great Britain and the current debate appears to have been initiated by a series of papers presented at a symposium on "Penetration through Insect Cuticles."*

How diametrically opposite conclusions can be reached by different investigators doing what appears to be sound basic research is one of the enigmas in the present status of insect toxicology. Probably the different points of view are reconcilable. It does not appear likely that any insecticide will follow one route exclusively and one would expect different insecticides to vary with respect to pathways taken. In the absence of spiracles, setae, and conjunctivae, an insecticide is going to follow different penetration pathways than where these favored portals of entry are present. A compound is distributed according to its affinity for various phases; a comparatively low concentration in a particular phase, e.g., hemolymph, need not mean that the phase plays no part in distributing the compound.

Under field conditions in insect pest control, the contact of insect with insecticide is often practically exclusively a contact of tarsi and dry insecticide residue. The toxicant can be expected to penetrate the localized tarsal portals of entry and it is difficult to visualize the rapid toxication that often takes place without the hemolymph playing a major role as a transport system.

The increased research prompted by the lively controversy regarding the pathways by which toxicants penetrate the cuticle and reach the central nervous system is bound to have the end result of greatly increasing our understanding of cuticle permeability.

In a review of literature on cuticle permeability, one is impressed by the many instances of migration of organic molecules away from not only liquid, but also solid deposits. Examples are the migration of solid insect wax from artificial beeswax films or insect cuticles (e.g., carabid beetles) to permeate a layer of silica aerogel for distances of as much as hundreds of microns, or the rapid lateral spread of insecticides, from waxy or crystalline deposits, on or in insect cuticle. Liquids can, of course, migrate even more rapidly, moving from one point of application to cover the surface of an entire insect, including the wings, within a few minutes. Adsorbed films of either solid or liquid organic substances are dynamic and potentially mobile, but are too often regarded by biologists as static systems.

Acknowledgments

The author gratefully acknowledges the assistance of Drs. E. B. Edney, M. S. Quraishi, and A. G. Richards, who read and criticized the manuscript and offered many useful suggestions.

* Organized by a British Physiochemical and Biophysical Panel in November 1969, and published in *Pesticide Science* 1:209–223, 1970.

References

Afifi, S. E. D., and Roan, C. C. (1961). *J. Kans. Entomol. Soc.* **34**, 87.

Ahearn, G. A. (1970). *J. Exp. Biol.* **53**, 573.

Ahmed, H., and Gardiner, B. G. (1970). *Pestic. Sci.* **1**, 217.

Alexander, P., Kitchener, J. A., and Briscoe, H. V. A. (1944a). *Trans. Faraday Soc.* **40**, 10.

Alexander, P., Kitchener, J. A., and Briscoe, H. V. A. (1944b). *Ann. Appl. Biol.* **31**, 156.

Alexandrov, W. J. (1935). *Acta Zool. (Stockholm)* **16**, 1.

Armstrong, G., Bradbury, F. R., and Standen, H. (1951). *Nature (London)* **167**, 319.

Armstrong, G., Bradbury, F. R., and Britton, H. G. (1952). *Ann. Appl. Biol.* **39**, 548.

Barlow, F., and Hadaway, A. B. (1958). *Bull. Entomol. Res.* **49**, 315.

Beament, J. W. L. (1945). *J. Exp. Biol.* **21**, 115.

Beament, J. W. L. (1954). *Symp. Soc. Exp. Biol.* **3**, 94.

Beament, J. W. L. (1955). *J. Exp. Biol.* **32**, 514.

Beament, J. W. L. (1957). *Proc. Int. Congr. Plant Protect., 2nd, 1956,* p. 96.

Beament, J. W. L. (1958a). *J. Exp. Biol.* **35**, 494.

Beament, J. W. L. (1958b). *J. Insect Physiol.* **2**, 199.

Beament, J. W. L. (1959). *J. Exp. Biol.* **36**, 391.

Beament, J. W. L. (1960). *Nature (London)* **186**, 408.

Beament, J. W. L. (1961a). *Biol. Rev. Cambridge Phil. Soc.* **36**, 281.

Beament, J. W. L. (1961b). *J. Exp. Biol.* **38**, 277.

Beament, J. W. L. (1964). *Advan. Insect Physiol.* **2**, 67.

Beament, J. W. L. (1965). *Symp. Soc. Exp. Biol.* **19**, 273.

Beament, J. W. L., Noble-Nesbitt, J. J., and Watson, J. A. L. (1964). *J. Exp. Biol.* **41**, 323.

Benezet, H. J., and Forgash, A. J. (1972). *J. Econ. Entomol.* **65**, 53.

Bennett, S. E., and Thomas, C. A., Jr. (1963). *J. Econ. Entomol.* **56**, 239.

Bikerman, J. J. (1958). "Surface Chemistry," 2nd rev. ed. Academic Press, New York.

Bohm, O. (1951). *Pflanzenschutzberichte* **7**(3-4), 33.

Brown, A. W. A. (1951). "Insect Control by Chemicals." Wiley, New York.

Brown, E. W., and Scott, W. O. (1934). *J. Pharmacol. Exp. Ther.* **50**, 373.

Brown, H. T., and Escombe, F. (1900). *Phil. Trans. Roy. Soc. London, Ser. B* **193**, 223.

Browning, T. O. (1954). *J. Exp. Biol.* **31**, 331.

Buck, J. B., and Keister, M. L. (1949). *Biol. Bull.* **97**, 64.

Buerger, A. A., and O'Brien, R. D. (1965). *J. Cell. Comp. Physiol.* **66**, 227.

Bursell, E. (1955). *J. Exp. Biol.* **32**, 238.

Bursell, E. (1964). *In* "The Physiology of Insecta" (M. Rockstein, ed.), Vol. I, pp. 323–357. Academic Press, New York.

Bursell, E., and Clements, A. N. (1967). *J. Insect. Physiol.* **13**, 1671.

Burt, P. E. (1970). *Pestic. Sci.* **1**, 88.

Burt, P. E., and Goodchild, R. E. (1969). *Rothamsted Rep.* Part 1, pp. 210–12.

Burt, P. E., and Lord, K. A. (1968). *Entomol. Exp. Appl.* **11**, 55.

Burt, P. E., Lord, K. A., Forrest, J. M., and Goodchild, R. E. (1971). *Entomol. Exp. Appl.* **14**, 255.

Buxton, P. A. (1932). *Biol. Rev. Cambridge Phil. Soc.* **7**, 275.
Chattoraj, A. N., and Sharma, V. P. (1964). *Beitr. Entomol.* **14**, 525.
Cloudsley-Thompson, J. L. (1955). *Discovery* **16**, 248.
Cloudsley-Thompson, J. L. (1956). *Ann. Mag. Nat. Hist. (Ser. 12)* **9**, 305.
Cloudsley-Thompson, J. L. (1968). "Spiders, Scorpions, Centipedes, and Mites." Pergamon, Oxford.
Collins, M. S. (1969). *In* "Biology of Termites" (K. Krishna and F. M. Weesner, eds.), Vol. 1, pp. 433–58. Academic Press, New York.
Crawford, C. S., and Cloudsley-Thompson, J. L. (1971). *Entomol. Exp. Appl.* **14**, 99.
David, W. A. L., and Gardiner, B. O. C. (1950). *Bull. Entomol. Res.* **41**, 1.
Davies, M. E., and Edney, E. B. (1952). *J. Exp. Biol.* **29**, 571.
Delye, G. (1969). *Bull. Soc. Entomol. Fr.* **74**, 51.
Den Boer, P. J. (1961). *Arch. Neer. Zool.* **14**, 283.
Dennell, R. (1946). *Proc. Roy. Soc., Ser. B* **133**, 348.
Dresden, D., and Krijgsman, B. J. (1948). *Bull. Entomol. Res.* **38**, 575.
Ebeling, W. (1939). *Hilgardia* **12**, 665.
Ebeling, W. (1961). *Hilgardia* **30**, 531.
Ebeling, W. (1971). *Annu. Rev. Entomol.* **16**, 123.
Ebeling, W., and Clark, W. R. (1962). "The Penetration of Liquids through Lipid Layers." Unpublished manuscript on file in the Dept. of Entomology, University of California, Los Angeles.
Ebeling, W., and Pence, R. J. (1957). *Ann. Entomol. Soc. Amer.* **50**, 637.
Ebeling, W., and Reierson, D. A. (1972). "Effect of Surfactant on the Penetration of Aqueous Solutions through the Cockroach Lipid Barrier." Unpublished manuscript on file in the Dept. of Entomology, University of California, Los Angeles.
Ebeling, W., and Wagner, R. E. (1959). *J. Econ. Entomol.* **52**, 190.
Edney, E. B. (1947). *Bull. Entomol. Res.* **38**, 263.
Edney, E. B. (1949). *Nature (London)* **164**, 321.
Edney, E. B. (1951). *J. Exp. Biol.* **28**, 91.
Edney, E. B. (1956). "The Water Relations of Terrestrial Arthropods." Cambridge Univ. Press, London and New York.
Edney, E. B. (1957). *Cambridge Monogr. Exp. Biol.* No. 5, pp. 1–108.
Edney, E. B. (1966). *Comp. Biochem. Physiol.* **19**, 387.
Edney, E. B. (1967). *Science* **156**, 1059.
Edney, E. B. (1968). *Amer. Zool.* **8**, 309.
Edwards, G. A. (1953). *In* "Insect Physiology" (K. D. Roeder, ed.), pp. 55–95. Wiley, New York.
El Bashir, S. (1967). *Entomol. Exp. Appl.* **10**, 111.
Fan, H. Y., Cheng, T. H., and Richards, A. G. (1948). *Physiol. Zool.* **21**, 48.
Fast, P. G. (1964). *Mem. Entomol. Soc. Canada* **37**, 1.
Fast, P. G., and Brown, A. W. A. (1962). *Ann. Entomol. Soc. Amer.* **55**, 663.
Fine, B. C., Godin, P. J., Thain, E. M., and Marks, T. B. (1967). *J. Sci. Food Agr.* **18**, 220.
Finlayson, L. R. (1932). *Rep. Entomol. Soc. Ontario* **63**, 56.
Fisher, R. W. (1952). *Can. J. Zool.* **30**, 254.
Forgash, A. J., Cook, B. J., and Riley, R. C. (1962). *J. Econ. Entomol.* **55**, 544.
Gabler, H. (1939). *Z. Angew. Entomol.* **26**, 1–62.
Gerolt, P. (1969). *J. Insect Physiol.* **15**, 563.
Gerolt, P. (1970). *Pestic. Sci.* **1**, 209.

Gibbs, K. E., and Morrison, F. O. (1959). *Can. J. Zool.* **37**, 633.

Gluud, A. (1968). *Zool. Jahrb., Abt. Anat. Ontog. Tiere* **85**, 191.

Glynne-Jones, G. D., and Edwards, G. A. (1952). *Bull. Entomol. Res.* **43**, 67.

Govaerts, J., and Leclercq, J. (1946). *Nature (London)* **157**, 483.

Gunn, D. L. (1933). *J. Exp. Biol.* **10**, 274.

Hadaway, A. B., and Barlow, F. (1951). *Nature (London)* **167**, 854.

Hadaway, A. B., Barlow, F., Grose, J. E. H., Turner, C. R., and Flower, L. S. (1970). *Bull. W.H.O.* **42**, 353.

Hadley, N. F. (1970). *J. Exp. Biol.* **53**, 547.

Hafez, M., El-Ziady, S., and Hafnawy, T. (1970). *J. Parasitol.* **56**, 154.

Harris, E. J. (1956). "Transport and Accumulation in Biological Systems." Butterworth, London.

Hayes, W. P., and Liu, Y. S. (1947). *Ann. Entomol. Soc. Amer.* **40**, 401.

Hetrick, L. A. (1961). *Fla. Entomol.* **44**, 53.

Hoffman, R. A., Roth, A. R., Lindquist, A. W., and Butts, J. S. (1952). *Science* **115**, 312.

Holdgate, M. W., and Seal, M. (1956). *J. Exp. Biol.* **33**, 82.

Hoskins, W. M. (1932). *J. Econ. Entomol.* **25**, 1212.

Hoskins, W. M. (1940). *Hilgardia* **13**, 307.

Hurst, H. (1950). *J. Exp. Biol.* **27**, 238.

Ingram, R. L. (1955). *Ann. Entomol. Soc. Amer.* **48**, 481.

Jackson, L. L., and Baker, G. L. (1970). *Lipids* **5**, 239.

Johnson, C. G. (1942). *Biol. Rev. Cambridge Phil. Soc.* **17**, 151.

Kalmus, H. (1944). *Nature (London)* **153**, 714.

Kalra, R. L. (1970). *Bull. W.H.O.* **42**, 623.

Kanungo, K. (1965). *J. Insect Physiol.* **11**, 557.

Kennedy, C. H. (1927). *Ann. Entomol. Soc. Amer.* **20**, 87.

King, G. (1944). *Nature (London)* **154**, 575.

Klinger, H. (1936). *Arb. Physiol. Angew. Entomol. Berlin-Dahlem* **3**, 49 and 115.

Knülle, W. (1965). *Z. Vergl. Physiol.* **49**, 586.

Knülle, W. (1967). *J. Insect Physiol.* **13**, 333.

Knülle, W., and Spadafora, R. R. (1969). *J. Stored Prod. Res.* **5**, 49.

Kühnelt, W. (1928). *Zool. Jahrb., Abt. Anat. Ontog. Tiere* **50**, 219.

Kühnelt, W. (1939). *Proc. Int. Congr. Entomol., 7th, 1938* p. 797.

Kurihara, N., Nakajima, E., and Shindo, H. (1970). *In* "Biochemical Toxicology of Insecticides" (R. D. O'Brien and I. Yamamoto, eds.), pp. 41–50. Academic Press, New York.

Lafon, M. (1943). *Ann. Sci. Natur., Zool. Biol. Anim.* [11] **11**, 114.

Landmann, W., Lovegren, N. V., and Feuge, R. O. (1960). *J. Amer. Oil Chem. Soc.* **37**, 1.

Langmuir, I. (1938). *Science* **87**, 493.

Langmuir, I., and Schaefer, V. J. (1943). *Publ. No. 21*, pp. 17–39. Amer. Ass. Advan. Sci., Washington, D.C.

Lees, A. D. (1946). *Parasitology* **37**, 1.

Lees, A. D. (1947). *J. Exp. Biol.* **23**, 379.

Lees, A. D. (1948). *Discuss. Faraday Soc.* **3**, 187.

LeRoux, E. J., and Morrison, F. O. (1954). *J. Econ. Entomol.* **47**, 1058.

Lewis, C. T. (1962). *Nature (London)* **193**, 904.

Lewis, C. T. (1965). *J. Insect Physiol.* **11**, 683.

Locke, M. (1956). *Quart. J. Microsc. Sci.* **99**, 29.

Locke, M. (1961). *J. Biophys. Biochem. Cytol.* **10**, 589.

Locke, M. (1964). *In* "The Physiology of Insecta" (M. Rockstein, ed.), Vol. III, pp. 379–470. Academic Press, New York.

Locke, M. (1965). *Science* **147**, 295.

Lord, K. A. (1948). *Biochem. J.* **43**, 72.

Lovegren, N. V., and Feuge, R. O. (1954). *J. Agr. Food Chem.* **2**, 558.

Lovell, J. B. (1963). *J. Econ. Entomol.* **56**, 310.

Ludwig, D. (1937). *Physiol. Zool.* **10**, 342.

McIndoo, N. E. (1916). *J. Agr. Res.* **7**, 89.

McIntosh, A. H. (1957). *Ann. Appl. Biol.* **45**, 189.

Majumder, S. K., Narasimhan, K. S., and Subrahamanyan, V. (1959). *Nature (London)* **184**, 1165.

Mansingh, A. (1965). *J. Econ. Entomol.* **58**, 162.

Matsumura, F. (1963). *J. Insect Physiol.* **9**, 207.

Matsumura, F., and Brown, A. W. A. (1961). *Mosquito News* **21**, 192.

Matsumura, F., and Brown, A. W. A. (1963). *Mosquito News* **23**, 26.

Matsubara, H. (1960). *Bochu-Kagaku* **25**, 138.

Mead-Briggs, A. R. (1956). *J. Exp. Biol.* **33**, 737.

Mellanby, K. (1932). *Proc. Roy. Soc., Ser. B* **111**, 376.

Mellanby, K. (1935). *Biol. Rev. Cambridge Phil. Soc.* **10**, 317.

Mitchell, J. W., Smale, B. C., and Metcalf, R. L. (1960). *Advan. Pest Contr. Res.* **3**, 359.

Mitchell, P. A. (1956). "A Textbook of General Physiology." McGraw-Hill, New York.

Moore, R. F., Jr., Hopkins, A. R., Taft, H. M., and Anderson, L. L. (1967). *J. Econ. Entomol.* **60**, 64.

Moore, W. (1921). *Minn., Agr. Exp. Sta., Tech. Bull.* **2**, 1.

Moriarty, F., and French, M. C. (1971). *Pestic. Biochem. Physiol.* **1**, 286.

Müller, A. (1930). *Proc. Roy. Soc., Ser. A* **127**, 417.

Munson, S. C., Padilla, G. M., and Weissmann, M. L. (1954). *J. Econ. Entomol.* **47**, 578.

Nair, M. R. G. K. (1957). *Indian J. Entomol.* **19**, 37.

Nair, M. R. G. K. (1959). *Indian J. Entomol.* **21**, 246.

Nakajima, E., Shindo, H., Kurihara, N., and Fujita, T. (1969). *Radioisotopes* **18**, (9), 37.

Nelson, D. R. (1969). *Nature (London)* **221**, 854.

Nicoli, R. M. (1958). *Indian J. Malariol.* **12**, 469.

Noble-Nesbitt, J. J. (1969). *J. Exp. Biol.* **50**, 745.

Noble-Nesbitt, J. J. (1970a). *Nature (London)* **225**, 753.

Noble-Nesbitt, J. J. (1970b). *Pestic. Sci.* **1**, 204.

O'Brien, R. D. (1961). *Advan. Pest Contr. Res.* **4**, 75.

O'Kane, W. C., Westgate, W. A., Glover, L. C., and Lowry, P. R. (1930). *N.H. Agr. Exp. Sta., Tech. Bull.* **39**, 1.

O'Kane, W. C., Walker, G. L., Guy, H. G., and Smith, O. J. (1933). *N.H., Agr. Exp. Sta., Tech. Bull.* **54**, 1.

Okasha, A. Y. K. (1971). *J. Exp. Biol.* **55**, 435.

Olson, W. P. (1970). *Comp. Biochem. Physiol.* **35**, 273.

Olson, W. P., and O'Brien, R. D. (1963). *J. Insect Physiol.* **9**, 777.

Oppenoorth, F. J. (1958). *Nature (London)* **181**, 425.

Oppenoorth, F. J., and Dresden, D. (1958). *Indian J. Malariol.* **12**, 309.

Orgell, W. H. (1957). *Proc. Iowa Acad. Sci.* **64,** 189.

Ossowski, L. L. (1944). *Endeavour* **3,** 32.

Parkin, E. A. (1944). *Ann. Appl. Biol.* **31,** 84.

Pence, R. J. (1956). *J. Econ. Entomol.* **49,** 553.

Pepper, J. H., and Hastings, E. (1943). *J. Econ. Entomol.* **36,** 633.

Perry, A. S. (1964). *In* "The Physiology of Insecta" (M. Rockstein, ed.), Vol. 3, pp. 45–378. Academic Press, New York.

Pfaff, W. (1952). *Hofchen-Briefe (Engl. ed.)* **3,** 93.

Poonawalla, Z. T., and Quraishi, M. S. (1970). *Can. Entomol.* **102,** 1136.

Pradhan, S., Nair, M. R. G. K., and Krishnaswami, S. (1952). *Nature (London)* **170,** 619.

Quraishi, M. S. (1970). *Can. Entomol.* **102,** 1189.

Quraishi, M. S., and Poonawalla, Z. T. (1969). *J. Econ. Entomol.* **62,** 988.

Ramsay, J. A. (1935a). *J. Exp. Biol.* **12,** 355.

Ramsay, J. A. (1935b). *J. Exp. Biol.* **12,** 373.

Reiser, R., Chadbourne, D. S., Kuiken, K. A., Rainwater, C. F., and Ivy, E. E. (1953). *J. Econ. Entomol.* **46,** 337.

Richards, A. G. (1941). *Trans. Amer. Entomol. Soc.* **67,** 161.

Richards, A. G. (1951). "The Integument of Arthropods." Univ. of Minnesota Press, Minneapolis.

Richards, A. G. (1957). *J. Insect Physiol.* **1,** 23.

Richards, A. G. (1958). *Ergeb. Biol.* **20,** 1.

Richards, A. G., and Anderson, T. F. (1942a). *J. Morphol.* **71,** 135.

Richards, A. G., and Anderson, T. F. (1942b). *J. N.Y. Entomol. Soc.* **50,** 147.

Richards, A. G., and Cutkomp, L. K. (1946). *Biol. Bull.* **90,** 97.

Richards, A. G., and Fan, H. Y. (1949). *J. Cell. Comp. Physiol.* **33,** 177.

Richards, A. G., and Korda, F. H. (1950). *Ann. Entomol. Soc. Amer.* **43,** 49.

Richards, A. G., and Weygandt, J. L. (1945). *J. N.Y. Entomol. Soc.* **53,** 153.

Richards, A. G., Clausen, M. B., and Smith, M. N. (1953). *J. Cell. Comp. Physiol.* **42,** 395.

Richardson, C. H., and Shepard, H. H. (1930). *J. Agr. Res.* **41,** 337.

Ricks, M., and Hoskins, W. M. (1948). *Physiol. Zool.* **21,** 258.

Rideal, E. K., and Tadayon, J. (1954a). *Proc. Roy. Soc., Ser. A* **225,** 346.

Rideal, E. K., and Tadayon, J. (1954b). *Proc. Roy. Soc., Ser. A* **225,** 357.

Roy, D. N., Ghosh, S. M., and Chopra, R. N. (1943). *Ann. Appl. Biol.* **30,** 42.

Sarkaria, D. S., and Patton, R. L. (1949). *Trans. Entomol. Soc. Amer.* **75,** 71.

Sawicki, R. M., and Lord, K. A. (1970). *Pestic. Sci.* **1,** 213.

Scheie, P. O., Smith, T., and Greer, R. T. (1968). *Ann. Entomol. Soc. Amer.* **61,** 1346.

Sebba, F., and Briscoe, H. V. A. (1940). *J. Chem. Soc., London* p. 106.

Sellers, L. G., and Guthrie, F. E. (1971). *J. Econ. Entomol.* **64,** 352.

Sharma, V. P., and Chattoraj, A. N. (1964). *Beitr. Entomol.* **14,** 533.

Sikes, E. K., and Wigglesworth, V. B. (1931). *Quart. J. Microsc. Sci.* **74,** 165.

Slifer, E. H. (1950). *Ann. Entomol. Soc. Amer.* **43,** 173.

Slifer, E. H. (1954). *Biol. Bull.* **106,** 122.

Slifer, E. H., Prestage, J. J., and Beams, H. W. (1959). *J. Morphol.* **105,** 145.

Spiller, D. (1958). *N. Z. Entomol.* **2,** 1.

Steinbrecht, R. A., and Müller, B. (1971). *Z. Zellforsch. Mikrosk. Anat.* **117,** 570.

Sternberg, J., and Kearns, C. W. (1952). *J. Econ. Entomol.* **45,** 497.

Sternburg, J., Kearns, C. W., and Bruce, W. N. (1950). *J. Econ. Entomol.* **43,** 214.

Tahori, A. S., and Hoskins, W. M. (1953). *J. Econ. Entomol.* **46,** 302 and 829.

Taylor, R. L., Herrmann, D. B., and Kemp, A. R. (1936). *Ind. Eng. Chem., Ind. Ed.* **28,** 1255.

Treherne, J. E. (1956). *J. Physiol. (London:* **133,** 171.

Treherne, J. E. (1957). *J. Insect Physiol.* **1,** 178.

Tulloch, A. P. (1970). *Lipids* **5,** 247.

Vinson, S. B., and Brazzel, J. R. (1966). *J. Econ. Entomol.* **59,** 600.

Vinson, S. B., and Law, P. K. (1971). *J. Econ. Entomol.* **64,** 1387.

Volmer, M. (1932). *Trans. Faraday Soc.* **28,** 359.

Volmer, M. (1938). *Chem. Zentralbl.* p. 2568.

Wagner, R. E., and Ebeling, W. (1959). *J. Econ. Entomol.* **52,** 208.

Warth, A. H. (1956). "The Chemistry and Technology of Waxes." Van Nostrand-Reinhold, Princeton, New Jersey.

Way, M. J. (1950). *Quart. J. Microsc. Sci.* **91,** 145.

Webb, J. E., and Green, R. A. (1945). *J. Exp. Biol.* **22,** 8.

Weber, H. (1933). "Lehrbuch der Entomologie." Fisher, Jena.

Wharton, G. W., and Kanungo, K. (1962). *Ann. Entomol. Soc. Amer.* **55,** 483.

Wiesmann, R. (1949). *Mitt. Schweiz. Entomol. Ges.* **22,** 257.

Wigglesworth, V. B. (1933). *Quart. J. Microsc. Sci.* **76,** 269.

Wigglesworth, V. B. (1938). *J. Exp. Biol.* **15,** 248.

Wigglesworth, V. B. (1941). *Nature (London)* **147,** 116.

Wigglesworth, V. B. (1942). *Bull. Entomol. Res.* **33,** 205.

Wigglesworth, V. B. (1944). *Nature (London)* **153,** 493.

Wigglesworth, V. B. (1945). *J. Exp. Biol.* **21,** 97.

Wigglesworth, V. B. (1947). *Proc. Roy. Entomol. Soc., Ser. B* **134,** 163.

Wigglesworth, V. B. (1948a). *Quart. J. Microsc. Sci.* **89,** 197.

Wigglesworth, V. B. (1948b). *Biol. Rev. Cambridge Phil. Soc.* **23,** 408.

Wigglesworth, V. B., and Beament, J. W. L. (1950). *Quart. J. Microsc. Sci.* **91,** 429.

Wilcoxon, F., and Hartzell, A. (1931). *Contrib. Boyce Thompson Inst.* **3,** 1.

Wilcoxon, F., and Hartzell, A. (1933). *Contrib. Boyce Thompson Inst.* **5,** 115.

Winston, P. W. (1967). *Nature (London)* **214,** 383.

Winston, P. W., and Beament, J. W. L. (1969). *J. Exp. Biol.* **50,** 541.

Winston, P. W., and Nelson, V. E. (1965). *J. Exp. Biol.* **43,** 257.

Winteringham, F. P. W., and Hewlett, P. S. (1964). *Chem. Ind. (London)* p. 1512.

Winteringham, F. P. W., Harrison, A., and Bridges, P. M. (1955). *Biochem. J.* **61,** 359.

Witt, P. N. (1947). *Z. Naturforsch. B* **2,** 361.

Woods, W. C. (1929). *Bull. Brooklyn Entomol. Soc.* **24,** 116.

Zacher, F., and Künicke, G. (1931). *Arb. Biol. Reichsanst. Land- Forstwirts. Berlin-Dahlem* **18,** 201.

Chapter 5

RESPIRATION—AERIAL GAS TRANSPORT

P. L. Miller

I. Introduction

Since 1964, our knowledge of some aspects of the physiology of insect breathing has increased, while in other areas there has been little progress. The latter will be discussed only briefly, but more detail will be given where there is new information. Some useful reviews and books which deal wholly or in part with the subject have appeared since 1964 by Wigglesworth (1965), Miller (1966a), Hinton (1968a, 1969), Guthrie and Tindall (1968), Chapman (1969), Bursell (1970), Steen (1971), Mill (1972), and Whitten (1972).

Animals in terrestrial habitats require efficient respiratory systems which allow rapid exchanges of oxygen and carbon dioxide while restricting water loss. The insect epidermis is a highly adaptive tissue; one of its important functions is the formation of a tracheal system which can bring oxygen in the gaseous state to within a few microns of the respiratory sites. Tracheae terminate in tracheoles, normally believed to be blind-ending tubes, which in active tissue may lie among the mitochondria. The tracheal system shows both a conservative metameric organization and extensive modifications to meet the differing needs of organs. This is most marked in wing-bearing segments. Tracheae also sometimes serve nonrespiratory functions.

Gases move independently within tracheae and tracheoles down concentration gradients by diffusion, or by bulk flow in larger tracheae down pressure gradients. Bulk flow may be caused by locomotory activity, or by specialized pumping strokes. It may also occur in the absence of muscular movements when the tracheal system is kept more or less completely closed, a partial vacuum developing which periodically sucks air in through slightly open spiracles. No active transport (i.e., by chemical means) of gases is known and respiratory pigments are absent from most insects, although some chironomid larvae, *Gastrophilus* larvae, and *Anisops* contain hemoglobin.

Tracheal systems similar to those of the insects are found in several arthropod classes, and they have been evolved more than once. The most elaborate respiratory system outside the insects is probably that of the Solifugids where longitudinal trunks occur which are ventilated by muscular pumping movements, and spiracle valves open and close in phase with pumping so that there is probably unidirectional ventilation, as in some insects.

II. The Organization and Evolution of the Tracheal System

Physiological specializations arise as developments and modifications of existing systems. Evolutionary questions are therefore pertinent to physiologists. Their answers help to explain the structure and organization of systems in contemporary insects. The tracheal system in pterygote insects is monophyletic and the basic elements can be recognized in Thysanura (Barnhart, 1961). Tracheae preserve a segmental organization linked to a segmental (or sometimes intersegmental) disposition of spiracles. Although the gross morphology of holometabolous larvae may differ fundamentally from that of their adults, the larval tracheal system is sometimes preserved in modified form in the adult rather than being replaced by a completely new system (Whitten, 1957, 1968). Pupae must be able to breathe and they depend on the continued functioning of some of their tracheae throughout development.

Tracheae join at nodes across intersegmental boundaries. At ecdysis the tracheae of one metamere dissociate from those of the next at the nodes. Nodes may have a similar significance for segmental organization, as do the intersegmental membranes. There may be many nodes or tracheal fusions between the tracheae of adjacent segments, particularly in the thorax. Longitudinal tracheal trunks are found in most insects: they run the length of the insect joining together all parts of the tracheal system. But some regions may become partially and secondarily isolated from the remainder—for example, in connection with flight autoventilation (Section IV,D,2) or heat conservation.

Major tracheal branching patterns are often found to be similar in all members of a family or order and they are sometimes useful for establishing phylogenetic relationships (Landa, 1969; Whitten, 1959, 1960, 1972). Minor branching patterns show much flexibility and variation, and they may not be constant from individual to individual of the same species. Major branching is clearly genetically determined, but minor branching patterns result from the interactions of genetic plans with local needs. Generally the system anticipates demands to be made (e.g., developing flight muscle) or outlasts them (in degenerating flight muscle), but it is also capable of responding to experimentally imposed demands; for example, areas deprived of their normal tracheal supply can reorganize neighboring tracheae to meet their needs (Wigglesworth, 1954, 1959). The tracheal system may provide useful material for developmental biologists who wish to analyze the mode of expression of a genetically planned but flexible system.

Spiracles occur either on or between segments. Many insect embryos possess 12 pairs, but only 10 pairs normally remain after hatching. There is much speculation about which two disappear. A tendency to reduce further the number of functional spiracles occurs in many holometabolous adults as body shape becomes shorter and fatter. Dipterous larvae, too, show a reduction, often leaving only two (amphipneustic) or one (metapneustic) pair functional, while in some aquatic larvae all spiracles are closed (apneustic).

Matsuda (1970) has recently summarized the arguments concerning the allocation of spiracles to segments. The discussion concerns mainly spiracle 1 which is mesothoracic in embryological origins in *Lepisma*, in many Hemimetabola, and some Coleoptera, although the final position may be intersegmental. In most Holometabola, however (Lepidoptera, Diptera, Hymenoptera), spiracle 1 arises on the prothorax in the embryo and may then move posteriorly in development. An earlier migration from meso- to prothorax may be undetectable. A prothoracic spiracle would be particularly important for supplying cephalic tracheae, so that its loss in Hemimetabola is surprising. An alternative possibility is that the embryological pro- and mesothoracic spiracles fuse to produce the existing spiracle 1. Spiracle 1 in many species (e.g., Dictyoptera, Odonata, Orthoptera, and some Coleoptera) has two separate orifices leading to two atria, but it is covered by one valve (Miller, 1960; Poonawalla, 1966). The motor innervation of spiracle 1 in locusts stems from the pro- and mesothoracic ganglia, although in most other species examined it comes from the prothoracic ganglion alone. Sensory neurons from the valve and peritreme run into the mesothoracic ganglion (Section V,A). These facts are consistent with the hypothesis of a dual origin for this spiracle, but there is no supporting embryological information.

The evolution of a closable valve which reduces water loss must have been important in enabling insects to colonize drier habitats. Spiracle valves or similar structures have been independently evolved a number of times in arthropods (e.g., in spiders, scorpions, solifugids). They occur in all pterygotes (although secondarily lost in some) and in Thysanura (Noble-Nesbitt, 1971). These valves are controlled by one or more muscles. Perhaps the primitive state was with two muscles, an opener and a closer, but in many spiracles only the closer is present and a mechanical elastic system brings about opening when the closer relaxes. Less often, the closer muscle is replaced [e.g., in *Thermobia* (Noble-Nesbitt, 1973) or the solifugid, *Galeodes*], or supplemented by cuticular elasticity (e.g., in *Blaberus*—abdominal spiracles), and control is exercised mainly by the opener muscle.

In some insects, tracheae and air sacs have come to serve functions other than respiration. For example, sacs allow internal organs to change in volume during an instar without disturbing the outer contours of the insect (Clarke, 1957, 1958; Tonapi, 1958). They reduce the weight of large insects and may serve functions in connection with development and ecdysis (Allegret and Barbier, 1966). They may act as sound resonators (Pringle, 1954) or heat insulators (Church, 1960; Weis-Fogh, 1964a). Tracheoles may possibly be directly involved in the excitation of firefly lanterns (Peterson and Buck, 1968) or in water movement (Oschman and Wall, 1969). Many other functions probably wait discovery.

III. The Distribution and Structure of Respiratory Systems

A. GENERAL FEATURES OF TRACHEAL SYSTEMS

Tracheae arise at spiracular atria, branch repeatedly, and terminate as tracheoles supplying tissues. In *Rhodnius* the summed cross-sectional area of the branches remains constant (Locke, 1958), but in silkworm larvae the summed cross-section at the origin of the tracheoles (of which there are $1\frac{1}{2}$ million in a fifth-stage larva) may be only one sixty-second of its maximal value (Nunnome, 1944). In dragonflies, too, there is a reduction in total cross-sectional area in flight-muscle tracheae between the primary tube and the tracheoles (Fig. 1) (Section IV,A) (Weis-Fogh, 1964b).

The largest tracheae of big insects may be several millimeters in diameter.

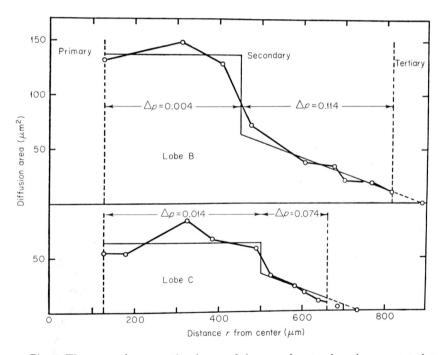

Fig. 1. The summed cross-sectional area of the secondary tracheae in representative lobes of two flight muscles in *Aeshna*. The secondary tracheae start from a primary trachea which runs axially through the muscle (Section IV,D). The values of the oxygen partial pressure difference (Δp) are shown for conditions during flight (Section IV,A). (From Weis-Fogh, 1964b.)

Tracheae taper to a diameter of 1–2 μm and there is then sometimes a further reduction of bore at the point where the tracheoles start: they in turn taper to 0.5–0.1 μm and may run for several hundred μm.

Tracheal volume varies greatly in different insects and at different stages of development. The tracheal system may occupy 42% of the volume of *Locusta* at the beginning of an instar, but only 3.8% towards the end of the instar as organs grow (Clarke, 1957, 1958). The tracheal volume of a 5 gm *Hyalophora* pupa is 250–400 mm^3 (5–10% of the body volume) (Kanwisher, 1966; Levy and Schneiderman, 1966a). This figure may increase as fuel reserves are used and water is lost.

The richness of tracheation of organs is clearly related to metabolic demands. An attempt to quantify the tracheation of an organ was made by Weis-Fogh (1964b) who introduced the concept of the *hole fraction* which is expressed as the summed cross-sectional area of the tracheae leading to an organ divided by the total surface area of that organ. In insect flight muscle the hole fraction lies between 10^{-1} and 10^{-3}. Figures for other organs or tissues are unknown but would provide interesting information, particularly if they were compared with metabolic rates.

B. TRACHEAL EPITHELIA

Tracheae, air sacs, and tracheoles are ensheathed by epithelial cells. Large tracheae are encircled by several cells, and both simple and septate desmosomes occur between the adjoining membranes of neighboring cells (Beaulaton, 1964). Reinforced junctions of this type may help to resist the expansion of the developing intima. Small tracheae lie invaginated in a single cell, while tracheoles lie within the tracheoblasts which form them. The tracheole is surrounded by a plasma membrane which does not initially appear to be joined to the outer plasma membrane of the tracheoblast. Tracheal cells often contain many microtubules (Smith, 1968) arranged parallel to the length of the trachea. These may have a supporting function or play a role in tracheal formation. There are few mitochondria but sometimes an abundance of glycogen granules.

Tracheoles often indent the membranes of the cells they supply, thus becoming functionally intracellular and ending close to mitochondria. This occurs in the flight muscles of many insects (but not in dragonflies or cockroaches) and in some other tissues, for example, fat-body, luminescent organs, testes, anal papillae of mosquitoes, and tymbal muscles of cicadas (Edwards *et al.,* 1958; Smith, 1961, 1963, 1968). Indenting tracheoles remain ensheathed in their own tracheoblast as well as in the membrane of the indented cell. Basement membranes of tracheoblast and of cell supplied may fuse, anchoring the tracheole firmly. The development of flight-muscle tracheation and the invasion of the T-system canals of the muscle cells by tracheoblasts has been described by Brosemer *et al.* (1963), Bücher (1966), and Beinbrech (1970). The factors controlling tracheal development including hormones, tissue factors, and lactate have recently been examined by Pihan (1972).

C. THE STRUCTURE OF TRACHEAE AND TRACHEOLES

The tracheal intima is continuous with the outer epicuticle of the body surface. It comprises a triple-layered structure containing cuticulin (Locke, 1969; Wigglesworth, 1970). Cuticulin is thought to be a stabilized lipid, occurring as a polyester, associated with protein. In the trachea the intima may bear on its luminal surface a wax monolayer. A protein epicuticle layer may underlie the intima (Locke, 1969) and in large tracheae under the taenidium there may be a continuous endocuticular layer which reduces longitudinal flexibility of the tube. The intima is buckled either into a helix or into rings and a "stiff, elastic, refractile material containing stabilized lipid" (Wigglesworth, 1970) is deposited into the buckles to form the taenidium. The taenidium is thought to comprise mainly protein, perhaps tanned, and chitin, the latter organized into fibers running circumferentially. The taenidium in large tracheae is often darkened. Beaulaton (1964) has described two sizes of fibril, 80–100 Å and 50 Å in diameter, associated with the taenidium; the fibrils are organized into four separate regions. Locke (1958) suggested that buckling results from expansion of the newly deposited intima which is restrained by the surrounding epithelial cells. Locke's hypothesis is consistent with the observation that intertaenidial size and separation increase more or less linearly with tracheal bore in a given species (with some exceptions), but many types of taenidia occur and the complex details of the surface patterning suggest that a more active part is played by the cells (Buck, 1962; Whitten, 1972).

The intima is often thrown into complex patterns of tubercles, major and minor buckles, papillae, bristles, and hairs (Richards, 1951; cf. Locke and Krishnan, 1971). Additional stout taenidial bars may be added which strengthen large tracheae (cf. Amos and Miller, 1965), while in some Buprestid beetles the prothoracic tracheae are unusual in possessing large buckles every 0.5 mm separated by a reticulated cuticular pattern (Tonapi, 1968). The tracheae of mature physogastric queen termites are derived from the tracheal system in the young alate female by extension and possibly by growth. The annular buckles become separated and the whole trachea increases considerably in diameter and length. The tracheae arise in a sheaf from a highly modified spiracle with no closing apparatus and show an unusually low frequency of branching. The system in alate males and females shows marked sexual dimorphism (Bordereau, 1971; P. L. Miller, unpublished).

The tracheolar intima is similar to that of the trachea, but there is probably no wax layer (Beament, 1964) and no underlying protein epicuticle (Fig. 2). The taenidial folds are either annular or helical (several parallel

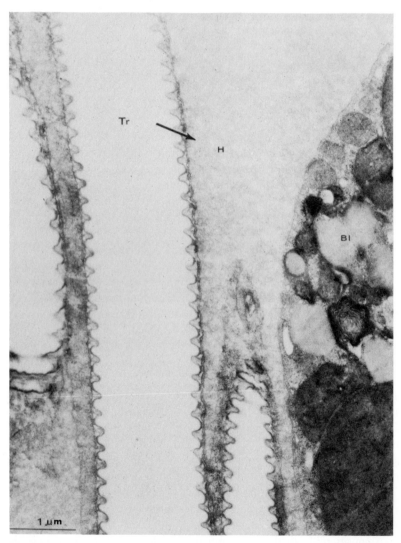

Fig. 2. Electron micrograph of large tracheoles from the leg of a pupal *Sarcophaga*. The tracheoles are unusual in lying in the hemolymph, and they possess a very thin cytoplasmic layer. Bl, blood cell; H, hemocoel; Tr, tracheole. (From Whitten, 1968.)

helices may be found) and they usually have no taenidial infilling. However, Whitten (1968) described a taenidial thread in the tracheoles of pupae of Diptera. The intima of tracheoles is pierced by pores 30 Å in diameter which persist in the mature tracheole and may have significance for liquid movements across the tracheole wall (Locke, 1966). In many insects tracheolar linings are not shed at ecdysis, but old tracheoles are glued onto new tracheae by an osmophilic material (Wigglesworth, 1959; Whitten, 1968).

A clear homogeneous layer is seen in electron micrographs under the intima of tracheae and tracheoles; Whitten (1969, 1972) has made the interesting suggestion that it may comprise the rubbery protein resilin. Most trachea display considerable elasticity, springing back to their resting length after stretch, and sections of resilin elsewhere appear homogeneous in electron micrographs. However, no direct tests for resilin in tracheae have been made so far.

D. Air Sacs

The tracheal system includes deformable regions which permit variations of tracheal volume to occur. Such regions are normally in the form of air sacs which appear to be dilated tracheae with a reduction or complete absence of the taenidium. Sacs are usually very thin walled and flexible; they may be held expanded by slight negative pressure in the hemolymph (cf. Davey and Treherne, 1964).

Sacs may occur throughout the body cavity and also in appendages such as legs, wings, elytra, and antennae. In the locust and many other Hemimetabola, they appear in the first instar and their number increases with successive ecdyses. In the pentatomid, *Halys,* however, they are first formed in the last larval instar (Bahadur and Bhatnagar, 1971). In insects such as flies, wasps, bees, ants, dragonflies, and Scarabaeid beetles they are extremely abundant: but in Cerambycid beetles and Belostomatid water bugs, for example, sacs are few or absent, their place being taken by many deformable parallel-sided tracheae. In *Oryctes* the sacs are each clothed in a layer of fat which may help in heat conservation during or after flight (Nair and Karnavan, 1966).

The sac systems of *Drosophila* and *Mycetophila* are shown in Fig. 3. In the Hymenoptera a good correlation between the develpment of air sacs and the size and activity of various species has been found (Tonapi, 1958).

E. Spiracular Structures

1. *Valves and Musculature*

In this account spiracles are numbered from 1 to 10, spiracles 1 and 2 being thoracic and the remainder abdominal, although spiracle 3 sometimes

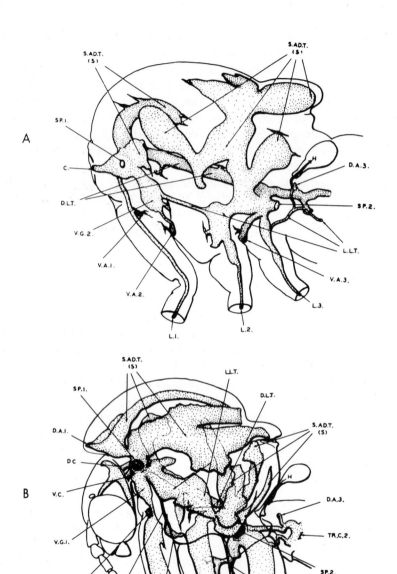

supplies metathoracic flight muscles. The innervation is described in Section V,A.

The closable valve comprises either a single or a pair of external lids which are pulled down onto or across the atrial orifice, or an internal lever which shuts off the tracheae where they emerge from the atrium. When spiracle muscles are inactive, most valves open by cuticular elasticity, but the abdominal spiracles of *Blaberus* are *closed* by elasticity. In the former case, maintained closure depends on muscular tension normally resulting from a stream of motor impulses in the nerves: a myogenic mechanism may also play a part (Section V). In locust abdominal spiracles closer muscles are short and broad while the openers are long and narrow. In *Blaberus* abdominal spiracles the opener is many times larger than the closer (Fig. 4C).

Spiracle 1 of *Sphodromantis* (Fig. 4B) possesses two closer muscles which act synergistically on separate parts of the valve. They share the same innervation but their different insertions allow either the dorsal orifice only, or both orifices, to open and close according to the motor impulse frequency (Miller, 1971b). The closer of spiracle 1 in *Periplaneta* is also divided partially into two, but its action has not been examined. In *Schistocerca* spiracle 1 has separate opener and closer muscles. At times the opener remains continuously contracted and the closer periodically shuts the valve against the opener by expanding a cuticular spring (Section V) (Miller, 1960).

The spiracles of *Hyalophora* have an internal closing mechanism. A strong closer muscle pulls a lever across the lower end of the atrium and it is opposed by an elastic ligament (Beckel, 1958). In other Lepidoptera an opening muscle may be present. On the other hand, all the spiracles of *Thermobia* possess opener muscles and are closed by cuticular elasticity alone. In spiracle 1 (Fig. 4A) a set of muscle fibers opens both the external lips, while in other spiracles muscles open an internal slit (Noble-Nesbitt, 1973). Anatomical descriptions of many other types of spiracle exist and in some there may be three or four muscles, but little is known of their action (Maki, 1938).

The closing mechanism is secondarily lost in some terrestrial species (e.g., larvae of all Diptera and some Scarabaeidae and Geotrupidae) as well as in many aquatic species. Insects with no closing mechanisms live in wet or moist habitats and often possess a sieve plate covering the spiracle (Section III,E,2). Some stratiomyid larvae shut off their large posterior valveless

---

Fig. 3. The tracheal systems of the pterothorax of (A) *Drosophila*, and (B) *Mycetophila*. Both systems arise by growth of the larval and pupal tracheal epithelium around the existing system. Wide sacs with many fusions are found in Diptera where oxygen supply in flight is largely by diffusion. SP., spiracles; S.ADT., secondary adult sacs; D.L.T., dorsal longitudinal trunk; V.A., ventral tracheal anastomosis; L, leg. (From Whitten, 1968.)

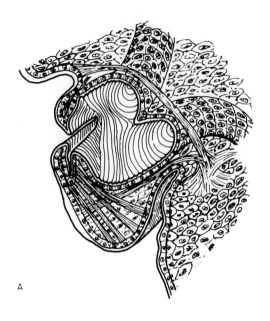

A

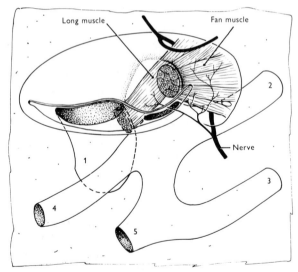

B

Fig. 4. Diagrams of spiracles representing different types of valve musculature.
(A) *Thermobia* spiracle 1; sectional view of the valves showing the opener muscles;
no closer muscle is present (Noble-Nesbitt, 1973). (B) *Sphodromantis* spiracle 1;
the two closer muscles are seen by transparency through the valve, and their shared
innervation is indicated (Miller, 1971b). (C) *Blaberus giganteus* spiracle 7; the
large opener muscle divided into two parts and the much smaller closer muscle
are shown.

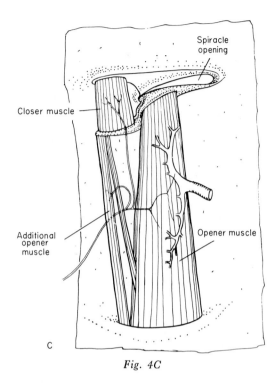

Fig. 4C

spiracles between the posterior tergite and sternite when their habitat dries out (Hinton, 1953). In the Nigerian species *Cyrtopus fastuosus* the closing system shows CO_2-sensitivity like that of spiracle valves, and when closed it reduces total water loss from the larva in a dry atmosphere to 0.14 ± 0.01 mg/gm/hour (Miller, 1970).

2. Sieve Plates

The spiracles of many insects, for example, Diptera (Keilin, 1944; Hassan, 1944), Coleoptera (Imms, 1957; Lotz, 1962; Hinton, 1967a; Richter, 1969), Lepidoptera (Nunnome, 1944), and aquatic Hemiptera (Hamilton, 1931; Miller, 1961), are covered either by sets of interlocking bristly filters or by sieve plates. These structures may overlie a closable valve. The sieve plate is a continuous but porous cuticular structure which is normally composed of stout trabeculae covering the spiracular opening: the trabeculae bear small branches at right angles which in turn hold the "membrane," a delicate structure of fine struts bounding small pores. Continuous channels or aeropyles join the atrial airspace to the outside. Richter (1969) believes that in Coleoptera the sieve plates of Scarabaeidae evolved from separate branching and interlocking atrial spines attached to the sides of the atrium and to the valve margins, such as are found in Cerambycidae and Dytiscidae.

Doubt about the porous nature of the sieve plate is best resolved by examination with the stereoscan electron microscope (Hinton, 1967a; Parsons, 1972) and by attempts to measure the airflow through it. In *Hydrocyrius* (Belostomatidae) under a pressure of 10 cm water, air flows through spiracle 1 at 1.5 ml/minute/mm² and through spiracle 3, with larger pores, at about 100 ml/minute/mm² (Miller, 1961). In *Anisops* (Notonectidae), air flows through abdominal spiracles at about 10 ml/minute/mm² under the same pressure (Miller, 1964b). Some values for pore size in different sieve plates are given in Table I.

The sieve plates of adult Scarabaeids are divided down the middle by two lips which open at emergence to allow the pupal tracheae to be withdrawn. The lips are bristly and may allow air to pass between them. In other insects an ecdysial opening may lie to one side of the plate, or on a protruding bulla. In Diptera a separate tube is formed which opens beside the plate to allow the shed cuticle to pass through and then closes to leave an ecdysial scar (Hinton, 1947).

Sieve plates are often described as structures which exclude dust from entering the tracheal system (Chapman, 1969). While this may be correct, they are not better developed on inspiratory than expiratory spiracles (see Section VI,C,7) (cf. Connell and Glynne Jones, 1953). They may serve the following additional functions: (1) prevention of entry of water into spiracles. The open spiracles of aquatic species are protected either by com-

TABLE I

PORE SIZES IN THE SIEVE PLATES OF DIFFERENT SPECIES

Family	Species	Spiracle	Pore size	Reference
..abaeidae	*Lepidoderma albo-hirtum* (larvae)	Abdominal	Slits: 4 μm × 0.1 −0.5 μm Pores: 0.2–0.3 μm in diameter	Hinton, 1967a
Saturniidae	*Bombyx* (larvae)	Abdominal	6 × 3 μm	Nunnome, 1944
Notonectidae	*Anisops pellucens*	Abdominal	2–10 μm²	Miller, 1964b
Belostomatidae	*Hydrocyrius columbiae*	Spiracles 1, 2 and 4–10	0.5 μm²	Miller, 1961
		Spiracle 3	>3 μm²	

plex arrays of bristles (e.g., Dytiscid adults) or by sieve plates (e.g., Belostomatidae). But in adult water-beetles and bugs most spiracles are additionally protected by overlying sclerites, elytra, or extra sets of long bristles (Notonectidae). Sieve plates protect the spiracles of terrestrial species against occasional wetting (rain). Many Scarabaeid adults and larvae burrow in rotting vegetable matter or in dung and their spiracles need to be well protected against wetting by surface-active materials, as also are some eggs laid in similar situations (Hinton, 1969). (2) Sieve plates may help to prevent the entry of parasites. Acarines are sometimes found in considerable numbers in the tracheae of large insects (cf. Amos and Miller, 1965) and *Acarapis* is a well-known hazard to beekeepers, the mites being attracted to the bees' spiracles (Sachs, 1952). (3) Sieve plates provide increased frictional drag and so reduce the bulk movement of air through the spiracle. But diffusion which is dependent on the sum of the pore perimeters and not on total pore area is not impeded (Keister and Buck, 1961). The restriction of bulk air movements resulting from body movements may help to conserve water in inactive species, but in more active species it conflicts with the needs of ventilation. Spiracle 2 is of major significance for the flight-muscle tracheal supply in most beetles and this spiracle usually lacks bristles or plates which might impede autoventilation (see Section VI,D,2). Spiracle 3 in *Hydrocyrius,* which again is important for the flight muscles, has a much coarser filter than other spiracles. Alternatively, the presence of lips dividing the plates of some spiracles (Scarabaeidae and abdominal spiracles of Notonectidae and Belostomatidae) may provide a bypass which can be opened by strong ventilatory movements. (4) It is unlikely that the diffusion of water vapor is affected differently from that of oxygen by sieve plates, filters, or internal felt chambers (cf. Imms, 1957). When spiracle valves are closed for most of the time, the tracheal system probably contains air saturated with water vapor: the drop in partial pressure of water vapor and oxygen is then greatest across the spiracle. A sieve plate might reduce water loss preferentially only if water vapor diffused as polymers or was adsorbed by the plate, but there is no evidence for either.

F. RESPIRATORY SYSTEMS OF THE EGG SHELL

Many terrestrial and a few aquatic species have very elaborate structures in the eggshell or chorion which may allow gas exchange to take place in air or water. An understanding of the fine structure has depended on careful light and electron microscopy (Hinton, 1960a,b,c, 1961, 1962, 1963) supplemented by use of the scanning electron microscope (Hinton, 1967b, 1968b, 1969, 1970; Hartley, 1971) and on the examination of ultra-thin sections of chorion (Smith and Telfer, 1970; Smith *et al.,* 1971). The chorion is a proteinaceous structure cross-bonded by sulfur and hydrogen bonds and it is formed by the follicular cells of the mother. In many species an air layer is held in the shell by a complex branching structure which communicates with the exterior through aeropyles. According to Hinton, in the egg of muscid flies there is a central zone pierced by small holes and bearing struts which pass both outward and inward. The former branch at right angles to produce an anastomosing network on the outside of the chorion; the inwardly directed struts fuse with a continuous layer next to the vitelline membrane. A wax layer is present at this level in many eggs (Wigglesworth and Beament, 1950, 1960), but it appears to be absent in *Musca* eggs, which desiccate rapidly in dry air. The gas layer held between the struts makes contact with the atmosphere through the many pores, 0.5–1.0 μm in diameter, in the superficial networks which are scattered over the whole surface except at the hatching lines. The chorion of *Calliphora* (Fig. 5) is similar.

The 50–60-μm-thick chorion of *Hyalophora* has a lamellar organization consisting of helicoidally arranged microfibrils like those of cuticle (Smith *et al.,* 1971). Above an inner trabeculate layer, 0.4 μm thick, a system of canals runs through the chorion. The canals become narrow and contorted as they pass through a central zone. They are filled with extracellular microfilaments, 70 Å in diameter. Some canals are believed to be formed by cytoplasmic extensions of the follicular cells.

In the tettigoniid, *Homorocoryphus,* the chorion is 10 μm thick, and composed of an inner region (6.5 μm thick), pierced by many small vertical air-filled channels, 0.2–0.3 μm in diameter (Hartley, 1971). A middle air-filled pillar zone connects with wider (0.5 μm) pores leading to the surface. Between 7.9 and 11.4% of the surface consists of pores.

Many terrestrial eggs are laid in situations which are likely to be temporarily flooded. Hinton has suggested that the chorionic aeropyle and meshwork systems may in some species act as plastrons if they provide an adequate area of air–water interface. In *Rhodnius* there are 185 aeropylar openings with an aggregate area of 536 μm^2—too small to act significantly as a plastron; but in the egg of the staphylinid, *Oxypus,* the comparable area is 3000–4000 μm^2/mg tissue and this can support 25% of normal respiration when the egg is submerged. Other eggs have more extensive plastrons, for example, *Antheraea* has 4×10^5 μm^2/mg, which compares favorably with the 10^5–10^6 μm^2/mg of adult aquatic insects with efficient plastrons (Hinton, 1970). However, since plastron-bearing eggs are stationary and the water surrounding them may become stagnant, the benefit gained from a plastron may be short lived.

Direct measurements of oxygen uptake and rates of development in *Homorocoryphus* eggs have suggested that the respiratory needs in air could

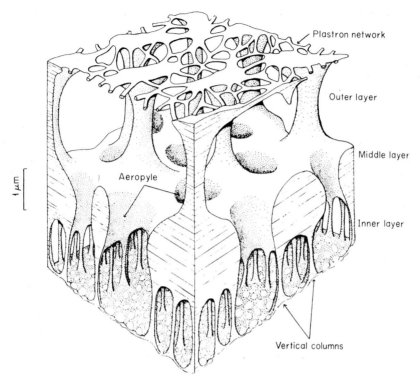

Fig. 5. Reconstruction of part of the chorion (eggshell) of *Calliphora* between the hatching lines. The three layers are shown with penetrating aeropyles leading to the superficial plastron. (From Hinton, 1963.)

be adequately met if the pore area was reduced by 10^5 (assuming all gas exchange to be by the pores) (Hartley, 1971). Hartley suggests that the pores are not primarily for gas exchange in air or water, but act to trap a layer of water after chance wetting which is then taken in by the egg. Egg development requires considerable water uptake and the chorion stretches by 1.7 times. Stretching is presumed to be allowed for by an increase in pore diameter. When submerged, the peripheral parts of the canals become water-logged and development is delayed (cf. Wigglesworth and Beament, 1960).

This conclusion does not extend to the eggs of many other species whose aeropyles are not wetted even under considerable pressures. Hinton argues that this property is correlated with resistance to wetting by surfactant materials which might be encountered, for example, by the eggs of Cordylurids placed in cow dung, which resist wetting for 30 minutes under a pressure of 300 mm Hg, or of *Drosophila* laid on rotting fruit, which are not wetted after 30 minutes at 600–1000 mm Hg.

Some eggs must resist periodic submersion and also desiccation in air. The latter is assisted by a reduction of the cross-sectional area of the diffusion pathway between the plastron and tissues. The development of one or more horns provides an answer to the problem (Hinton, 1960c) (Fig. 6). The distal part of the horn bears many pores opening on the surface (the plastron); these communicate with the main chorionic air space by ducts running through the center of the horn. The horn also allows the pore area to remain above a shallow water layer, like a snorkel. Horns have been developed independently in seven families of Diptera, six of Hemiptera, and one of Hymenoptera (Hinton, 1969). Horns show a basic similarity in structural organization with each other and some resemblance to the aeropyle system of some eggs, to the sieve plates covering the spiracles of some insects, and to the spiracular gills of others. These similarities suggest a common developmental mechanism, part of the synthesis probably depending on extracellular self-assembly.

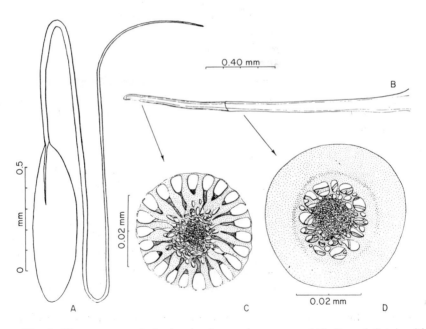

Fig. 6. The structure of the horns of some insect eggs. (A) Egg of *Sepsis* with a long horn. (From Hinton, 1960c.) (B), (C), and (D) Egg of *Nepa*. In (C), a transverse section of the horn near the apex through the plastron is shown. (D) is a section below the plastron showing the central air-filled meshwork. (From Hinton, 1961.)

G. Pupal Horning

Horns developed on the pupae of some Diptera bear resemblance to the horns on eggs. When the third-stage larval cuticle of higher Diptera hardens and darkens to form the puparium, two small areas remain clear. An extension from the pupal spiracle, the horn, is thrust through each to unite the pupal tracheal system with the atmosphere. The horn is a porous structure and opens into a felt chamber (Keister, 1953). Respiratory horns normally occur in pairs—*Ptychoptera* is exceptional in having only one—and are usually larger and more complex in the Orthorrhapha where they possess two closing devices (Satchell, 1948). Horns of Cyclorrhapha are small, simpler, and possess one closing device. When the horns of *Ophyra* are blocked with wax the pupa dies (Roddy, 1955), but in *Phormia* adequate supplies of oxygen reach the pupa either through the horns or, if blocked, through the old larval spiracles (Keister, 1953). In some Cyclorrhapha, the horns never break through the puparium and this led Keister to believe that, in general, they are not essential for respiration in higher Diptera. *Pseudolimnophila,* which pupates in mud, has long, highly specialized horns equipped with distal flaps which enable the pupa to float when the habitat is flooded (Hinton, 1954). These may be intermediate between the spiracular gills of aquatic species (Hinton, 1968a) and the horns of terrestrial species.

IV. The Movement of Gases in the Tracheal System

A. Diffusion

Useful discussions of diffusion in animal tissues have been provided by Krogh (1941), Weis-Fogh (1964b), Bursell (1970), Steen (1971), and Mill (1972).

Gaseous diffusion alone meets the needs of small insects or of the inactive stages of larger species. All insects depend on liquid diffusion for the final stages between tracheoles and mitochondria. Muscular movements and hemolymph circulation shorten the liquid diffusion pathway, and intracellular convention too may play a part. Since gaseous diffusion may be several hundred thousand times faster than liquid diffusion, the value of bringing air-filled tracheoles close to respiring tissues is apparent. For similar reasons, tracheae play an important role in apneustic aquatic larvae.

Hemolymph circulation is probably important for gas exchange, particularly in poorly tracheated tissues. Tracheae clustered round the heart may provide a pathway for gas movement between hemolymph and the atmosphere (Wigglesworth, 1965). More CO_2 than oxygen is transported by the hemolymph since the permeability coefficient of CO_2 in tissues is much higher than for oxygen. A fraction of the CO_2 produced may therefore leave the insect across the cuticle and not via the tracheal system. However, when the cuticle is thick, CO_2 loss by this route is minimal (Schneiderman and Williams, 1955), although if the cuticle is thin, as in *Calliphora* larvae, 10% of the total CO_2 loss may be transcuticular (Fraenkel and Herford, 1938), or in *Phormia* larvae, 2.5% (Buck and Keister, 1956).

In *Tenebrio* and *Cossus* larvae Krogh (1920) calculated that, at rest with the spiracles open, the difference in partial pressure of the air and at the ends of the tracheoles need not exceed 0.02 atm. After exercise, at the distal end of grasshopper leg tracheoles, the difference was greater but still within the limits of the system. Thus, Krogh was able to demonstrate that diffusion was an adequate process for supplying oxygen to the tissues.

Weis-Fogh (1964b) reached the same conclusion for active flight muscle with very high rates of oxygen consumption. He analyzed in detail the pathway between spiracle and tracheole in *Aeshna*. The first part (the primary tube) is ventilated, and the subsequent air diffusion pathway is up to 1 mm long. With an oxygen consumption of 1.8 ml/gm muscle/minute, he concluded that the difference in oxygen partial pressure between the primary and the tracheole lies between 0.05 and 0.13 atm in different muscles. Thus, no alternative process need be postulated.

Weis-Fogh has shown that the dense mesh of tracheoles surrounding a dragonfly flight muscle makes it reasonable to regard each fiber as being surrounded by air. The maximum distance for tissue diffusion is about 10 μm and tracheoles must therefore be not more than 20 μm apart. Since the tracheoles do not indent dragonfly muscles, this defines the maximum diameter for the fiber. In fact, the fibers are tubular with a central core of nuclei and radially arranged mitochondria (Smith, 1966, 1968) so that a slightly greater diameter is permissible. However, even in the damselfly *Enallagma* the mitochondria all appear to be less than 10 μm from the nearest tracheole. In the most active flight muscles known (in Diptera and Hymenoptera), the tracheoles indent the muscle fibers and lie within 2–3 μm of each other, although the theoretical maximum is 6–8 μm.

In many insects the terminal part of the tracheole is filled with a liquid (Wigglesworth, 1930, 1932). When the tissue is active, or is made hypoxic, the liquid disappears from the tracheole and gas advances towards the tip. A balance between capillary forces and tissue osmotic forces is believed to determine the level of the liquid in the tracheole. Tissue activity causes the release of metabolites which raise the osmotic pressure and liquid is withdrawn from the tracheole; as the metabolites are dispersed the liquid level returns to normal. Thus a simple feedback system depending on physical principles is believed to determine the filling of the tracheoles, working perhaps via the colloid osmotic pressure of the tracheoblasts (Wigglesworth, 1953). Quantitative aspects of the system are unknown. Surfactant materials like those found at the mammalian lung surface, if present, would reduce the capillary forces, while the osmotic force is dependent partly on the composition of the tracheolar fluid. Beament (1964) has argued that active transport may play a part, particularly if the capillary forces are large. The tracheoblasts, however, do not resemble cells known to be involved in active transport elsewhere.

It is postulated that active transport is also involved in the withdrawal of liquid from the tracheal system after ecdysis (pneumatization). Air may first appear in tracheae remote from spiracles and the tracheal system of apneustic insects becomes air filled while the insects are submerged. Beament has suggested that when the tracheal intima is tanned, a wax layer is made to present a hydrofuge surface to the lumen. The consequent low adhesion of water to the walls allows gas bubbles to appear when a slight reduction of pressure occurs. This may be brought about by the active uptake of liquid by the cells. Tanning is believed to be under hormonal control in some insects and this seems to extend to tracheae too, since when bursicon from the last abdominal ganglion is prevented from reaching some regions in *Locusta* the tracheae in those parts remain liquid filled (Vincent, 1971).

B. PASSIVE SUCTION VENTILATION

Passive suction ventilation occurs in the absence of muscular movements when a partial vacuum develops in tracheae and air is sucked in through slightly open spiracles. The tracheae of submerged mosquito larvae collapse as oxygen is consumed and they reinflate instantly when the spiracles make contact with air at the water surface. Only a very brief exposure to the air is therefore necessary. The selective advantage of suction ventilation in most cases is not, however, in the acceleration of the gas exchange but in the reduction of water loss. This may explain its occurrence in organisms as different as fruits and plant-storage organs, insect pupae, and fleas (Schneiderman, 1960; Levy and Schneiderman, 1966c). In many insects suction ventilation is associated with the discontinuous release of CO_2, which was thought to reflect the existence of periodic respiration (Punt, 1950, 1956; Punt et al., 1957). However, in the pupae of *Agapema* and *Hyalophora*, continuous oxygen uptake proceeds while CO_2 is released in bursts (Buck and Keister, 1955; Schneiderman and Williams, 1953, 1955). Normal gas exchange occurs if such pupae are made active by wounding, or if one or more spiracles are held open. On the other hand, blockage of all spiracles abolishes the bursts of CO_2 release and all exchange becomes minimal. To explain the situation, Buck (1958) suggested that a partial vacuum occurs in the tracheae due to the consumption of oxygen, while CO_2 forms bicarbonate in the hemolymph and tissues. Air is sucked in through slightly open spiracles and this effectively bars the outward diffusion of CO_2 and water vapor. Accumulating CO_2 and bicarbonate eventually cause the spiracles to open fully and a burst of CO_2 diffuses out.

Kanwisher (1966) made continuous measurements for up to 6 weeks of the CO_2 and water vapor release. His technique was to measure the thermal conductivity of the gas mixture with a heated thermistor (see also Edwards, 1970). In addition he used an infrared gas analyzer sufficiently sensitive to measure the rate of CO_2 release between the bursts. He showed that in a dry atmosphere similar amounts of CO_2 and water were lost (Fig. 7). With

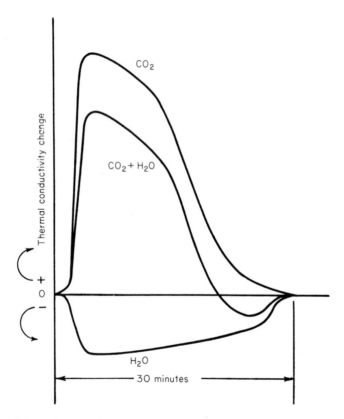

Fig. 7. Thermal conductivity recordings of CO_2 and water-vapor output during a single burst from a pupal *Hyalophora* in diapause. (From Kanwisher, 1966.)

fat as a fuel, water losses were equaled by water produced from metabolism and in high humidities there might even be a gain of water, although pupae normally living under such conditions do not show burst cycles (Sláma, 1960).

Accurate measurements of a number of parameters in *Hyalophora* by Schneiderman and his colleagues have caused some modifications to be made to Buck's hypothesis, and have contributed much important information about insect respiration (Schneiderman, 1960; Levy and Schneiderman, 1966a,b,c; Schneiderman and Schechter, 1966; Brockway and Schneiderman, 1967). Observations of the spiracle over long periods have confirmed that the valve plays a key role. The duration of the whole cycle is very variable and depends on temperature, maturation, and on individual differences. A burst cycle is divisible into three stages, summarized in Fig. 8. In the first, rising pCO_2 causes full valve opening which lasts for 15–30 minutes. Valve opening is triggered by a pCO_2 of 6.4% when the pO_2 is 3.5%. About 350 mm³ of CO_2 may be released in a burst, and since the total tracheal volume is only about 400 mm³ in a 5 gm pupa, most of the CO_2 must be derived from tissue bicarbonate and only about 10% from tracheal CO_2 (Buck and Keister, 1958). During the burst the tracheal pCO_2 falls from 6.4 to 3.5% and the spiracles then reclose. Carbonic anhydrase occurs in some insect tissues, but it is not associated with the tracheal system so the possible role it plays in CO_2 release is unknown (Buck and Friedman, 1958; Edwards and Patton, 1967).

In the second stage, the constriction phase, the spiracles appear to close fully. The pO_2 falls to 3.5% and the intratracheal pressure falls to —4 mm Hg; the pupa shortens slightly. Oxygen consumption continues and it is assumed that the spiracles are not hermetically sealed, and that the partial vacuum causes air to enter even though the exit of CO_2 and nitrogen is barred.

In the third stage the valve starts to make small movements, opening slightly for a few seconds and then reclosing (fluttering). The pO_2 stays at about 3.5% and the pCO_2 gradually rises. Each small opening is accompanied by a steplike rise in pressure toward atmospheric. Small oscillations of pressure, termed microcycles, then continue between —0.03 and —0.15 mm Hg. During this stage, some outward diffusion of nitrogen occurs, but CO_2 escape, with a shallower partial pressure drop, continues to be barred by the inward movement of air. This stage is terminated by the next CO_2 burst (stage 1).

Pupae made to breathe 60% O_2 continue to produce CO_2 in bursts but the pressure falls by as much as —10 mm Hg between bursts and there are no microcycles. By perfusing the ganglia, or the remainder of the insect separately, with gases, it has been shown that the spiracles are directly sensi-

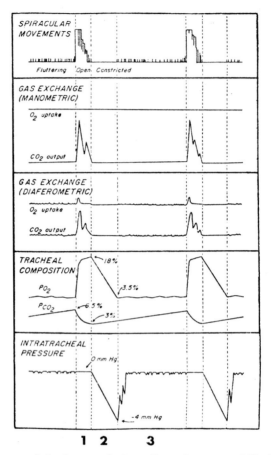

Fig. 8. Summary of the burst cycle in a diapausing pupa of *Hyalophora*. Stages 1, 2, and 3, described in the test, are indicated. (From Levy and Schneiderman, 1966c.)

tive to CO_2, opening fully in 14% regardless of the pO_2 (Burkett and Schneiderman, 1967). In contrast, when the ganglia are treated with 30% CO_2 there is no spiracle opening. However, ganglia perfused with 3.5% O_2 cause spiracle-valve fluttering. Thus CO_2 may act on spiracle muscles (see Section V,C) while hypoxia seems to act on the ganglia.

The pupae of temperate species may spend much of the winter in diapause at subzero temperatures. Burkett and Schneiderman (1968) have found that the spiracles continue to control CO_2 bursts at $-5°C$ while small flutters of the valve occur at 1–30 times/hour during stage 3. At between -5 and $-10°C$, burst production ceases and the spiracles remain closed, although they continue to show CO_2 sensitivity until the temperature is lowered to between $-10°C$ and $-16°C$ when they stay closed.

Discontinuous release of CO_2 has been noted in many other species. For example, it occurs in *Periplaneta* (Wilkins, 1960) and in *Schistocerca* (Hamilton, 1964) (Fig. 9). In *Schistocerca* (and probably in *Periplaneta*) it depends on periodic ventilation (see Section IV,C,1), but again its role may be in water retention. In fleas, Herford (1938) noted the collapse and reinflation of tracheae at regular intervals, the latter corresponding to spiracle opening. The flea spiracles are probably chemically controlled like those of *Hyalophora* (Wigglesworth, 1935) and a similar type of passive suction ventilation may occur in these species.

Discontinuous CO_2 release has been measured in the prepupa of the wood-boring beetle *Orthosoma*, but normally the pCO_2 of the logs inhabited by the beetles is high enough to abolish burst cycles and the spiracles remain open (Paim and Beckel, 1963). The pupae can survive for 10 days in 97% CO_2 and feeding larvae of this species congregate in areas where the pCO_2 is as high as 30–50%, although they are narcotized if the pCO_2 exceeds 50%. Mated females, too, are attracted to areas of high pCO_2 for oviposition (Paim and Beckel, 1964).

In summary, passive suction ventilation allows the internal CO_2 level to reach a high value before it triggers the opening of the spiracles; the time during which escape of CO_2 takes place is thus limited. This, in turn, restricts the time available for water vapor loss. The burst cycle is therefore interpreted as a means of conserving water. A high internal pCO_2 is made possible because (a) much CO_2 is banked as bicarbonate, (b) the inward draught of air during the constriction and flutter periods impedes the outward movement of CO_2, and (c) the spiracle triggering level is high—apparently higher than that needed to keep the spiracle open.

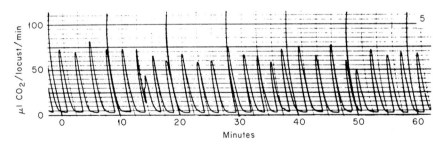

Fig. 9. Continuous diaferometric recordings of CO_2 output from a single adult male *Schistocerca* shortly after ecdysis. The mean output is 19.61 μl of CO_2/minute, and each CO_2 burst corresponds to the appearance of ventilation. (From Hamilton, 1964.)

C. Muscular Ventilation

1. Occurrence and Form of Ventilatory Movements

Ventilating movements commonly occur in the abdomen of large insects where they take the form of dorsoventral or longitudinal pumping strokes. In locusts, both inspiration and expiration are actively performed by muscles, but in many other species (e.g., cockroaches) expiration is muscular while inspiration is dependent on cuticular elasticity, enhanced sometimes by resilin. For example, in the abdomen of *Orcytes* and *Melolontha,* pairs of resilin ribs tend to keep the tergum and sternum in the inspiratory position (Andersen and Weis-Fogh, 1964). Pumping movements act on compressible regions of the tracheal system driving air in and out of open spiracles. Air sacs remote from the pump can be affected through the displacement of hemolymph.

Pumping can be recorded with a kymograph or electromechanical transducer when the movement of a sclerite is registered against a fixed point in space. A superior technique employs a Hall generator which measures the dorsoventral excursion of the sternum with reference to the tergum of the same segment (R. Hustert, personal communication). Measurements of the total volume of air pumped have been made by Weis-Fogh (1967) by placing the abdomen of a locust in a small plethysmograph connected to a spirometer. He separately measured the volume of air expelled from spiracle 10, and was thus able to make simultaneous calculations of the total and unidirectionally pumped volumes (Section IV,C,7). An inactive *Schistocerca* pumps about 40 liters (from 11 to 72) per kg per hour, and with 5% CO_2 this rises to 240–280 liters.

Some values for the ventilatory frequencies, amplitudes, and volumes of a number of species are given in Tables 189, 190, and 192 of Altman and Dittmer (1971). However, values from different species measured by several workers are not always comparable since conditions may not be adequately standardized. Moreover, the values given in Table 192 for amplitudes of ventilatory strokes need to be corrected by multiplying each by 10^6.

The ratio of stroke volume to total tracheal volume varies within an individual, since growing organs reduce the tracheal volume. Thus, a resting *Schistocerca* normally exchanges about 5% of the tracheal air with each stroke and 20% with hyperventilation, but in gravid females the figures are likely to be larger.

Ventilation may occur continually at normal temperatures (e.g., sometimes in mature locusts) or periodically (e.g., in *Byrsotria,* Myers and Retzlaff, 1963), or it may occur only after periods of activity (*Periplaneta*). Hamilton (1964) continuously measured the CO_2 output from *Schistocerca* and found that during the first day after the last ecdysis, and in old males, the output was periodic. Each burst coincided with the appearance of ventilation and he interpreted this as a means of conserving water (Fig. 9). Similarly, *Byrsotria* pumps 3–15 times every 7 minutes (Myers and Retzlaff, 1963).

2. Auxiliary Pumping Systems

Additional pumping activity synchronized with abdominal movements occurs in a number of insects during respiratory stress. Some forms of auxiliary pumping are listed in Table II. In *Schistocerca* three additional forms of movement supplement the principal dorsoventral pump in the abdomen; these are abdominal telescoping movements, and head and prothoracic pumping (Miller, 1960). Head protraction (inspiration) and retraction (expiration) occur precisely in phase with abdominal strokes and are brought about by the complex neck musculature, part of which acts on the cervical sclerites. The musculature receives innervation from the subesophageal, prothoracic, and mesothoracic ganglia, and inspiratory and expiratory bursts of motor impulses have been recorded coming from each ganglion (P. S. Mills, unpublished). Contrary to an earlier report (Miller, 1960), the subesophageal ganglion need not be present for bursts to be formed in the pro- and mesothoracic ganglia, although severance of the neck connectives abolishes almost all the movement. The movements can be evoked in all except the first instar, and the presence of the metathoracic ganglion is always essential.

Prothoracic pumping occurs commonly in several Prionine beetles (Miller, 1971a), but it has not been recorded in other Cerambycidae. Many Cerambycids do, however, stridulate by making nodding movements with the prothorax. Miller (1971c) suggested that such stridulation evolved from Prionine prothoracic pumping, but R. A. Crowson (personal communication) has pointed out that the Disteniinae and Philinae possess stridulatory files and are probably close to the original Cerambycid stock. Prionines may therefore have secondarily lost their stridulatory files, but retained the movements as a ventilatory mechanism.

Many other movements may be intermittently coupled to strong ventilatory pumping. Some of these may help to resist expiratory pressures and make the expulsion of air more effective (cf. Walcott and Burrows, 1969; Miller, 1971a), but others do not seem to have respiratory significance (e.g., palp twitching in locusts) and may serve other functions (see Section IV,C,5).

3. Neuromuscular Mechanisms

The dorsoventral and longitudinal inspiratory and expiratory muscles of the abdomen of *Schistocerca* have recently been examined physiologically and anatomically (Lewis, 1973; R. Hustert, personal communication). Each of the inspiratory dorsoventral muscles is innervated by the median and trans-

TABLE II

AUXILIARY PUMPING SYSTEMS

Species	Segments	Form of movements	Inspiratory muscles	Expiratory muscles	Innervation	Stimulus	Reference
Schistocerca gregaria	Abdominal	Longitudinal	189 (Seg. 4)	182 (Seg. 4)	Lateral nerves from 1st abdominal ganglion	CO_2	Lewis, 1973
	Prothoracic	Longitudinal	—	59, 60	Pro-, mesothoracic ganglia	CO_2	Miller, 1960
	Head pumping[a]	Longitudinal	50, 52 and others	49, 55, 57 and others	Subesophageal, pro- and mesothoracic ganglia	CO_2	P. Mills, unpublished
Hydrocyrius columbiae and other Belostomatidae	Prothoracic	Longitudinal	?	?	?	Preflight warm up	Miller, 1961
Melolontha melolontha	Prothoracic	Longitudinal	?	?	?	CO_2	Miller, 1971a
	Metathoracic[b]	Dorsoventral	—	Sternoepisternal, pleurosternal, or M.72 (Larsen, 1966)	Metathoracic ganglion	Accompanies abdominal ventilation	Miller, 1971a
Mallodon downesi and *Macrotoma palmata* (Prioninae)	Metathoracic	Dorsoventral	—	Sternoepisternal, pleurosternal, or M.72 (Larsen, 1966)	Metathoracic ganglion	Accompanies abdominal ventilation	Miller, 1971a
Sphodromantis lineola	Mesothoracic[b]	Dorsoventral	—	Pleurosternal or M.57 ("spiracle 2 dilator" of Levereault, 1938)	Axons in N1 from metathoracic ganglion join mesothoracic transverse nerve and then run to M.57	Accompanies abdominal ventilation	P. L. Miller, unpublished

[a] Also noted in *Chorthippus* by du Buisson (1924), and occurs in other Acrididae.

[b] Thoracic pumping has been noted in several other insects, e.g., *Hydrophilus*, *Dytiscus* (Brocher, 1931), *Dixippus*, and *Embia* (du Buisson, 1924).

verse nerves of the segmental ganglion, while the dorsoventral expiratory muscles, and the longitudinal expiratory and inspiratory muscles are all innervated by paired lateral nerves (Fig. 10). In dragonfly larvae Mill and Hughes (1966) found median nerves running to inspiratory muscles. The innervation of *Melanoplus* probably follows a similar plan (Tyrer, 1971).

With each inspiratory stroke, a burst of motor impulses can be recorded in two axons in the transverse nerve, while motor bursts in several axons in the lateral nerves accompany expiration. In inspiratory bursts spikes appear at constant frequency, but in expiratory bursts they sometimes wax and wane (Figs. 11 and 12). The duration of inspiratory bursts is not related to the length of the ventilatory cycle, but expiratory burst length is approximately proportional to cycle length (Fig. 13). In many species pauses

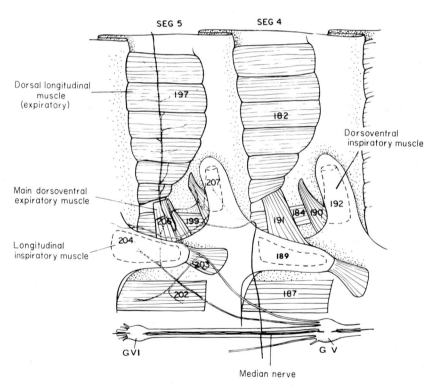

Fig. 10. The main ventilatory muscles and their innervation in abdominal segments 4 and 5 of *Schistocerca*. The median nerve of GV (the second separate abdominal ganglion) innervates the dorsoventral inspiratory muscle (207) of segment 5. The anterior lateral nerve of GV innervates principally the dorsal longitudinal muscles (197); the posterior lateral nerve supplies the longitudinal inspiratory (204) and dorsoventral expiratory muscles (206). (Based on Lewis, 1973.)

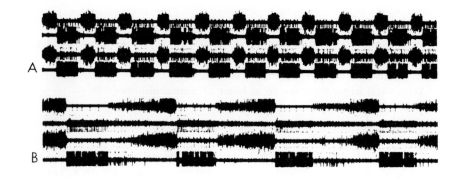

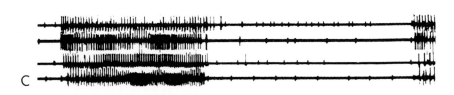

Fig. 11. Oscillograms of motor impulses in nerves to ventilatory muscles in *Schisto-cerca*. (A) Simultaneous recordings from four efferent nerve trunks: second and fourth lines are from median nerves of metathoracic and first abdominal ganglia, respectively, which supply inspiratory muscles in segments 3 and 4. The first and third lines are from lateral nerves of the same two ganglia which supply the dorso-ventral expiratory muscles. All activity shown is efferent and the CNS is intact. (B) The same after cutting between mesothoracic and metathoracic ganglia. The frequency of ventilation declines but the coordination is unimpaired. (C) Simulta-neous recordings from the median nerves of four adjacent ganglia (metathoracic and first three abdominal) showing one ventilatory cycle commencing with an inspira-tory burst in which two axons in each median nerve are active. Small spikes in the interburst period are efferent to spiracle muscles. Horizontal scales, 1 second. (From Lewis, 1973.)

commonly occur between the end of one inspiration and the beginning of the following expiration. In *Schistocerca* the expiratory stroke may be divided into two phases, separated by a stationary interval during which expiratory motor nerves continue to fire (Miller, 1965). As the amplitude of ventilation increases, more units are recruited, larger units tending to participate later in the expiratory burst (Lewis, 1973; Hinkle and Camhi, 1972). Inspiratory bursts frequently overlap preceding expiratory bursts, but the reverse seldom occurs.

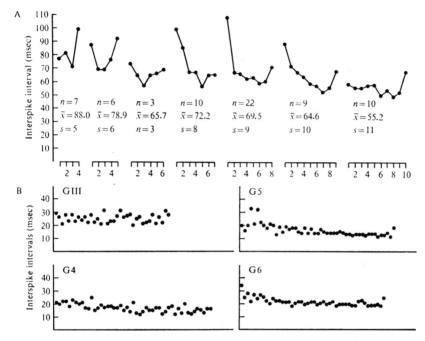

Fig. 12. Sequential interval histograms of motor impulses in abdominal nerves during a ventilatory cycle in *Schistocerca*. (A) A single excitatory axon to the main dorsoventral expiratory muscles; (B) median nerve axon firing during inspiration. The "parabolic" shape of the expiratory burst contrasts with the steady-rate firing of the inspiratory burst. (From Lewis, 1973.)

In intact resting locusts R. Hustert (personal communication) has recognized three patterns of ventilatory movement. In the first, termed the plateau cycle, there is a marked compression phase at the end of the expiratory stroke before the ensuing inspiratory stroke starts. In the second pattern, inspiration follows immediately after the expiratory stroke. Several cycles of strong ventilation are often followed by a long pause or quiescent phase when the third pattern may appear. This takes the form of miniature inspirations with no corresponding bursts in expiratory muscles.

4. The Pacemaker System

Four aspects concerning ventilatory pacemakers will be considered: (1) their localization, (2) the distribution of their activity to motor neurons, (3) their regulation, and (4) the origin of their activity.

1. Early attempts to localize pacemakers depended on making cuts in the CNS and examining subsequent activity. They gave rise to a variety of inter-

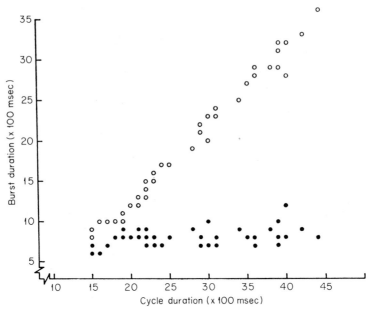

Fig. 13. The duration of the ventilatory cycle in *Schistocerca* plotted against the duration of expiratory bursts (open circles) and of inspiratory bursts (closed circles) recorded in abdominal efferent nerves. As the cycle length increases, expiratory bursts lengthen while inspiratory bursts show little change. CNS intact. (From Lewis, 1973.)

pretations in different species (e.g., Fraenkel, 1932; Schreuder and de Wilde, 1952; Miller, 1960; Myers and Retzlaff, 1963).

Isolated abdominal ganglia of *Schistocerca* can each generate a rhythm, although this may not appear until the ganglion is treated with CO_2. The rhythm generated by the isolated metathoracic ganglion, to which the first three abdominal ganglia are fused, resembles the ventilatory rhythm of the intact insect; no more anterior separated ganglion seems able to generate a ventilatory rhythm (Miller, 1966a). In *Periplaneta* the metathoracic, second abdominal, and sometimes the third abdominal ganglion can each generate a rhythm when isolated (Farley *et al.*, 1967). These observations therefore suggest that there are several available pacemakers for ventilation. However, pumping strokes are generally well synchronized throughout the abdomen and one pacemaker is assumed to coordinate the whole system.

Lewis examined the timing of motor bursts in nerves leading to pumping muscles from different ganglia, in hyperventilating *Schistocerca*. He found much variation from one preparation to another and even within one insect.

No ganglion consistently leads, but on the average the metathoracic ganglion fires earlier than abdominal ganglia. He interprets this to mean that the metathoracic ganglion initiates the stroke, and relays activity to other ganglia which in turn produce motor bursts. However, variable delays within abdominal ganglia may mask the order in which they are excited. The delays become much less variable when ventilation is fast (Lewis et al., 1973).

R. Hustert (personal communication) finds that strong ventilation is well synchronized throughout the locust abdomen, but that when isolated ventilatory cycles occur, or when there are miniature inspirations, delays of 200–400 mseconds may occur between the activity of adjacent segments. The movements are therefore peristaltic in nature and probably depend on a different intersegmental coordinating system from that which controls strong ventilation. Likewise, Lewis (1973) has found that in locusts with the metathoracic ganglion removed, ventilation in the posterior segments commonly involves long delays between activity in neighboring segments (see below and Fig. 14).

2. Pacemaker activity is thought to be distributed to motor neurons in other ganglia by interneurons. Bursts of impulses in a single neuron in each abdominal connective of Schistocerca precede expiratory efferent bursts sometimes by more than 100 mseconds. The delay between a connective burst and an expiratory burst is shortened when ventilation is rapid. The bursts continue unchanged in deafferented preparations, and individual impulses which pass through ganglia without delay have been tracked throughout the abdominal cord. They are therefore believed to occur in coordinating interneurons and not in motor or sensory axons. They do not occur when the metathoracic ganglion is removed. The suggested organization of interneurons which coordinate ventilation is shown in Fig. 14. Similar bursts have been recorded in the cord anterior to the metathoracic ganglion during hyperventilation and they are believed to coordinate auxiliary pumping in anterior segments.

Farley et al. (1967) recorded bursts of spikes propagating posteriorly at 3 meters/second in abdominal connectives of Periplaneta. The bursts arose from the metathoracic ganglion and were thought to coordinate ventilation in posterior segments as in the locust. Less often, the second abdominal ganglion appeared to take over the pacemaker function, and at such times impulses were recorded in connectives propagating anteriorly.

3. Command interneurons may act on pacemakers to control their output. In Schistocerca the head and anterior thoracic ganglia boost ventilation when treated with CO_2 by acting on the metathoracic ganglion (Miller, 1960). The brain is thought to be particularly important in this respect, and Huber (1960) showed that brain stimulation in crickets can either stimulate or depress ventilation. In Periplaneta, Rounds (1968) made localized injections

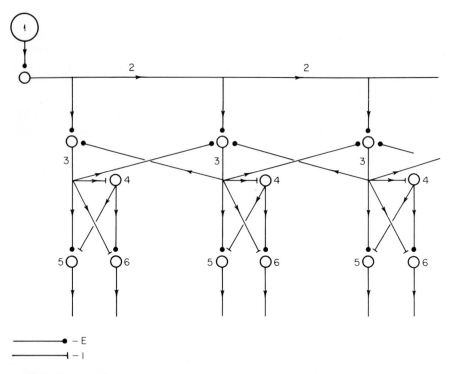

Fig. 14. A model of a neural arrangement which explains some aspects of the coordinated ventilatory output in *Schistocerca*. 1, Pacemaker cell which produces rhythmic bursts; 2, a coordinating interneuron which runs throughout the abdominal cord and is excited by the pacemaker (1). Bursts of activity in cell 2 excite inter-neurons 3 in each segmental ganglion; these in turn bilaterally excite the expiratory motor neurons, 5, and simultaneously inhibit interneurons, 4, and inspiratory motor neurons, 6. Cell 4 is tonically active when not inhibited by cell 3. A second interseg-mental coordinating system runs parallel to that provided by cell 2 and takes the form of lateral connections between cells 3. It coordinates ventilation when cell 2 is inactive. E, excitatory synapse; I, inhibitory synapse (cf. Lewis *et al.,* 1973).

of KCl solutions and concluded that the brain and mesothoracic ganglia played important roles in controlling ventilation.

Command neurons themselves may act as CO_2 receptors. Some neurons in the ganglia of *Helix* and *Aplysia* and in the vertebrate respiratory center are depolarized by CO_2, while others may be hyperpolarized (Walker and Brown, 1970). CO_2 is believed to act by increasing the membrane conduc-tance to chloride ions through a change in pH. Whether a cell is depolarized or hyperpolarized by CO_2 may therefore depend on its resting potential and its chloride equilibrium potential. Pacemaker cells in *Aplysia* are generally much less responsive to CO_2 than other neurons (Chalazonitis, 1963; Brown and Berman, 1970). If insect neurons show the same responses to CO_2 as are found in *Aplysia* cells, then command interneurons and not pacemakers probably act as the sensing system which responds to low CO_2 levels (1–2%) and stimulates ventilation. This agrees with conclusions reached from the local applications of CO_2 to parts of the CNS (Miller, 1960).

4. The mechanism of the ventilatory pacemaker itself is not understood. Patterned endogenous output may arise from the interaction of several cells, as has been suggested for the vertebrate respiratory center (Salmoirhagi, 1963) and the insect flight system (Wilson, 1968). In some arthropods cardiac ganglia bursts are thought to arise from positive coupling between a number of spontaneously active cells (Lang, 1971). In other cases, burst production may be the property of a single neuron. Chen *et al.* (1971) have isolated single cells from *Aplysia* ganglia which produce spontaneous bursts similar to those in the intact ganglion. In crab ganglia Mendelson (1971) has recorded from single cells which produce sinusoidally, oscillating potential changes without generating impulses. These cells may drive antagonistic sets of motor neurons which cause the scaphognathites to beat and so bring about ventilation of the gill chamber. Many rhythmically active systems in arthropods resemble the insect ventilatory system and it is possible that the principles which apply both to the origin of activity and its distribution up and down the nerve cord will be found to be similar.

Intracellular records from the metathoracic neuropile region of crickets (Bentley, 1969) or from the cell bodies of motor and interneurons in locust metathoracic ganglia (Burrows, 1974) promise to provide much new information about insect ventilating systems. Some ventilatory neurons receive a barrage of excitatory synaptic potentials which depolarize the cell and produce a burst of impulses. A burst of inhibitory synaptic potentials may then hold the cell away from the firing threshold until the next cycle is initiated. Many cells are coupled to ventilation and there is much variety of activity among them.

5. Interactions between Ventilation and Other Systems

Several types of repetitive activity may take place at the same or related frequencies either because they are all driven by one oscillator, or because their separate oscillators have become coupled to a common frequency.

Bentley (1969) described many flight and leg motor neurons in cricket ganglia which were coupled to the ventilatory rhythm, and Hoyle (1964) made similar observations in locusts. Such coupling may sometimes occur because of extensive deafferentation or during hyperventilation; it was not usually seen in the metathoracic ganglia of almost intact *Schistocerca* which were pumping normally (Burrows, 1974). Coupling of wing and leg motor neurons does occur in hyperventilating intact beetles, but it is absent when ventilation is normal (Miller, 1971a). Auxiliary forms of ventilation (Section IV,C,2) may have evolved through the chance coupling of other motor systems to ventilation, together with a lowering of the thresholds for interaction. Hinkle and Camhi (1972) have shown how large fast units to the longitudinal muscles of the locust abdomen may be affected simultaneously by two oscillators. In flight they fire at the wingbeat frequency, but only during the latter part of each abdominal expiratory stroke. Ventilation seems to cause fluctuations in the threshold sensitivity of the neurons to the flight oscillator.

In crickets different forms of activity may take place at one of two common frequencies—a fast and a slow. The slow frequency may include activities such as walking, the chirp frequency of the male's calling song, and ventilation (Kutsch, 1969). Moreover, the responsiveness of the female cricket to the male's song also seems to vary according to the ventilatory cycle. Stout and Huber (1972) found an auditory interneuron which fired in response to male chirps only at certain phases of ventilation. For effective communication it would appear to be necessary for calling male and listening female to ventilate in phase with each other. In fact, ventilation in other insects can be paced by externally applied and repeated stimuli. For example, flashes of light at suitable frequencies can entrain ventilation in *Sphodromantis* (Miller, 1971b), while mechanical or electrical stimuli can entrain cockroach (Farley and Case, 1968) and dragonfly larva (Mill, 1970) ventilation.

6. Proprioceptive Input

The insect abdomen contains a variety of proprioceptors including campaniform sensilla, chordotonal organs, stretch receptors, and multiterminal neurons (Finlayson, 1968).

Rhythmic bursts of impulses in afferent nerves coincident with ventilation have been recorded in cockroaches (Farley *et al.*, 1967), *Dytiscus* and *Locusta* (Hughes, 1952), and in *Schistocerca* (Lewis, 1973). Kehler *et al.* (1970) recorded ventilatory bursts from ventral phasic mechanoreceptors in *Blaberus* and *Periplaneta;* they corresponded to longitudinal and to dorsoventral pumping movements. Florentine (1967, 1968) noted phasic activity in response to vibrations in laterally situated receptors in cockroaches, and suggested that they might be affected by sclerite movements during ventilation. Wing-stretch receptors are stimulated by ventilation in the locust (Möss, 1971). The scolopophorous organ of *Lethocerus* responds to ventilation (Barber and Pringle, 1966), and also to wing movements; a depth-detecting function has also been suggested for it (Miller, 1961).

The part played by receptors which respond phasically to ventilation is not understood. Their input may be important for tonic adjustments of frequency or amplitude, perhaps with respect to variations in the abdominal loading caused by food or developing eggs. Or it may be necessary for the maintenance of normal frequencies, as in the flight system. However, the ventilatory system also receives feedback from the tracheal tensions of CO_2 and oxygen to which the CNS responds directly.

The responsiveness of the cockroach (Farley and Case, 1968), dragonfly larva (Mill, 1970), and slowly ventilating mantis systems (Miller, 1971b) to phasic input, which can reset the oscillator, suggests that cycle-to-cycle adjustments are important in normal ventilation. Such control might be necessary, for example, when the abdomen is used as a rudder in flying locusts (Camhi, 1970; see also Pearson, 1972). The responsiveness of the system also enables the insect to flush out the tracheae in anticipation of sudden action.

7. Unidirectional Ventilation

When spiracles remain open, ventilation pumps air in and out of them tidally. In many insects spiracle valves make movements synchronized with ventilation—some open with inspiration, others with expiration. In a few species, such movements have been shown to pass air unidirectionally through the insect (e.g., Fraenkel, 1932; Bailey, 1954; Weis-Fogh, 1967). In an actively ventilating *Schistocerca,* air enters spiracles 1–4, which open during inspiration, and it leaves from spiracles 5–10, which open toward the end of expiration. In quiescent insects only spiracles 1 and 10 may function (Miller, 1960). Weis-Fogh (1967) has shown that about 80% (33 liters/kg/hour) of the total volume of air ventilated by the abdomen passes unidirectionally when at rest. However, when pumping is strongly stimulated during the first 5 minutes of flight, or just after a flight, the actual volume pumped unidirectionally remains about the same, but it now constitutes only 18% of the total volume. The spiracles therefore act much less efficiently as one-way valves.

Similar coordination of the spiracles occurs in *Sphodromantis* with spiracle 1 playing a major role. At rest, 15–20 liters of air/kg/hour enter spiracles 1, and about 45% is pumped unidirectionally, while 55% is tidal. However, when ventilation is stimulated, unidirectional ventilation becomes more efficient and 60–70 liters may pass through spiracle 1, of which 95% circulates unidirectionally and only 5% is tidal (Fig. 15) (P. L. Miller, unpublished).

From morphological evidence it was suggested that part of the air inspired by *Schistocerca* entered the dorsal orifice of spiracle 1, circulated round the head, and was then passed posteriorly down the ventral longitudinal trunks. Some tracheal anastomoses become blocked in the adult and this probably prevents short-circuiting of the flow outlined above (Miller, 1960). Direct

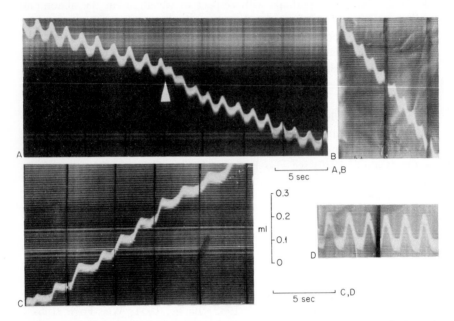

Fig. 15. Photographic records of direct measurements of air pumped through spiracles 1 in *Sphodromantis*. Each small oscillation corresponds to one ventilatory cycle while the slope of the line indicates the volume of air passed unidirectionally through the spiracle. (A) Resting ventilation, much of the ventilation is tidal and a small fraction is passed unidirectionally. At the arrow the insect was stimulated mechanically and for the two subsequent cycles all air pumped through spiracles 1 was unidirectional; (B) shows hyperventilation in which a high percentage of the total exchange through spiracle 1 is unidirectional (anterior-to-posterior); (C) reversed unidirectional ventilation in which the air is passed from posterior to anterior; spiracle 1 is mainly expiratory and only a small fraction of the exchange is tidal; (D) exchange through spiracle 1 when the valves are mechanically held open; all exchange is tidal.

measurements indicate that a similar circulation takes place in *Sphodromantis*. In many Mantodea the head and forelegs are separated from spiracle 1 by the lengthened prothorax so that the problem of ensuring adequate cephalic ventilation may be greater. A small thermistor, heated 1°C above ambient, was sealed into various prothoracic tracheae and the ventilatory phase which produced maximum cooling was noted (Fig. 16). When ventilation is strongly unidirectional, only the dorsal orifice of spiracle 1 (Fig. 4B) opens with inspiration; air is sucked into trachea 1 and passes anteriorly to the head and foreleg air sacs. Expiration drives air from the anterior sacs down trachea 2, bypassing spiracle 1 which is firmly closed throughout

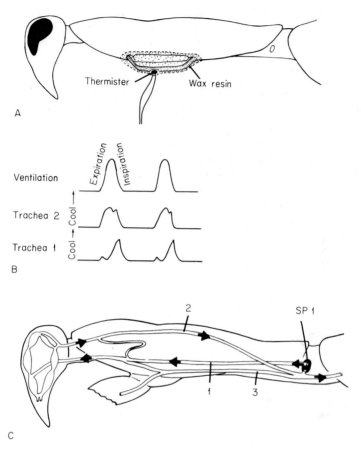

Fig. 16. Measurements of the route followed by unidirectional ventilating airstreams in *Sphodromantis*. (A) shows a thermistor inserted into a prothoracic trachea; (B) indicates the form of thermistor records from trachea 2 (leading *from* the head) and trachea 1 (leading *to* the head) during ventilation; (C) indicates the routes followed by airstreams in the principal longitudinal tracheae in the prothorax.

expiration. It then passes posteriorly to the abdomen (P. L. Miller, unpublished).

Unidirectional ventilation is found in aquatic animals such as some crustacea, molluscs, and fish. Among terrestrial species it is known only in birds and insects. Recent measurements with heated thermistors have suggested the routes followed by circulating air in birds' lungs (Bretz and Schmidt-Nielsen, 1971). Synchronized spiracle movements have been seen in Orthoptera, Dictyoptera, Odonata, Hymenoptera, and Coleoptera. Unidirectional ventilation is probably therefore a widespread phenomenon in larger insects and its advantages may be twofold: It may improve the efficiency of gaseous exchange by reducing or abolishing the dead space and by ensuring adequate ventilation of the head; secondly, it may reduce water loss (Buck, 1962).

8. The Reversal of Unidirectional Ventilation

Unidirectional airstreams normally pass in a posterior direction through insects. In some genera, e.g., *Arphia, Chortophaga, Dissosteira,* and *Periplaneta,* the direction of the current can be temporarily reversed (McArthur, 1929; McGovran, 1932). In *Sphodromantis* and *Schistocerca* temporary reversals may occur after treatment with 10% CO_2 or when anterior tracheae are blocked (Fig. 15C). Reversal is sometimes achieved within one ventilatory cycle in *Sphodromantis* and abdominal spiracles come to open with inspiration while spiracle 1 opens with expiration (see Section V,D). The volume of air pumped may be as great as normal and 90% may be unidirectional (P. L. Miller, unpublished).

9. The Compression Phase

Expiratory spiracles usually open toward the end of the expiratory stroke. During the earlier part, all spiracles may be closed and hemolymph movements inflate arthrodial membranes, although no significant compression of tracheal gases occurs (Wigglesworth, 1965). Watts (1951) recorded expiratory pressures of up to 10 mm Hg in *Schistocerca,* while Weis-Fogh (1967) measured up to 15 mm Hg during expiration and —4 mm Hg during inspiration, in hyperventilating locusts.

In *Blaberus giganteus* abdominal spiracles open sometimes for only 40–60 mseconds at the end of expiration and they emit a blast of air. This may at times have significance for the distribution of noxious chemicals. A compression phase may ensure that air is expelled from a larger proportion of the tracheal system by allowing pressures to equalize before the spiracles open (McCutcheon, 1940; Hrbráček, 1949). It may also have significance for hemolymph circulation.

D. Gas Exchanges during Flight

1. Tracheal Specializations

The metabolic rates of some flying insects are not exceeded by those of any other organisms. Metabolism during flight is entirely aerobic so that high rates of oxygen delivery and of CO_2 and water removal are essential. The transition between rest and flight is usually abrupt, although in some insects flight is preceded by a period of warming up. Systems must therefore be able to switch from low to high rates of delivery within a wingbeat cycle.

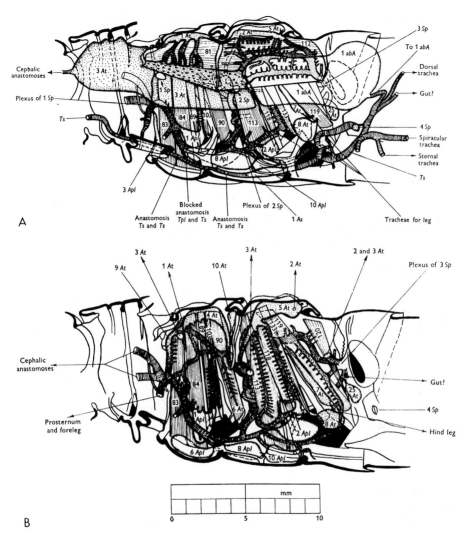

Fig. 17. The tracheal anatomy of the right side of the thorax of *Schistocerca.*
(A) Longitudinal section in the plane of symmetry showing the large central airsacs
(3 *At*) which connect directly to flight-muscle tracheae and spiracles, and the large
abdominal sacs (unstippled, I *abA*) with no direct connections to the flight system;
(B) the same after removal of the dorsal longitudinal muscles showing the primary
and secondary supplies to many of the flight muscles. (From Weis-Fogh, 1964a.)

The tracheal system is much modified to deliver oxygen at high rates to
active flight muscles, but the modifications do not necessarily obscure the
basic metameric pattern (Whitten, 1968).

Weis-Fogh (1964a) regards the path between spiracle and tracheole as divisible into three parts: a *primary* trachea, ventilated in flight, passes either axially through the center of the muscle (centro-radial plan), or down one side (latero-radial plan), or is expanded into a broad sac along one side of the muscle (latero-linear plan). The primary trachea normally expands into an air sac beyond the muscle and the sacs of neighboring muscles may be confluent (Miller, 1962). *Secondary* tracheae leave the primary at right angles. In the first subalar muscle of *Aeshna* they occur every 21 μm and taper from a diameter of 7 to one of 1.5 μm. In the same muscle each secondary gives off about 25 *tertiary* branches which end in tracheoles encircling the muscle fibers. Some aspects of the system in *Schistocerca* are shown in Fig. 17. In large insects such as *Lethocerus,* secondary tracheae terminate beyond the muscle surface in small air sacs and thus can be ventilated, but in most insects gas movements in secondary and tertiary tubes depend on diffusion alone (Section IV,A). In many insects the tracheoles indent the muscle fibers (Section III,B).

2. Autoventilation

The term means the ventilation of flight muscle tracheae directly by flight movements. It takes place mainly as the result of dorsoventral movements of the nota and lateral movements of the pleura with each wing beat. Weis-Fogh (1967) demonstrated that such movements do bring about a considerable exchange of air between thoracic sacs and the outside air through open spiracles. A flying *Schistocerca* increases its oxygen uptake 24 times, whereas abdominal ventilation increases by only 4 times as compared with resting values (Table III). Locusts with abdominal ventilation abolished are still able to fly. Weis-Fogh measured the volume changes during a single wingbeat (Fig. 18A) and from this calculated the potential airflow through open thoracic spiracles (Fig. 18B). A total of 950 liters/kg/hour could be exchanged in strong flight, and 750 liters in steady flight. About one third of this is believed to be available for pterothoracic ventilation, some of the remainder affecting abdominal air sacs. Weis-Fogh also measured the pressure changes caused by wingbeats and by abdominal pumping in the thorax and found them to be small, probably because the spiracles were open. He concluded that autoventilation exchanges about 250 liters with the outside air, while abdominal pumping can exchange 144 liters in steady flight. Of the 144 liters, 70 are pumped in and out of pterothoracic spiracles, 44 are pumped tidally through other regions, and the remaining 33 are pumped unidirectionally through the insect (Table III).

The pterothoracic tracheae are considerably isolated from the remainder of the system in *Schistocerca* (Weis-Fogh, 1964a), and in other insects some anastomoses being occluded. This probably makes autoventilation more effective in flying insects which do not depend on an abdominal pump.

TABLE III

MEASUREMENTS OF THE FLIGHT VENTILATION SYSTEM IN *Schistocerca gregaria*[a]

Measurement	Method	Values
Metabolic rate	Measurement of heat production, of oxygen consumption, etc.	65 kcal/kg/hour or 13.7 liters O_2/kg/hour
Volume ventilated by the abdomen in steady flight (after first 10 minutes)	Abdomen enclosed in a small plethysmograph joined to a spirometer	144(39–245) liters air/kg/hour
Volume of air ventilated in unidirectional anterior-to-posterior flow, during steady flight	Spiracles 4–9 waxed: spiracle 10 intubated and output led to capillary with kerosene droplet	33 liters/kg/hour
Volume change in pterothorax caused by flight movements	Dead locust (HCN); thorax in sealed vessel; thoracic spiracles and soft cuticle waxed; abdomen in separate vessel and air movement through abdominal spiracles measured when wings were moved by a lever	25 μl/stroke = 760 liters/kg/hour
Pressure changes in pterothorax resulting from abdominal ventilation during flight	Hypodermic needle in hemolymph adjacent to air sacs; pressure transducer	2.22–3.7 mm Hg (peak-to-peak)
Pressure changes in pterothorax resulting from flight movements		1.48–1.85 mm Hg (during abdominal expiration) 0.75–1.11 mm Hg (during abdominal inspiration)
Pressure changes in air sacs resulting from abdominal ventilation during recovery from CO_2		+14.8 mm Hg (expiration); −3.7 mm Hg (inspiration)
Volume of pterothoracic tracheal system in a standard locust	Injection	100–150 μl
Actual volume of air ventilating flight-muscle tracheae in flight resulting from flight movements		250 liters/kg/hour
Total volume of air ventilating the flight muscle tracheae in flight		320 liters/kg/hour (250 from thoracic movements and 70 from abdominal movements)
Gas content of pterothoracic air sacs in flight	Samples extracted from spiracle 2 and analyzed	10–15% oxygen; 4.8–7.9% CO_2

[a] From Weis-Fogh, 1964b, 1967.

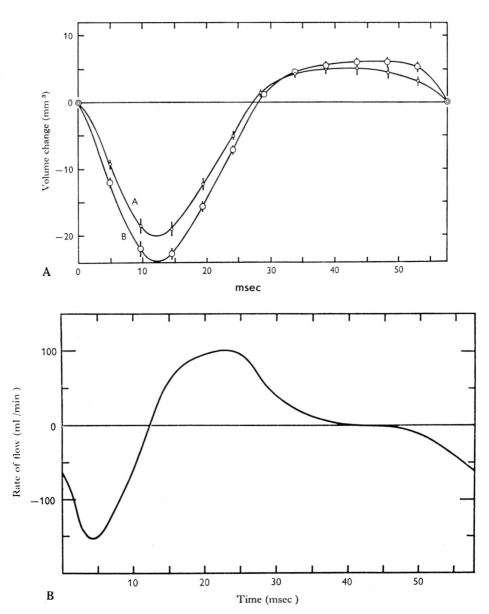

Fig. 18. (A) The change in thoracic volume caused by the wing movements of a standard *Schistocerca* during a wingstroke. Two plots are shown for slightly differing amplitudes of wing movement; (B) the potential rate of air flow through thoracic spiracles of *Schistocerca* caused by the wing movements during a standard wingstroke. Negative values indicate that air is being sucked in through the spiracles. (From Weis-Fogh, 1967.)

Spiracles which supply flight muscles normally remain open in flight. Spiracle 2 in many Acrididae has a lever system actuated by tension in flight muscles which causes wider gaping of the valves. In flying *Schistocerca* spiracles 1 and 4–10 remain synchronized with ventilation and they probably account for the 33 liters pumped unidirectionally. However, spiracle 1 closes for only a small fraction of the ventilatory cycle (Miller, 1960). Abrupt sensory stimulation may cause reflex spiracle opening, often before flight movements appear, in dragonflies (Miller, 1962), mosquitoes (Krafsur *et al.*, 1970), and other insects.

Adult dragonflies pump weakly in flight with the abdomen and depend to a large extent on efficient autoventilation (Weis-Fogh, 1967). Wingbeats cause relatively large lateral movements of the pleura which are joined loosely to each other, anterior to the forewings, at the dorsal carina. Pleural movements take place twice per wingbeat cycle and in *Aeshna cyanea,* with a wingbeat amplitude of 57°, produce a potential rate of ventilation of 800 liters/kg/hour, or with an amplitude of 90°, a ventilation of 1600 liters. Efficient pterothoracic ventilation is required in dragonflies since they have high metabolic rates but no indenting tracheoles (Sections III,B and IV,A).

Many Dictyoptera, Orthoptera, Hemiptera, Isoptera, Coleoptera, and Lepidoptera depend on autoventilation for much of their oxygen supply in flight. In Coleoptera two types of supply system have been recognized (Miller, 1966b). In the first, exemplified by Scarabaeidae and Buprestidae, air sacs are abundant in the pterothorax and strong abdominal pumping, at least in larger species, supplements autoventilation. Some of the most massive flying insects known (e.g., *Goliathus regius*) depend on this system. In the second system, exemplified by Cerambycidae, there are few or no air sacs and even in some large species abdominal pumping is weak or absent in flight. Flight autoventilation is supplemented by draught or ram ventilation in which air is ducted through large tracheae as a result of a Bernoulli effect. In *Petrognatha,* two pairs of giant rigid tracheae pass through the metathorax between spiracles 2 and 3 (Fig. 19). Large numbers of deformable secondary tracheae leave them and invade the flight muscles. Direct measurements with a small manometer show that in a wind speed of 5 meters/second, 3900 liters of air/kg/hour are ducted through the four giant tracheae. Metanotal movements acting on the secondaries probably pump air in and out of the primaries: these alone must be adequate for short periods to enable the beetle to become airborne. Moreover, tethered flight continues in the laboratory for a few seconds after the wind tunnel is turned off.

In Diptera and Hymenoptera, the volume changes produced by flight movements are probably too small to bring about effective exchanges between the tracheae and the atmosphere, although they may make a significant contribution to the movement of gases and hemolymph within the

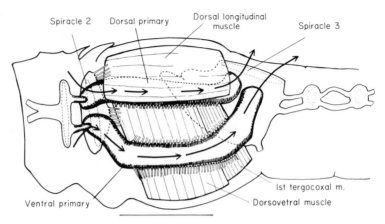

Fig. 19. Medial section of the pterothorax of *Petrognatha gigas* to show the two pairs of giant primary tracheae which run between spiracles 2 and 3 and through which air is ducted during flight, providing a form of draught ventilation. Horizontal scale, 10 mm. (Based on Miller, 1966b.)

pterothorax (Weis-Fogh, 1964b, 1967). In Hymenoptera, the abdomen acts as an efficient telescoping pump with a high frequency and large amplitude. During flight it probably ventilates pterothoracic air sacs effectively, although wasps with the abdomen removed can still "fly" (Krogh, 1941). Active unidirectional ventilation occurs in the mesothorax of flying bees (Bailey, 1954).

Most Diptera do not have a strong abdominal pump, although ventilatory movements may appear in some species (e.g., Krafsur *et al.,* 1970). A system of pterothoracic air sacs with wide connections and many anastomoses (Fig. 3) probably enables most flies to depend on diffusion alone for their oxygen supply in flight (Weis-Fogh, 1964b; cf. Wigglesworth, 1963; Whitten, 1968, 1972).

V. Control of the Spiracles

A. INNERVATION

Anatomical studies have indicated that median nerves innervate spiracle muscles in Odonata, Orthoptera, Dictyoptera, Plecoptera, and Lepidoptera (see Horridge, 1965). Physiological evidence is available from dragonflies, cockroaches, mantids, locusts, and moths (Miller, 1966a). Typically, a median nerve leaves from a ganglion mid-dorsally and splits into transverse nerves which run to the spiracles and elsewhere. Transverse nerves are commonly joined by branches from the posterior ganglion. Plotnikova (1968,

1969) stained with methylene blue some of the neurons in *Locusta* ganglia whose axons run in median nerves. A vegetative or trophic function has been ascribed to the median nerve system (Voskresenskaya, 1968), and neurosecretory axons have been identified running in median and transverse nerves (Brady and Maddrell, 1967; Smalley, 1970). Finlayson (1966) described a "sensory" nerve cell lying on a nerve trunk and sending two processes into the CNS and one into a spiracle muscle in the tsetse fly, while Whitten (1963) suggested that a similarly situated neuron in the fly might act as a CO_2 receptor. Peripherally situated neurosecretory cells (possibly derived from sensory cells) occur in connection with the abdominal transverse nerves of *Carausius,* and their spontaneous activity has been recorded (Finlayson and Osborne, 1968, 1970). Bhatia and Tonapi (1968) have suggested that cockroach spiracles are influenced by neurosecretory substances.

The innervation of the spiracles of *Schistocerca* is summarized in Fig. 20. The opener and closer muscles of spiracles 1 and 3 each receive two motor axons which run in the appropriate median nerves. The axons themselves split at the division of the median nerve into left and right transverse nerves, as is known in other species (Case, 1957). The opener muscle of spiracle 1 receives an additional excitatory axon from the mesothoracic ganglion (Miller, 1965). Impulses in the prothoracic nerves to the opener each produce small 1–5 mV depolarizations which facilitate readily. The mesothoracic axon to the opener produces larger 10–20 mV depolarizations which

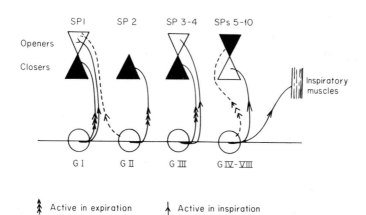

Fig. 20. A summary of the innervation of the spiracle muscles of *Schistocerca gregaria.* Spiracle muscles active during expiration are shown blacked in. Motor nerves active with expiration are shown with triple arrows. In each spiracle, the closer muscle is innervated by the median nerve of the appropriate segment (continuous lines). Median nerves also supply the opener muscles of spiracles 1, 3, and 4; in the abdomen the median nerves additionally supply the dorsoventral inspiratory muscles. Lateral nerves supplying opener muscles are shown as broken lines. Sp, spiracle; GI, II, and III, thoracic ganglia; GIV–VIII, abdominal ganglia.

show less facilitation, and many fibers are triply innervated. In *Locusta* the innervation of the opener of spiracle 1 is the same, but the action of the axons is slightly different, the prothoracic axons producing larger depolarizations than the mesothoracic (Miller, 1971d).

The closer muscle of spiracle 2 of *Schistocerca* receives two excitatory axons from the mesothoracic median nerve. They produce 10–60 mV depolarizations and Hoyle (1959) has termed one axon "fast" and the other "slow." All fibers appear to be doubly innervated. In spiracles 4–10 the median nerve in each segment supplies two axons to the closer muscle, while the opener muscle is supplied from the anterior lateral nerve trunk (Lewis, 1973).

In *Blaberus* the innervation of abdominal spiracles is similar to that in *Schistocerca* and normal activity is controlled by the two unpaired axons to the closer and by a single paired axon to the opener. In spiracle 1, however (which has only a closer muscle), the two excitatory median nerve axons are joined by a third paired inhibitory axon, also from the prothoracic ganglion, which is believed to accelerate muscle relaxation when the excitatory axons become silent (Miller, 1969). An inhibitory nerve has also been reported in the spiracles of *Hyalophora* (van der Kloot, 1963).

Mechanoreceptors occur in the valves and in the vicinity of the spiracles. In many insects their stimulation has been shown to cause reflex spiracle closing. In the cockroach (Case, 1957) and dragonfly (Miller, 1962) afferent neurons run from bristle receptors to the ganglion posterior to that from which the spiracular motor nerves emerge.

B. THE ACTION OF CO_2 AND HYPOXIA

The local action of CO_2 on spiracles has been described by many authors (e.g., Hazelhoff, 1926; Schreuder and de Wilde, 1952; Case, 1957; Krafsur and Graham, 1970). The failure to find sensory structures associated with the spiracle muscle (Beckel, 1958), together with the experiments of Hoyle (1960), have demonstrated the direct action of CO_2 on the muscle. CO_2 lowers muscular tension in spiracle 2 of *Schistocerca* and the valve opens as a result of elasticity, despite a maintained discharge of action potentials from the CNS. CO_2 may act on the muscle membrane by increasing the conductance to chloride ions, as in neurons. This would tend to repolarize a partially depolarized membrane and bring about relaxation. In *Hyalophora,* van der Kloot (1963) produced results in agreement with this hypothesis, but according to Hoyle (1960) the junctional potentials are reduced in *Schistocerca* and the membrane may be partially depolarized, at least in high CO_2 concentrations. Hoyle suggested that CO_2 might act both at the membrane and on the coupling mechanism (see Aidley, 1967).

In insects with one-muscle spiracles whose movements are not closely linked to ventilation, the peripheral action of CO_2 probably plays an important part in control of the valve (e.g., dragonflies and Diptera). However, CO_2 also acts on the ganglia, not only to stimulate ventilation, but also to adjust the motor discharge to the spiracle muscles. In spiracle 1 of *Schistocerca*, CO_2 causes the three axons to the opener muscle to discharge at appropriate moments in the ventilatory cycle (Section V,D), or, in the absence of ventilation, CO_2 reduces the tonic discharge to the closer muscle and accelerates that to the opener. In *Sphodromantis* and *Blaberus*, CO_2 acts initially to open the valve by local action on the muscle; it then reaches the CNS and causes a reduction in the tonic discharge which may allow the valve to open further. Finally, it stimulates ventilation which causes synchronized bursts of impulses to appear in the motor nerves to the spiracles (cf. Hazelhoff, 1926). Although the tonic discharge can be suppressed, the motor neurons are still able to produce expiratory bursts.

In dragonflies the tonic discharge to the spiracles is more readily reduced by oxygen lack than by CO_2. Ten percent oxygen in nitrogen causes a reduction, while about 2% may suppress the discharge altogether (Miller, 1964a). The spiracles of *Hyalophora* are directly affected by CO_2, whereas oxygen lack causes them to open by acting on the CNS (see Section IV,B) (Burkett and Schneiderman, 1967). By causing a reduction of the tonic discharge, hypoxia enables the spiracle muscle to be more sensitive to CO_2. A priming effect of hypoxia on the spiracle sensitivity to CO_2 has been observed in several species (Case, 1956; Schneiderman, 1960; Krafsur and Graham, 1970), and it may be brought about by this means.

Several other factors, such as starvation, age, water balance, temperature, and wounding (Miller, 1964a; Krafsur, 1971) can affect spiracle control, possibly by acting through alterations of the tonic discharge. Different species show different degrees of spiracle control (Krafsur, 1971) and not all spiracles on one insect have the same threshold; for example, resting fleas use only spiracles 3 and 10 (Wigglesworth, 1935), and adult mosquitoes use only 1 and 2 (Krafsur *et al.*, 1970). Quiescent dragonflies, locusts, mantids, and cockroaches may use only the first and last spiracles or a variable number of those in between. Differing rates of tonic discharge may again be responsible for some of this variation.

C. SPIRACULAR RESPONSES TO HUMIDITY AND WATER BALANCE

The importance of spiracle valves for reducing water loss suggests that humidity and water balance play a part in their control, but this aspect has been little studied. Tsetse flies are known to adjust their spiracle control according to the state of their water reserves and possibly also in response to the prevailing humidity (Bursell, 1957, 1970). At high temperatures, how-ever, they open their spiracles temporarily to encourage evaporative cooling (Edney and Barrass, 1962). Mosquitoes respond to low humidities by in-creasingly strict spiracle control: the valves open less often and less fully (Krafsur, 1971). Spiracle control is also stricter in starved or old mosquitoes. The valves may be controlled by the water balance of the insect as well as responding to humidity by means of receptors (Krafsur, 1971). The tonic discharge to the spiracles of dragonflies rises in frequency when the insects are deprived of water, or are perfused with hypertonic solutions. The CO_2 threshold of the spiracles rises in consequence (Miller, 1964a).

In *Locusta* the rate of transpiration is reduced at low humidities by a decrease of the frequency and amplitude of ventilation. Loveridge (1968) has shown that normal water losses through the spiracles amount to 2.9 mg/gm/hour, and these increase to 6.2–8.0 mg with hyperventilation. Thus, ventilation is a major cause of water loss, and its reduction can achieve a saving of 5.0 mg/gm/hour. Central neurons can apparently adjust the threshold of their response to CO_2 according to the humidity. The occur-rence of periodic ventilation in teneral *Schistocerca* was described as a water-saving mechanism by Hamilton (1964) (Section IV,C,1).

Structural modifications may also help to reduce water losses from spira-cles, and these are most highly developed in some xerophilous arthropods (Lewis, 1963; cf. Delye, 1965). Spiracles may be sunk, or they may lie con-cealed under sclerites. Abdominal spiracles of many Coleoptera and some Hemiptera (Scudder, 1963) lie under the elytra or wings, which fit tightly along the sides and provide an opening only at the tip of the abdomen (Ahearn, 1970). These mechanisms help to stabilize an air layer above the spiracle opening and so reduce water losses by convection; they are compar-able to the protected stomata of desert plants. At 25°C the respiratory water loss in the desert tenebrionid, *Eleodes armata,* is only 2% of the total, al-though at 40°C it amounts to 20% of the total (Ahearn, 1970).

D. Spiracle Movements Coordinated with Ventilation

Unidirectional ventilation, resulting from synchronized spiracle activity, has been described in Section IV,C,7. Synchronized activity is brought about by coordinating interneurons, possibly similar to those postulated in Section IV,C,4, which run from a ventilation center to the spiracle motor neurons. They cause the motor neurons to produce bursts of impulses or they inhibit their tonic activity at appropriate phases of the ventilatory cycle. In *Schistocerca* spiracle 1, a high-frequency burst slams the spiracle shut at the onset of expiration; both closer and opener neurons may continue to fire throughout expiration, but at the start of inspiration the valve swings open as the closer impulses abruptly cease. The mesothoracic axon may continue to fire throughout inspiration. This pattern of activity, summarized in Fig. 21, may be due to the activity of two or three coordinating interneurons (Miller, 1965, 1967). One or two are excitatory and fire during expiration (some degree of coupling between the firing of spiracles 1 and 2 suggests that their motor neurons are driven by a common interneuron), while the third is inhibitory and may act with inspiration. When all are inactive, a tonic discharge results from pacemaker activity possibly within the spiracle motor neurons.

Irregular bursts can occur in the spiracle nerves of an isolated prothoracic segment of *Sphodromantis* in response to CO_2. Activity synchronized with ventilation in the intact insect can be explained by postulating a single coordinating interneuron which triggers bursts at appropriate moments in the cycle, but there is no direct evidence for this unit (Miller, 1971b).

Reversal of the ventilating airstream in *Sphodromantis* is caused by a change in coupling of the spiracles to the ventilatory pump (Section IV,C,8). In spiracle 1, reversal coincides with attenuation and sometimes with disappearance of the expiratory bursts to the closer muscle, whereas inspiratory firing is accelerated and the valve now closes only with inspiration (Fig. 22).

E. Independent Action by Denervated Spiracles

In several species, denervated spiracles have been observed to close spontaneously (Hazelhoff, 1926; Wigglesworth, 1935, 1941; Beckel and Schneiderman, 1957; Miller, 1960; Hoyle, 1961). In most cases a continuous tension

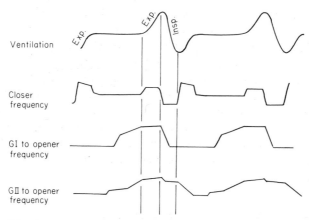

Fig. 21. The frequencies of action potentials in the nerves to the opener and closer muscles of spiracle 1 of *Schistocerca* during a ventilatory cycle. The different contributions of the pro- (GI) and mesothoracic (GII) nerves to the activity of the opener before and during inspiration should be noted. (Based on Miller, 1965.)

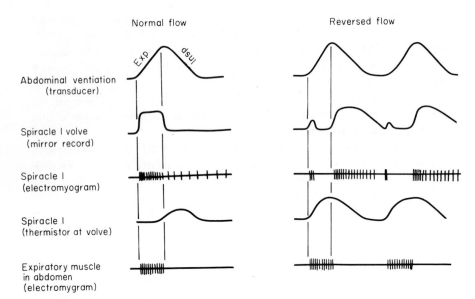

Fig. 22. The activity of spiracle 1 of *Sphodromantis* during normal unidirectional and reversed ventilation. Second line, mirror records from the valve with closing upwards. Fourth line, the thermistor records the timing of air movements through the valve during inspiration, with normal flow, and during expiration with reversed flow.

is exerted by a closer muscle against a cuticular spring, although CO_2 can lead to a reduction of tension and temporary opening of the spiracle. Hoyle (1961) suggested that closure in the denervated spiracle 2 of *Schistocerca* is caused by potassium ions in the hemolymph which depolarize the muscle, and produce a contracture. Spiracle closing may follow denervation after a delay, possibly due to water loss through the open spiracle, or to a change in excitability of the muscle (Horridge, 1965).

In *Hyalophora* pupae, van der Kloot (1963) recorded slow spontaneous potential changes and spikes in the denervated spiracle muscle. The repolarizations reached higher values after treatment with CO_2 and these allowed the muscle to relax and the valve to open. Van der Kloot believed the activity to be myogenic and he suggested that in the intact spiracle, control is exercised by the interaction of excitatory and inhibitory nerves with local myogenic activity.

F. NONRESPIRATORY FUNCTIONS OF SPIRACLES

Several insects use spiracles for ejecting and sometimes frothing noxious defensive fluids (Roth and Eisner, 1962). Quinones are expelled from spiracle 4 in the cockroach *Diploptera punctata*, and this spiracle, which possesses both opener and closer muscles, remains closed except when the insect is attacked. *Blaberus* and *Byrsotria* have tracheal swellings like those of *Diploptera* near spiracle 4, but their function is unknown. *Romalea* and *Poekilocerus* froth defensive secretions from spiracle 1 and some Arctiid moths may do the same. The Madagascar cockroach *Gromphadorhina* hisses through spiracle 4 by means of violent abdominal expiratory movements during which other spiracles remain closed (Guthrie and Tindall, 1968; M. Burns, personal communication). Male *Gromphadorhina* also hiss during courtship and an odor of unknown function is released. Queen bees were thought to pipe through spiracles, but piping is now known to be produced by wing-muscle vibrations (Simpson, 1964). Pheromones may be released from the tracheal system via spiracles, but no definite examples seem to be known.

Acknowledgments

I am grateful to Dr. M. Burns, Dr. M. Burrows, Dr. R. A. Crowson, Mr. R. Hustert, Dr. G. W. Lewis, Dr. P. J. Mill, Mrs. P. Mills, and Dr. J. Noble-Nesbitt for permission to make use of their unpublished work and comments.

References

Ahearn, G. A. (1970). *J. Exp. Biol.* **53**, 573.
Aidley, D. J. (1967). *Advan. Insect Physiol.* **4**, 1.
Allegret, P., and Barbier, R. (1966). *Bull. Soc. Zool. Fr.* **91**, 31.
Altman, P. L., and Dittmer, D. S. (1971). "Respiration and Circulation," Biological Handbooks. Fed. Amer. Soc. Exp. Biol., Bethesda, Maryland.
Amos, W. B., and Miller, P. L. (1965). *Entomologist* **98**, 88.
Andersen, S. O., and Weis-Fogh, T. (1964). *Advan. Insect Physiol.* **2**, 1.
Bahadur, J., and Bhatnagar, B. S. (1971). *Acta Entomol. Bohemoslov.* **68**, 15.
Bailey, L. (1954). *J. Exp. Biol.* **31**, 589.
Barber, S. B., and Pringle, J. W. S. (1966). *Proc. Roy. Soc., Ser. B* **164**, 21.
Barnhart, C. S. (1961). *Ann. Entomol. Soc. Amer.* **54**, 177.
Beament, J. W. L. (1964). *Advan. Insect Physiol.* **2**, 67.
Beaulaton, J. (1964). *J. Microsc. (Paris)* **3**, 91.
Beckel, W. E. (1958). *Proc. Int. Congr. Entomol., 10th, 1956* **2**, 87.
Beckel, W. E., and Schneiderman, H. A. (1957). *Science* **126**, 352.
Beinbrech, G. (1970). *Zool. Anz., Suppl.* **33**, 401.
Bentley, D. R. (1969). *J. Insect Physiol.* **15**, 677.
Bhatia, S. S., and Tonapi, G. T. (1968). *Experientia* **24**, 1224.
Bordereau, C. (1971). *Arch. Zool. Exp. Gen.* **112**, 33.
Brady, J., and Maddrell, S. H. P. (1967). *Z. Zellforsch. Mikrosk. Anat.* **76**, 389.
Bretz, W. L., and Schmidt-Nielsen, K. (1971). *J. Exp. Biol.* **54**, 103.
Brocher, F. (1931). *Arch. Zool. Exp. Gen.* **74**, 25.

Brockway, A. P., and Schneiderman, H. A. (1967). *J. Insect Physiol.* **13**, 1413.

Brosemer, R. W., Vogell, W., and Bücher, T. (1963). *Biochem. Z.* **338**, 854.

Brown, A. M., and Berman, P. R. (1970). *J. Gen. Physiol.* **56**, 543.

Bücher, T. (1966). *In* "Aspects of Insect Biochemistry" (T. W. Goodwin, ed.), p. 15. Academic Press, New York.

Buck, J. B. (1958). *Biol. Bull.* **114**, 118.

Buck, J. B. (1962). *Annu. Rev. Entomol.* **7**, 27.

Buck, J. B., and Friedman, S. (1958). *J. Insect Physiol.* **2**, 52.

Buck, J. B., and Keister, M. L. (1955). *Biol. Bull.* **109**, 144.

Buck, J. B., and Keister, M. L. (1956). *Physiol. Zool.* **29**, 137.

Buck, J. B., and Keister, M. L. (1958). *J. Insect Physiol.* **1**, 327.

Burkett, B. N., and Schneiderman, H. A. (1967). *Science* **156**, 1604.

Burkett, B. N., and Schneiderman, H. A. (1968). *Nature (London)* **217**, 95.

Burrows, M. (1974). *Phil. Trans.* (in press).

Bursell, E. (1957). *Proc. Roy. Entomol. Soc. London, Ser. A* **32**, 21.

Bursell, E., ed. (1970). "An Introduction to Insect Physiology." Academic Press, New York.

Camhi, J. M. (1970). *J. Exp. Biol.* **52**, 519 and 533.

Case, J. F. (1956). *Physiol. Zool.* **29**, 163.

Case, J. F. (1957). *J. Insect Physiol.* **1**, 85.

Chalazonitis, N. (1963). *Ann. N.Y. Acad. Sci.* **109**, 451.

Chapman, R. F. (1969). "The Insects, Structure and Function." Amer. Elsevier, New York.

Chen, C. F., von Baumgarten, R., and Takeda, R. (1971). *Nature (London)* **233**, 27.

Church, N. S. (1960). *J. Exp. Biol.* **37**, 171 and 186.

Clarke, K. U. (1957). *Proc. Roy. Entomol. Soc. London, Ser. A* **32**, 67.

Clarke, K. U. (1958). *Proc. Int. Congr. Entomol., 10th, 1956* **2**, 205.

Connell, J. U., and Glynne-Jones, G. D. (1953). *Bull. Entomol. Res.* **44**, 291.

Davey, K. G., and Treherne, J. E. (1964). *J. Exp. Biol.* **41**, 513.

Delye, G. (1965). *Insectes Soc.* **12**, 285.

du Buisson, M. (1924). *Bull. Cl. Sci., Acad. Roy. Belg.* [5] **10**, 373 and 645.

Edney, E. B., and Barrass, R. (1962). *J. Insect Physiol.* **8**, 469.

Edwards, G. A., Ruska, H., and de Harven, E. (1958). *J. Biophys. Biochem. Cytol.* **4**, 107.

Edwards, L. J. (1970). *Ann. Entomol. Soc. Amer.* **63**, 627.

Edwards, L. J., and Patton, R. L. (1967). *J. Insect Physiol.* **13**, 1333.

Farley, R. D., and Case, J. F. (1968). *J. Insect. Physiol.* **14**, 591.

Farley, R. D., Case, J. F., and Roeder, K. D. (1967). *J. Insect Physiol.* **13**, 1713.

Finlayson, L. H. (1966). *J. Insect Physiol.* **12**, 1451.

Finlayson, L. H. (1968). *Symp. Zool. Soc. London* **23**, 217.

Finlayson, L. H., and Osborne, M. P. (1968). *J. Insect Physiol.* **14**, 1793.

Finlayson, L. H., and Osborne, M. P. (1970). *J. Insect Physiol.* **16**, 791.

Florentine, G. J. (1967). *J. Insect Physiol.* **13**, 215.

Florentine, G. J. (1968). *J. Insect. Physiol.* **14**, 1577.

Fraenkel, G. (1932). *Z. Vergl. Physiol.* **16**, 371.

Fraenkel, G., and Herford, G. V. B. (1938). *J. Exp. Biol.* **15**, 266.

Guthrie, D. M., and Tindall, A. R. (1968). "The Biology of the Cockroach." Arnold, London.

Hamilton, A. G. (1964). *Proc. Roy. Soc., Ser. B* **160**, 373.

Hamilton, M. A. (1931). *Proc. Zool. Soc. London* p. 1067.

Hartley, J. C. (1971). *J. Exp. Biol.* **55**, 165.

Hassan, A. A. G. (1944). *Trans. Roy. Entomol. Soc. London* **94**, 103.

Hazelhoff, E. H. (1926). "Regeling der Ademhaling bij Insecten en Spinnen." Proefschrift, Rijks-Universiteit de Utrecht.

Herford, G. M. (1938). *J. Exp. Biol.* **15**, 327.

Hinkle, M., and Camhi, J. M. (1972). *Science* **175**, 550.

Hinton, H. E. (1947). *Trans. Roy. Entomol. Soc. London* **98**, 449.

Hinton, H. E. (1953). *Trans. Soc. Brit. Entomol.* **11**, 209.

Hinton, H. E. (1954). *Proc. Roy. Entomol. Soc. London, Ser. A* **29**, 135.

Hinton, H. E. (1960a). *J. Insect Physiol.* **4**, 176.

Hinton, H. E. (1960b). *Quart. J. Microsc. Sci.* **101**, 313.

Hinton, H. E. (1960c). *Phil. Trans. Roy. Soc. London, Ser. B* **243**, 45.

Hinton, H. E. (1961). *J. Insect Physiol.* **7**, 224.

Hinton, H. E. (1962). *Quart. J. Microsc. Sci.* **103**, 243.

Hinton, H. E. (1963). *J. Insect Physiol.* **9**, 121.

Hinton, H. E. (1967a). *Aust. J. Zool.* **15**, 947.

Hinton, H. E. (1967b). *J. Insect Physiol.* **13**, 647.

Hinton, H. E. (1968a). *Advan. Insect Physiol.* **5**, 65.

Hinton, H. E. (1968b). *J. Insect Physiol.* **14**, 145.

Hinton, H. E. (1969). *Annu. Rev. Entomol.* **14**, 343.

Hinton, H. E. (1970). *Sci. Amer.* **223**, 84.

Horridge, G. A. (1965). *In* "Structure and Function in the Nervous Systems of Invertebrates" (T. H. Bullock and G. A. Horridge, eds), Chapters 17 and 21, pp. 965 and 1165. Freeman, San Francisco, California.

Hoyle, G. (1959). *J. Insect Physiol.* **3**, 378.

Hoyle, G. (1960). *J. Insect. Physiol.* **4**, 63.

Hoyle, G. (1961). *J. Insect Physiol.* **7**, 305.

Hoyle, G. (1964). *In* "Neural Theory and Modeling" (R. F. Reiss, ed.), p. 346. Stanford Univ. Press, Stanford, California.

Hrbráček, J. (1949). *Vest. Cesk. Spolecnosti Zool.* **13**, 136.

Huber, F. (1960). *Z. Vergl. Physiol.* **43**, 359.

Hughes, G. M. (1952). *Nature (London)* **170**, 531.

Imms, A. D. (1957). "A General Textbook of Entomology," 9th ed. Methuen, London.

Kanwisher, J. W. (1966). *Biol. Bull.* **130**, 96.

Kehler, J. G., Smalley, K. N., and Rowe, E. C. (1970). *J. Insect Physiol.* **16**, 483.

Keilin, D. (1944). *Parsitology* **36**, 1.

Keister, M. L. (1953). *J. Morphol.* **93**, 573.

Keister, M. L., and Buck, J. B. (1961). *J. Insect Physiol.* **7**, 51.

Krafsur, E. S. (1971). *Ann. Entomol. Soc. Amer.* **64**, 93, 97, and 294.

Krafsur, E. S., and Graham, C. L. (1970). *Ann. Entomol. Soc. Amer.* **63**, 691.

Krafsur, E. S., Willman, J. R., Graham, C. L., and Williams, R. E. (1970). *Ann. Entomol. Soc. Amer.* **63**, 684.

Krogh, A. (1920). *Pfluegers Arch. Gesamte Physiol. Menschen Tiere* **179**, 95 and 113.

Krogh, A. (1941). "Comparative Physiology of Respiratory Mechanisms." Univ. of Pennsylvania Press, Philadelphia (reprint, 1959).

Kutsch, W. (1969). *Z. Vergl. Physiol.* **63**, 335.

Landa, V. (1969). *Acta Entomol. Bohemoslov.* **66**, 289.

Lang, F. (1971). *J. Exp. Biol.* **54**, 815.

Larsen, O. (1966). *Opusc. Entomol., Suppl.* **30**, 1.

Levereault, P. (1938). *Univ. Kans. Sci. Bull.* **25**, 577.

Levy, R. I., and Schneiderman, H. A. (1966a). *J. Insect Physiol.* **12**, 83.

Levy, R. I., and Schneiderman, H. A. (1966b). *J. Insect Physiol.* **12**, 105.

Levy, R. I., and Schneiderman, H. A. (1966c). *J. Insect Physiol.* **12**, 465.

Lewis, G. W. (1973). "Neurophysiology of Insect Respiratory Movements". D. Phil. Thesis, Oxford University.

Lewis, G. W., Miller, P. L., and Mills, P. S. (1973). *J. Exp. Biol.* **59**, 149.

Lewis, J. G. E. (1963). *Entomol. Exp. Appl.* **6**, 89.

Locke, M. (1958). *Quart. J. Microsc. Sci.* **99**, 29 and 373.

Locke, M. (1966). *J. Morphol.* **118**, 461.

Locke, M. (1969). *J. Morphol.* **127**, 7.

Locke, M., and Krishnan, N. (1971). *Tissue & Cell,* **3**, 103.

Lotz, G. (1962). *Z. Morphol. Oekol. Tiere* **50**, 726.

Loveridge, J. P. (1968). *J. Exp. Biol.* **49**, 15.

McArthur, J. M. (1929). *J. Exp. Zool.* **53**, 117.

McCutcheon, F. H. (1940). *Ann. Entomol. Soc. Amer.* **33**, 35.

McGovran, E. R. (1932). *J. Econ. Entomol.* **25**, 271.

Maki, T. (1938). *Mem. Fac. Sci. Agr., Taihoku Imp. Univ.* **24**, 1.

Matsuda, R. (1970). *Mem. Entomol. Soc. Can. No.* 76, p. 1.

Mendelson, M. (1971). *Science* **171**, 1170.

Mill, P. J. (1970). *J. Exp. Biol.* **52**, 167.

Mill, P. J. (1972). "Respiration in the Invertebrates." Macmillan, New York.

Mill, P. J., and Hughes, G. M. (1966). *J. Exp. Biol.* **44**, 297.

Miller, P. L. (1960). *J. Exp. Biol.* **37**, 224, 237, and 264.

Miller, P. L. (1961). *J. Insect Physiol.* **6**, 243.

Miller, P. L. (1962). *J. Exp. Biol.* **39**, 513.

Miller, P. L. (1964a). *J. Exp. Biol.* **41**, 331 and 345.

Miller, P. L. (1964b). *Nature (London)* **201**, 1052.

Miller, P. L. (1965). *In* "Physiology of the Insect Central Nervous System" (J. E. Treherne and J. W. L. Beament, eds.), p. 141. Academic Press, New York.

Miller, P. L. (1966a). *Advan. Insect Physiol.* **3**, 279.

Miller, P. L. (1966b). *J. Exp. Biol.* **45**, 285.

Miller, P. L. (1967). *J. Exp. Biol.* **46**, 349.

Miller, P. L. (1969). *Nature (London)* **221**, 171.

Miller, P. L. (1970). *Entomol. Mon. Mag.* **105**, 233.

Miller, P. L. (1971a). *J. Insect Physiol.* **17**, 395.

Miller, P. L. (1971b). *J. Exp. Biol.* **54**, 587 and 599.

Miller, P. L. (1971c). *J. Entomol. (A)* **46**, 63.

Miller, P. L. (1971d). *Proc. Int. Congr. Entomol., 13th, 1968* **2**, 26.

Möss, D (1971). *Z Vergl. Physiol.* **73**, 53.

Myers, T. B., and Retzlaff, E. (1963). *J. Insect Physiol.* **9**, 607.

Nair, R. S. S., and Karnavan, G. K. (1966). *Nature (London)* **211**, 1207.

Noble-Nesbitt, J. J. (1971). *Proc. Int. Congr. Entomol., 13th, 1968* **1**, 422.

Noble-Nesbitt, J. J. (1973). In preparation.

Nunnome, Z. (1944). *Bull. Sericult. Exp. Sta. Jap.* **12**, 17 and 41.

Oschman, J. L., and Wall, B. J. (1969). *J. Morphol.* **127**, 475.

Paim, U., and Beckel, W. E. (1963). *Can. J. Zool.* **41**, 1133 and 1149.

Paim, U., and Beckel, W. E. (1964). *Can. J. Zool.* **42**, 59, 295, and 327.

Parsons, M. C. (1972). *Can. J. Zool.* **50**, 865.

Pearson, K. G. (1972). *J. Exp. Biol.* **56**, 173.

Peterson, M. K., and Buck, J. B. (1968). *Biol. Bull.* **135**, 335.

Pihan, J. C. (1972). *Bull. Soc. Zool. France* **97**, 351.

Plotnikova, S. I. (1968). *In* "Neurobiology of Invertebrates" (J. Salánki, ed.), p. 59. Plenum, New York.

Plotnikova, S. I. (1969). *In* "Les problèmes actuelles de la structure et de la fonction du système nerveux des insectes" (G. J. Bey-Bienko, ed.), Vol. 53, p. 36. Acad. Sci. U.S.S.R. (in Russian).

Poonawalla, Z. T. (1966). *Ann. Entomol. Soc. Amer.* **59**, 807.

Pringle, J. W. S. (1954). *J. Exp. Biol.* **31**, 525.

Punt, A. (1950). *Physiol. Comp. Oecol.* **2**, 59.

Punt, A. (1956). *Physiol. Comp. Oecol.* **4**, 121 and 132.

Punt, A., Parsler, W. J., and Kuchlein, J. (1957). *Biol. Bull.* **112**, 108.

Richards, A. G. (1951). "The Integument of Arthropods." Univ. of Minnesota Press, Minneapolis.

Richter, P. O. (1969). *Ann. Entomol. Soc. Amer.* **62**, 869 and 1388.

Roddy, L. R. (1955). *Ann. Entomol. Soc. Amer.* **48**, 407.

Roth, L. M., and Eisner, T. (1962). *Annu. Rev. Entomol.* **7**, 107.

Rounds, H. D. (1968). *Comp. Biochem. Physiol.* **24**, 653.

Sachs, H. (1952). *Z. Bienenforsch.* **1**, 148.

Salmoirhagi, G. C. (1963). *Ann. N.Y. Acad. Sci.* **109**, 571.

Satchell, G. H. (1948). *Parasitology* **39**, 43.

Schneiderman, H. A. (1960). *Biol. Bull.* **119**, 494.

Schneiderman, H. A., and Schechter, A. N. (1966). *J. Insect Physiol.* **12**, 1143.

Schneiderman, H. A., and Williams, C. M. (1953). *Biol. Bull.* **105**, 320.

Schneiderman, H. A., and Williams, C. M. (1955). *Biol. Bull.* **109**, 123.

Schreuder, J. E., and de Wilde, J. (1952). *Physiol. Comp. Oecol.* **2**, 355.

Scudder, G. G. E. (1963). *Can. J. Zool.* **41**, 1.

Simpson, J. (1964). *Z. Vergl. Physiol.* **48**, 277.

Sláma, K. (1960). *J. Insect Physiol.* **5**, 341.

Smalley, K. N. (1970). *J. Insect Physiol.* **16**, 241.

Smith, D. S. (1961). *J. Biophys. Biochem. Cytol.* **10**, 123.

Smith, D. S. (1963). *J. Cell Biol.* **16**, 323.

Smith, D. S. (1966). *J. Cell Biol.* **28**, 109.

Smith, D. S. (1968). "Insect Cells, Their Structure and Function." Oliver & Boyd, Edinburgh.

Smith, D. S., and Telfer, W. H. (1970). *In* "Insect Ultrastructure" (A. C. Neville, ed.), p. 117. Blackwell, Oxford.

Smith, D. S., Telfer, W. H., and Neville, A. C. (1971). *Tissue & Cell* **3**, 477.

Steen, J. B. (1971). "Comparative Physiology of Respiratory Mechanisms." Academic Press, New York.

Stout, J. F., and Huber, F. (1972). *Z. Vergl. Physiol.* **76**, 302.

Tonapi, G. T. (1958). *Indian J. Entomol.* **20**, 245.

Tonapi, G. T. (1968). *Naturwissenschaften* **55**, 44.

Tyrer, M. (1971). *J. Exp. Biol.* **55**, 305 and 315.

van der Kloot, W. G. (1963). *Comp. Biochem. Physiol.* **9**, 317.

Vincent, J. F. V. (1971). *J. Insect Physiol.* **17**, 625.

Voskresenskaya, A. K. (1968). *In* "Neurobiology of Invertebrates" (J. Salánki, ed.), p. 367. Plenum, New York.

Walcott, B., and Burrows, M. (1969). *J. Insect Physiol.* **15,** 1855.

Walker, L. J., and Brown, A. M. (1970). *Science* **167,** 1502.

Watts, D. T. (1951). *Ann. Entomol. Soc. Amer.* **44,** 527.

Weis-Fogh, T. (1964a). *J. Exp. Biol.* **41,** 207.

Weis-Fogh, T. (1964b). *J. Exp. Biol.* **41,** 229.

Weis-Fogh, T. (1967). *J. Exp. Biol.* **47,** 561.

Whitten, J. M. (1957). *Quart. J. Microsc. Sci.* **98,** 123.

Whitten, J. M. (1959). *Syst. Zool.* **8,** 130.

Whitten, J. M. (1960). *J. Morphol.* **107,** 233.

Whitten, J. M. (1963). *Ann. Entomol. Soc. Amer.* **56,** 755.

Whitten, J. M. (1968). *In* "Metamorphosis: A Problem in Developmental Biology" (W. Etkin and L. I. Gilbert, eds.), p. 43. North-Holland Publ., Amsterdam.

Whitten, J. M. (1969). *J. Morphol.* **127,** 73.

Whitten, J. M. (1972). *Annu. Rev. Entomol.* **17,** 373.

Wigglesworth, V. B. (1930). *Proc. Roy. Soc., Ser. B* **106,** 229.

Wigglesworth, V. B. (1932). *Proc. Roy. Soc., Ser. B* **109,** 354.

Wigglesworth, V. B. (1935). *Proc. Roy. Soc., Ser. B* **118,** 397.

Wigglesworth, V. B. (1941). *Proc. Roy. Entomol. Soc. London Ser. A* **16,** 11.

Wigglesworth, V. B. (1953). *Quart. J. Microsc. Sci.* **94,** 507.

Wigglesworth, V. B. (1954). *Quart. J. Microsc. Sci.* **95,** 115.

Wigglesworth, V. B. (1959). *J. Exp. Biol.* **36,** 632.

Wigglesworth, V. B. (1963). *Nature (London)* **198,** 106.

Wigglesworth, V. B. (1965). "The Principles of Insect Physiology," 6th ed. Methuen, London.

Wigglesworth, V. B. (1970). *Tissue & Cell* **2,** 155.

Wigglesworth, V. B., and Beament, J. W. L. (1950). *Quart. J. Microsc. Sci.* **91,** 429.

Wigglesworth, V. B., and Beament, J. W. L. (1960). *J. Insect Physiol.* **4,** 184.

Wilkins, M. B. (1960). *Nature (London)* **185,** 481.

Wilson, D. M. (1968). *Advan. Insect Physiol.* **5,** 289.

Chapter 6

RESPIRATION: AQUATIC INSECTS

P. J. Mill

I. Introduction

A gas-filled tracheal system, as described in Chapter 5 of this volume, is found in all aquatic insects, with the exception of the early larval stages of a few dipterans such as *Chironomus* (Pause, 1918) and *Simulium* (Wagner, 1926), and of the lepidopteran *Acentropus* (Nigmann, 1908).

In many larvae with a closed tracheal system, that is, one in which none of the spiracles is functional, respiration occurs exclusively over the general body surface (cutaneous respiration) and the subepidermal tracheal plexus is well developed. This plexus may be generally distributed (Hagemann, 1910; Miall, 1895; Nigmann, 1908; Zavřel, 1920) or more or less confined to areas where the cuticle is thin (Burtt, 1936; Lübben, 1907; Schoenemund, 1925).

Many dipteran larvae possess small blood-filled sacs which have been referred to as "blood-gills," but the general consensus of opinion is that most of these structures are not true gills (Alterberg, 1934; Fox, 1920; Thorpe, 1933a; Wigglesworth, 1933, 1953) and it has been suggested that their primary function is in the control of chloride ion uptake (Koch, 1938).

In a large number of aquatic larvae with a closed tracheal system, gills, containing abundant ramifications of the tracheal system (tracheal gills), have evolved. On the other hand, a wide variety of aquatic insects possesses an open tracheal system. In many cases, they are entirely reliant on coming to the surface to breathe, or on tapping the air spaces of submerged aquatic plants (Ege, 1915b). Some carry bubbles or films of air (physical or gas gills) when they are submerged (Ege, 1915a) and their spiracles then communicate directly with this air. A gas gill may be temporary, when the animal has to return periodically to the surface of the water to replenish it (Portier, 1911). Alternatively, it may be permanent (plastron), involving the development of either the body hairs (Brocher, 1912a,b; Davis, 1942; Thorpe and Crisp, 1947a) or cuticular modifications of outgrowths from the vicinity of the spiracles (spiracular gills) (Hinton, 1953). Nevertheless, even in larvae which possess tracheal gills or breathe atmospheric air, cutaneous respiration is in many cases still important (Thorpe, 1933b; Wigglesworth, 1933). For instance, Fraenkel and Herford (1938) have shown that, in well-aerated water, *Culex* larvae (which bear functional spiracles) take up about half of their oxygen requirements through the body wall.

In any animal total metabolism (M), expressed as oxygen consumption/unit time, is related to volume (V), according to the equation

$$M = kV^b \tag{1}$$

where k and b are constants for any given individual and the value of k depends on the shape of the body. If the body is of uniform specific gravity, weight (W) is directly proportional to volume, and hence Eq. (1) can be rewritten

$$M = kW^b \tag{2}$$

If total metabolism is linearly related to surface area (S) then Eq. (2) becomes

$$M = kW^{2/3} \tag{3a}$$

or, in terms of metabolic rate (R), defined as oxygen consumption/unit weight/unit time

$$R = kW^{-1/3} \tag{3b}$$

This relationship is known as the surface law.

If, on the other hand, total metabolism is linearly related to weight, Eq. (2) becomes

$$M = kW \tag{4a}$$

or

$$R = k \tag{4b}$$

(Keister and Buck, 1964; Mill, 1972).

Since the specific gravity of most animals is not homogeneous, a value of $\frac{2}{3}$ for b does not necessarily indicate a surface relationship (Gilchrist, 1956). Similarly, if b differs markedly from $\frac{2}{3}$ it does not necessarily mean that the surface law does not apply and it has been pointed out that too rigorous an application of the relationship between b and the validity of the surface law has led to some confusion (Edwards, 1958).

When a diffusion gradient is established across the body wall or between the environment and a gas gill, the adjacent layer of water must also be considered as a barrier. The latter is known as the "boundary layer" (Krogh, 1919; von Ruttner, 1926; Paganelli et al., 1967; Feldmeth, 1970a) and its thickness decreases with increasing flow of water over the body. However, even with a flow of water there will always be a zone close to the body which remains relatively stagnant, with the water velocity increasing asymptotically with distance from the surface (Paganelli et al., 1967) (Fig. 1a). Tracheal gills may have a dual function: the uptake of oxygen and, by their activity, the reduction of the boundary layer. The effective thickness of the boundary layer (ΔX) has been defined as "that thickness of perfectly still water which has the same diffusion resistance as the actual boundary layer" (Rahn and Paganelli, 1968). Paganelli et al. (1967), using a model of a gas gill, used Fick's law of diffusion for the steady state to define ΔX.

$$\Delta X = \frac{(A)(\alpha_0)(D_0)(\Delta P_0)}{J_0}$$

where A is the membrane area (cm^2), α_0, solubility of oxygen in water ($cm^3/cm^3/atm$), D_0, diffusion coefficient of oxygen in water ($cm^2/second$), ΔP_0, oxygen pressure difference between the environment (outside the boundary layer) and the inside, and J_0, observed inward diffusion of oxygen ($cm^3/second$). They plotted computed values of ΔX against water flow (Fig. 1b). In still water, the effective boundary layer came to about 300 μm, reducing to about half this at a flow rate of 1 cm/second. It continued to

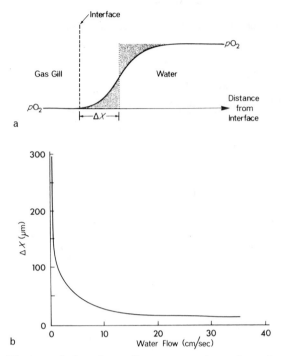

Fig. 1. (a) Diagram of the pO_2 gradient next to the surface of a gas gill. The effective thickness of the boundary layer (ΔX) is such that the shaded areas are equal. (b) The relationship between the effective thickness of the boundary layer (μm) and the current velocity (cm/second). (After Paganelli *et al.*, 1967.)

decrease asymptotically with increase in rate of flow and reached a value of about 10 μm at 35 cm/second.

Aquatic animals can be divided broadly into two categories on the basis of their ability to regulate their metabolic rate (Prosser and Brown, 1961). "Respiratory conformers" show a decrease in their metabolic requirements, and hence in their respiratory rate, as the oxygen concentration of the environment decreases. "Respiratory regulators" maintain a steady metabolic rate over a wide range of environmental oxygen concentrations. The lowest oxygen tension at which an animal is capable of regulating its metabolism is often referred to as the critical level (P_c). Below this it becomes a conformer. At 10°C, an oxygen content of 7.9 ml/liter is in equilibrium with air and so, at this temperature, respiratory regulators will have a critical level below 7.9.

In many cases, the available evidence is overwhelming in favor of a respiratory function for tracheal gills; in others they have been shown to be of special importance under adverse conditions; but in certain cases it is dubious whether, in the animal's normal environment, they are at all essential for respiration.

General accounts of aquatic respiration in insects are given by Wigglesworth (1953) and Edwards (1953), who summarize much of the earlier work, and Mill (1972), while excellent reviews on specific aspects of gas gills are those of Thorpe (1950), Crisp (1964), and Hinton (1968).

II. Closed Tracheal Systems

A. CUTANEOUS RESPIRATION

A number of aquatic larvae with a closed tracheal system do not possess gills. In many cases, e.g., *Chironomus,* most of the body wall is richly tracheated and the cuticle is uniformly thin. Such animals are probably the most likely to have a metabolic rate linked to surface area, and indeed Fox (1920) has shown that the larva of *Chironomus* absorbs oxygen over its general body surface.

In third- and fourth-instar larvae of *Chironomus riparius* a double log plot of metabolic rate (mm³ O_2/mg dry wt/hour) against dry weight produces a linear relationship with a slope of about —0.30 (i.e., total oxygen consumption varies as the 0.70 power of the dry weight) (Edwards, 1958) (Fig. 2). However, a similar logarithmic transform of metabolic rate against wet weight shows an alteration in slope between the two instars, due to a different relationship between dry weight and wet weight. A double log plot of dry weight against wet weight for fourth-instar larvae gives a straight

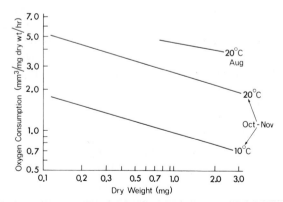

Fig. 2. A double log transform of the relationship between oxygen consumption (mm³/mg dry wt/hour) and dry weight (mg) in larvae of *Chironomus.* (After Edwards, 1958.)

line with a slope of 1.186. Thus, in this instar the total oxygen consumption varies as the 0.83 power (i.e., 0.70×1.186) of the wet weight. Furthermore, Edwards (1958) showed that volume is approximately proportional to wet weight, not to dry weight, and that the estimated surface area varies as the 0.70 power of the wet weight. Thus, these larvae appear not to have a metabolic rate which is directly proportional to surface area or to body weight. These results are contrary to the view of Bertalanffy (1951) who states that the metabolic rate of insect larvae (not necessarily aquatic ones) is weight proportional. The Q_{10} of *C. riparius* is 2.6 over the range $10°-20°C$, but it is of particular interest to note that, when the metabolic rates of "summer" and "winter" larvae are measured at 20°C, the former show a higher rate (Edwards, 1958) (Fig. 2). This may be because animals belonging to the overwintering generation undergo a period of arrested development (Lees, 1955). Erman and Helm (1970) have looked at the metabolic rate of the larvae of eight species of chironomids and they found that when they plotted $\log_{10}$ O_2 consumption (μl/animal/hour) against $\log_{10}$ body length (mm) they obtained a straight line with a slope of 2.26.

Tubifex shows an interesting response to changes in environmental oxygen tension (pO_2), exposing more of its body above the surface of the mud as the pO_2 decreases. The larvae of a few species of *Tanytarsus* (Diptera) are found in fast-flowing water and they build tubes which they ventilate by oscillations of the body (Keilin, 1944).

In other larvae the tracheal supply to the body wall is primarily concentrated in specific zones. For example, in the larva of *Atrichopogon trifasciatus* (Diptera, Ceratopogoninae) there are eight elliptical regions ("respiratory organs") on the dorsal surface where the cuticle is very thin and the underlying hypodermis well developed. This hypodermis is richly supplied with tracheoles. The larva lives on the moist region of partially submerged stones in rivers and lakes, when it is covered with a thin film of water. Thus, these "respiratory organs" probably function in much the same way as lungs. If the larva becomes dry, each organ is pulled downward by a pair of muscles and is virtually enclosed by surrounding scales (Burtt, 1936). Physiological data for this animal do not appear to be available.

B. Tracheal Gills

Tracheal gills are thin outgrowths of the body wall which contain a rich tracheal plexus. They are almost exclusively confined to larval stages. However, they are found in some trichopteran pupae. Also, in some plecopterans and in the trichopteran *Hydropsyche*, they persist into the adult stage in a shriveled, nonfunctional condition (Imms, 1957). They occur most often on the abdomen, but are also found on the thorax and head.

In some instances there has been considerable controversy over the function of tracheal gills (Wigglesworth, 1953; MacNeill, 1960), especially since in many cases the larvae live quite successfully after they have been removed. Also, their absence may cause virtually no difference to the total oxygen consumption of the animal (Morgan and O'Neil, 1931). However, Koch (1936) has pointed out that although an abundant tracheal supply may not be a very good criterion on its own for determining function, if the structure has very low metabolic requirements, and thus takes up more oxygen from the environment than it needs for itself, then it will be an important site of gaseous exchange. In general, the thin cuticular covering of the gills, their rich supply of tracheal capillaries, and the absence of structures within them which possess a high metabolic rate all tend to argue in favor of this function.

1. Lateral Abdominal Gills

a. *Morphology and Occurrence.* Pairs of segmental, abdominal gills are the most common arrangement. In their simplest form they are flat, platelike outgrowths of the body wall (lamellae), as found in many ephemeropteran larvae. In *Baetis* and *Cloeon* there are seven pairs. In the former, each gill is a single lamella, but in *Cloeon* some are double. In other ephemeropterans the gills are biramous with simple (*Leptophlebia*) or multifurcate (*Habrophlebia*) branching and further elaboration occurs in *Ecdyonurus* and other dorsoventrally flattened larvae, where many of the lamellae shield a tuft of filamentous gills. In *Ephemera* and *Hexagenia,* the gills are biramous and lanceolate, except for the first pair which is vestigial. They are fringed with long filaments and are generally reflexed over the back. It has been demonstrated by Dodds and Hisaw (1924) that for different species of mayflies the area of the gills is inversely proportional to the oxygen content of their environment.

In all other orders the gills are filamentous. Among the Odonata, lateral abdominal gills are found in just two zygopteran families, the Epallagidae and Polythoridae (Calopterygidea), the members of which also possess caudal gills (Section II,A,2). The gills are simple in the Epallagidae, but are segmented in the Polythoridae (Calvert, 1911; Needham, 1911; Ris, 1912; Lieftinck, 1962). Abdominal gills are also found in just one family of plecopterans, the Eustheniidae (Tillyard, 1923). Among the Neuroptera they are found in larvae belonging to the family Sialoidea (suborder Megaloptera) and may be simple, as in *Corydalis* and *Chauliodes* (eight pairs), or segmented, as in *Sialis* (seven pairs). In addition to the lateral gills, *Sialis* has a single, segmented, terminal filament on the ninth segment. In the other suborder of the Neuroptera, the Planipennia, only the larvae of *Sisyra* are known to have gills and here they consist of seven pairs of jointed filaments

(Anthony, 1902). Finally, among Coleopterans, larvae of the Gyrinidae have ten pairs, and larvae of *Hydrocharis* and *Berosus* (Hydrophilidae) have seven pairs of hair-fringed filamentous gills (Imms, 1957).

Considerable elaboration occurs in some genera of mayflies. Thus, in *Caenis,* which has gills on the first six segments, the first pair are reduced, while the second pair are very large and cover the remainder, forming a branchial chamber. In *Prosopistoma* the wing sheaths and the highly developed pro- and mesothoracic terga help to form a branchial chamber which completely encloses the five pairs of gills.

b. Physiology. The general *Ephemera* and *Hexagenia* contain some of the largest ephemeropterans, all of which are adapted for burrowing. *Ephemera simulans* is found in streams and lakes where there is a rather coarse sediment (e.g., gravel) and can survive as long as the oxygen content of the interstitial water exceeds about 1.20 cm³/liter (Eriksen, 1968). Its filamentous gills are held reflexed over the top of the abdomen and beat continuously, the frequency of the rhythm being inversely proportional to the environmental oxygen concentration. The same is also true for *E. vulgata* (Wingfield, 1939) and this relationship was first observed for ephemerid larvae by Babák and Foustka (1907). It is now certain that the gills of burrowing species are directly concerned with the uptake of oxygen as well as with increasing the flow of water over the body to reduce the thickness of the "boundary layer." Eriksen (1963b) compared the oxygen uptake of anesthetized normal and gill-less larvae of *E. simulans* and *H. limbata* and found that, at an oxygen concentration of 6 cm³/liter, about 45 and 47%, respectively, of the total uptake is via the gills. In other words, the metabolic rate of normal larvae is about twice that of gill-less ones. These values of course exclude any contribution which the gills make in disturbing the "boundary layer."

This is substantiated by the earlier work of Morgan and Grierson (1932) on *H. recurvata*. They obtained a similar figure (43%) for oxygen uptake via the gills. Also they found that, while gill-less larvae survived through the winter, they were less active than normal larvae and rarely burrowed, and they did not survive in the spring. Morgan and Porter (1929) observed that the normal negative phototactic response was reversed in gill-less larvae.

Wingfield (1939) obtained a much higher value (73%) for the oxygen uptake across the gills in *E. vulgata*. However, none of the earlier workers used anesthetized larvae as a control and so gill and body movements will have had an effect on their results. Morgan and Grierson (1932) used an initial oxygen concentration in their experiments of 8–9 cm³/liter and it never fell below 5 cm³/liter. Although the gills were probably still beating at this concentration, they gave no information on this. Wingfield (1939) similarly used high oxygen concentrations (6.5–8 cm³/liter) and notes that the gills in *E. vulgata* still beat slowly. Thus, his figure does include the effect of the gills on the "boundary layer." A difference in technique which may be important in accounting for the different values is that Morgan and Grierson (1932) removed the gills from most of the larvae 14–16 weeks before carrying out any measurements, whereas Wingfield (1939) used the gill-less larvae after 3–4 hours.

Hexagenia limbata has much larger gills than *E. simulans* and only requires about a quarter the frequency of beat to produce the same respiratory current as the latter species. It is capable of producing a respiratory current faster than that of *E. simulans,* can tolerate an interstitial oxygen concentration down to 0.80 cm^3/liter, and is found in finer sediments than *E. simulans* (Eriksen, 1968). In *Ephemera danica* (Wautier and Pattée, 1955), *E. simulans,* and *H. limbata* (Eriksen, 1963a), the metabolic rate (cm^3 O$_2$/gm/hour) was invariably shown to be lower when the animal was provided with a substrate compared with experiments carried out in the absence of a substrate. Furthermore, the metabolic rate reached a minimum when the substrate particle size was similar to that of the animal's natural environment. Eriksen (1963a) added confirmation to this by carrying out substrate preference experiments with *E. simulans.* The larvae showed a marked preference for substrate particles of the same size at which their metabolic rate was lowest.

In the absence of a substrate, experiments on the metabolic rate of *E. vulgata* (Fox et al., 1937) indicated that this animal is a respiratory conformer (Fig. 3). However, Eriksen (1963b) has shown that if similar experiments are carried out in the presence of an optimal substrate, *E. simulans* and *H. limbata,* at least, are respiratory regulators, both with a critical level in the region of 1 cm^3/liter (Fig. 3).

The work of Fox et al. (1937) showed that the larvae of *Leptophlebia* ($P_c = 2.5$) and *Cloeon* ($P_c = 1.5$) are both regulators (Fig. 3). Since both live in static water, where changes in the environmental oxygen content are to be expected, the ability to regulate would be of considerable use.

The lateral gills in *Cloeon* beat rhythmically at low oxygen concentrations, but only intermittently at higher concentrations. At 10°C the critical level in gill-less larvae is only 3 ml/liter, compared with the 1.5 ml/liter of normal animals (Wingfield, 1937). Above 3 ml/liter the metabolic rate is about the same in both (Fig. 3). This indicates that the gills are not normally essential in terms of oxygen uptake and only become important at low oxygen tensions, when they act either to increase oxygen uptake directly and/or by reducing the thickness of the boundary layer. Above a concentration of 3 ml O$_2$/liter, increase in oxygen uptake over the general body surface is sufficient to cope with normal demands. Wingfield's (1939) evidence favors the latter since, when the water was mechanically stirred, gill-less larvae had the same metabolic rate as normal animals at all oxygen concentrations.

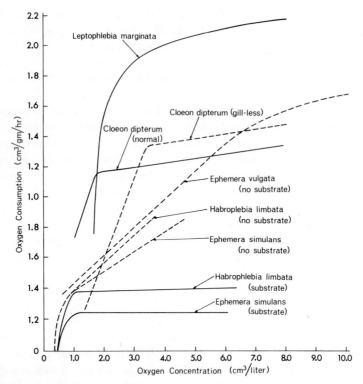

Fig. 3. The relationship between oxygen consumption (cm³/gm/hour) and the environmental oxygen concentration (cm³/liter) for various ephemeropteran larvae. (*Leptophlebia marginata* and *Ephemera vulgata* at 10°C, after Fox *et al.,* 1937; *Cloeon dipterum* at 10°C, after Wingfield, 1937; *Ephemera simulans* and *Habrophlebia limbata* at 13°C, after Eriksen, 1963b.)

In *Baetis,* which lives in fast-flowing water, the gills do not beat and the metabolic rates of normal and gill-less larvae are identical over the range 5–8 ml O₂/liter. Wingfield (1939) suggested that it can thus be assumed that the gills are not involved in oxygen uptake. However, another possibility is that, at these oxygen concentrations, *Baetis* can take up sufficient oxygen over its general body surface to meet any increase in its metabolic requirements. Larvae of *Baetis rhodani* are less resistant to low environmental pO_2 than those of the pond-dwelling *Cloeon dipterum* (Fox *et al.,* 1935). Thus, *Baetis* larvae can only tolerate an oxygen concentration down to 4 ml/liter, which is about the minimum they are likely to experience in nature (Wingfield, 1939). Ambühl (1959) stated that *Baetis* and *Rhithrogena* depend on the water current for ventilation. Fox *et al.* (1937) have shown that while small *Baetis* larvae are conformers, large larvae are regulators ($P_c = 5$). However, these may have been larvae of different species.

Leptophlebia, Cloeon, and *Baetis* are not burrowers and so it is unlikely that the presence of a substrate would affect the results obtained for these animals by Fox *et al.* (1937). However, they all tend to live among plants and the microhabitat cannot yet be ruled out as being of importance in determining the metabolic rate.

Information on the removal of CO_2 from the body is somewhat sparse. Morgan and Grierson (1932) have shown that gill-less larvae of the burrowing mayfly *H. recurvata* only eliminate half as much as normal larvae and they suggest that this buildup in CO_2 concentration in the body accounts for a change in their behavior from being photonegative to photopositive. This change can also be elicited in normal larvae of *Heptagenia interpunctata* by increasing the carbon dioxide concentration of the water (Wodsedalek, 1911). Possibly this increase in body CO_2 is the stimulus for the positive phototropism exhibited by many species prior to emergence, at which time the gills are presumably losing their respiratory function (Morgan and Grierson, 1932). Anisopteran dragonfly larvae come to the surface as a result of low environmental oxygen concentration (Section II,A,3).

2. Caudal Gills

a. Morphology and Occurrence. The presence of three (one median and two lateral) caudal gills is a diagnostic feature of zygopteran larvae. In a few cases, such as the semiterrestrial and terrestrial species of *Megalagrion* from Hawaii (Williams, 1936), the gills are somewhat reduced, but otherwise are generally well developed.

In young larvae the gills are ferular, with the median one quadrilateral and the lateral ones triangular (triquetral) in section (MacNeill, 1960). In the Calopteryginae the lateral gills retain this shape. However, together with the medial ones, they usually become lamellate or, in a few families, saccoid (Tillyard, 1917).

MacNeill (1958, 1960) has shown that in many species the gills are divided into two more or less distinct zones (proximal and distal). In such cases the distal zone is not discernible in early instars, but it increases in length by successive outgrowths of an organized structure present within the proximal part. Often, the additions produce a banded appearance (chevron bands). This is termed "protrusive growth" and zgopteran larvae can be divided into two types on the basis of whether protrusive growth is absent (simplex type) or present (duplex type) (MacNeill, 1958). In duplex gills the distal zone has a thinner cuticle than the proximal zone and a relatively small number of spines. In early instars there is no easily identifiable division between the two zones, but later on in lamellate gills it may consist of an abrupt change in the type of armature, a small node, or a larger constriction (MacNeill, 1960). Duplex saccoid gills only occur in three genera of the family Protoneuridae and in all cases the distal zone is extremely small.

It has been suggested that the caudal gills (or lamellae) of zygopteran larvae are used for swimming, or at least that they aid swimming by acting as rudders (Réamur, 1734–1742; von Rosenhof, 1749; Bodine, 1918) ; while Dufour (1852) and Roster (1886) suggested that they served for both respiration and locomotion. However, swimming is a comparatively rare event, even in the better swimmers (Schmidt, 1919). It is achieved by lateral movements of the body and is virtually unimpaired by removal of the gills. Furthermore, saccoid gills would appear to hinder, rather than aid, swimming movements (MacNeill, 1960). It has been suggested that the primitive method of respiration in dragonfly larvae is by means of lateral abdominal gills, as are found in a few species (Section II,A,1) (Snodgrass, 1954), but the observation of Robert (1958) that *Epallage fatime* lives in streams which sometimes dry up, leaving stagnant pools, has led him to suggest that lateral abdominal gills may be a secondary adaptation to conditions of low oxygen tension (also Corbet, 1962). Nevertheless they are undoubtedly fairly archaic structures (Corbet, 1962; Lieftinck, 1962).

MacNeill (1960) argues strongly that protrusive growth, and hence the duplex gill, is an evolutionary development which increases the respiratory efficiency of the caudal gills, both by increasing their surface area and by providing a region with an especially thin cuticle.

An alternative means of increasing the surface area of the gills is seen in the development of the saccoid form, which presumably renders protrusive growth unnecessary. This may explain why duplex saccoid gills are only found in a few species and why, when they do occur, the distal region is always very small (MacNeill, 1960). In most cases, however, saccoid gills would be extremely vulnerable and they are only found in those species which live under rocks and in rock crevices in fairly fast-flowing water, where they are reasonably well protected.

Removal of the caudal gills in *Coenagrion* results in the development of a rich supply of tracheoles in the body wall (Harnisch, 1958), which is also indicative of a respiratory function.

It is tempting to suggest that in zygopteran larvae general cutaneous respiration is still important because of the vulnerability of external caudal lamellae; whereas anisopteran larvae, with invulnerable internal gills, have little need of cutaneous respiration (MacNeill, 1960).

b. Physiology. In general it seems that, while gill-less larvae can survive well at high concentrations of oxygen, the gills as well as the general body surface are involved in oxygen uptake and become increasingly important for survival as the environmental oxygen tension (pO_2) decreases. Pennak and McColl (1944) demonstrated that large gill-less larvae of *Enallagma civile* and *E. cyathigerum* obtain sufficient oxygen through their general body surface over a wide range of pO_2. However, they were only able to extract oxygen from the water down to an oxygen concentration of about 14.5% saturation, compared with 2.4% saturation for normal larvae. They also concluded that the rectal epithelium is of no significance in oxygen uptake in older larvae, and in young larvae is, at best, only of minimum importance (Section II,A,3). These animals both live in ponds and Pennak and McColl (1944) suggest that there may be differences between larvae occupying different habitats in, for example, the relative respiratory significance of the various regions of the body (gills, body surface, rectal surface). In *Agrion pulchellum* it has been shown that 32–45% of the total oxygen uptake is via the caudal gills (Koch, 1934) and in *Coenagrion* as much as 60% (Harnisch, 1958).

Under conditions of low oxygen concentration a number of species have been observed to make ventilatory movements—*Copera* (Lieftinck, 1940), *Ceriagrion* (Gardner, 1956), *Platycnemis* (Corbet, 1962).

In *Calopteryx virgo* and *C. splendens* removal of the gills has no effect at high concentrations of oxygen, but increases the mortality at low concentrations (Zahner, 1959). As the oxygen content decreases, larvae of these two species spread the gills and wing sheaths and move the body from side to side. They then come to the surface and lie along it or expose the head and thorax, or possibly leave the water completely. Gill-less larvae lie along the surface with the tip of the abdomen protruding (Zahner, 1959). Larvae of *Lestes viridis* also lie along the surface under conditions of low oxygen concentration (Robert, 1958). At low oxygen tensions, larvae of *Pyrrhosoma nymphula* prevented from reaching the surface showed periods of rapid swimming alternating with periods of rest, when the gills were spread out and the body moved from side to side as in *Calopteryx* (Lawton, 1971).

Pyrrhosoma nymphula, which lives in ponds, has a metabolic rate (μl/larva/hour) which increases with increase in size and temperature [$Q_{10} = 2.20$ (5°–10°C) and 3.12 (10°–16°C)]. The temperature effect does not appear to show any metabolic acclimatization. These larvae showed a slight decrease in metabolic rate with decrease in oxygen concentration down to 50% saturation and so are only fairly good respiratory regulators. Below this critical level, further decrease in oxygen concentration caused a marked decline in metabolic rate (Lawton, 1971).

3. Rectal Gills

a. Morphology and Occurrence. Rectal tracheal gills are confined to the anisopteran dragonflies. They lie in the anterior region of the rectum, which has become enlarged to form a branchial chamber. Basically there are six main gill folds, each of which runs the full length of the branchial chamber. Each is supported by a series of short cross-folds on either side. This has been called the simplex system by Tillyard (1917), who has provided a classification of the various types. In some families (e.g., Cordulegastrinae) the gill folds consist of an undulated membrane which is not subdivided (undulate type), but in most Gomphinae the edges of the gills are divided into long papillae (papillate type).

A number of families have the so-called duplex system, in which only the cross-folds function as gills. This also can be subdivided on the basis of the characteristics of the individual gills. In the implicate type (only found in the Brachytronini), consecutive lamellate gills, arranged on either side of a reduced main fold, overlap each other. In the foliate type (confined to the Aeshnini) the bases of the gills are constricted. This type can be further divided. Thus, *Aeshna* has crinkled lamellate gills (normal foliate) while in *Anax* the distal portion of each gill is swollen and bears a number of papillae (papillofoliate). Finally, there is the lamellate type, restricted to the Libellulidae. Here, the main folds have completely disappeared and each gill, as the name suggests, is a simple lamella with a broad base. In the Synthemini there are only 12 lamellae in each longitudinal row (archilamellate), but all others have many more (neolamellate) (Tillyard, 1917).

In all cases, the gills are covered by an extremely thin layer of cuticle. This encloses an epithelial syncytium through which run the tracheal capillaries. The capillaries do not end blindly, but are looped. At their bases the walls of the gills are separated by extensions of the hemocoele which contain the trachae surrounded by hypobranchial tissue (Sadones, 1896; Tillyard, 1916) and the epithelium of one or both walls has become well developed to form a basal pad(s) (Tillyard, 1917).

Immediately posterior to the branchial chamber there is a highly muscular region of the rectum called the vestibule. In contrast to the branchial chamber, the circular muscles of the vestibule are extremely well developed and it has a number of dilator muscles attached to it. The branchial chamber and vestibule are guarded by valves. Water is pumped in and out of the branchial chamber via the anus and vestibule and the mechanism by which this is achieved is discussed in Section IV,B.

b. Physiology. The cuticle of anisopteran dragonfly larvae is fairly thick and it is doubtful whether it is of much importance in the exchange of respiratory gases. The respiratory function of the gills, which are contained in the branchial chamber, appears not to have been questioned, although there is little direct evidence for this. There is a marked tidal ventilatory current, the anus serving as both expiratory and inspiratory opening (e.g., Tillyard, 1917; Hughes and Mill, 1966) (Section IV,B).

In larvae which burrow in mud, care must be taken not to draw sediment into the branchial chamber during inspiration. Thus, the posterior region of the abdomen is often curved dorsally so that the anus protrudes above the surface (e.g., *Cordulegaster*), while some tropical gomphids have a respiratory tube formed by elongation of the last abdominal segment (Corbet *et al.*, 1960).

The frequency of ventilation increases with decrease in oxygen concentration and with increase in temperature (Matula, 1911). It has also been shown that the rate of carbon dioxide output in *Aeshna umbrosa* increases with increase in temperature, although the larvae eventually acclimatize. Young larvae have a higher rate of output than older larvae (Sayle, 1928). Larvae of *Anax imperator* show an almost linear increase in rate of oxygen consumption with increase in environmental pO_2 (Klekowski and Kamler, 1968).

When larvae of *Aeshna grandis* are subjected to an oxygen concentration of only 2.5 cm³/liter at 17°–18°C, they come to the surface to breathe, exposing the anus and pumping air in and out of their branchial chamber (Notatmung) except toward the end of the last instar, when they expose their thoracic spiracles (Wallengren, 1914b). The latter is similar to the preemergence behavior of anisopteran larvae (Lucas, 1900; Wesenberg-Lund, 1913; Calvert, 1929; Byers, 1930) which presumably experience an oxygen lack due to increased metabolism and water temperature (Lutz and Jenner, 1960; Corbet, 1962). Robert (1958) has shown that larvae of *Brachytron pratense* resubmerge if the temperature falls. This behavior is not initiated by increased environmental carbon dioxide tension (Wallengren, 1914b). However, presumably lowered environmental pO_2 causes a rise in tissue pCO_2.

The metabolic rate ($\mu l\ O_2$/gm dry wt/hour) of larvae of *Anax junius* varies with body weight and temperature. A double log plot of metabolic rate against dry body weight gave values for b of -0.32 at 13°C and -0.24 at 20°C. However, at 27°C and 34°C b does not differ significantly from zero and thus the metabolic rate is weight independent at higher temperatures (Petitpren and Knight, 1970) (Fig. 4).

It has been suggested that rectal inspiration is also of importance in zygopteran larvae (von Rosenhof, 1749) and a number of authors have reported pumping of water in and out of the rectum via the anus (Balfour-Browne, 1909; Calvert, 1915; Tillyard, 1916b, 1917; Gericke, 1919; Lieftinck, 1962). Folds or gills are certainly present in the rectum and have been described, for example, in *Calopteryx maculata*, *Hetaerina americana* and *Argia putrida* (Calvert, 1915), and *Austrolestes lede* (Tillyard, 1916b). However, their tracheal supply is poor or absent (Cullen, 1918; Jamiesen, 1918; Carroll, 1918). Tillyard (1917) uses the term blood-gills and suggests that they

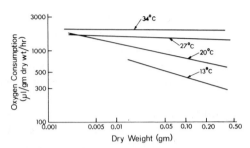

Fig. 4. A double log transform of the relationship between oxygen consumption (μl/gm dry wt/hour) and dry weight (gm) in larvae of *Anax junius* at different temperatures. The slope is -0.32 at $13°C$ and -0.24 at $20°C$, but does not differ significantly from zero at the two higher temperatures. (The correlation coefficients are -0.5995 at $13°C$, -0.7259 at $20°C$, -0.1245 at $27°C$, and -0.0504 at $34°C$.) (After Petitpren and Knight, 1970.)

are concerned with the elimination of carbon dioxide. Bodine (1918) claimed that the rectum is the major site of respiratory exchange, with cutaneous respiration doubtful and the caudal gills only acting as rudders. Contrary to this, Tillyard states that zygopteran larvae respire primarily through the caudal gills and use the rectum as an auxillary method of respiration (Tillyard, 1915, 1917). The reader is referred to Section II,A,2 for more current views of the relative roles of cutaneous and gill respiration.

4. Other Gills

a. Morphology and Occurrence. A number of plecopterans have tufts of finger-like gills which occur in a variety of places on the head, thorax, and abdomen, and also on the coxae. Similar tufts are found on the ventral surface of the abdomen of *Corydalis* and *Neuromus* (Megaloptera) (Imms, 1957).

In all of those trichopteran larvae which construct portable cases, as well as in a few of those that do not, filamentous gills occur in segmental groups in various positions on the abdomen. *Macronema zebratum,* for example, has about 60 gills (Morgan and O'Neil, 1931). In case-bearing caddis larvae there is an increase in the number of gill filaments with increase in body size. Larvae without cases do not show this relationship, but there is an indication that these animals have an inverse relationship between the number of filaments and their normal environmental pO_2 (Dodds and Hisaw, 1924).

Phalacrocerca (Diptera) and *Nymphula stratiotalis* and *N. maculalis* (Lepidoptera) also possess filamentous gills, on most body segments in the former and laterally on the meso- and metathoracic and abdominal segments in *Nymphula* (Buckler, 1875; Miall, 1895; Welch, 1916). *Nymphula macu-*

lalis has over 400 gills (Welch, 1916). In *Dicranota* (Diptera) there are two ventral pairs on the last abdominal segment (Imms, 1957). In two genera of mayflies, *Jolia* and *Oligoneuria,* tracheal gills are found on the head. Finally, *Peltodytes* (Coleoptera) has a number of jointed filaments on the dorsal surface of the thorax and abdomen, and another coleopteran (*Hygrobia*) has ventral gills near the base of each leg and on the first three abdominal segments (Imms, 1957).

  b. Physiology. It was suggested as long ago as 1734 by Réaumur (1734–1742) that the tracheal gills of *Phryganea* (Trichoptera) larvae are respiratory organs and this interpretation was accepted by Dufour (1847) for a hydropsychid larva and by Houghton (1868) for some other species of trichopterans. Sleight (1913), however, came to the conclusion that this was not the case for all caddis fly larvae, some species of which rely exclusively on cutaneous respiration. Morgan and O'Neil (1931), working on *Macronema zebratum,* found that removal of the entire complement of tracheal gills had little or no effect on oxygen consumption in this animal. Indeed gill-less larvae still built cases and pupated.

  Larvae of the caddis fly *Pycnopsyche guttifer* live in slow-flowing water (mean velocity 5.0 cm/second) near the edges of streams. A closely related species, *P. lepida,* occupies a similar habitat, except that during its last two instars it moves into more rapidly flowing water, with a mean velocity of 14.5 cm/second (Cummins, 1964; Feldmeth, 1970a). Anesthetized animals of both species have a fairly steady metabolic rate at current velocities down to about 2 cm/second. Below this it begins to decrease, presumably due to the formation of a thick boundary layer which would retard diffusion of respiratory gases (Feldmeth, 1970a) (Fig. 5a).

  Normal animals of many species produce a ventilatory current through their cases by rhythmic abdominal undulations (van Dam, 1937; Philipson, 1954; Leader, 1970). In *Pycnopsyche guttifer* and *P. lepida* the frequency increases with decrease in current velocity (Feldmeth, 1970a), presumably to compensate for the latter (Fig. 5b). Philipson (1954) demonstrated a similar relationship in a fast-water species, *Hydropsyche instabilis* (no case), and a slow-water species, *Polycentropus flavomaculatus,* using stirring of the water instead of current flow. In the former, fast stirring inhibits the movements completely. However, these two species behave differently with regard to changes in environmental pO_2 over the range 2–6 ml O_2/liter. Thus, *H. instabilis* shows little change in frequency as the pO_2 is lowered to about 4 ml/liter, but with gentle stirring the frequency tends to increase. Below 4 ml/liter, animals in fast-stirred water start to show undulations which increase in frequency with decrease in pO_2, but animals in less well-stirred water or in still water exhibit a marked decline in frequency. On the other

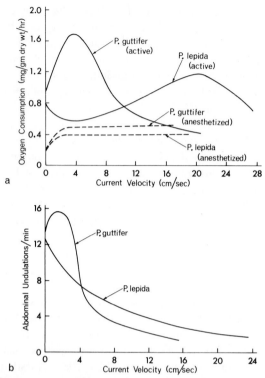

Fig. 5. (a) The relationship between oxygen consumption (mg/gm dry wt/hour) and current velocity (cm/second) in normal and anesthetized larvae of *Pycnopsyche guttifer* and *P. lepida*. (b) The relationship between the frequency of ventilation and current velocity in *P. guttifer* and *P. lepida*. (After Feldmeth, 1970a).

hand, *P. flavomaculatus* always exhibits an increase in frequency with decrease in environmental pO_2. A similar relationship is found in *Limnophilus flavicornis* (Fox and Sidney, 1953) and in *Phyrganea grandis* (van Dam, 1937). In the latter, ventilation is intermittent in air-saturated water. With decreasing pO_2 or increasing temperature the ventilatory frequency increases and the pauses become shorter, eventually disappearing altogether. Environmental pCO_2 has no effect over the range 0–29% saturation (van Dam, 1937). Ambühl (1959) observed that the frequency of ventilatory movements in *Hydropsyche angustipennis* (no case) and *Halesus digitatus* depend on environmental pO_2 and current velocity. In the littoral and sublittoral larvae of *Philanisus plebeius*, initial exposure to low environmental pO_2 causes an increase in the frequency of ventilation. However, this adapts in 5–10 minutes. The adapted frequency increases slightly down to about 50% saturation, but below this there is a rapid fall-off and complete inhibition at low values of pO_2. In animals which may often be left stranded out of the water this ability to become quiescent is presumably of survival value (Leader, 1970).

The maximum metabolic rate (and locomotor activity) of *Pycnopsyche guttifer* and *P. lepida* occurs in the velocity range in which each species is found in the field (Feldmeth, 1970a) (Fig. 5a). However, these differences in metabolic rate are apparently determined, not by specific genetic adaptations, but by acclimation to the different current velocities (Feldmeth, 1970b). The rate of oxygen uptake in *Philaniscus plebeius* is maximal in normal seawater at 25°C, at which temperature the frequency of ventilation also reaches a maximum and the efficiency of the ventilatory mechanism in terms of oxygen extraction is about 25%.

The real effectiveness of the ventilatory current can be expressed by the difference between the metabolic rates of anesthetized and maximally active normal animals (Fry, 1947). In the above two species of *Pycnopsyche,* the activity recorded was not maximal, but normal, and Feldmeth (1970a) suggests the term "routine scope for activity" for the difference between the metabolic rates of anesthetized and normally active animals.

The aquatic caterpillars of *Nymphula stratiotalis* and *N. maculalis* live in cases made of leaves and possess numerous filamentous gills (Buckler, 1875; Welch, 1916). Both species produce a current through their cases by rapid undulations of the anterior end of the body for short periods at regular intervals; dorsoventrally in *N. stratiotalis* (Buckler, 1875), horizontally in *N. maculalis* (Welch and Sehon, 1928). In the latter species, the passage of an artificial current through the case, or the removal of part of the case, reduces the frequency of the undulation periods. The frequency of these periods can be increased by decreasing the oxygen content of the water and below 1.0 cm³/liter the undulations are continuous. Also, the frequency increases with increase in temperature, again becoming continuous at about 30.5°C. Below 4.5°C the undulations cease (Welch and Sehon, 1928). This evidence certainly indicates a respiratory function for the gills of *Nymphula*.

Morgan and O'Neal (1931) have shown that the tracheal gills of *Macroneme zebratum* are important for the removal of carbon dioxide from the body. Normal larvae eliminate carbon dioxide more rapidly than gill-less ones and consume more oxygen in the presence of an excess concentration of carbon dioxide.

III. Open Tracheal Systems

Insects with functional spiracles face the primary danger of flooding of their tracheal system. There are two main categories into which these insects can be divided. On the one hand, there are those which need to maintain contact with the surface or with the air spaces of plants, and at best can only make short dives; on the other, there are those which carry air (gas gills) around with them and which have evolved varying degrees of independence of atmospheric oxygen.

A. Total Reliance on Atmospheric Oxygen

1. Surface Breathers

The principal problems faced by surface breathers are how to prevent water from entering the spiracles and how to break through the surface film. They are solved by having part of the cuticle, or a group of hairs, around the spiracle with a greater affinity for air than for water and with which the water thus has a high "contact angle" (Brocher, 1909). Thus, when the structure comes into contact with the surface film "the cohesion of the water is greater than the adhesion to the body" (Wigglesworth, 1953) and the spiracular region remains dry. Such regions of the cuticle or hairs are said to be "hydrofuge." In the case of a ring of hairs around the spiracle, their hydrofuge nature will result in them curving over the spiracle when the animal is submerged and spreading out because of the surface tension effect when the animal is at the surface (Wigglesworth, 1953) (Fig. 6).

Aquatic dipteran larvae provide examples of hydrofuge areas of cuticle around the spiracles. This is achieved by the oily secretion of the peristigmatic glands in this region, first described towards the end of the last century (Leydig, 1859; Batelli, 1879; Viallanes, 1885). Their function in this respect was first suggested by Gazagnaire (1886) and they have since been described in many species (Dolley and Farris, 1929; Keilin et al., 1935; Phillips, 1939).

In many species of Tipulida and Anisopodidae (Diptera) the spiracles open onto a postabdominal disk. When the animal submerges, the disk is withdrawn and the surrounding fleshy lobes close over it to prevent entry of water (Keilin, 1944).

A number of dipteran larvae have their functional spiracles situated on the end of a projection from the body, the "postabdominal respiratory siphon," as in the Culicidae (Diptera). Again the spiracles can be closed by the fleshy perispiracular valves (Keilin, 1944). In Eristalis and Ptychoptera, both of which are bottom dwellers, the siphon is very extensible (Grobben, 1876; Miall, 1895) and a two-centimeter larva of Eristalis can extend its siphon to about 12 centimeters (Réamur, 1734–1742). The siphon is telescopic and is apparently extended by the contraction of circular muscles and withdrawn by the relaxation of these muscles and the invagination of

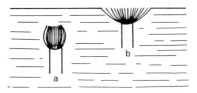

Fig. 6. A spiracle surrounded by a group of hydrofuge hairs. (a) In equilibrium below the surface of the water; (b) in equilibrium at the surface. (After Wigglesworth, 1953.)

the distal part takes place by means of retractor muscles (Miall, 1895). Krogh (1943) has shown that the body wall is virtually impervious and that respiration only occurs via these posterior spiracles.

Hydrofuge hairs surrounding the spiracles have been described in the larva of *Hedriodiscus truquii* (Diptera) which lives in thermal springs (Stockner, 1971) but otherwise seem to be fairly well restricted to insects with gas gills, such as *Notonecta* (Hemiptera) (Hoppe, 1911; Brocher, 1913). Also, in some insects which have a gas gill, communication with the atmosphere is made via a respiratory tube. Thus, in *Nepa* (Hemiptera) the region around the posteriorly situated spiracles is developed to form a tube, and in *Hydrophilus* (Coleoptera) the antennae have become modified to perform a similar function (Miall, 1895; Wigglesworth, 1953).

A double log plot of the rate of oxygen uptake against body weight for larvae of *Hedriodiscus truquii* showed a linear relationship up to a weight of 40 mg. Beyond this, the respiratory rate did not increase (Stockner, 1971).

2. Utilization of Plant Air Spaces

Larvae, pupae, and young adults of the *Donaciinae* (Coleoptera) live in mud on the roots of aquatic plants (Böving, 1910), an environment which has a low oxygen content (Varley, 1937), and Houlihan (1969) points out that the larvae are able to tolerate a high environmental pCO_2. According to Beadle and Beadle (1949), this is characteristic of many aquatic insects. The functional spiracles of the larva are at the end of a long, pointed posterior siphon, which it uses to penetrate the gas spaces in the roots (Deibel, 1911; Varley, 1937; Houlihan, 1969).

Böving (1910) observed that, prior to pupation, the larva constructs a cocoon on a root. While doing this, it bites a hole in the root and leaves a space in the floor of the cocoon above the hole, and it was suggested by Ege (1915b) that the pupa and adult continue to obtain their oxygen from the plant. The hole in the root eventually becomes covered over by a scar, but oxygen still passes through this into the cocoon (Houlihan, 1970). In *Donacia simplex* pupation occurs around October. The pupal stage is short, but the adult stays in the cocoon until June. Houlihan (1970) demonstrated that the respiratory rate ($\mu l/O_2$/mgm/hour) shows little change with increase in temperature over the range 4–14°C in adults taken from the cocoons in January, but those removed progressively later in the year started to show a marked increase in oxygen uptake above 8°C. Emerged, inactive adults showed a similar increase in the rate of oxygen uptake with increase in temperature, but the overall rate was much higher than in those taken from their cocoons. Houlihan (1970) suggests that the adults are thus adapted to a low oxygen tension, brought about by the formation of scar tissue which slows down the diffusion of oxygen into the cocoon. Several other insects obtain their supply of oxygen in a similar manner. The larvae of the dipterans *Chrysogaster* and *Notiphila* occupy the same type of habitat as *Donacia;* and they and the larvae of other dipterans, such as *Mansonia* and *Taeniorhynchus richiardii,* have similar respiratory siphons. In the pupae of *Notiphila* and *Mansonia* the siphons are thoracic (Wesenberg-Lund, 1918; Edwards, 1919; Keilin, 1915, 1944; Varley, 1937). *Lissorhaptrus* (weevil) larvae also pierce plants (Varley, 1937).

Alternatively, some insects, such as *Elmis* (Coleoptera) (Brocher, 1912b) and the pupating larva of *Hydrocampa* (Lepidoptera) (Ege, 1926), bite into the air spaces. However, see Section III,B,2 with reference to *Elmis.*)

B. Gas Gills

It has been known for some time that many insects carry air with them when they submerge and it was first suggested by Comstock (1887) that these bubbles extract oxygen from the water. Brocher (1910) thought that in *Dytiscus* the air had a purely hydrostatic function and was not involved in any way with respiration, but both Babák (1912) and Wesenberg-Lund (1912) considered it to be of both respiratory and hydrostatic importance. Ege (1915a) worked on this problem with a variety of species (dytiscids, corixids, and *Notonecta*) and demonstrated that the air did indeed have an important role to play in respiration. In all of the species at which he looked, the trapped air acted, not only as a store of oxygen, but also as a physical or gas gill during winter. This was also the case in the summer for *Hydrophorus* and the smaller species of *Corixa;* but in *Notonecta, Dytiscus,* and various other dytiscids only the stored, or primary, supply of oxygen was used if the animals were at all active.

Ege (1915a) worked exclusively with animals in which the air "store" steadily diminished in size, ultimately depending on the temperature and the amount of activity, forcing the animal to return to the surface to replenish its supply. These are therefore compressible gills.

On the other hand, many insects have evolved a gas gill which is incompressible and this allows them to remain below the surface of the water indefinitely. The incompressible gas gill was called the "plastron" by Thorpe and Crisp (1947a), a term first employed by Brocher (1912b) to cover all forms of gas gill.

In both types of gas gill, when the oxygen tension (pO_2) in the gill falls below that of the water, oxygen will diffuse into it, the rate depending on the difference in tension. Nitrogen is extremely important in these gills and is primarily responsible for their functioning (Ege, 1915a; Thorpe and Crisp 1947b; Rahn and Paganelli, 1968). The initial gas concentrations in the gas film will be the same as in air (i.e., about 20% oxygen, $<1\%$ carbon dioxide, and 79% nitrogen). Carbon dioxide is extremely soluble in water and diffuses out of the store as fast as it is produced by the animal. Thus, gas gills can be thought of solely in terms of oxygen and nitrogen.

1. Compressible Gas Gills (Temporary Air Stores)

a. Structure and Occurrence. Compressible gas gills are normally held in place by body hairs and/or consist of bubbles in the subelytral space. Thus, among the Coleoptera, some members of the Dryopoidea and the Hydrophilidae have a film of air held in place by a hair pile. The hairs are not so regular and densely packed as in the case of the plastron (Section III,B,2). In the Dryopoidea there are between 6×10^4 and 8×10^5 hairs/cm^2; in the Hydrophilidae 1×10^4 to 1×10^7/cm^2. In a number of hydrophilids there are two sets of hairs (long and short). The long hairs hold a bubble which decreases in volume fairly quickly when the animal dives, whereas the smaller hairs hold a gas film which is relatively permanent as long as the animal does not dive too deep. These two layers have been termed the "macroplastron" and "microplastron," respectively (Thorpe and Crisp, 1949). When the macroplastron has collapsed, the long hairs which supported it help to protect the inner layer from wetting. *Dryops* and some hydrophilids can crawl about on submerged plants, enveloped by an air bubble from which only the animal's extremities project.

Many hemipterans (Nepidae, Corixidae, Notonectidae, Pleidae, Naucoridae, Belostomatidae) carry a layer of air supported by an extensive hair pile and often supplemented by an air bubble in the subelytral space (Brocher, 1912b; Ege, 1915a; Thorpe, 1950). *Gerris* and *Velia* both have a macroplastron and a microplastron supported by long and short hairs, respectively, and some adult trichopterans also have these two layers (Thorpe, 1950).

Among the Lepidoptera the larvae of *Palustra* and *Diacrisia* (Arctiidae) carry air in two lateral tracts of both large and small hairs (Myers, 1935; Thorpe, 1950). The larvae of some pyralids (e.g., *Cataclysta*) and *Hydrocampa* have longitudinally ribbed or flanged spines and, in the former at least, these are thought to retain an air film (Lübben, 1907b; Thorpe, 1950). All of these animals belong to group III of Thorpe and Crisp (1949) and Thorpe (1950).

In a wide range of other insects found on the surface or in the vicinity of water, hydrofuge hairs are present. They are not dense and probably serve to prevent wetting if the animal accidentally falls into the water. In other words they have more of a "rainproofing" effect. These belong to group IV (Thorpe and Crisp, 1949; Thorpe, 1950).

b. *Function*. In a compressible gill, the pressure will always be the same as the hydrostatic pressure. Since oxygen is being consumed by the insect, the oxygen tension in the gill will fall. When it becomes lower than that in the water (i.e., immediately if the water is in equilibrium with the air) the difference in tension (ΔP_0) will cause an inward diffusion of oxygen. However, the fall in total pressure in the gill caused by the ΔP_0 produces a difference in the nitrogen tension between the gill and the water (ΔP_N) and nitrogen diffuses out. Immediately below the surface, where the gill is at approximately atmospheric pressure, $\Delta P_0 = \Delta P_N$ if the water is in equilibrium with the air (Rahn and Paganelli, 1968).

The primary oxygen supply is about 0.19 of the volume of the air store (Ege, 1915a). The lifetime of the gill based on this stored oxygen alone (T_s) is given by the ratio of the initial volume of oxygen (V_{io}) to the rate of its consumption by the animal (q) (Rahn and Paganelli, 1968).

$$T_s = V_{io}/q \qquad (5)$$

If nitrogen diffused out of the store at the same rate that oxygen diffused in, the oxygen taken up from the water would be equal to the initial volume of nitrogen (V_{iN}). This is about 0.79 of the total volume and hence would increase the gill lifetime by about four times (Ege, 1915a). Thus, under these conditions, the additional lifetime provided by uptake from the water (T_w) is the ratio of the initial volume of nitrogen to the animal's rate of oxygen consumption.

$$T_w = V_{iN}/q \qquad (6)$$

The total gill lifetime (T) is, of course, $T_s + T_w$.

However, Ege (1915a) pointed out that oxygen diffuses into the gill faster than nitrogen diffuses out, and so the value of T_w will be increased by the ratio between the "invasion coefficients" of these two gases. The invasion coefficient can be defined as "the quantity of gas crossing a unit area of gas–liquid interface under unit difference of gas tension between two phases" (Thorpe, 1950). Equation (6) becomes

$$T_w = \frac{V_{iN}}{q} \frac{i_0}{i_N} \qquad (7)$$

(where i_0 is the invasion coefficient of oxygen, and i_N, the invasion coefficient of nitrogen). Ege determined values for the invasion coefficients of oxygen and nitrogen and obtained a ratio of 3.2. Thus oxygen uptake from the water (V_{wo}) is $0.8 \times 3.2 = 2.56$ times the total volume of the gill, compared with a storage level of only 0.2 of the total volume.

The ratio between the total volume of oxygen utilized (V_0) and the volume of the oxygen store is called the "gill factor." Another way of expressing this is as the ratio between total gill lifetime and the gill lifetime due to the oxygen store alone (Rahn and Paganelli, 1968), i.e.,

$$\text{Gill factor} = V_0/V_{i0} = T/T_s$$

From Ege's data this gives

$$\text{Gill factor} = \frac{V_0}{V_{i0}} = \frac{2.76 \times \text{gill volume}}{0.2 \times \text{gill volume}} = 13.8$$

Another way of looking at this is to consider the amount by which the gill can extend its lifetime over and above that due to the oxygen store alone. This is given by V_{wo}/V_{i0} or T_w/T_s. From the above data $V_{wo}/V_{i0} = 2.56/0.2 = 12.8$. In other words, the gill can extract 12.8 times as much oxygen from the water as it possesses initially in its store (Ege, 1915a).

Rahn and Paganelli (1968) have described a mathematical model for a compressible gill, operating under conditions where the water is in equilibrium with the air at 1 atm at all depths. They showed that, if reduction in volume occurs solely by decreasing the height of the gill (i.e., keeping its area constant), when the animal is just below the surface of the water (where $\Delta P_0 = \Delta P_N$) the oxygen tension in the gill falls rapidly at first but soon reaches a steady state, which is then maintained until the volume reaches zero. This is in marked contrast to what happens if the height remains constant while the area decreases. Under the latter conditions a steady state is never reached, the oxygen tension in the gill falling continuously. Since, in both cases, the volume of oxygen shows a steady decrease with time, it is suggested that the metabolic demands of an insect are met both by utilization of the oxygen store and by diffusion of oxygen from the water. They estimated that the stored oxygen contributes less than 5% to the total amount utilized. Ege's data gives a value of about 7%. Assuming this to be negligible, Rahn and Paganelli (1968) derived equations to determine ΔP_0 and total gill lifetime (T).

$$\Delta P_0 = \frac{\Delta X}{i_0} \frac{V_0}{A} \tag{8}$$

and

$$T = \frac{\Delta X}{i_N A} \frac{V_{iN}}{\Delta P_N} \tag{9}$$

(where i_0 and i_N are equal to the product of the solubility and the diffusion coefficient of oxygen and nitrogen, respectively). Combining Eqs. (8) and (9)

$$T = \frac{V_{iN}}{q} \frac{i_0}{i_N} \tag{10}$$

They used values of 4.07×10^{-5} and 2.04×10^{-5} (mm Hg)$^{-1}$ for the solubilities (Hodgman, 1958) and 2.28×10^{-5} and 2.08×10^{-5} cm^2 second^{-1} for the diffusion coefficients (Gertz and Loeschke, 1954) of oxygen and nitrogen, respectively, in water at 20°C. From this data the ratio $i_0/i_N =$ 2.19. From Eqs. (5) and (10)

$$\text{Gill factor} = \frac{T}{T_s} = \frac{V_{iN}}{V_{i0}} \frac{i_0}{i_N} = \left(\frac{0.79}{0.21}\right) \times 2.19 = 8.28$$

This value is lower than that derived from Ege's data for two reasons. First, Ege used a higher value for the ratio of the invasion coefficients. Second, the approximation used by Rahn and Paganelli to derive T (i.e., $T = T_w$ to within 5%) leads to a slight underestimation of the gill factor.

The rate of decrease in volume of the bubble just below the surface in air-saturated water is given by Crisp (1964)

$$\frac{dV}{dt} = Ai_0(\Delta P_O) - q + Ai_N(\Delta P_N) \tag{11}$$

(where ΔP_N is equal and opposite to ΔP_O).

The effect of increasing oxygen consumption in Rahn and Paganelli's constant area model is to increase ΔP_O and so cause a decrease in the steady-state level of gill pO_2. This has the effect of increasing ΔP_N and hence the diffusion rate of nitrogen increases and the gill lifetime is shortened. However, increase in depth has very little effect on the steady-state level of pO_2, although the gill lifetime is still shortened. There are two reasons for the latter. First, the dive produces a transient increase in gill pO_2 above the water pO_2, resulting in oxygen *loss* from the gill. Second, and more important, increase in depth increases ΔP_N and so nitrogen diffuses out faster.

Rahn and Paganelli (1968) used an O_2–N_2 diagram based on the O_2–CO_2 diagram of Fenn *et al.* (1946) to demonstrate the relationship between these two gases in compressible and incompressible (plastron) gills. In Fig. 7a pN_2 is plotted against pO_2 in atmospheres. The 45° diagonals give the proportions of these gases at the stated pressures and the dotted line gives their relative proportions in air.

Part of this figure is reproduced in Fig. 7b and point A indicates the partial pressures of the gases in air at 1 atm or in air-saturated water at all depths. If the animal remains just below the surface, its metabolism will

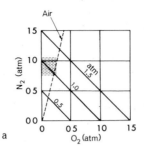

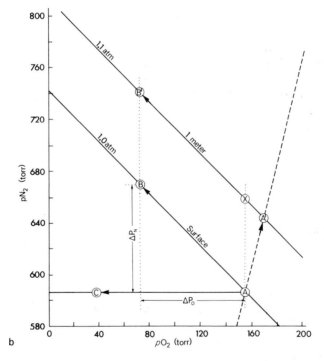

Fig. 7. O₂–N₂ diagrams. (a) This shows the relationship between oxygen and nitrogen up to a total pressure of 1.5 atm. The dashed line indicates their relative proportions in air. The shaded area is enlarged in (b). (b) A indicates the relative proportions of oxygen and nitrogen in air at 1.0 atm. In a compressible gas gill the proportions of these gases change during a dive in the direction of B, provided the animal remains just below the surface of the water. If the animal dives to 1 meter the initial proportions are given by A′ and they change in the direction of B′. To right of X, the gas gill loses oxygen to the water. In an incompressible gas gill the change in the proportions of the two gases is in the direction of C at all depths. (After Rahn and Paganelli, 1968.)

cause a fall in pO_2 and a corresponding rise in pN_2 to keep the pressure in the gill constant. This will cause a change along the line AB until the steady state is reached at B. The length AB increases with increase in metabolic rate (q) and/or decrease in gill surface area (A). Thus, B will be further to the left in animals such as *Dytiscus,* with a high value of q/A, compared with the smaller dytiscids, such as *Hyphydrus,* in which this ratio is low (Ege, 1915a; Rahn and Paganelli, 1968). ΔP_O and ΔP_N are given by the vertical and horizontal differences from A, respectively, and are equal to each other.

If the animal dives, the pressure to which the gill is subjected increases to a new level A' and the steady state will then be achieved when B' is reached. If the metabolic rate is the same as before, B' will lie vertically above B and, since ΔP_O remains the same under this condition, whereas ΔP_N has increased, gill lifetime will be shortened. Note that between A' and X oxygen will be lost to the surrounding water because A' lies to the right of A (Rahn and Paganelli, 1968).

Some values of gill pO_2 at the start, and at various times after the start, of a dive are given by Ege (1915a). At the time of diving, gill pO_2 varied between 16.0% and 19.6% in dytiscids and *Notonecta,* and in a specimen of *Corixa geoffroyi* he recorded 20.3% immediately after breathing. The tension of this gas fell rapidly after the animals dived (Table I). Gill pCO_2 was low (maximum recorded was 2.8%) and remained so after diving. A type of gill which is intermediate between the compressible type and the plastron is found in the elmid beetle *Potamodytes tuberosus.* This beetle lives in fast-flowing water, clinging to the substrate and facing upstream. It possesses a large air bubble supported on its widespread front legs and drawn out posteriorly by the current. Technically, this is a compressible structure, and it does decrease in size rapidly if the current velocity is reduced. However, in the normal situation of a fast current of air-saturated water, it is a permanent gas gill (Stride, 1955). The explanation of this can be given in terms of the Bernouilli equation (Stride, 1958).

$$p + \tfrac{1}{2}\rho V^2 = \text{constant}$$

(where p = pressure, ρ, a gravitational constant, and V, velocity). Thus, increase in flow velocity will produce a region of reduced pressure. In the case of *Potamodytes,* the water velocity will increase as it is deflected in the region of the animal. Thus, the air-saturated water will experience a drop in pressure, become supersaturated, and give up oxygen and nitrogen to the bubble. If the bubble is removed, it will slowly grow again until it reaches an optimum size for the current velocity. Stride (1955, 1958) demonstrated that the bubble pressure falls below atmospheric pressure when a critical current velocity is exceeded.

TABLE I

THE OXYGEN TENSION (pO_2) IN GAS GILLS DURING A DIVE[a]

Air	pO_2 about 20%	
At start of dive	pO_2 16.0–19.6%	

Species	Time after diving (minutes)	pO_2 (%)
Dytiscidae		
Colymbetes fuscus	1	15.4
Acilius sp.	2	11.9
Colymbetes sp.	5	9.2
Dytiscus marginalis	5	3.0
Ilybius sp.	20	0.7
Notonectidae		
Notonecta sp.		
Dorsal space	{ 1	11.0
	2	5.3
Canals (ventral surface)	{ 1	15.4
	2	3.9
Corixidae		
Corixa geoffroyi		
Passive	{ 10	9.4
	15	5.8
Active	5	6.5

[a] Data from Ege (1915a).

2. Incompressible Gas Gills (Plastron Respiration)

a. Structure and Occurrence. A plastron is a thin film of air which may be held in place by hairs or by cuticular modifications.

The "hair plastron" is found on the adults of certain coleopterans, *Aphelocheirus aestivalis* (Hemiptera) and *Acentropus niveus* (Lepidoptera). The best-known example is that of *Aphelocheirus* (Ussing, 1910; Larsen, 1924; Szabo-Patay, 1924; Thorpe and Crisp, 1947a). The adult has nine pairs of spiracles, two thoracic and seven abdominal. The first three have single openings into deep longitudinal grooves, while the others have numerous small openings along canals which radiate out from the center, forming a "rosette." The grooves, canals, and openings are all lined with fine hairs, as is also the whole of the ventral and much of the dorsal surface of the animal. There are between 2 and 2.5×10^8 hairs/cm² (Fig. 8A). The animal is heavier than water, in contrast to those animals which carry air bubbles, and avoids regions of low pO_2 by means of its tactile sensilla, visual sense, and specialized pressure receptors (Thorpe and Crisp, 1947c).

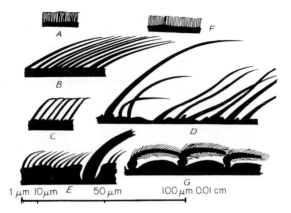

Fig. 8. Diagrams of sections through the hair pile of various insects. (A) *Aphelocheirus aestivalis* (abdominal sternum); (B) *A. aestivalis* (pressure sense organ); (C) *Haemonia mutica* (thoracic sternum); (D) *Donacia simplex* (abdominal sternum); (E) *Elmis maugei* (thoracic sternum); (F) *Stenelmis crenata* (metathoracic sternum); (G) *Phytobius velatus* (abdominal sternum). (From Crisp and Thorpe, 1948.)

The plastron of *Aphelocheirus* is extremely efficient and is only equalled in a few elmid beetles [*Stenelmis crenata* (Fig. 8) and *Cylloepus barberi*] and in the weevil *Phytobius velatus* (Thorpe and Crisp, 1949). *Phytobius* is covered by scales which touch or overlap each other over most of the body (Fig. 8). The scales are covered with a dense hair pile (1.8 to 2.0 $\times$ 10^8 hairs/cm²). All of these animals belong to group I of Thorpe and Crisp (1949) and Thorpe (1950) which is characterized by a hair density in excess of 10^8/cm². In the rest of the animals which possess a hair plastron, the hairs tend to be longer and not so dense as in those just described, and they have been assigned to a second group (group II of Thorpe and Crisp, 1949; Thorpe, 1950). However, like the animals in group I, members of group II do not normally need to keep returning to the surface and may remain submerged for months (Brocher, 1912b; Davis, 1942; Thorpe and Crisp, 1949). They are characterized by a hair density of 3×10^6 to 1.6×10^7/cm² (Fig. 8C,E).

Included in group II are most of the aquatic elmid beetles (e.g., *Elmis* and *Riolus*). *Elmis maugei* and *Riolus capreus* have eight pairs of spiracles which open into a lateral groove on each side. These grooves communicate with the subelytral space and the plastron. The latter covers most of the lateral and lateroventral surfaces of the animal. The hairs hold a thin film of air, but a thicker layer is often present. This extra layer (macroplastron) is soon lost by the Ege effect when the animal submerges. It can only be maintained by frequent visits to the surface or by using some of the air in the subelytral space. However, under well-oxygenated conditions the microplastron is sufficient for the animal's needs (Thorpe, 1950).

Also in this group is another beetle, *Haemonia mutica* (Donaciinae), the plastron of which covers the whole of the ventral surface and the antennae. The hair density is greater on the latter. The hairs are bent at their tips, as in *Aphelocheirus,* but are much longer (Thorpe and Crisp, 1949).

The hemipteran *Hydrocyrius columbiae* (Belostomatidae) has a film of hair held in place on the ventral surface by small hairs, bent at their tips, at a density of 1 to 2×10^6/cm². Longer hairs, ending in flattened blades, lie over the surface of the short hairs. This ventral air film communicates via three pairs of "bridges" with a subelytral bubble. When the animal comes to the surface the air supply is renewed via a posterior retractile siphon. All of the spiracles except the first communicate with the gas gill (Miller, 1961).

The lepidopteran *Acentropus niveus* has a wingless form which remains submerged for the whole of its adult life of 3 or more days (Berg, 1941). Its plastron consists of three- and four-pronged scales and it probably also belongs to group II (Thorpe, 1950). Thorpe and Crisp (1949) have commented that in group II the plastron is functionally but not structurally perfect.

The other way in which the plastron may be held in place is by means of cuticular modifications, and these are invariably associated with spiracular gills. The spiracular gill is an outgrowth of the spiracular atrium and/or of the region immediately surrounding it (Hinton, 1966b) (Fig. 9). With the exception of those in some of the Chironomidae, all spiracular gills possess a plastron (Hinton, 1968). They are confined to the pupal stages of certain Coleoptera and Diptera, except in the Torridinicolidae, where the larvae also bear them, and the Sphaeridae and Hydroscaphidae, where they are confined to the larvae (Hinton, 1968).

In most cases the plastron consists of vertical struts which divide near their tips to provide horizontal branches which are fused with the branches of adjacent struts. The effect is to form an open hydrofuge network, supported by struts, enclosing an air-filled lumen. Other cuticular branches may traverse the lumen. This type is found in the marine tipulids *Geranomyia* and *Dicranomyia,* in all other dipterans except the tipulids mentioned below, and in the Torridincolidae (Coleoptera) (Hinton, 1967b,c, 1968) (Fig. 10a).

There are several other types (Hinton, 1968). In the tipulids *Antocha, Lipsothrix* (semiaquatic), and some species of *Orimargula* the gill bears grooves ("plastron lines") roofed over by an open network (Fig. 10b,c). In *Lipsothrix* the roofing is flat, but it is arched in the other two genera (Hinton, 1966a, 1967a). In *Orimargula hintoni* the plastron consists of

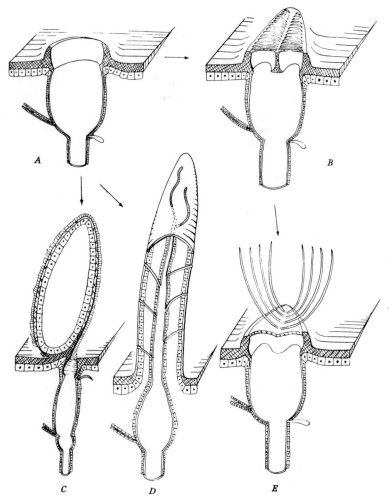

Fig. 9. Diagrams to show the origin of different types of spiracular gills. (A,B) Spiracles; (C) simulid type of spiracular gill; (D) tipulid type of spiracular gill; (E) possible stage in the development of the type of spiracular gill found in *Psephenoides.* (From Hinton, 1966b.)

numerous, evenly spaced tubercles (Hinton, 1968). The spiracular gills of *Psephenoides* (Coleoptera) consist of tubular outgrowths (up to 40 in *P. gahani* (Hinton, 1966b) (Fig. 10d). In the Eubriinae, Sphaeridae, and Hydroscaphidae (Coleoptera) the dilated spiracle forms the plastron (Hinton, 1966b, 1967c).

b. Function. In the compressible gas gill described in Section III,B,1, the gill must always be at hydrostatic pressure and so nitrogen will diffuse out

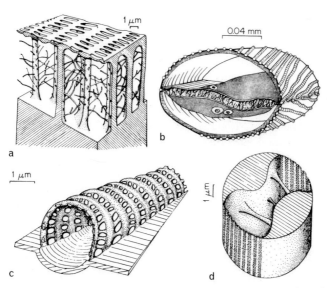

Fig. 10. Diagrams of some of the types of plastron. (a) From *Aphrosylus celtiber;* (b) base of spiracular gill branches of *Antocha vitripennis* to show the plastron lines, aeropyles, and spiracular atrium; (c) part of plastron line of *Antocha bifida;* (d) branch of a spiracular gill of *Psephenoides volatilis* showing the plastron network. [(a) from Hinton, 1967b; (b) from Hinton, 1957; (c) from Hinton, 1966a; (d) from Hinton, 1966b).]

and result in the eventual loss of the gill. In the case of the plastron, however, which is incompressible, while gill pO_2 will decrease and so establish a ΔP_0, gill pN_2 must remain constant and so nitrogen will not diffuse out. With reference to the O_2–N_2 diagram in Fig. 6b, the proportion of the two gases changes from A along the direction of the horizontal line AC. The position of C, like that of B, becomes further to the left as the metabolic rate/gill surface area ratio increases. Gill pO_2 can never exceed 155 mm Hg and it can be seen that its minimum value is zero, at which level gill pN_2 = total gas pressure = 595 mm Hg (or 0.79 atm) (Rahn and Paganelli, 1968). In *Coxelmis novemnotata* Davis (1942) has shown that the spaces between the hairs are such that the surface tension forces could withstand a much lower gill pressure than actually occurs.

The "diffusion resistance" of oxygen (R) along a closed path has been defined as the ratio of the tension difference (ΔP_0) to its rate of uptake by the tissues (q) (Thorpe and Crisp, 1947b), i.e.,

$$R = \Delta P_0/q$$

To obtain a standard measure for the efficiency of respiration, Thorpe and Crisp (1947b) used the term $1/q_bR$ (where q_b is the basal metabolic rate) as a "respiratory index." They found that the respiratory index of the adult of *Aphelocheirus aestivalis* is very similar to that of the fifth-instar larva (which does not possess a plastron). However, the diffusion resistance of the larval cuticle is 3.7×10^4 which gives a permeability of 1.6×10^{-7} cm^3O_2/second/cm^2. In contrast, the thicker adult cuticle has a permeability of only about a quarter of this value and diffusion through it can only just provide the animal with its basal metabolic requirements (6 mm^3O_2/hour/ animal) (Thorpe and Crisp, 1947b). This indicates that the adult *Aphelo-cheirus* obtains a large part of its oxygen supply via its plastron.

It is of interest to know whether the whole or only part of the plastron is functional since, if there is a significant difference in gill pO_2 along the length of the plastron as a result of the diffusion of oxygen through it, ΔP_0 will decrease with increase in distance from the spiracles and so only the region of the plastron close to the spiracles will be fully utilized (Crisp and Thorpe, 1948; Crist, 1964).

The average drop in oxygen tension across the water–air interface $(\Delta P_0)_{av}$ is given by the ratio between the metabolic rate (q) and the product of the area of the plastron and the invasion coefficient of oxygen:

$$(\Delta P_0)_{va} = q/Ai_o$$

For efficiency $(\Delta P_0)_{av}$ should be considerably less than 0.2 atm, the pO_2 of air-saturated water. For hair plastrons $(\Delta P_0)_{av}$ lies between 0.55×10^{-2} (*Aphelocheirus*) and 10.8×10^{-2} (*Haemonia*) when $q = q_b$, the basal metabolic rate. In *Hydrophilus*, which has a compressible gill, $(\Delta P_0)_{av} =$ 12 to 17×10^{-2} when $q = q_b$ (Thorpe and Crisp, 1949). For the spiracular gill plastron of *Simulium ornatum* $(\Delta P_0)_{av}$ lies between 8×10^{-2} and 16×10^{-2} for actively moving pharate adults, but for the early period of adult pharate life it is 3 to 6×10^{-2} (Hinton, 1968). Thus, all the plastrons investigated show a favorable ratio between oxygen uptake and the area of the plastron.

The relationship between $(\Delta P_0)_{av}$ and the actual drop in pressure $(\Delta P_0)_x$ at distance x from the spiracle (Fig. 11) is given by

$$(\Delta P_0)_x = \Delta P_{0av} n x_1 \cosh n(x_1 - x)/\sinh n x_1 \qquad (12)$$

(Thorpe and Crisp, 1974b; Crisp and Thorpe, 1948; Crisp, 1964) (where

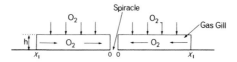

Fig. 11. Diagrammatic view of a flat plastron to show the diffusion pathways of oxygen. x_1 is the maximum extent of the plastron. (After Crisp and Thorpe, 1948.)

x_1 is the maximum extent of the plastron; n, $(i_0/Dh)^{1/2}$; D, the diffusion constant of oxygen within the plastron; h, the height of the plastron). From Eq. (12) it can be seen that $(\Delta P_0)_x$ is maximal at the spiracle, where $x = 0$

$$(\Delta P_0)_{x=0} = \Delta P_{0av} \, nx_1/\tanh nx_1$$

and minimal at the outer edges of the plastron, where $x = x_1$

$$(\Delta P_0)_{x=x_1} = \Delta P_{0av} \, nx_1/\sinh nx_1$$

Thus, the functional efficiency of the plastron, in terms of the uniformity of oxygen uptake over its surface, is dependent solely on the function $nx_1 = (i_0x_1^2/Dh)^{1/2}$ (Crisp, 1964). Diffusion alone has been shown to be adequate to account for the uptake of oxygen by the plastron (Thorpe and Crisp, 1947b). When vaues of the relative drop in partial pressure across the plastron $[(\Delta P_0)_x/\Delta P_{0av}]$ are plotted against relative distance from the spiracle (x/x_1) it can be seen that there is little deviation from unity when nx_1 is less than 1.0 and so under this condition the entire plastron is functional (Fig. 12). However, when nx_1 exceeds 1.0 the deviations become more and more marked and thus less and less of the plastron can be utilized (Crisp and Thorpe, 1948; Crisp, 1964). In the case of the tubular spiracular gills of some species of *Psephenoides* h is replaced by $r/2$ (where r is the radius) in the above expression (Crisp, 1964; Hinton, 1968). Hence

$$nx_1 = (2i_0x_1^2/Dr)^{1/2}$$

nx_1 for the hair plastron of *Haemonia* (Coleoptera–Donaciinae) is 0.47 (Thorpe and Crisp, 1949) and values for the plastrons of spiracular gills range from 0.19 in *Antocha bifida* to 0.97 in the longest gill branches of *Simulium ornatum* (Hinton, 1968) and, hence, they are all fully utilized. The principal diffusion barriers are at the water–air interface and between the tracheoles and the tissues (Thorpe and Crisp, 1947b).

There is a conflict between the need to provide a large interface for oxygen uptake and to supply the maximum resistance to water invasion. For the hair plastron the result has been the development of a dense hair pile, which still leaves sufficient space between the hairs to provide an adequate water–air interface and to allow a reasonably free diffusion path to the spiracles. In its most advanced form (e.g., in *Aphelocheirus*) the hairs are evenly spaced, short, and bent over almost horizontally at the tip and the latter increases the resistance of the plastron to wetting (Thorpe, 1950).

The hair plastron can theoretically be destroyed by wetting in two ways: by lowering the surface tension or by increasing the pressure on it to cause the hairs to bend. In *Aphelocheirus*, wetting occurs at a surface tension of about 26 dynes/cm (e.g., 10 to 12% butyl alcohol), and Thorpe and Crisp

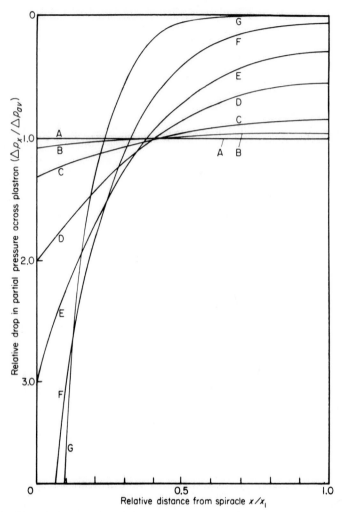

Fig. 12. The relative drop in oxygen tension across the plastron interface ($\Delta P_x/\Delta P_{av}$) plotted against the relative distance from the spiracle (x/x_1). Each curve is derived from equation (12) $nx_1 = 0.1$ (A), 0.5 (B), 1.0 (C), 2.0 (D), 3.0 (E), 5.0 (F), and 10.0 (G). (From Crisp, 1964.)

(1947a) made use of this to check on the respiratory function of the plastron. They wet various amounts of the plastron and showed that the treated animals exhibited the same behavioral patterns as those subjected to lowered environmental pO_2. In *Haemonia mutica,* which belongs to group II, wetting of the body plastron occurs in 7 to 9% butyl alcohol, and wetting of the antennal plastron in 9 to 10% (Thorpe, 1950).

Insects with spiracular gill plastrons are found in flowing water where the surface tension normally exceeds 70 dynes/cm. Like the hair plastrons of *Aphelocheirus*, the plastrons of *Antocha* and *Simulium* are not wetted until the surface tension falls to about 25 dynes/cm. This value corresponds to a reduction in the contact angle to 55° to 60°.

Thorpe and Crisp (1947a) showed that the hairs of *Aphelocheirus* (group I) start to bend at an excess pressure of 3.5–5.0 atm (1 atm is equivalent to a depth of about 10 meters), thus causing collapse of the plastron, whereas, in the structurally less efficient plastron of *Haemonia* (group II) only an additional 0.5–2.0 atm is required (Thorpe, 1950). (The microplastrons found in some animals assigned to group III can typically only withstand an additional 0.5 atm). *Hydrocyrius* can withstand a pressure of 220 cm Hg (about 3 atm) for at least 5 minutes without wetting of the plastron, although some flattening of the hairs occurs and the tracheae are flattened at this pressure (Miller, 1961).

However, Hinton (1968) points out that the time for which the plastron is subjected to excess pressure is of importance. In the spiracular gill plastron of *Geranomyia unicolor,* wetting occurs immediately at 1.4 atm, but at 1 atm it takes 10 minutes, and at 0.3 atm over 2 hours. In all spiracular gill plastrons, which are structurally more rigid than hair plastrons, increase in pressure will cause wetting long before any structural deformation occurs (Hinton, 1968). In contrast to Thorpe and Crisp (1947a), Hinton (1968) has shown that this is also the case for the hair plastron of *Aphelocheirus*. While the resistance to wetting varies in different species, all those with spiracular gill plastrons that have been investigated have a wide margin of safety (Hinton, 1968). Decrease in surface tension causes wetting at lower pressures. There are five ways in which structural deformation is prevented in spiracular gill plastrons. In many tipulids (e.g., *Geranomyia*) the rigidity of the vertical struts is sufficient in itself; in *Psephenoides* and the Eubriinae there are, in addition, strong cuticular struts across the lumen, and in the Blepharoceridae parts of the tissue have become cuticularized. The tipulids *Antocha* and *Orimargula* have an internal pressure of over 4 atm and in the Simuliidae the surrounding water has free access to the lumen and so will equalize any pressure differences (Hinton, 1968).

Insects possessing a plastron must live in well-oxygenated water, such as is provided by fast currents, the edges of lakes, and the intertidal zone (Hinton, 1968). This is extremely important since, under conditions of low environmental pO_2, the plastron will work in reverse and extract oxygen from the tissues rather than from the water. *Aphelocheirus* can only survive down to an oxygen concentration of 8.6% of the total gases, and below this it shows signs of asphyxia (Thorpe, 1950). In many cases, plastron-bearing animals are subjected to large fluctuations in the water level of their environment and they are often left dry for considerable periods. The plastron, however, is rigid and does not collapse under these conditions, and so it serves to prevent water loss without impairing the inward diffusion of oxygen (Hinton, 1968).

IV. Ventilatory Currents

Ventilatory currents provide the respiratory surfaces with a continual supply of well-oxygenated water and reduce the diffusion barrier imposed by the establishment of a boundary layer.

In most cases, ventilatory currents are produced either by undulations of the body or by rhythmic movements of the gills themselves (Section II). Leg movements may also be used to produce currents, as in hemipterans (Comstock, 1887). Thus, in *Notonecta,* de Ruiter *et al.* (1951) demonstrated that leg movements and active or passive movement through the water increased the rate of oxygen uptake. Finally, anisopteran dragonfly larvae ventilate their branchial chamber by muscular pumping movements. However, in only a few cases do we have any detailed knowledge of the nature and mode of production of ventilatory currents.

A. Ephemeropteran Larvae

In *Ecdyonurus dispar, Ephemera danica,* and *Leptophlebia marginata* the ventilatory currents flow from the sides and/or below the animal across the gills toward the dorsal surface and then backward along the dorsal midline (Eastham, 1936, 1937, 1939). However, this is not the case in *Cloeon dipterum,* where the current passes backward over the dorsal surface of the abdomen and outward between the gills (Eastham,· 1932), or in *Caenis horaria,* in which the ventilatory current passes between the gills and across the dorsal surface of the abdomen from one side to the other (Eastham, 1934) (Fig. 13).

Common to all species is an anterior-to-posterior metachronal rhythm, each gill starting to move slightly in advance of the next posterior one, and in all except *Caenis* the gills on opposite sides of the same segment beat in unison.

Each gill in *Ecdyonurus dispar* consists of an anterior lamella and a posterior bunch of filaments. The second to sixth pairs of lamellae produce the ventilatory current which irrigates their own surfaces and the bunches of filaments. Each lamella is slightly curved, with its posterior surface convex, and projects laterally when at rest. During ventilation it moves backward and upward and then forward and downward, its tip traveling in an ellipse

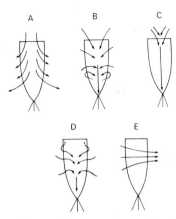

Fig. 13. Diagram of the ventilatory currents flowing over the abdomen of (A) *Cloeon dipterum;* (B) *Leptophlebia marginata;* (C) *Ephemera vulgata;* (D) *Ecdyonurus venosus;* and (E) *Caenis horaria.* (Redrawn from Eastham, 1932.)

(Fig. 14a). The backward movement is the effective stroke and the convex posterior surface leads, gradually increasing its angle to the direction of movement. During the forward recovery stroke the lamella is "feathered" so that it makes virtually no angle with its path of motion. Thus, each lamella generates a current upward and backward. The lag between successive pairs of lamellae results in the lower border of each lamella coming into contact with the posterior surface of the next posterior one toward the start of the effective stroke. The angle between them steadily decreases, forcing the intergill water upward and inward (Fig. 14b). As they move forward, there is a corresponding suction phase drawing in water from below and the side.

Ecdyonurus lives on stones in fast-flowing water. It is dorsoventrally flattened and its head is usually oriented upstream. In this position the gills are relatively protected from the force of the water. The net effect of their movements is to draw water over themselves from the fairly still zone around the sides and below the body, and then force it posteriorly over the dorsal surface of the abdomen (Eastham, 1937).

Ephemera danica has a similar ventilatory current, except that water is only drawn toward the dorsal surface of the abdomen anteriorly. There are seven pairs of bifurcate gills, each narrow lamella being fringed with numerous filaments, but only the second to sixth produce the current. At rest, each gill branch is held reflexed over the dorsal abdominal surface. The effective stroke is a downward and inward movement, during which an undulatory wave passes from the base of the gill branch to its tip, thereby setting up a current in this direction. Another undulation passes posteriorly

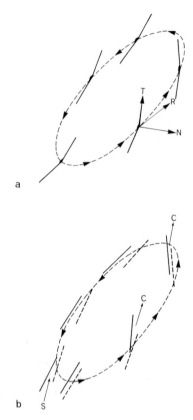

Fig. 14. Ecdyonurus venosus. (a) The elliptical path of a left gill as seen from behind. T,N,R, tangential, normal and resultant forces to the surface of the gill. Arrows indicate direction of movement. (b) The paths of two adjacent left gills viewed from behind. The solid line is the anterior of the two. C, compression; S, suction. (After Eastham, 1937.)

across the gill surface, generating a posterior flow (Fig. 15). The net effect is that each gill branch produces a current directed diagonally backward towards the dorsal midline. The gill branch then moves upward and outward. It presents an appreciable angle to its own path of motion in both directions and so the basic oscillatory movement probably contributes little or nothing to the current.

The filaments of the two branches of each gill and of consecutive gills are arranged so that they overlap at all times and are so close together that there is no flow between them. However, the two second gills are twisted spirally such that, during the effective stroke, water is drawn in laterally at their bases and flows round to their inner surfaces. The backward and downward

Fig. 15. *Ephemera vulgata*. Diagrams to show the undulatory movements of a gill. (A) The axial undulation; (B) the lateral aspect of a left gill; a,t,r, axial, transverse, and resultant forces. (From Eastham, 1939.)

stroke of consecutive pairs of gills results in a posteriorly directed current. The intersegmental phase lag is about one-eighth of an oscillation (i.e., 45°), which means that when the second pair is at one end of an oscillation, the sixth pair is at the other end (Eastham, 1939).

Larvae of *Ephemera* burrow in fairly fine sediment and Eastham (1939) suggests that the reason for the virtual absence of lateral currents is that they would damage the walls of the burrow.

Like *Ephemera, Leptophlebia marginata* has seven pairs of bifurcate gills, the first and last of which are not used in the production of the ventilatory current. The gill branches are simple and lanceolate. As in *Ecdyonurus* they draw water in towards the dorsal midline, but they differ from this animal in one very noticeable respect. Successively more posterior gills are held at lower angles to the abdomen so that the anterior ones sample the water dorsolaterally, the posterior ones ventrolaterally.

The lamellae move through an ellipse with the effective stroke directed backward and inward (Fig. 16a). During the effective stroke both lamellae of each pair present a large angle to their path of motion, the anterior one being inclined towards the midline and so producing a posteromedian current. During the return stroke, both lamellae are "feathered," as in *Ecdyonurus*. However, the posterior lamella of each pair lags behind the anterior one and does not reach so far forward during the return stroke (Fig. 16b). Thus, during the effective stroke the anterior lamella meets the posterior one at an angle which is open toward the midline. This angle is progressively reduced and the interlamellar water squeezed inwards. As in *Ephemera,* each gill is about one-eighth oscillation in advance of the one next behind it; in other words, there is a phase lag of 45° (Eastham, 1936).

Cloeon dipterum, like *Leptophlebia,* lives in still water. However, the lamellate gills of *Cloeon* are all held laterally in the same plane. They draw water posteriorly along the dorsal midline and then force it outward and

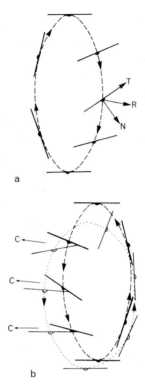

Fig. 16. Leptophlebia marginata. The elliptical path of the anterior lamella of the second pair of gills viewed from behind. In (b) the posterior lamella is also included (dotted line). T,N,R, tangential, normal; and resultant forces to the surface of the gill. C, compression. (After Eastham, 1936.)

upward between themselves. The last pair do not beat, but serve to deflect the current laterally (Eastham, 1932).

In *Ecdyonurus, Ephemera,* and *Cloeon,* the backwardly directed current is enhanced by the alternate phase of suction and compression produced by each pair of gills moving slightly in advance of the pair next behind. This strengthening does not occur in *Leptophlebia* because of the differences in gill attitude.

Caenis horaria differs markedly from the above species. It lives partly buried in mud and its second pair of gills is expanded into large plates which cover and protect the third to sixth pairs, forming a sort of branchial chamber. During ventilation these plates lift to an angle of 30° to 40° to the body surface to allow the other gills to oscillate. The ventilatory current is completely lateral, either from left to right or from right to left. The flow pulsates with weak reversals between each forward pulse. Each gill is a flat lamella with numerous marginal filaments. The effective stroke is downward and backward when the filaments are straight and close together so that water cannot pass between them. During the upward and forward return stroke the marginal filaments lag behind and spread out to allow water to pass between them. The gill also pivots and follows an elliptical path (Fig. 17).

To produce a flow from left to right, the underside of the gill faces the right on the effective downstroke and faces the left on the upward return stroke, thus producing a screw effect. In both directions the gill presents a marked angle to its direction of motion, but the return stroke is ineffective in producing any current for the reasons outlined above. There is a delay of about one-third oscillation between successive gills (i.e., a phase lag of about 120°) and so the sixth gill is in the same position as the ipsilateral third gill. In this animal, the members of each pair of gills are out of phase by about one-third oscillation also. When the current is from left to right, the left gill of each pair is in advance of the right gill. The lateral flow produced by the individual gills is enhanced by the lateral peristaltic wave caused by the phase lag between the members of a pair, and this in turn is reinforced by the longitudinal metachronal rhythm (Eastham, 1934).

The direction of the current is reversed by stopping the oscillatory movement and then starting it again, changing the direction of the elliptical path, the attitude of the gills, and the phase relationship between the two sides. When swimming, the gill movements cause the body to spin on its longitudinal axis, gill-less larvae swimming dorsal side up (Eastham, 1934).

B. ANISOPTERAN (ODONATA) LARVAE

In anisopteran dragonfly larvae the gills are arranged in six longitudinal sets in a branchial chamber formed from the hindgut (Tillyard, 1916a,

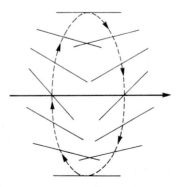

Fig. 17. Caenis horaria. Elliptical path of any gill viewed from behind with the current flowing from left to right. (After Eastham, 1934.)

1917) (Section II,B,3) (Fig. 18). The branchial chamber can be closed at its ends by the pre- and postbranchial valves. Behind it is a muscular vestibule, inserted onto which are six rows of dilator muscles (Snodgrass, 1954). The vestibule communicates with the exterior via the anus and this opening is controlled by the anal valve (Tonner, 1936; Hughes and Mill, 1966; Mill and Pickard, 1972b).

The first phase of normal ventilation is expiration and this can be recognized by the upward movement of the abdominal sterna, especially of the more posterior segments. Snodgrass (1954) states that there is also some compression of the terga, especially during strong expiratory movements. At the same time as the sterna are being raised, the anal valve is opened to about one-third of its maximum extent (Mill and Pickard, 1972b) (Figs. 19, 20). The rising sterna produce a pressure of 2 to 4 cm H_2O in the branchial chamber and this forces the contained water out through the anal valve (Hughes and Mill, 1966) (Fig. 21).

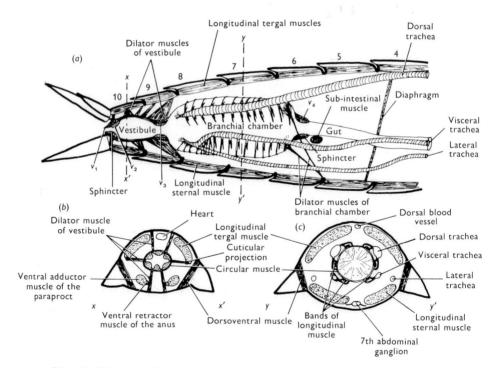

Fig. 18. Diagrams of the respiratory system in an anisopteran dragonfly larva of the aeshnid type. (a) Longitudinal section; (b,c) transverse sections through the regions indicated. v_1 and v_2 are the anal valve, v_3 is the postbranchial valve and v_4 is the prebranchial valve. (From Hughes and Mill, 1966.)

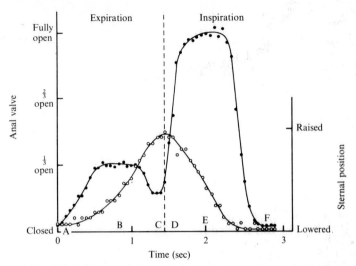

Fig. 19. The relationship between opening of the anal valve and sternal movement, measured from successive cine frames, during a single cycle of ventilation in *Anax*. (From Mill and Pickard, 1972b.) A–E refer to the photographs in Fig. 20.

When they reach their uppermost position the sterna start to fall again, signifying the start of inspiration. At the same time the anal valve opens fully, this often being preceded by a brief closing movement (Mill and Pickard, 1972b) (Figs. 19, 20). The net effect of these actions is to cause a negative pressure of about 0.5 cm H_2O in the branchial chamber and so water is drawn in (Hughes and Mill, 1966) (Fig. 21). It is thought that this is aided by active dilation of the vestibule.

The effect of the narrow anal valve opening, coupled with high branchial chamber pressure, during expiration, and the wide opening and low negative pressure during inspiration, is to prevent too much mixing of inspired with expired water.

Apart from the musculature directly associated with the hindgut, other abdominal muscles are obviously involved. In *Aeshna* and *Anax* each abdominal segment from the fourth to the ninth has three pairs of dorsoventral muscles (Wallengren, 1914a; Whedon, 1918; Steiner, 1929; Mill, 1965). Wallengren (1914a) thought that all of these, and also the dorsoventral oblique muscles, are expiratory and help to lift the sterna, but Whedon (1918) suggested that the only expiratory muscles are the middle or "respiratory" pair of dorsoventrals. It was confirmed by Mill and Hughes (1966) that the so-called respiratory dorsoventral muscles are the ones normally concerned with raising the sterna, although in certain circumstances other dorsoventral muscles may be involved (Mill, 1970). The respiratory dorso-

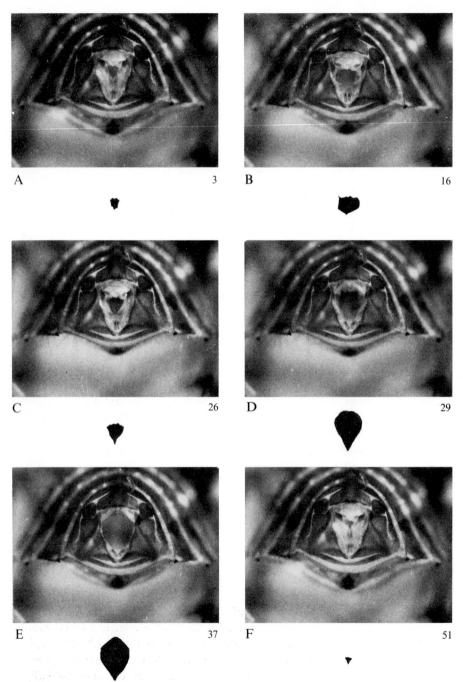

Fig. 20. Cine frames of the anal valve and abdominal sterna viewed from behind, illustrating different points in a single cycle of ventilation in *Anax*. The silhouettes below each frame are of the open valve area. A–F refer to the corresponding points in Fig. 19, and the frame numbers are also given. Camera speed 18 frames/second. (From Mill and Pickard, 1972b.)

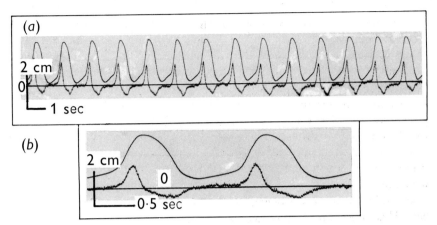

Fig. 21. Recordings of normal ventilation in *Aeshna*. The upper trace shows the dorsoventral movements of the sterna (upward indicates lifting). The lower trace is the pressure recorded in the branchial chamber in relation to the outside (zero) pressure. (From Hughes and Mill, 1966.)

ventral muscles differ from the other two pairs in that they are richly tracheated and contain numerous mitochondria, regularly oriented on either side of the Z-lines (Mill and Lowe, 1971). In *Libellula* two pairs of richly tracheated, segmental muscles are invariably concerned with lifting the sterna.

Running transversely across the abdomen between segments 4 and 5 and segments 5 and 6 are a muscular diaphragm and a subintestinal muscle, respectively. The diaphragm was first described by Amans (1881), who suggested that it was involved in ventilation, defecation, and extension of the labial mask; the subintestinal muscle was described by Wallengren (1914a). It was suggested by Snodgrass (1954) that the sterna return to their resting position solely as a result of the natural elasticity of the tergal plates. Tonner (1936) thought that contraction of the subintestinal muscle forces the sterna down, but that the diaphragm only starts to contract later and thus has little effect, except in widening the posterior segments. However, earlier Matula (1911) and Whedon (1918) were of the opinion that both muscles aid the downward movement of the sterna, and it has been shown that this is indeed the case, with both muscles contracting synchronously (Mill and Hughes, 1966). However, tergal elasticity is still thought to play an important role.

Normal ventilation is a repetitive event which may continue for some time. It is interrupted by other forms of ventilation and by periods of quiescence. Frequencies in the range 12–57/minute have been recorded (40–114/minute in *Libellula*) (Hughes and Mill, 1966), the frequency increasing with increase in temperature (Babák and Foustka, 1907). Wallengren (1914b) has shown that a 4-centimeter long larva changes about 83% of the contents of its branchial chamber and vestibule with each oscillation. This amounts to about 0.05 cm³. Thus, at a frequency of 20/minute, the larva pumps about 1 cm³ of water/minute. When the environmental pO_2 falls below about 2.5 cm³/liter at 17° to 18°C the larva comes to the surface and pumps air instead of water. This has been called *Notatmung,* or "emergency ventilation," by Wallengren (1914b).

Tonner (1936) distinguished two other types of ventilation. In "gulping ventilation" (*Schlukatmung*) the anal valve opens and the vestibule is filled with water. The anal valve then closes and the postbranchial valve opens. The vestibule contracts strongly, forcing the water into the branchial chamber. This procedure is repeated several times in rapid succession, thus "pumping up" the branchial chamber. The ensuing expiratory phase is rather slow. In many cases the water is retained in the branchial chamber for several seconds, when it is churned about by the action of the intrinsic musculature. This latter is "chewing ventilation" (*Kauatmung*).

Hughes and Mill (1966) have recorded maintained positive pressures in the branchial chamber. They are of two types. In one, the pressure increases in a series of jerks similar to normal ventilatory cycles, but at a higher frequency which produces a temporal summation effect. Coincident with each pressure increase, the sterna show a slight expansion. When the maximum is reached (5 to 6 cm H_2O) the pressure starts to fall slowly and then drops abruptly, the sterna rising back to their resting level at the same time (Fig. 22). This is probably gulping ventilation. In the other type, the pressure increases smoothly and fairly rapidly up to as high as 35 cm H_2O, and is accompanied by lifting of the sterna. The pressure may fall immediately. but is usually maintained for several seconds (Fig. 23).

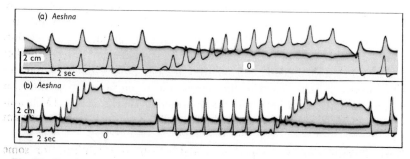

Fig. 22. Simultaneous movement and pressure recordings during gulping ventilation. The upper trace is the record of sternal movement (upward indicates lifting), the lower trace is the pressure record in the branchial chamber in relation to the outside (zero) pressure. (a) Oscilloscope recording; (b) oscilloscope recording illustrating a more maintained pressure. (From Hughes and Mill, 1966.)

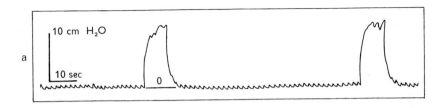

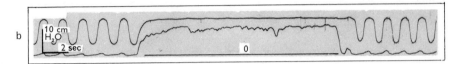

Fig. 23. (a) Pen recording of the changes in branchial chamber pressure in *Aeshna*. The small oscillations are normal ventilatory cycles; (b) oscilloscope recording of a similar pattern in *Anax* (lower trace), with sternal movement on the upper trace (upward indicates lifting.) (From Hughes and Mill, 1966.)

Finally, the jet-propulsive mechanism of locomotion also subserves respiration. This is basically a rapid ventilatory movement, but in addition some longitudinal contraction of the abdomen occurs (Wallengren, 1914a; Tonner, 1936) and the legs are moved posteriorly to lie alongside the body. Pressures of up to 40 cm H_2O have been recorded, with a rate of increase of about 1000 cm/second as against 6–12 cm/second for normal ventilation (Hughes, 1958; Hughes and Mill, 1966).

V. The Control of Ventilation

The effect of various environmental parameters, such as pO_2, pCO_2, and temperature, on the initiation and frequency of ventilation has been mentioned in previous sections. Here we are concerned primarily with the neural control of ventilation, for which information is only available for the pumping mechanism of aeshnid dragonfly larvae.

Ventilation in aeshnid larvae consists of two phases, expiration and inspiration, both of which involve abdominal skeletal muscles (Section IV,B). The segmental, expiratory "respiratory dorsoventral muscles" are innervated by the corresponding second segmental nerves, while the inspiratory "diaphragm" and "subintestinal transverse muscle" receive their nerve supply from the unpaired median nervous system (Mill, 1965).

Rhythmic bursts of activity have been recorded in the nerves to all of these muscles. The expiratory bursts in the second segmental nerves of *Aeshna* normally consist of a single unit, although a second may sometimes be present, whereas in *Anax* there are almost invariably three units. However, in both genera only one unit has ever been shown to produce an electrical effect in the respiratory dorsoventral muscle. There is always a 1:1 relationship between nerve and muscle potentials and the firing pattern shows a characteristic increase in frequency which causes facilitation of the muscle potentials (Mill and Hughes, 1966; Mill, 1970) (Fig. 24).

The other segmental nerves also show activity synchronized with the expiratory bursts in the second nerves, but neither this nor the other active units in the second nerves has been seen to instigate any muscular contraction in dissected preparations. Some units could be inhibitory, others could serve to increase the tone in certain muscles (Mill, 1970), and indeed, in chronic preparations, synchronous activity has been recorded from the anterior dorsoventral muscles on a few occasions (Mill and Pickard, 1972a).

The expiratory bursts on opposite sides of the same segment are in phase and there may or may not be a 1:1 relationship between the action potentials. On the other hand, the activity spreads from behind forward. The expiratory bursts first appear from the last (eighth) abdominal ganglion and then from successively more anterior ganglia (as far forward as the fourth) with an intersegmental delay of about 100 mseconds (Fig. 25). Thus, there is a pacemaker in the eighth ganglion. This posterior-to-anterior rhythm is also seen in the control of swimmeret movements in decapod crustaceans (Hughes and Wiersma, 1960), in contrast to the anterior-to-posterior metachronal rhythm of the gills of ephemeropteran larvae (Eastham, 1932).

The bursts cease simultaneously in all segments (Mill and Hughes, 1966).

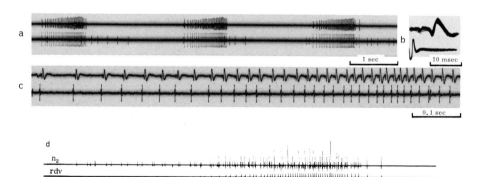

Fig. 24. Recordings from the second lateral nerve on one side of the seventh abdominal segment and the corresponding respiratory dorsoventral muscle. (a) Aeshna: Lower trace—nerve (lat); upper trace—muscle (rdv); (b) same preparation with superimposed sweeps from a single burst; (c) single burst from same preparation; (d) Anax: Upper trace–nerve (n_2); lower trace—muscle (rdv). Note that in both cases only one unit in the nerve produces muscle potentials. [(a–c) from Mill and Hughes, 1966; (b) from Mill, 1970.)]

Fig. 25. Expiratory bursts recorded from right respiratory dorsoventral muscles of various segments, as indicated, showing the posterior–anterior rhythm. Upper trace—sternal movement(s) (upward indicates lifting). Chronic recording. (From Pickard and Mill, 1972).

However, this occurs before the sterna are fully raised, and records of the tension developed by individual muscles show that tension starts to fall as soon as electrical activity ceases. The continued upward movement of the sterna could be due to the inertia effect of water leaving the branchial chamber or to the sterna "clicking" into a stable position after being lifted past a critical level (Pickard and Mill, 1972).

Inspiration is closely coupled to expiration and as soon as the dorsoventral muscles start to relax the diaphragm and subintestinal muscle contract (Fig. 26). Recordings from these muscles show in-phase bursts of activity. The bursts are fairly uniform, with possibly a slight overall increase in frequency toward the end (Mill and Hughes, 1966; Mill, 1970). Alternate expiratory and inspiratory bursts have been recorded from the fifth segmental nerves of the last abdominal ganglion, which innervates all of the musculature of the tenth segment (Mill, 1963, 1970). A summary diagram of normal ventilation is given in Fig. 27.

Jet propulsion also involves the respiratory dorsoventral muscles and recordings from them show short bursts of activity in which the frequency is high and the muscle potentials large (Fig. 28). Other abdominal muscles are also involved and synchronous activity has so far been recorded from the anterior and posterior dorsoventral muscles (Mill and Pickard, 1972a).

Electrical stimulation of a first segmental nerve has no effect during an expiratory burst. However, if the stimulus is applied between bursts, it often elicits a normal expiratory burst which then resets the rhythm (Mill and Hughes, 1966). Furthermore, it was shown by Mill (1970) that repetitive electrical stimuli, delivered at a frequency slightly greater than that of ventilation and starting in an interexpiratory period, are capable of causing the

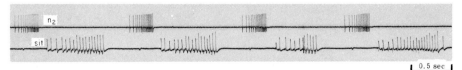

Fig. 26. Recording showing the alternation of expiratory and inspiratory activity in *Aeshna*. Upper trace—a second lateral nerve of the seventh abdominal ganglion (n_2). Lower trace—the subintestinal transverse muscle (sit). (From Mill, 1970.)

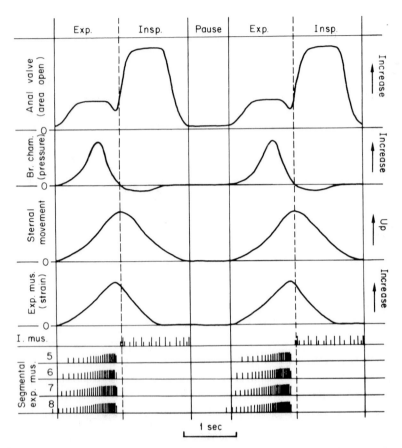

Fig. 27. Summary diagram of normal ventilation in the aeshnid dragonfly larva showing, from the bottom, expiratory and inspiratory muscle activity; the strain produced by a single expiratory muscle; sternal movement; branchial chamber pressure; and opening of the anal valve. (From Mill, 1972, after Mill and Pickard, 1972b.)

ventilatory cycle to become phase-locked with this new frequency. A similar effect has been shown by Miller (1971) in a mantid (*Sphodromantis lineola*), while in *Periplaneta americana* only a decrease in frequency can be obtained (Farley and Case, 1968). However, the effect of proprioceptive feedback on the control mechanism is probably very small.

Matula (1911) demonstrated that removal of the head ganglia caused an increase in ventilatory frequency, whereas removal of the prothoracic ganglion caused a decrease (as also did removal of the legs). He revealed also that removal of the head ganglia had no effect on the relationship of

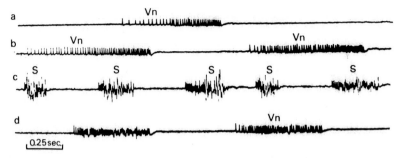

Fig. 28. Continuous recording from a respiratory dorsoventral muscle of *Anax* during normal ventilation (Vn) and swimming (S). Chronic recording. (From Mill and Pickard, 1972a.)

ventilatory frequency with temperature (Section IV,B). Furthermore, he showed that removal of either head or prothoracic ganglia negated the frequency increase associated with decrease in environmental pO_2. From this he concluded that, in addition to the abdominal ventilatory centers, there are centers in the head and prothoracic ganglia sensitive to lowering of the environmental pO_2. However, Wallengren (1913) found that the frequency increase caused by removal of the head ganglia was only temporary and these animals were still sensitive to pO_2 changes, but he agreed that the prothoracic ganglion must contain a center which controls ventilation according to environmental oxygen tension.

It has been suggested by Mill and Hughes (1966) that the expiratory center in the last abdominal ganglion is activated by a command interneuron or interneurons from a higher center in the head or thoracic ganglia. Although normally the center in the eighth ganglion fires first, with successively more anterior centers then firing in sequence, it has been seen in some preparations that one of the centers is firing out of turn (Pickard and Mill, 1972). Thus, the centers as far forward as the sixth at least presumably receive a direct input from the command system. A possible neural arrangement which accounts for the known parameters in a single ganglion is shown in Fig. 29, and this also demonstrates how the system may oscillate between expiration and inspiration. Some of the details of the control of ventilation in these animals may well have parallels in the crayfish swimmeret control system (Ikeda and Wiersma, 1964; Wiersma and Ikeda, 1964).

VI. Respiratory Pigments

Of the four respiratory pigments, only hemoglobin has been found in insects and this in only a very few species. Presumably this is a result of the

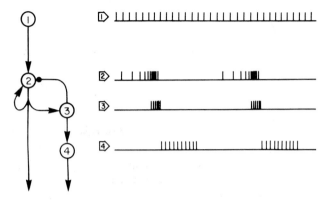

Fig. 29. A diagrammatic representation of one way in which an oscillating system of the type found in the dragonfly larva could work. (From Mill, 1972.)

tracheal system method of supplying the tissues with oxygen, the vascular system thus having little respiratory significance.

A. Hemoglobin as a High Affinity Pigment

Hemoglobin was first recorded in larvae of *Chironomus* by Lankester (1867), but it was not for several decades after this that any work was carried out on the function of this pigment. *Chironomus plumosus* (and probably the other chironomids as well) has the smallest known hemoglobin, consisting of only two unit molecules (MW = 31,400) and with a sedimentation constant of 2.0. Fox (1945), working on *C. riparius,* demonstrated that it has a high affinity for oxygen, with a p_{50} of 0.6 mm Hg. at 17°C. There is no Bohr effect and the temperature effect is very small indeed ($p_{50} = 0.5$ at 10°C).

The oxygen uptake of normal larvae has been compared with that of larvae in which the hemoglobin was rendered functionless by treatment with carbon monoxide to produce carboxyhemoglobin (Harnisch, 1936; Ewer, 1942). Harnisch (1936) concluded that in *C. thummi* the hemoglobin is functional at all oxygen tensions and that its main function is to increase the recovery rate after anaerobis. In contrast, Ewer (1942), working on *C. plumosus,* found that the hemoglobin only transports oxygen at low environmental tensions of this gas, a condition presumably experienced at times in the animal's natural habitat (they live in mud burrows in still water). He showed that at 17°C the pigment only functions below about 40% air saturation (i.e., at 3 ml O_2/liter) and down to 15% air saturation oxygen uptake is independent of environmental pO_2. This is in agreement with the earlier work of Leitch (1916) who found that larvae of *Chironomus* sp.

only appear to use their hemoglobin below an environmental pO_2 of about 7 mm Hg at 17°C.

Walshe (1950) investigated the behavior of *C. plumosus* larvae in glass U-tubes. She found that in well-aerated water they spent about 50% of their time ventilating the tube, with pauses for filter feeding or rest. A decrease in environmental pO_2 caused an increase in the amount of time spent ventilating and, below 10% air saturation, feeding ceased and the animal ventilated continuously. Under completely anaerobic conditions the larvae became immobile. In contrast, larvae treated with carbon monoxide stopped feeding at a much higher environmental pO_2 (26% air saturation) and became immobile at about 9% saturation. Thus, hemoglobin helps to keep the larvae aerobic, and therefore active, at low oxygen tensions. Since both environmental oxygen and carbon dioxide have an effect Walshe (1950) suggested that they are pH sensitive.

If the larvae undergo a period of anaerobis they only build up a small oxygen debt and this is repaid when the water is aerated. Initially there is a large increase in oxygen uptake as the larvae become active, but this soon subsides, remaining slightly above normal for about 1 hour. In larvae in which this activity is suppressed by lowering the temperature, the initial excessive increase in oxygen consumption does not occur and the animals take twice as long to repay their debt (Walshe, 1947a) (Fig. 30). According to Walshe (1947a) the hemoglobin is only involved during the initial increase, at which time the water was not fully aerated in her experiments. Furthermore, Walshe (1950) has shown that *Chironomus* larvae can repay

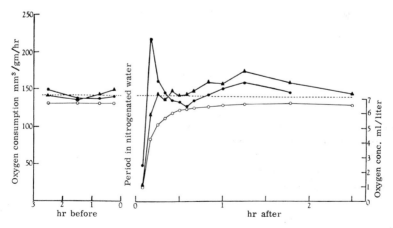

Fig. 30. Rate of oxygen consumption of *Chironomus plumosus* larvae at 17°C before and after 16 hours in anaerobic conditions. ●, normal animals; ▲, carbon monoxide-treated animals; ○, oxygen concentration of water. The dotted line is the average oxygen uptake before the period of anaerobis. (From Walshe, 1947a.)

their oxygen debt in water which is only 7% saturated. Thus, *Chironomus* hemoglobin is used to transport oxygen at low tensions of environmental oxygen.

A number of authors have suggested a storage function for this pigment (Miall and Hammond, 1900; Pause, 1918; Lindroth, 1942). Leitch (1916) showed that its storage capacity is limited to about 12 minutes and Walshe (1950) confirmed this with a figure of about 9 minutes for the storage capacity in a resting animal. Thus, a storage function is not significant during long periods of anaerobis.

The larva of another chironomid which possesses hemoglobin, *Tanytarsus brunnipes,* unlike that of *Chironomus,* is very intolerant of low environmental pO_2 (Thienemann, 1923). In spite of this, its hemoglobin does not transport oxygen above 25% air saturation at 17°C, and even below this level contributes little to the metabolic rate. In fact, at all tensions below air saturation, its oxygen uptake varies with environmental pO_2 (Walshe, 1947b). *Tanytarsus* also lives in mud tubes but in well-aerated water, and nothing is known about its normal metabolic rate. This may be of considerable importance in reaching an understanding of the role of this pigment (Walshe, 1947b) since the polychaete *Nereis* has a lower metabolism when it is in its tube (Hyman, 1932) and Eriksen (1963a) has shown that burrowing ephemeropteran larvae have a much lower metabolic rate when they are provided with a normal substrate in which to burrow.

The chironomids are unique in having a low molecular weight respiratory pigment in the plasma; in all other cases pigments of low molecular weight are contained in cells. This may be possible here because insects do not use ultrafiltration for excretion, a method which has been shown to remove low molecular weight proteins from plasma.

B. Hemoglobin as a Low Affinity Pigment

Larvae of *Gastrophilus* and the adults and larvae of the notonectids *Anisops* and *Buenoa* have hemoglobin contained in richly tracheated groups of abdominal cells (Hungerford, 1922; Poisson, 1926; Bare, 1929; Keilin, 1944; Miller, 1966).

The hemoglobin of *Anisops pellucens* differs from that of chironomid larvae in that it has a very low affinity for oxygen ($p_{50} = 28$ mm Hg at 24°C) (Fig. 31). It shows no Bohr effect, but there is a considerable temperature effect, increase in the latter shifting the oxygen dissociation curve to the right.

When *Anisops* and *Buenoa* dive they show a period of neutral buoyancy (Bare, 1926, 1929; Jaczewski, 1936; Kaiser, 1940) and this is possible almost entirely because of the storage function of the hemoglobin (Bare, 1929;

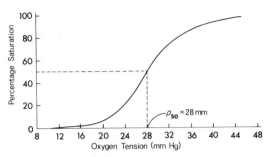

Fig. 31. Oxygen dissociation curve for the hemoglobin of intact adult *Anisops pelucens* at 24°C, p_{50} about 28 mm Hg. (After Miller, 1966).

Hutchinson, 1953; Hungerford, 1958; Miller, 1964a,b, 1966). The adults possess a compressible gas gill consisting of a ventral air film connected to a bubble lying between the bases of the legs and in the subelytral space (Section III,B,1). Unlike *Notonecta,* however, there is no air film on the wings of the adults and so they are less buoyant and have a much smaller water–air interface (de Ruiter *et al.,* 1952; Miller, 1966).

In adults of *A. pellucens* the average duration of a dive is about 5 minutes and this is reduced to about 1 minute in animals treated with carbon monoxide. Thus, the hemoglobin provides a store of oxygen which can last for about 4 minutes (Miller, 1966). This compares with 1–2 minutes in *A. debilis* (Miller, 1964b) and 4 minutes in *Gastrophilus* larvae (Keilin and Wang, 1946). The environmental pO_2 does not affect the duration of a dive and so the gas gill is probably acting mainly as an oxygen store also. This has led Miller (1966) to suggest the following sequence of events during a dive by *Anisops.* When the animal dives it is at first active and the gas gill starts to decrease in size, and so it becomes less buoyant. At the onset of the phase of neutral buoyancy, activity is minimal and the hemoglobin starts to give up its oxygen. This passes into the abdominal tracheae and hence into the gas gill to maintain the neutral buoyancy. From here, most oxygen is probably taken up via the thoracic spiracles. At a gill pO_2 of about 20 mm Hg the hemoglobin will be unloaded and the animal will start to lose its gas gill again. This decreases its buoyancy, resulting in more activity and the animal soon has to return to the surface to replenish its stores (Fig. 32).

VII. Other Functions of the Respiratory System

Compressible gas gills are normally sufficiently large to make the animal positively buoyant and it is decrease in this buoyancy, caused by the shrinking gill, that provides the stimulus for the animal to return to the surface

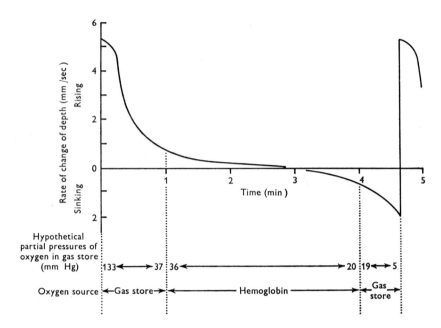

Fig. 32. Diagram to show the rate of change of depth during a dive of adult *Anisops pellucens* and its utilization of oxygen. Hypothetical values of gill pO_2 are also given. (From Miller, 1966.)

(Both, 1934). The rather unusual case of *Anisops* and *Buenoa* has been dealt with in the previous section.

In the plastron-bearing adults of *Aphelocheirus* there is a pair of sense organs, each consisting of an oval plate bearing hydrofuge hairs which are longer and less abundant than those of the plastron. Interspersed with these are sensory hairs. These organs are sensitive to uniform and differential pressure changes (Thorpe and Crisp, 1947c). They abut against the spiracular rosettes of the second abdominal segment and so are in close contact with the tracheal system. The trachea from these same spiracles each connect with a small air sac (one on each side) and Thorpe and Crisp (1947c) suggest that these sacs damp out fluctuations in the gas pressure in the tracheal system caused by body movements or changes in environmental gas tension.

Three pairs of similar sense organs are present in *Nepa* (Baunacke, 1912; Hamilton, 1931), but Thorpe and Crisp (1947c) have shown that their response is not so finely graded as in *Aphelocheirus* and they only respond to differential, not absolute, pressure.

Corethra and *Mochlonyx* (Diptera-Nematocera) larvae contain tracheal air sacs, which have a hydrostatic function (Krogh, 1911; Koch, 1918; Damant, 1924; Wigglesworth, 1953). In *Corethra* there are two of these sacs and they are about all that is left of the tracheal system. The gas inside the air sacs equilibrates with the gas mixture in the environment (Krogh, 1911) and their size can be varied (Damant, 1924). Thus, the buoyancy of the animal can be altered to enable it to float at different levels in the water (Wigglesworth, 1953).

In the larvae of the mayfly *Hexagenia* the gills appear to have a balancing role as well as being important sites for gaseous exchange (Morgan and Grierson, 1932), and this probably applies to other larvae with lateral abdominal gills. Finally, as described in Section V, anisopteran larvae use an exaggerated form of ventilation to produce jet-propulsive swimming.

VIII. Conclusions

In general it would seem that tracheal gills are of particular importance for respiration in those insects which live in still water or in otherwise stagnant conditions. The gills may have a dual role. In all cases they provide a large surface-to-volume ratio; but abdominal gills often have the additional function of increasing the flow of water over the animal and so reducing the thickness of the stagnant boundary layer. Those insects which live almost exclusively in well-aerated conditions also often have tracheal gills, but do not move them. Although a large amount of oxygen may be taken up over the gills in such animals the gills are not normally necessary for survival. Certain paradoxes exist, such as in *Ecdyonurus* larvae, which live in apparently well-aerated conditions and yet use their gills to produce a ventilatory current. However, the region immediately around them does tend to be fairly still and perhaps an appreciable boundary layer develops in the absence of gill movements.

Increase in surface area has been brought about in many ways. For instance, some trichopterans have a very large number of gills, in some mayflies each lamella shields a filamentous gill, and in most zygopteran dragonfly larvae the gills are very long or are saccoid.

Instead of using the gills to produce ventilatory currents, a number of insects, especially tube-dwellers, undulate their bodies. Also leg movements are important in some notonectids, and in anisopteran dragonfly larvae water is pumped in and out of a special branchial chamber.

Environmental oxygen tension tends to be more important in most cases than carbon dioxide tension for providing ventilatory stimuli. Thus low pO_2 causes the gills to start beating in many ephemeropteran larvae, *Phyrganea* larvae start to ventilate their tubes and aeshnid larvae come to the surface and start to breath air instead of water. However, in *Corixa geoffroyi* and *Notonecta,* the ventilatory leg movements are initiated by an increase in environmental pCO_2 and not by a decrease in pO_2. Temperature is also important and in many cases the frequency of ventilation increases with increase in temperature.

Hydrostatic stimuli are important in animals with gas gills. *Notonecta glauca* has been shown to rise because of a decrease in buoyancy as the gill becomes smaller and it can be made to rise by increasing the air pressure above the water (Both, 1934). Botjes (1932) suggested that *Corixa* rises for the same reason, whereas Wrede and Kramer (1926) thought that *Naucoris cimicoides* rises because of the low oxygen tension that ultimately develops in the gill.

References

Alterberg, G. (1934). *Biol. Zentralbl.* **54**, 1.

Amans, P. (1881). *Rev. Sci. Natur. Montpellier* [ser. 3] **1**, 63.

Ambühl, H. (1959). *Schweiz. Z. Hydrol.* **21**, 133.

Anthony, M. H. (1902). *Amer. Natur.* **26**, 615.

Babák, E. (1912). *Pfluegers Arch. Gesamte Physiol. Menschen Tiere* **147**, 349.

Babák, E., and Foustka, O. (1907). *Arch. Gesamte Physiol. Menschen Tiere* **119**, 530.

Balfour-Browne, F. (1909). *Proc. Zool. Soc. London* p. 253.

Bare, C. O. (1926). *Ann. Entomol. Soc. Amer.* **19**, 93.

Bare, C. O. (1929). *Univ. Kans. Sci. Bull.* **18**, 265.

Batelli, A. (1879). *Bull. Soc. Entomol. Ital.* **11**, 77.

Baunacke, W. (1912). *Zool. Jahrb., Abt. Anat. Ontog. Tiere* **34**, 179.

Beadle, L. C., and Beadle, S. F. (1949). *Nature (London)* **164**, 235.

Berg, K. (1941). *Vidensk. Medd. Dan. Naturh. Foren. Kbh.* **105**, 59.

Bertalanffy, L. (1951). *Amer. Natur.* **85**, 111.

Bodine, J. H. (1918). *Proc. Acad. Natur. Sci. Philadelphia* **70**, 103.

Both, M. P. (1934). *Z. Vergl. Physiol.* **21**, 167.

Botjes, J. O. (1932). *Z. Vergl. Physiol.* **17**, 557.

Böving, A. G. (1910). *Int. Rev. Gesamten Hydrobiol. Hydrogr.* **7**, Biol. Suppl., 1.

Brocher, F. (1909). *Ann. Biol. Lacust.* **4**, 89.

Brocher, F. (1910). *Ann. Biol. Lacust.* **4**, 383.

Brocher, F. (1912a). *Ann. Biol. Lacust.* **5**, 5.

Brocher, F. (1912b). *Ann. Biol. Lacust.* **5**, 136.

Brocher, F. (1913). *Zool. Jahrb., Abt. Allg. Zool. Physiol. Tiere* **33**, 225.

Buckler, W. (1875). *Entomol. Mon. Mag.* [1] **12**, 160.

Burtt, E. T. (1936). *Proc. Roy. Entomol. Soc. London, Ser. A* **11**, 61.

Byers, C. F. (1930). *Univ. Fla. Publ., Biol. Sci. Ser.* **1**, 1.

Calvert, P. P. (1911). *Entomol. News* **22**, 49.

Calvert, P. P. (1915). *Entomol. News* **26**, 385.

Calvert, P. P. (1929). *Proc. Amer. Phil. Soc.* **68**, 227.

Carroll, M. (1918). *Proc. Acad. Natur. Sci. Philadelphia* **70**, 86.

Comstock, J. H. (1887). *Amer. Natur.* **21**, 577.

Corbet, P. S. (1962). "A Biology of Dragonflies." Witherby, London.

Corbet, P. S., Longfield, C., and Moore, N. W. (1960). "Dragonflies." Collins, London.

Crisp, D. J. (1964). *Recent Progr. Surface Sci.* 2, 377.

Crisp, D. J., and Thorpe, W. H. (1948). *Discuss. Faraday Soc.* 3, 210.

Cullen, A. M. (1918). *Proc. Acad. Natur. Sci. Philadelphia* 70, 75.

Cummins, K. W. (1964). *Ecol. Monogr.* 34, 271.

Damant, G. C. C. (1924). *J. Physiol. (London)* 59, 345.

Davis, C. (1942). *Proc. Linn. Soc. N.S.W.* 67, 1.

Deibel, J. (1911). *Zool. Jahrb., Abt. Anat. Ontog. Tiere* 31, 107.

de Ruiter, L., Wolvekamp, H. P., and van Tooren, A. J. (1951). *Acta Physiol. Pharmacol. Neer.* 1, 657.

de Ruiter, L., Wolvekamp, H. P., van Tooren, A. J., and Vlasblom, A. (1952). *Acta Physiol. Pharmacol. Neer.* 2, 180.

Dodds, G. S., and Hisaw, F. L. (1924). *Ecology* 5, 262.

Dolley, W. L., and Farris, E. J. (1929). *J. N.Y. Entomol. Soc.* 37, 127.

Dufour, L. (1847). *Ann. Sci. Natur.* 8, 341.

Dufour, L. (1852). *Ann. Sci. Natur., Zool.* [3] 17, 65.

Eastham, L. E. S. (1932). *Nature (London)* 130, 58.

Eastham, L. E. S. (1934). *Proc. Roy. Soc., Ser. B* 115, 30.

Eastham, L. E. S. (1936). *J. Exp. Biol.* 13, 443.

Eastham, L. E. S. (1937). *J. Exp. Biol.* 14, 219.

Eastham, L. E. S. (1939). *J. Exp. Biol.* 16, 18.

Edwards, F. W. (1919). *Entomol. Mon. Mag.* [3] 5, 83.

Edwards, G. A. (1953). *In* "Insect Physiology" (K. D. Roeder, ed.), pp. 55–95. Wiley, New York.

Edwards, R. W. (1958). *J. Exp. Biol.* 35, 383.

Ege, R. (1915a). *Z. Allg. Physiol.* 17, 81.

Ege, R. (1915b). *Vidensk. Medd. Dan. Naturh. Foren. Kbh.* 66, 183.

Ege, R. (1926). *In* "Physiological Papers Dedicated to August Krogh, London." p. 25

Eriksen, C. H. (1963a). *J. Exp. Biol.* 40, 447.

Eriksen, C. H. (1963b). *J. Exp. Biol.* 40, 455.

Eriksen, C. H. (1968). *Can. J. Zool.* 46, 93.

Erman, D. C., and Helm, W. T. (1970). *Hydrobiologia* 36, 505.

Ewer, R. F. (1942). *J. Exp. Biol.* 18, 197.

Farley, R. D., and Case, J. F. (1968). *J. Insect Physiol.* 13, 591.

Feldmeth, C. R. (1970a). *Comp. Biochem. Physiol.* 32, 193.

Feldmeth, C. R. (1970b). *Physiol. Zool.* 43, 185.

Fenn, W. O., Rahn, H., and Otis, A. B. (1946). *Amer. J. Physiol.* 146, 637.

Fox, H. M. (1920). *J. Gen. Physiol.* 3, 565.

Fox, H. M. (1945). *J. Exp. Biol.* 21, 161.

Fox, H. M., and Sidney, J. (1953). *J. Exp. Biol.* 30, 235.

Fox, H. M., Simmonds, B. G., and Washbourn, R. (1935). *J. Exp. Biol.* 12, 179.

Fox, H. M., Wingfield, C. A., and Simmonds, B. G. (1937). *J. Exp. Biol.* 14, 210.

Fraenkel, G., and Herford, G. V. B. (1938). *J. Exp. Biol.* 15, 266.

Fry, F. E. J. (1947). *Publ. Ont. Fish. Res. Lab.* 68, 1.

Gardner, A. E. (1956). *Proc. London Ent. Nat. Hist. Soc.* 1954–5, 109.

Gazagnaire, J. (1886). *C.R. Acad. Sci.* 102, 1501.

Gericke, H. (1919). *Zool. Jahrb., Abt. Allg. Zool. Physiol. Tiere* 36, 157.

Gertz, K. H., and Loeschke, H. H. (1954). *Z. Naturforsch. B* **9**, 1.

Gilchrist, B. M. (1956). *Hydrobiologia* **8**, 54.

Grobben, C. (1876). *Sitzungsber. Akad. Wiss. Wien* **72**, 433.

Hagemann, J. (1910). *Zool. Jahrb., Abt. Anat. Ontog. Tiere* **30**, 373.

Hamilton, M. A. (1931). *Proc. Zool. Soc. London* p. 1067.

Harnisch, O. (1936). *Z. Vergl. Physiol.* **23**, 391.

Harnisch, O. (1958). *Biol. Zentralbl.* **77**, 300.

Hinton, H. E. (1953). *Trans. Soc. Brit. Entomol.* **11**, 209.

Hinton, H. E. (1957). *Proc. Roy. Soc., Ser. B* **147**, 90.

Hinton, H. E. (1966a). *Proc. Roy. Entomol. Soc. London Ser. A* **41**, 107.

Hinton, H. E. (1966b). *Phil. Trans. Roy. Soc. London, Ser. B* **251**, 211.

Hinton, H. E. (1967a). *Proc. Roy. Entomol. Soc. London Ser. A* **42**, 35.

Hinton, H. E. (1967b). *J. Mar. Biol. Ass. U.K.* **47**, 485.

Hinton, H. E. (1967c). *Aust. J. Zool.* **15**, 955.

Hinton, H. E. (1968). *Advan. Insect Physiol.* **5**, 65.

Hodgman, C. D. (1958). *In* "Handbook of Chemistry and Physics" (C. D. Hodgman, ed.), 40th ed. Chem. Rubber Publ. Co., Cleveland, Ohio.

Hoppe, J. (1911). *Zool. Jahrb., Abt. Allg. Zool. Physiol. Tiere* **31**, 189.

Houghton, W. (1868). *Pop. Sci. Rev.* **7**, 287.

Houlihan, D. F. (1969). *J. Insect Physiol.* **15**, 1517.

Houlihan, D. F. (1970). *J. Insect Physiol.* **16**, 1607.

Hughes, G. M. (1958). *J. Exp. Biol.* **35**, 567.

Hughes, G. M., and Mill, P. J. (1966). *J. Exp. Biol.* **44**, 317.

Hughes, G. M., and Wiersma, C. A. G. (1960). *J. Exp. Biol.* **37**, 657.

Hungerford, H. B. (1922). *Can. Entomol.* **54**, 262.

Hungerford, H. B. (1958). *Proc. Int. Congr. Entomol., 10th, 1956* Vol. 1, p. 337.

Hutchinson, G. E. (1953). *In* "The Elementary Chemical Composition of Marine Organisms" (A. P. Vinogradov, ed.) Yale Univ. Press, New Haven, Connecticut.

Hyman, L. H. (1932). *J. Exp. Zool.* **61**, 209.

Ikeda, K., and Wiersma, C. A. G. (1964). *Comp. Biochem. Physiol.* **12**, 107.

Imms, A. D. (1957). "A General Textbook of Entomology" (O. W. Richards and R. G. Davies, rev. eds.), 9th ed. Methuen, London.

Jaczewski, T. (1936). *Ann. Mus. Zool. Pol.* **11**, 171.

Jamieson, J. P. (1918). *Proc. Acad. Natur. Sci. Philadelphia* **70**, 81.

Kaiser, E. W. (1940). *Dan. Sci. Invest. Iran No.* 3, p. 139.

Keilin, D. (1915). *Bull. Sci. Fr. Belg.* [7] **49**, 15.

Keilin, D. (1944). *Parasitology* **36**, 1.

Keilin, D., and Wang, Y. L. (1946). *Biochem. J.* **40**, 855.

Keilin, D., Tate, P., and Vincent, M. (1935). *Parasitology* **27**, 257.

Keister, M. L., and Buck, J. (1964). *In* "The Physiology of Insecta" (M. Rockstein, ed.), Vol. 3, pp. 617–658. Academic Press, New York.

Klekowski, R. Z., and Kamler, E. (1968). *Pol. Arch. Hydrobiol.* **15**, 121.

Koch, A. (1918). *Mitt. Zool. Inst. Univ. Munster* **1**, 11.

Koch, H. J. (1934). *Natuurwetensch. Tijdschr.* (*Ghent*) **16**, 75.

Koch, H. J. (1936). *Mem. Acad. Roy. Belg., Cl. Sci.* **16**, 3.

Koch, H. J. (1938). *J. Exp. Biol.* **15**, 152.

Krogh, A. (1911). *Skand. Arch. Physiol.* **25**, 183.

Krogh, A. (1919). *J. Physiol.* (*London*) **52**, 391.

Krogh, A. (1943). *Entomol. Medd.* **23**, 49.

Lankester, E. R. (1867). *J. Anat. Physiol. Norm. Pathol. Homme Anim.* **2**, 114.

Larsen, O. (1924). *Ark. Zool.* [1] **16**, No. 16, 1.

Lawton, J. H. (1971). *J. Anim. Ecol.* **40**, 385.

Leader, J. P. (1970). *J. Insect Physiol.* **17**, 1917.

Lees, A. D. (1955). "The Physiology of Diapause in Arthropods." Cambridge Univ. Press, London and New York.

Leitch, I. (1916). *J. Physiol. (London)* **50**, 370.

Leydig, F. (1859). *Arch. Anat. Physiol., Leipzig* p. 149.

Lieftinck, M. A. (1940). *Treubia* **17**, 281.

Lieftinck, M. A. (1962). *Proc. Int. Congr. Entomol., 11th, 1960* Vol. 3, p. 274.

Lindroth, A. (1942). *Zool. Anz.* **138**, 244.

Lübben, H. (1907a). *Z. Wiss. Insekten Biol.* **3**, 174.

Lübben, H. (1907b). *Zool. Jahrb., Abt. Anat. Ontog. Tiere* **24**, 71.

Lucas, W. J. (1900). "British Dragonflies (Odonata)." Upcott & Gill, London.

Lutz, P. E., and Jenner, C. E. (1960). *J. Elisha Mitchell Sci. Soc.* **76**, 192.

MacNeill, N. (1958). *Entomol. Gaz.* **9**, 102.

MacNeill, N. (1960). *Proc. Roy. Irish Acad., Sect. B* **61**, 115.

Matula, J. (1911). *Pfluegers Arch Gesamte Physiol. Menschen Tiere* **138**, 388.

Miall, L. C. (1895). "The Natural History of Aquatic Insects." Macmillan, London.

Miall, L. C., and Hammond, A. R. (1900). "The Structure and Life History of the Harlequin Fly." Oxford Univ. Press, London and New York.

Mill, P. J. (1963). *Comp. Biochem. Physiol.* **8**, 83.

Mill, P. J. (1965). *Proc. Zool. Soc. London* **145**, 57.

Mill, P. J. (1970). *J. Exp. Biol.* **52**, 167.

Mill, P. J. (1972). "Respiration in the Invertebrates." Macmillan, London.

Mill, P. J., and Hughes, G. M. (1966). *J. Exp. Biol.* **44**, 297.

Mill, P. J., and Lowe, D. A. (1971). *J. Insect Physiol.* **17**, 1947.

Mill, P. J., and Pickard, R. S. (1972a). *Odonatologica* **1**, 41.

Mill, P. J., and Pickard, R. S. (1972b). *J. Exp. Biol.* **56**, 537.

Miller, P. L. (1961). *J. Insect Physiol.* **6**, 243.

Miller, P. L. (1964a). *Nature (London)* **201**, 1052.

Miller, P. L. (1964b). *Proc. Roy. Entomol. Soc. London, Ser. A* **39**, 166.

Miller, P. L. (1966). *J. Exp. Biol.* **44**, 529.

Miller, P. L. (1971). *J. Exp. Biol.* **54**, 599.

Morgan, A. H., and Grierson, M. C. (1932). *Physiol. Zool.* **5**, 230.

Morgan, A. H., and O'Neil, H. D. (1931). *Physiol. Zool.* **4**, 361.

Morgan, A. H., and Porter, M. C. M. (1929). *Anat. Rec.* **44**, 221.

Myers, J. G. (1935). *Proc. Roy. Entomol. Soc. London Ser. A* **10**, 65.

Needham, J. G. (1911). *Entomol. News* **22**, 145.

Nigmann, M. (1908). *Zool. Jahrb., Abt. Syst., Oekol. Geogr. Tiere* **26**, 489.

Paganelli, C. V., Bateman, N., and Rahn, H. (1967). *Underwater Physiol., Proc. Symp., 3rd, 1966* p. 452.

Pause, J. (1918). *Zool. Jahrb., Abt. Allg. Zool. Physiol. Tiere* **36**, 339.

Pennak, R. W., and McColl, C. M. (1944). *J. Cell. Comp. Physiol.* **23**, 1.

Petitpren, M. F., and Knight, A. W. (1970). *J. Insect Physiol.* **16**, 449.

Philipson, G. N. (1954). *Proc. Zool. Soc. London* **124**, 547.

Phillips, M. E. (1939). *Ann. Entomol. Soc. Amer.* **32**, 325.

Pickard, R. S., and Mill, P. J. (1972). *J. Exp. Biol.* **56**, 527.

Poisson, R. (1926). *Arch. Zool. Exp. Gen.* **65**, 181.

Portier, P. (1911). *Arch. Zool.* [5] **8**, 89.

Prosser, C. L., and Brown, F. A. (1961). "Comparative Animal Physiology," 2nd ed. Saunders, Philadelphia, Pennsylvania.

Rahn, H., and Paganelli, C. V. (1968). *Resp. Physiol.* 5, 145.

Réaumur, R. A. F. (1734–1742). "Memoires pour Servir à l'Histoire des Insectes," Vol. 3.

Ris, F. (1912). *Tijdschr. Entomol.* 55, 157.

Robert, P. A. (1958). "Les Libellules (Odonates)." Delachaux et Niestlé S. A., Neuchâtel.

Roster, D. A. (1886). *Bull. Soc. Entomol. Ital.* 18, 239.

Sadones, J. (1896). *Cellule* 11, 273.

Sayle, M. H. (1928). *Biol. Bull.* 54, 212.

Schmidt, E. (1919). *Zool. Anz.* 50, 235.

Schoenemund, E. (1925). *Arch. Hydrobiol.* 15, 339.

Sleight, C. S. (1913). *J. N.Y. Entomol., Soc.* 21–22, 4.

Snodgrass, R. E. (1954). *Smithson. Misc. Collect.* 123, 1.

Steiner, L. F. (1929). *Ann. Entomol. Soc. Amer.* 22, 297.

Stockner, J. G. (1971). *J. Fish. Res. Bd. Can.* 28, 73.

Stride, G. O. (1955). *Ann. Entomol. Soc. Amer.* 48, 344.

Stride, G. O. (1958). *Proc. Int. Congr. Entomol., 10th, 1956* Vol. 2, p. 335.

Szabo-Patay, J. (1924). *Ann. Mus. Natur. Hung.* 21, 33.

Thienemann, A. (1923). *Verh. Int. Ver. Limnol. Kiel Stuttgart* 1, 108.

Thorpe, W. H. (1933a). *Nature (London)* 131, 549.

Thorpe, W. H. (1933b). *Proc. Int. Congr., Entomol., 5th, 1932* p. 345.

Thorpe, W. H. (1950). *Biol. Rev. Cambridge Phil. Soc.* 25, 344.

Thorpe, W. H., and Crisp, D. J. (1947a). *J. Exp. Biol.* 24, 227.

Thorpe, W. H., and Crisp, D. J. (1947b). *J. Exp. Biol.* 24, 270.

Thorpe, W. H., and Crisp, D. J. (1947c). *J. Exp. Biol.* 24, 310.

Thorpe, W. H., and Crisp, D. J. (1949). *J. Exp. Biol.* 26, 219.

Tillyard, R. J. (1915). *Proc. Linn. Soc. N.S.W.* 40, 422.

Tillyard, R. J. (1916a). *J. Linn. Soc. London, Zool.* 33, 1.

Tillyard, R. J. (1916b). *Proc. Linn. Soc. N.S.W.* 41, 388.

Tillyard, R. J. (1917). "The Biology of Dragonflies." Cambridge Univ. Press, London and New York.

Tillyard, R. J. (1923). *Trans. Proc. N.Z. Inst.* 54, 197.

Tonner, F. (1936). *Z. Wiss. Zool.* 47, 433.

Ussing, H. (1910). *Int. Rev. Hydrobiol.* 3, 115.

van Dam, L. (1937). *Zool. Anz.* 118, 122.

Varley, G. C. (1937). *Proc. Roy. Entomol. Soc. London Ser. A* 12, 55.

Viallanes, H. (1885). *Ann. Sci. Natur.* [6] 17, 1.

von Rosenhof, R. (1749). "Der Wasser-Insecten zweyte Klasse. Der Monatlich-herausgegebenen Insecten-Belustigung," Part 2. Nürnberg.

von Ruttner, F. (1926). *Naturwissenschaften* 14, 1237.

Wagner, W. (1926). *Zool. Jahrb., Abt. Allg. Zool. Physiol. Tiere* 42, 441.

Wallengren, H. (1913). *Lunds Univ. Arsskr., Afd. 2* [N.S.] 9, No. 16, 1.

Wallengren, H. (1914a). *Lunds Univ. Arsskr., Afd. 2* [N.S.] 10, No. 4, 1.

Wallengren, H. (1914b). *Lunds Univ. Arsskr., Afd. 2* [N.S.] 10, No. 8, 1.

Walshe, B. M. (1947a). *J. Exp. Biol.* 24, 329.

Walshe, B. M. (1947b). *J. Exp. Biol.* 24, 343.

Walshe, B. M. (1950). *J. Exp. Biol.* 27, 73.

Wautier, J., and Pattée, E. (1955). *Bull. Mens. Soc. Linn. Lyon* [N.S.], No. 7, 178.

Welch, P. S. (1916). *Ann. Entomol. Soc. Amer.* **9,** 159.

Welch, P. S., and Sehon, G. L. (1928). *Ann. Entomol. Soc. Amer.* **21,** 243.

Wesenberg-Lund, C. (1912). *Int. Rev. Biol., Suppl.*

Wesenberg-Lund, C. (1913). *Int. Rev. Gesamten Hydrobiol. Hydrogr.* **6,** 155.

Wesenberg-Lund, C. (1918). *Vidensk. Medd. Dan. Naturh. Foren. Kbh.* **69,** 277.

Whedon, A. D. (1918). *Trans. Amer. Entomol. Soc.* **44,** 373.

Wiersma, C. A. G., and Ikeda, K. (1964). *Comp. Biochem. Physiol.* **12,** 509.

Wigglesworth, V. B. (1933). *J. Exp. Biol.* **10,** 1.

Wigglesworth, V. B. (1953). "The Principles of Insect Physiology," 5th ed. Methuen, London.

Williams, F. X. (1936). *Proc. Hawaii. Entomol. Soc.* **9,** 273.

Wingfield, C. A. (1937). *Nature (London)* **140,** 27.

Wingfield, C. A. (1939). *J. Exp. Biol.* **16,** 363.

Wodsedalek, J. E. (1911). *Biol. Bull.* **21,** 265.

Wrede, F., and Kramer, H. (1926). *Pfluegers Arch. Gesamte Physiol. Menschen Tiere* **212,** 15.

Zahner, R. (1959). *Int. Rev. Gesamten Hydrobiol.* **44,** 51.

Zavřel, J. (1920). *Bull. Int. Acad. Prague* **22,** 120.

Chapter 7

RESPIRATION: SOME EXOGENOUS AND ENDOGENOUS EFFECTS ON RATE OF RESPIRATION*

Margaret Keister and John Buck

* Reprinted, with permission, from the first edition.

I. Introduction

Because of ease of measurement, respiration is the most commonly used measure of intensity of aerobic metabolism, and the index of choice in exploring or comparing agents or conditions that affect metabolic rate and in describing changes in metabolism during growth and development. Many such studies have been made on insects but few have progressed beyond the descriptive stage to the question of the relative contributions of various tissues and cellular processes to overall metabolism; still fewer have dealt mechanistically with the actual control of respiration. For technical reasons, rate of oxygen uptake is much more frequently measured than rate of carbon dioxide release, but for periods in which the respiratory quotient is stable there is no generally valid theoretical reason for favoring one measurement over the other, and most of the considerations discussed below apply to either.

A. Bases for Expressing Metabolic Rate

Much attention has been given to the question of how best to express respiratory rate. Since the prime consumer of oxygen and producer of carbon dioxide is the tissue protoplasm, largely protein, perhaps the least equivocal basis is volume of gas exchanged per unit protein nitrogen content. This excludes most storage and skeletal material. However, micro-Kjeldahl digestion is a somewhat tedious procedure with the small samples usually available from insects, so a nitrogen basis is little used.

Tissue dry weight, sometimes after lipid extraction, is much used for expressing vertebrate respiratory rate, based apparently on the rationale that tissues vary in water content more than in content of solid matter. Such a basis would be inadvisable for whole insects because of the relative increase in exoskeletal tare and (in the case of lipid-extracted weight) because of the widely different importance of fat as fuel in different insects. In addition, the idea may involve two misconceptions: that water is metabolically inert, and that tissue hydration necessarily changes with changing body water content. Experimental data on the relative constancy of tissue hydration are given by Von Brand et al. (1957), and the cogency of the general argument can be illustrated by the following hypothetical example: A 100 mg insect consists of 70 mg water and 30 mg dry material, of which 15 mg is skeleton and 15 mg tissue, assumed to hydrate equally. The hydration of the tissue is then $35/(35 + 15) = 70\%$. During desiccation the insect loses 40% of its total weight, of which 5 mg is metabolic loss in dry weight (tissue only). Hence the water loss is 35 mg. But the remaining 10 mg of tissue (dry weight) is still hydrated by half the remaining 35 mg of water, so its water content is $17.5/(17.5 + 10) = 64\%$. Hence, although the original total water content has decreased 50%, tissue hydration has fallen less than 10%.

In summary, unit whole body live weight appears to be as good a basis as any for expressing insect respiratory rates.

B. The "Surface Law"

A priori it might be thought that total body uptake of oxygen would increase directly with body weight, each increment of tissue adding its intrinsic respiratory requirements to the whole. However, many years ago, in making comparisons between the metabolic rates of different sizes of adult mammals of one species and between adults of different species, it was discovered that respiratory rate usually does not increase linearly with body weight but at a slower rate. This could perhaps be attributed at least partly to the disproportionate increase in inert supporting tissues and storage materials that commonly occurs with increasing size. Nevertheless, the fact that the rate of increase is often close to the $\frac{2}{3}$ power of the weight led to the hypothesis that metabolic rate is proportional to body surface area—i.e., to the square of a linear dimension rather than to the cube. This hypothesis, beside being supported by direct measurement, has the specious theoretical basis that surface area of a spherical cell changes as weight$^{2/3}$ (square of radius or diameter) and that many sites of metabolic transfer (alveoli, villi, capillaries) are area limited. Much has therefore been made of the fact that, in contrast, postembryonic insect respiration often varies directly with body weight rather than body surface.

The facts themselves are quite clear. If one deals with physiologically homogeneous populations (comparable respiratory quotient, age, stage, activity, nutritional state, and percent storage and skeletal material) large enough to span several orders of magnitude of weight, a plot of respiratory rate against body weight on log-log coordinates usually yields a straight line. In mammals, as in the familiar mouse-to-elephant array, the slope of the line corresponds to a power of weight much less than 1 [usually about 0.75 according to best contemporary estimates (cf. Kleiber, 1961)]. Among insects a similar exponent has been claimed for adult cockroaches and phasmids (Enger and Savalov, 1958), *Drosophila* prepupae (Ellenby, 1939, 1945a, 1953, 1956), and certain aquatic insects (Balke, 1957; R. W. Edwards, 1958). It should be noted that, in most of these instances, surface was not measured directly. Furthermore, the computed areas were usually derived from body weight rather than from formulas such as those developed by Simanton (1933) for aphids and cockroaches, by Ellenby (1945b) for *Drosophila* pupae, and by Slifer (1954) for grasshoppers and locusts, which have been checked by direct measurement.

In contrast, a number of substantial investigations of insects indicates direct proportionality between rate of respiration and total body live weight. Such findings have been reported for lepidopterous pupae (Teissier, 1931) as well as for extensive arrays of adult Diptera and Coleoptera (Kittel, 1941) (Fig. 1). Extraneous factors such as feeding and age can cause substantial changes in gross weight that are not mirrored in respiration (Clarke, 1957). Moreover, data from eggs and pupae have to be viewed in general with caution, since there may be as great as tenfold differences in respiratory rate at different ages, with essentially no change in weight.

The early larval stadia usually involve some tissue differentiation, and, at least in the Holometabola, part of the last larval period is devoted to various changes in preparation for pupation—for example, cessation of feeding, activation of imaginal discs, changes in water content, and, in some

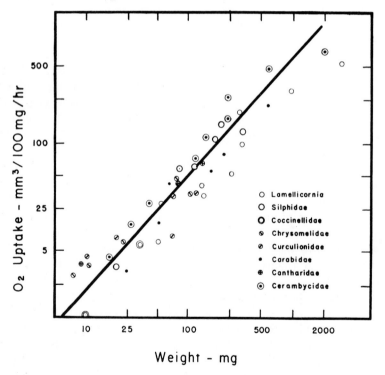

Fig. 1. Relation of respiratory rate to body weight in several families of beetles. (Redrawn from Kittel, 1941.)

insects, spinning. However, it is known that the definitive number of cells in many organs is attained early in larval life, with subsequent growth consisting primarily of cell enlargement (Trager, 1935; Abercrombie, 1936). It seems, therefore, possible that part of larval life might approximate a steady state of growth in which increment of tissue would involve an equivalent increment of metabolism. Hence larval growth might present an even better testing ground for the surface law than the comparison of adult insects of different species.

Bialascewicz (1937) did in fact find that the rate of oxygen uptake per unit weight of the *Bombyx* larva remains constant during the first 5 days of the fifth stadium while body weight concurrently more than triples. Similarly Janda (1961) reported that in the leaf wasp *Croesus* the rate of respiration is nearly proportional to body weight over the more than hundredfold range of larval weight during development. However, these authors also reported that the larvae consumed relatively enormous amounts of leaves per day—70% of total body weight in *Bombyx* and more than 100% in *Croesus*. This means that gross body weight at any one time must have included a considerable tare of unincorporated foodstuff; consequently, a direct proportionality between respiration and body weight (or indeed any other relationship) was fortuitous, and gave no valid basis for special deductions in regard to tissue metabolic rate or its relation to bodily dimensions. In this connection Teissier (1931) was able to vary the slope of the log respiration versus log body weight line in *Tenebrio* between 0.7 and 0.97 (i.e., from practically area proportionality to practically weight proportionality) by nutritional means. The same conclusion presumably applies to any developmental study in which care was not taken to allow the insects to empty the gut before measurement. On this basis the weight proportionality of respiratory rate reported by Müller (1943a) for *Tenebrio* larvae, Müller (1943b) for *Dixippus,* and Sláma (1959) for six species of sawflies must be accepted with reservation.

A good deal of effort has gone into curve fitting and the derivation of exponents and constants from respiratory rate data (e.g., Zeuthen, 1947; Kamioka, 1950; Edwards, 1953; Ludwig, 1956) but the results have descriptive rather than analytical value [see also Kleiber (1961) for a sophisticated view of the situation in mammals]. Even though respiratory rate in at least some insects shows an impressive and useful correlation with body weight, and even though tracheal surface area, in the few forms in which it has been estimated, likewise appears to increase directly with weight rather than $W^{2/3}$ (Sattel, 1956; Balke, 1957), it is necessary to resist the temptation to relate respiratory rate causally to body dimension. The relative independence of respiratory rate in relation to ambient oxygen tension (see Section IV,A) and the potentialities for sustained and often sudden many-fold increase or decrease in metabolic rate during flight or diapause (see Section II,B,C) show that respiratory control can be only tenuously related to body dimensions per se. The whole trend of contemporary thought, in fact, is to locate metabolic control at a subcellular or enzymic level.

II. Steady-State Respiration

A. Resting States

Control levels in insect respirometry are ordinarily determined in an atmosphere saturated with water vapor on animals which are deprived of food, and are, insofar as possible, motionless.* Whether this state [Krogh's (1941) "standard metabolism"] truly corresponds to the "basal metabolism" of the mammal (defined as rate of heat production in complete muscular and mental rest in conscious postabsorptive state) is less important than its duration and reproducibility. Table I gives some "normal" respiratory rates for a number of representative species. It is interesting that 1 mm^3/mg live weight/hour at 25°C could serve as a typical value for resting metabolic level in many of these forms.

It would be expected that basal respiratory rate would be approached in long inanition. In such instances the rate often appears to fall gradually, but since it is usually computed on the basis of initial weight rather than corrected continuously for substrate loss it is by no means certain that there is actually any decrease in rate per unit of active tissue. In some insects, moreover, respiration remains practically constant over many hours of inanition (e.g., Szwajsowna, 1916; Buck and Keister, 1949; Buck et al., 1952). In the desert locust, however, respiration falls steadily during at least the first 50 hours of starvation, the gut being empty after 9 hours (Clarke, 1957), and the decrease is about five times that to be expected on the basis of decrease in weight of respiring tissue (computation from Clarke's weight loss data, assuming an RQ of 1.0).

Even though an insect may make no visible movement for long periods it cannot be concluded that this resting respiration represents the minimum level of metabolism necessary to sustain life, or that it corresponds to sleep in vertebrates. Darkness is known to reduce the activity in certain diurnal species (Kittel, 1941; R. W. Edwards, 1958) and reduces grasshopper respiratory rate to the same extent as does decapitation (Bodine, 1922). In *Phormia,* ligating off the larval brain or decapitating the adult causes about a

---

* It is questionable whether insects constrained in bags, tubes, cages, or capsules (Butler and Innes, 1936; Lühmann and Drees, 1952; Will, 1952; Krüger, 1958; Allen, 1959a,b), and hence likely to struggle, attain the metabolic level of voluntary immobility.

5% fall in respiratory rate per unit weight (Keister and Buck, 1961). Decapitation also depresses respiration in silk moths (Raffy, 1934) and *Musca vicina* (Galun, 1960), although decapitation in *Culex* causes an *increase* in respiratory rate [i.e., release of inhibition (Galun, 1960)]. The results on the whole suggest that central nervous integrity may determine a metabolic rate somewhat above the maintenance level. Caution needs to be exercised in concluding that metabolism is steady unless respiratory rate has been measured over at least 24 hours. Various authors have described diurnal or other rhythms (e.g., Heller, 1930; Janda and Mrciak, 1957; Edelman, 1949, 1950; A. G. Richards, personal communication, 1963). In addition there are various types of puzzling fluctuations in rate, familiar to anyone who has measured respiration in individual insects, that suggest the possibility of relatively sudden metabolic regulation independent of any externally visible activity. For example, *Culex* adults immobilized at 12°C are reported to show "bursts of activity" (Galun, 1960) and the same is suggested by long-continued respirometry of motionless *Phormia* adults at 25°C (J. B. Buck and M. L. Keister, unpublished). The ability of some insects to endure considerable periods of hypoxia without incurring an appreciable oxygen debt (see Section IV,C) likewise indicates a disconcerting ability to bank the metabolic fires.

Among other types of maintained low-level metabolism may be mentioned the catalepsy which is common in adult phasmids (von Buddenbrock and von Rohr, 1923). Although this is marked by rigid extension of some of the legs, it appears to occasion no significant increase in metabolic rate as judged from the accompanying rate of oxygen uptake in comparison with other insects of comparable size (Table I). Decreased respiratory rate during summer sleep (aestivation?) has been reported by Lühmann and Drees (1952), and the inactivity seen in long inanition or desiccation at room temperatures may be physiologically equivalent.

B. DIAPAUSE

The primary physiological interests in diapause have centered around its mechanisms of induction and release, certain peculiarities of its gas exchange, and the biochemical nature of its oxidative metabolism. For present purposes it is sufficient to note that diapause represents a long, experimentally useful period during which metabolic rate, on a unit weight basis, is extraordinarily low and stable. Although the low rate was formerly thought to be due to a drastic decrease in mass of organized tissue it now appears that histogenesis proceeds concurrently with histolysis and the low respiratory rate signals a quantitative limitation in rate due to low cytochrome c concentration rather than any qualitative change in the electron transport

TABLE I

Oxygen Uptake Rates of Representative Insects in Various Stages of Development[a]

Name	Stage	Weight	Temp. (°C)	O$_2$ uptake (mm^3/mg/hr)[b]	Remarks	Reference
Diptera						
Musca domestica	Adult	—	20	2	Taken from hibernation	Edwards (1946)
("mouche")	Adult	—	20	5	—	Parhon (1909)
Musca domestica	Adult	18 mg	25	3.6	—	Fullmer and Hoskins (1951)
Calliphora erythrocephala	Larva	—	27	0.9	—	Fraenkel and Herford (1938)
Phormia regina	Larva	55 mg	25	1	—	Buck et al. (1952)
Phormia regina	Prepupa	55 mg	25	1–1.5	—	Keister (1953); Park and Buck (1960)
Phormia regina	1-day pupa	—	25	0.4	—	Park and Buck (1960)
Phormia regina	2–3-day pupa	—	25	0.3–0.4	Low point in U-shaped curve	Park and Buck (1960); Keister (1953)
Phormia regina	6-day pupa	—	25	1.2	Just before eclosion	Keister (1953)
Phormia regina	Adult, M	35–40 mg	25	2–2.6	No anesthesia, chilled for setup	Keister and Buck (1961)
Phormia regina	Adult, M	35–40 mg	25	1.75	CO$_2$ anesthesia for setup	Buck and Keister (1949)
Phormia terranovae	Pupa, M, F	—	25	0.125–1.7	Min. and max. of U-shaped curve	Taylor (1927)
Sarcophaga sarracenioides	Pupa	—	25	0.1–0.8	Min. and max. of U-shaped curve	Taylor (1927)
Procladius choreus	Larva	—	20	0.58	—	Luferov (1958)
Procladius nigriventris	Larva	—	20	0.47	—	Luferov (1958)
Ablabesmyia monilis	Larva	—	25	1.5	—	Luferov (1958)
Pollenia rudis	Adult, M, F	—	24.5	1.0	Overwintering	Argo (1939)

Species	Stage	Weight	Temp.	Rate	Notes	Reference
Glyptotendipes paripes	Larva	—	18	0.5		Harnisch (1959)
Stictochironomus histrio	Larva	—	18	0.4		Harnisch (1959)
Prodiamesa olivacea	Larva	—	18	0.5		Harnisch (1959)
Psectrotanypus varius	Larva	—	18	0.3		Harnisch (1959)
Chironomus 3 spp.	Larva	—	18	0.2, 0.3		Harnisch (1959)
Drosophila melanogaster	Pupa	1.2–1.4 mg	25	0.6–1.1	Max. and min. in U-shaped curve	Clare (1925)
Drosophila melanogaster	Pupa	0.6–0.9 mg	22.5	0.44–1.5	Max. and min. in U-shaped curve	Bodine and Orr (1925)
Drosophila melanogaster	Pupa, M, F	0.88–1.22 mg	25	0.67–2	Max. and min. in U-shaped curve	Poulson (1935)
Lepidoptera						
Bombyx mori	Egg	0.49 mg	24	0.1	Fertilized, just laid	Hsueh and Tang (1944)
Bombyx mori	Larva	0.5 mg	23	1.3	Newly hatched	Itaya (1940)
Bombyx mori	Larva, F	0.9 gm	23	0.6	Early 5th-instar	Itaya (1940)
Bombyx mori	Larva, M	1.8 gm	23	0.8	Late 5th-instar	Itaya (1940); Kamioka (1950)
Bombyx mori	Pupa, M	1.28 gm	23	0.37	1st day after pupation	Itaya (1940)
Bombyx mori	Pupa, F	1.63 gm	23	0.36	1st day after pupation	Itaya (1940)
Bombyx mori	Pupa, F	0.9 gm	23	0.16	4th day after pupation	Itaya (1940)
Bombyx mori	Pupa, M	0.7 gm	23	0.39	7th day after pupation	Itaya (1940)
Bombyx mori	Adult, M	0.49 gm	23	1.1	Young, virgin	Itaya (1940)
Bombyx mori	Adult, F	0.95 gm	23	0.98	Young, virgin	Itaya (1940)
Samia cecropia	Egg	—	28	0.24	Day 1, wt excluding wt of shell	Melvin (1928)
Samia cecropia	Egg	—	28	1.27	Day 9, wt excluding wt of shell	Melvin (1928)

TABLE I (*Continued*)

Name	Stage	Weight	Temp. (°C)	O₂ uptake (mm³/mg/hr)[b]	Remarks	Reference
Platysamia cecropia	Larva	—	25	1.2	Mature, prior to spinning	Schneiderman and Williams (1953)
Tropaea luna	Egg	—	28	0.28	1st day	Melvin (1928)
Tropaea luna	Egg	—	28	1.96	7th day, before hatch	Melvin (1928)
Pyrausta ainsliei	Egg	—	28	0.96	1st day	Melvin (1928)
Pyrausta ainsliei	Egg	—	28	4.2	6th day, before hatch	Melvin (1928)
Deilephila euphorbiae	Larva	2–4.5 gm	26	1.3	Feeding	Heller (1938)
Galleria mellonella	Larva	200 mg	30	1.6	Spinning	Burkett (1962)
Galleria mellonella	Pupa, M, F	125 mg	30	0.5	2-day pupa	Burkett (1962)
Galleria mellonella	Pupa, M, F	113 mg	30	1.0	7-day, near eclosion	Burkett (1962)
Galleria mellonella	Adult, M, F	av. 65 mg	30	2.3	24-hours-old	Burkett (1962)
Galleria mellonella	Larva	20–40 mg	25	22.1	Actively feeding, food contains brewers' yeast	Beck (1960)
Galleria mellonella	Pupa, M, F	ca. 100 mg	30	0.45–1.4	Min. and max. in U-shaped curve	Taylor and Steinbach (1931)
Ephestia kühniella	Pupa, M, F	—	25	0.25–0.8	Min. and max. in U-shaped curve	Taylor (1927)
Vanessa io	Pupa	250–300 mg	25	0.2–0.5	Min. and max. in U-shaped curve	Schwan (1940)
Orthoptera						
Periplaneta americana	Adult, M, F	—	25	0.34	—	Richards (1963a)
Blattella germanica	Adult, M	55 mg	27	0.8	—	Harvey and Brown (1951)
Blatta orientalis (*Periplaneta*)	—	150 mg	25	0.3	—	Slater (1927)
Blatta orientalis	Adult, M	—	25	0.3	—	Gunn and Cosway (1942)
Cryptocercus punctulatus	—	—	5–7.5	0.03–0.05	—	Gilmour (1940b)

Species	Stage/Sex	Weight	Temp.	Rate	Condition	Reference
Locusta migratoria	2nd-instar, F	60 mg	28	0.7	Mid instar, food *ad libitum*	Clarke (1957)
Schistocerca gregaria	5th-stage	1.3 gm	24–25	0.2	—	Bodenheimer (1929)
Schistocerca gregaria	Adult, M	1.7 gm	24–25	0.7	Young	Bodenheimer (1929)
Schistocerca gregaria	Adult, F	2 gm	24–25	0.4	Young	Bodenheimer (1929)
Melanoplus differentialis	2nd- to 4th-instar, M, F	0.09–0.34 gm	21	0.4	Restrained	Rogers (1929)
Dixippus morosus	1st stage	—	24	0.8	—	Janda (1952)
Dixippus morosus	Adult	—	24	0.4	—	Janda (1952)
Coleoptera						
Tenebrio molitor	Larva	—	30	0.8	—	Ludwig et al. (1962)
Tenebrio molitor	Larva	—	21	0.3	—	Edwards (1946)
Tenebrio molitor	Adult	—	30	1.3	—	Ludwig et al. (1962)
Tenebrio molitor	Adult	100 mg	19.5–22.5	0.4–0.6	—	Schmalfusz et al. (1939)
Tribolium confusum	Adult, M, F	22 mg	26	1.6	—	D. K. Edwards (1958)
Tribolium confusum	Adult	—	29	1.6	—	Kennington (1957)
Passalus cornutus	Adult	—	17	0.03	Taken in winter	Edwards (1946)
Melanotus communis	Adult	—	21	1.9	—	Edwards (1946)
Melasoma populi	Adult	—	25	0.125	"Winter sleep"	Marzusch (1952)
Melasoma populi	Adult	—	25	0.9	Active, but fasted 2 days	Marzusch (1952)
Leptinotarsa decemlineata	Adult	—	25	0.8	Active, but fasted 2 days	Marzusch (1952)
Chrysomela haemoptera	Adult	—	25	0.6	"Winter sleep"	Lühmann and Drees (1952)
Chrysomela haemoptera	Adult	—	15	0.2	"Winter sleep"	Lühmann and Drees (1952)
Chrysomela haemoptera	Adult	—	25	0.2	"Summer sleep"	Lühmann and Drees (1952)
Chrysomela haemoptera	Adult	—	25	0.6	Feeding period	Lühmann and Drees (1952)
Galeruca tanaceti	Adult	—	25	0.4	"Summer sleep"	Lühmann and Drees (1952)
Galeruca tanaceti	Adult	—	25	0.9	Active feeding	Lühmann and Drees (1952)
Phytodecta ruffpes	Adult	—	25	0.1–0.2	"Winter sleep"	Lühmann and Drees (1952)
Phytodecta ruffpes	Adult	—	25	0.14	"Summer sleep"	Lühmann and Drees (1952)

TABLE I (*Continued*)

Name	Stage	Weight	Temp. (°C)	O$_2$ uptake (mm^3/mg/hr)^b	Remarks	Reference
Phytodecta rufipes	Adult	—	25	1.25	Active period	Lühmann and Drees (1952)
Rhizopertha dominica	Adult	1.4 mg	34	3.6	1 week old	Birch (1947)
Calandra oryzae	Adult	1.7 mg	29	3.2	1 week old	Birch (1947)
Hypothenemus hampei	Larva	—	20	1.7	—	Edwards and González (1953)
Hypothenemus hampei	Pupa	—	20	0.8	—	Edwards and González (1953)
Hypothenemus hampei	Adult	—	20	2.8	—	Edwards and González (1953)
Popillia japonica	3rd-instar	220 mg	25	0.45	—	Ludwig (1946)
Hymenoptera "Abeille" (*Apis mellifera*)	Adult	—	32	15.6	In 20 gm groups, in spring	Parhon (1909)
Apis mellifera	Larva, queen	—	35	4.6	2–3 days old	Melampy and Willis (1939)
Apis mellifera	Larva, worker	—	35	2.9	2–3 days old	Melampy and Willis (1939)
Apis mellifera	Larva, queen	—	35	1.4	At pupation	Melampy and Willis (1939)
Apis mellifera	Larva, worker	—	35	0.4	At pupation	Melampy and Willis (1939)
Apis mellifera	Pupa, queen	—	35	0.7	Mid-pupal period	Melampy and Willis (1939)
Apis mellifera	Pupa, worker	—	35	0.5	Mid-pupal period	Melampy and Willis (1939)
Apis mellifera	Pupa, queen	—	35	1.5	Late-pupal period	Melampy and Willis (1939)
Apis mellifera	Pupa, worker	—	35	1.0	Late-pupal period	Melampy and Willis (1939)
Apis mellifera	Larva, worker	20 mg	32	1.9	—	Allen (1959a)
Apis mellifera	Larva, drone	20 mg	32	2.1	—	Allen (1959a)
Apis mellifera	Larva, worker	140 mg	32	0.8	Just before cell sealed	Allen (1959a)
Apis mellifera	Larva, drone	140 mg	32	1.7	—	Allen (1959a)

Apis mellifera	Larva, drone	400 mg	32	0.7	Just before cell sealed	Allen (1959a)
Camponotus pennsylvanicus	Adult, worker	—	25	0.6	Collected and measured at 6750 ft	Kennington (1957)
Formica ulkei	Adult, worker	—	22	0.8	—	Dreyer (1932)
Croesus septentrionalis	Larva	0.6 mg	21	2.4	1st-instar, 1st day	Janda (1961)
Croesus septentrionalis	Larva	84 mg	21	0.8	5th-instar, 12 days old	Janda (1961)
Hemiptera						
Oncopeltus fasciatus	Egg	—	24.8	0.5	48–72 hours old	Argo (1939)
Oncopeltus fasciatus	Egg	0.3 mg	32	0.21	1st 24 hours	Rutschky and Joseph (1957)
Oncopeltus fasciatus	Egg	—	32	1.02	Peak uptake, end of 4th day	Rutschky and Joseph (1957)
Anasa tristis	Egg	—	28	0.27	1st day, shell wt subtracted	Melvin (1928)
Anasa tristis	Egg	—	28	1.96	7th day, shell wt subtracted	Melvin (1928)
Homoptera						
Aonidiella aurantii	Pupa, M	0.5 mg	25	0.06	—	Yust and Shelden (1952)
Pseudococcus citri	Adult	1.3 mg	24.6	1.5	—	Argo (1939)
Odonata						
Agrion sp.	Nymph	—	18	0.27	—	Harnisch (1959)
Isoptera						
Zootermopsis nevadensis	Nymph	—	3	0.037	"Basal," i.e., inactive at 3°	Gilmour (1940a)

[a] Additional values in control columns of Tables II and III.
[b] Values converted to mm³/mg live wt/hr, and corrected to NTP where necessary.

pathway in terminal oxidation (Harvey, 1962). Some sample measurements indicating the magnitude of the oxidative depression are assembled in Table II.

C. Flight

As with diapause, most of the work on flight has centered on biochemical and biophysical aspects in which respirometry is involved only as an incidental index to metabolic changes. The features of greatest interest in regard to respiratory rate are (1) that the factor by which respiratory intensity increases over the resting level is often very large (Table III); (2) that flight represents a period, sometimes of hours (e.g., Weis-Fogh, 1952), during which oxidative metabolism is proceeding at the maximum steady-state pace, for a given temperature, that can be maintained without anaerobic supplementation; and (3) that the physiological governor is partly biomechanical rather than enzymatic, since wing-beat frequency (and, therefore, muscle metabolism) rises temporarily when wings are shortened or density of ambient gas decreased (Chadwick and Williams, 1949).

The exact ratio between metabolic rates in rest and in flight is not yet a theoretically useful datum, but it is worthwhile pointing out that it depends to an important extent on the precision with which the resting level can be measured and the range within which it normally fluctuates. If the resting rate is at about the limit of sensitivity of the respirometer the ostensible increase can vary by a large factor.

III. Respiratory Rate in Relation to Temperature

The increase in respiratory rate with rising temperature is perhaps the most over-confirmed fact in insect physiology. Many of the purely descriptive reports must be passed over entirely to allow for consideration of work having analytical value.

A. Typical Temperature–Respiration Response

In most insects absolute respiratory rate (per individual or per unit weight) is low at low temperatures, increases rapidly through the mid-range temperatures, and then breaks sharply as lethal temperatures are approached. Figure 2 shows examples of such curves, and very similar data have been recorded for one or more stages of *Calliphora* (Fraenkel and Herford, 1940), *Phormia* (Keister and Buck, 1961), *Tenebrio* (Janda and Kocián, 1933), *Galleria* (Crescitelli, 1935; Bell, 1940; Burkett, 1962),

TABLE II

RESPIRATORY RATES DURING DIAPAUSE AND, WHERE AVAILABLE, DURING DEVELOPMENT

Name	Stage	Weight	Temp. (°C)	O₂ uptake (mm³/mg/hr)	Remarks	Reference
Bombyx mori	Egg	0.49 mg	25	0.1	27 hr old, onset of diapause	Chino (1958); cf. Hsueh and Tang (1944)
Bombyx mori	Egg		25	0.008	10 days old, in diapause	Chino (1958)
Bombyx mori	Egg	0.4 mg	25	0.7	Postdiapause, early eye-spot stage	Yokoyama and Keister (1953)
Platysamia cecropia	Pupa	4–6 gm	25	0.15	Fresh pupa, prediapause	Schneiderman and Williams (1953)
Platysamia cecropia	Pupa	4–6 gm	25	0.016	Diapausing 10 to 16 weeks	Schneiderman and Williams (1953)
Antheraea pernyi	Pupa	—	28	0.014	Diapausing, unchilled	Waku (1957)
Antheraea pernyi	Pupa	—	28	0.08	After 96 days' chilling, development started	Waku (1957)
Cephaleia abietis	Larva		20	0.09	Diapausing	Kirberger (1953)
Melanoplus differentialis	Egg	0.069 mg dry	25	1.5/mg dry wt	Prediapause	Bodine and Boell (1937)
Melanoplus differentialis	Egg	0.075 mg	25	0.59	Diapause	Bodine and Boell (1937)
Melanoplus differentialis	Egg	0.078 mg	25	1.5	Postdiapause	Bodine and Boell (1937)
Agapema galbina	Pupa	1.02 gm	25	0.02	Diapause	Buck and Keister (1955)

TABLE III

OXYGEN UPTAKE DURING FLIGHT AND AT REST

Name	Weight	Temp. (°C)	Oxygen uptake (mm³/mg/hr)		Remarks	Reference
			At rest	In flight		
Drosophila americana	1.25 mg	20	1.6	28	1st ½ hr after flight 4.3	Chadwick (1947)
Drosophila repleta	3.4 mg	25	1.7	21	—	Chadwick and Gilmour (1940)
Lucilia sericata	34 mg	27	1.8–2.7	87	"At rest" = moving, not flying	Davis and Fraenkel (1940)
Bee	100 mg	18	21	312	"At rest" = slowly moving, wings at rest	Kosmin et al. (1932)
Apis mellifica [sic]	76–114 mg	21	1.75	94	"Good" flyers	Jongbloed and Wiersma (1934)
Periplaneta americana	—	20	0.36	36	Not free flight: roach fixed, wings vibrating	Poláček and Kubišta (1960)
Schistocerca gregaria	1.5–3.3 gm	28	0.63	15	In 10-min. flight: 3.5 immediately after, 0.65 1 hr after flight	Krogh and Weis-Fogh (1951)
				10–30	Calculated from fuel consumption at flying speeds of 2.3–3.5 m/sec	Weis-Fogh (1952)
Thais cassandra	0.13 gm	20–25	4	216	Not true flight: wing buzz induced by nicotine injection	Raffy and Portier (1931)
Sphinx ligustri	1.4 gm	20–25	0.8	24		Raffy and Portier (1931)
Deilephila euphorbiae	0.7 gm	20–25	0.5	88		Raffy and Portier (1931)
Deilephila elpenor	0.6 gm	20–22	0.2	6	QO_2 not measured: calculated from CO_2 output on assumed RQ of 1	Kalmus (1929)
Mimas tiliae	0.22–0.42 gm	22+	0.75	55.8		Zebe (1954)
Antheraea pernyi	0.72–1.7 gm	21–23.5	0.73	—		Zebe (1954)
Antheraea pernyi	0.595 gm	22.6	—	29.4		Zebe (1954)
Odonestis pruni	0.25 gm	21.3	0.62	65.8		Zebe (1954)
Vanessa io	0.23 gm	21.3	0.55	—		Zebe (1954)
Vanessa io	0.18 gm	21.7	—	53.6		Zebe (1954)
Metopsilus porcellus	0.285 gm	21.4	0.66	—		Zebe (1954)
Metopsilus porcellus	0.243 gm	22.9	—	92.1		Zebe (1954)
Tabanus affinis	160 mg	18.9	0.8	22	Body weights with crop empty: flight respiration calculated from sugar consumption	Hocking (1953)
Simulium venustum	2.5 mg	20.6	4.4	27		Hocking (1953)
Apis	105 mg	22.2	3.2	60		Hocking (1953)
Drosophila	0.9 mg	25.0	5.0	33		Hocking (1953)

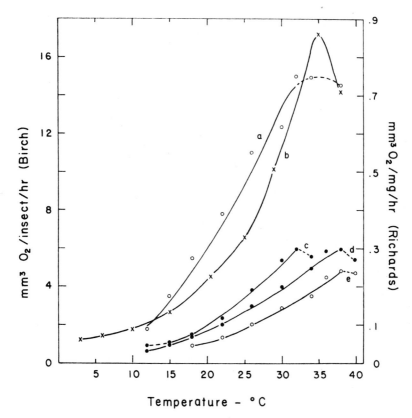

Fig. 2. Some "typical" respiration–temperature curves. (a) 12-day-old larvae of *Calandra;* (b) adult *Periplaneta,* means of both sexes combined; (c) 1-week-old adult *Calandra;* (d) 1-week-old adult *Rhizopertha;* (e) 18-day-old larvae of *Rhizopertha.* (Curves from *Calandra* and *Rhizopertha* redrawn from Birch, 1947; curve for *Periplaneta* from data in Table I, Richards, 1963a.)

Culex (Ellinger, 1915), *Periplaneta* (Dirken, 1922; Dehnel and Segal, 1956), *Dixippus* (von Buddenbrock and von Rohr, 1923), *Musca* and *Melanotus* (Edwards, 1946), and many other species. It should be noted that the *relative* increase in rate of respiration with increasing temperatures—that is, percent increase per degree—is highest at the low end of the temperature range.

It is clear that different forms have different break points at high temperatures, but no generalizations or correlations seem to have emerged as to response similarities or differences in relation to systematic position, size, stage, nutritional regimen, or habitat.

B. INTERPRETATION OF TYPICAL RESPONSE:
TEMPERATURE COEFFICIENTS AND CHARACTERISTICS

Rates of simple chemical reactions in general increase with rising temperature by a constant percent per degree over a wide range of temperature. As a rough generalization, in fact, the rate will double for every $10°$ rise in temperature ($Q_{10} = 2$; Van't Hoff rule). Q_{10}'s for biological processes, in contrast, vary widely and usually decrease as temperature rises. Scholander *et al.* (1953a) found Q_{10} values up to 50 in the $0°$ to $—5°C$ range for arctic insects (*Chironomus* larva) and calculated that if survival time were inversely proportional to respiratory rate an organism surviving 10 days at $0°C$ would have sufficient reserves to live for a thousand years at $—23°C$ and a million years at $—42°C$.

Such speculations are interesting but shed no light on why the rate of respiration changes so nonlinearly with temperature. Curves very similar to those in Fig. 2 have been obtained for single enzymic reactions, the progressive decrease in Q_{10} with increasing temperature and the eventual high-temperature inflection being ascribed to concurrent progressive denaturation of the enzyme. It would, however, be unjustified to refer the overall respiration-temperature curve directly to enzymic activity. For one thing, many enzyme systems undoubtedly contribute to overall "respiration." For another, gross locomotory activity, with its attendant oxygen demand, may follow a very similar course. In sum, then, the many Q_{10} data available for insect respiration are useful in comparing different organisms or temperatures but do not give any qualitative information about the underlying metabolism.

Temperature "characteristics" (activation energies in calories per mole) can be derived from the slopes of the straight lines that usually result from plotting logarithm of enzyme reaction rate against reciprocal of absolute temperature (Arrhenius plot). Such data have also been derived from whole insect respiration. The plots, themselves, whether straight lines over the entire range studied (Richards, 1957; Fig. 3 curve b), or involving intersecting lines (Fig. 3 curves a,c,d,e), are often of impressive regularity, but it is dubious indeed, as Richards (1957) states, whether it is justified at present to draw any conclusions therefrom. In other words, in a system in which overall gas exchange almost certainly involves a complex of sequential and parallel oxygen-requiring and carbon dioxide-producing reactions differing in intrinsic temperature coefficients, temperature thresholds, denaturation temperatures and rates, it is illusory to apply an analysis based on assumed single rate-limiting or "master" reactions. Richards' valiant and imaginative attempts to unravel some of the simultaneous temperature responses involved (Richards, 1957, 1958, 1959, 1963a,b; Richards and Suanraksa, 1962) serve to emphasize this point.

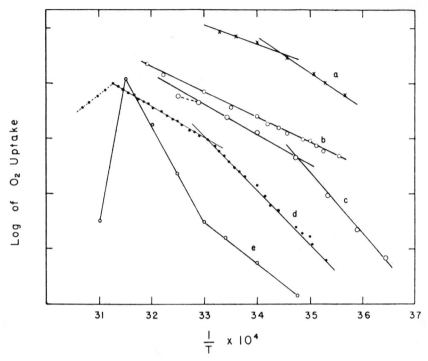

Fig. 3. Various Arrhenius plots of respiratory rates. (a) 1st-instar *Dixippus* (from Janda, 1952); (b), young larvae of *Oncopeltus* (from Richards, 1958); (c) *Drosophila* pupae (from Orr, 1925); (d) *Melanoplus* nymphs (from Rogers, 1929); (e) *Melasoma* adults (from Marzusch, 1952).

C. DEVIATIONS FROM THE STANDARD CURVE

It is not uncommon for a mid-range span of 15° or 20°C in the respiration-temperature curve to be essentially linear preceding a final prebreak increase in rate (region B, Fig. 4). Such an intercalated region of constancy in absolute metabolic increment per degree would be difficult if not impossible to explain in a unitary respiring system. However, in an overall gas exchange contributed to by at least two systems, the straight region could be visualized as the resultant of a temporary progressive slowing of the metabolic rate of system 1 superimposed on continuing increase in the rate of system 2, the final spurt (region C, Fig. 4) signaling the resumption of activity by system 1 and the eventual break (region D) marking the point at which irreversible damage to both systems attains dominance. To make this picture somewhat more concrete (though still speculative), system 1 could be locomotion and system 2 maintenance respiration.

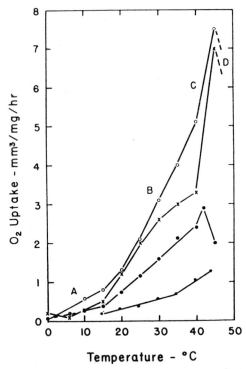

Fig. 4. Examples of respiration–temperature curves with distinctly different slopes in different parts of the temperature range (see text). Open circles, adult *Phormia;* crosses, adult *Galleria;* larger solid circles, *Galleria* larvae; smaller solid circles, third stage *Schistocerca.* (*Phormia* curves from Keister and Buck, 1961; *Galleria* curves from Burkett, 1962; *Schistocerca* curve from Bodenheimer, 1929.)

More extreme examples of similar polysystem interaction are perhaps seen in instances in which rate of respiration either fails to increase with temperature or actually decreases over part of the range. Most of these irregularities fall into two groups: those involving temperatures immediately above and below the rearing temperature, and those occurring considerably below the rearing temperature. As examples of the first group, (a) some strains of *Drosophila* pupae show plateaux at 18°–22°C and 27°–30°C, on either side of the rearing temperature of 25°C (Vernberg and Meriney, 1957); (b) a dung beetle shows no respiratory increase between 12° and 24°C and an ant none between 20° and 34°C (Slowtzoff, 1909); (c) bees exhibit a "zone of comfort" between 20° and 25°C in which respiration is lower than at temperatures on either side of this range (Woodworth, 1936). In all these instances, however, it must be supposed [as Parhon (1909) pointed out] that the insect is not a passive *esclave du milieu ambient,* but rather that changes in some controllable activity such as muscular contraction occur, or that body temperature is different from ambient. Precedent for such not necessarily visible activities is given by the process of flight preparation (Krogh and Zeuthen, 1941) and the body- and culture-temperature regulation of honey bees.

In the second group are cases of a definite plateau or break in the course of the curve at a temperature which is considerably lower than the rearing temperature and at which locomotion may be absent or restricted (Fig. 5). Such interruptions are seen, for example, in *Drosophila* pupae at 15°C (Orr, 1925), in cataleptic *Dixippus* between 8° and 15°C (von Buddenbrock and von Rohr, 1923), in *Pollenia* at 12.5°C (Argo, 1939), and in *Phormia* larvae between 10° and 15°C (Keister and Buck, 1961). It is curious that the figure 15°C occurs quite frequently as a "critical" temperature, not only in connection with respiratory rates but also as the hatching threshold of *Oncopeltus* eggs (Richards, 1959) and the threshold for spinning of full grown *Galleria* larvae (Burkett, 1962).

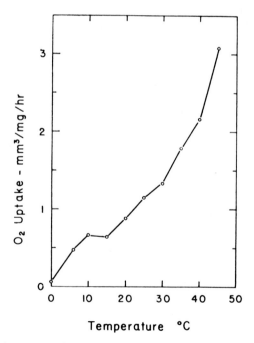

Fig. 5. Respiratory response of *Phormia* larvae to temperature, showing plateau at 10°–15°C. (From Keister and Buck, 1961).

D. Developmental Metabolism in Relation to Temperature

The concurrent effects of temperature upon rates of metabolism and of development during nonmotile stages are of considerable interest. From measurements of oxygen uptake throughout a given stage at several temperatures one can compute the respective costs of development. In the few such studies it is not surprising to find that not all insects behave alike, but there is also a disconcerting lack of agreement among investigations of the same insect.

In some species, temperature changes within the physiological range seem not to affect total oxygen uptake even though rate of development changes. *Tenebrio* pupae produce approximately the same total amount of carbon dioxide between 20.9° and 32.7°C, according to Krogh (1914). This conclusion is supported by integrating under the curves for oxygen uptake of *Tenebrio* pupae between 20° and 35°C presented by Lotz and Thiel (1962). However, integrating under the curves of Janda and Kocián (1933) for the same insect over the range 16°–37°C (Fig. 6) indicates a clear-cut minimum total oxygen uptake at 25°C. A nearly constant total carbon dioxide output was reported for the beetle *Acilius* (Krogh, 1916) and for oxygen uptake of *Drosophila* pupae between 14° and 25°C (Dobzhansky and Poul-

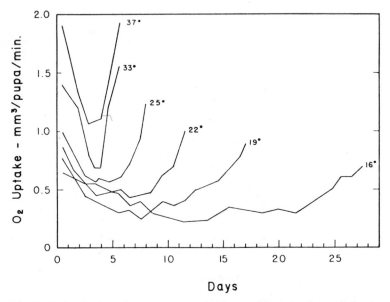

Fig. 6. Effect of temperature upon respiration and length of pupal development of *Tenebrio*. Integration under these curves indicates a minimum total pupal respiration at 25°C. (Redrawn from Janda and Kocián, 1933.)

son, 1935). However, Gromadska (1949) found the total carbon dioxide production of *Drosophila* pupae to be definitely lower at 25°C than at either 20° or 30°C. In several lepidopterous pupae the total oxygen uptake is minimal at some temperature within the physiological range. For *Ephestia* this temperature is around 20°C (Kozhantschikov and Maslowa, 1935) or 25°C (Gromadska, 1949); for *Galleria* 30°C (Crescitelli, 1935); for *Loxostege* around 25°C and for *Agrotis* around 32°C (Kozhantschikov and Maslowa, 1935).

These temperatures are not necessarily the same as those at which development is most rapid or mortality lowest, so there is some question as to whether they should be considered optima. However, Gromadska stated that for *Ephestia, Tribolium,* and *Drosophila* the physiological and ecological optimal temperatures agree. Crescitelli pointed out that the temperature of minimum oxygen uptake (30°C) is also the temperature at which the *Galleria* colonies were raised, and posed the question whether this was coincidental or indicated adaptation.

Richards and Suanraksa (1962) followed development and metabolism in the eggs of *Oncopeltus* and found that the total oxygen uptake was much greater at 17.5°C and 15°C than at 25°C. An even more interesting, though unexplained, result was the discovery that a fluctuating temperature, involving even short daily exposures to a favorable temperature, and giving an average temperature near the minimum for development, was more favorable than constant exposure to such a minimal temperature; e.g., eggs exposed to 15°C for 20 hours and to 30°C for 4 hours out of each day (average temperature 17.5°C) developed more rapidly and with lower mortality and lower total oxygen uptake than eggs exposed constantly to a fixed 17.5°C.

E. ADAPTATION TO TEMPERATURE

In any consideration of effects of temperature on respiration of insects, account should be taken of possible modifications due to rate of change in temperature and to past temperature history of the insects. Rogers (1929) reported that *Melanoplus* nymphs exposed to only one temperature other than the rearing temperature had higher rates of oxygen uptake at any given temperature than those exposed to a series of temperatures. No explanation was offered. When *Oncopeltus* eggs are transferred from 15° to 25°C, the respiratory rate "overshoots" slightly before settling down at the new level (Richards and Suanraksa, 1962). This transient phenomenon can easily be missed if the method of measurement involves a delay for temperature equilibration. Adults of *Tribolium,* transferred from 30° to 38°C, show a 2–3 day overshoot according to D. K. Edwards (1958). Nymphs of the dragonfly,

Aeschna, respond initially to temperature changes by increase or decrease in respiratory rate, but after a few days the rate returns to approximately the original level rather than to a new level characteristic of the experimental temperature (Sayle, 1928).

The subject of acclimatization of poikilotherms to temperature has been reviewed by Bullock (1954). A few instances among insects have been clearly demonstrated. When acclimatization does occur, the effect is usually a simple shift in the position of the respiration–temperature curve rather than a change in shape (Fig. 7). However, in some poikilotherms, there is a change in slope as well. The theoretical possibilities are discussed at length by Agrell (1947), who attempted to test them on seven species of insects kept for periods of 1 to 350 days at 5°, 18°, and 30°C. Unfortunately the results are complicated by starvation effects, the death of many of the insects after a few days at 30°C, and loss of vitality in the survivors. However, there was apparent compensational adjustment to 5°C in some species, as their respiration at any given temperature was higher than that of the 18°C-adapted insects. Certain beetles which have a seasonal quiescent period exhibit acclimatization to temperature. Thus "summer-sleeping" *Phytodecta* and *Galeruca* acclimated to 25°C respired more slowly at any given temperature than those acclimated to 15°C; but the rates of the two groups differed very little when measured at their respective adaptation temperatures (Lühmann and Drees, 1952; see also arrows in Fig. 7). A similar acclimatization is found in "winter-sleeping" *Melasoma* and *Leptinotarsa.* The actively feeding beetles of these species, however, showed no acclimatization (Marzusch, 1952). After 1 to 3 weeks at 10°, 16°, and 26°C, *Periplaneta* showed acclimatization, in that those that had been kept at the lowest temperature respired at higher rates at any given temperature than the warm-adapted ones. Nymphs acclimated to a greater degree than adults, and small adults more than large ones (Dehnel and Segal, 1956). D. K. Edwards (1958) produced a shift in the temperature-respiration curves of adults of *Tribolium* by keeping the beetles for months at 18° and 38°C. However, this result is a curious one since both the cold- and warm-adapted animals had lower rates at most temperatures than did animals that had been kept at 30°C.

In spite of instances such as these, insects in general are considered to have little ability to compensate for environmental temperature. Scholander *et al.* (1953b) compared similar species of arctic and tropic insects, all terrestrial, and concluded that there was little or no adaptation to temperature, or at least none that was evident in respiratory rates. Respiration of *Formica* varies directly and uniformly with temperature regardless of previous temperature history of the ants (Dreyer, 1932). In work with *Phormia* larvae, Keister and Buck (1961) found no evidence of a difference in response to

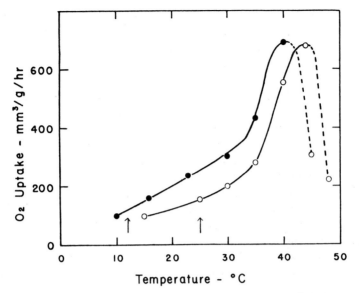

Fig. 7. Respiratory adaptation to temperature of *Melasoma*. Solid circles, respiration of beetles adapted to 12°C; open circles, respiration of beetles adapted to 25°C. The arrows call attention to nearly equal rates of the two groups at their respective adaptation temperatures. (Redrawn from Marzusch, 1952.)

temperature changes between larvae kept at 6°C for several days and those taken directly from 25°C cultures.

IV. Respiratory Rate in Relation to Oxygen Concentration

A. OXYGEN DEFICIENCY

Insects are often said to have a high degree of respiratory independence of ambient oxygen concentration. This is too sweeping a generalization.

Mayfly nymphs of various species show a complete spectrum from great sensitivity to changes in environmental oxygen concentration to great insensitivity (Fox *et al.*, 1937). In terrestrial insects there is a similarly wide variation. Among larvae at laboratory temperatures the critical oxygen concentration is below 7.5% in *Calliphora* (Fraenkel and Herford, 1938) and above 10% in *Phormia* (Keister and Buck, 1961). Among pupae the limiting concentration is below 1% in diapausing *Agapema* (Buck and Keister, 1955); about 2% in diapausing *Promethea* (Kurland and Schneiderman, 1959); below 5% in *Tenebrio* (Gaarder, 1918); about 5% in brainless cecropia (Schneiderman and Williams, 1954); 6–8% in *Musca* (Galun and Fraen-

kel, 1961); and above 10% in *Galleria* (Burkett, 1962) and *Phormia,* 1-
and 5-day-old pupae being more sensitive than those of intermediate age
(Park and Buck, 1960). Among adult insects, *Aedes* mosquitoes, both intact
and headless, have constant respiratory rates down to 3–4% oxygen (Galun,
1960); *Phormia* (Keister and Buck, 1961) and *Termopsis* (Cook, 1932)
to between 2 and 5%; and the coffee borer *Hypothenemus* down to 5%
(Edwards and González, 1952). The three stages of *Phormia* are compared
in Fig. 8.

In the above examples the measurements were made at single temperatures. Scattered experiments on the combined effects of temperature and
oxygen concentration (von Buddenbrock and von Rohr, 1923; Keister and
Buck, 1961) show, as expected, that at higher temperatures respiration cannot be maintained over as wide a range of oxygen concentration.

Throughout the animal kingdom it is exceedingly rare to find respiratory
rate increased by oxygen concentrations higher than atmospheric. However,
the oxygen consumption of diapausing silkworm eggs has been reported to
vary with oxygen concentration between 1 and 99%, although developing
eggs, young larvae, and pupae (all with higher respiratory demand in air)
show no respiratory increase in concentrations above that of air (Nittono
and Nakasone, 1956a,b,c). Similarly the respiration of grasshopper eggs was
said to be stimulated in 50 and 100% oxygen, though long exposure was
deleterious (Bodine, 1934b). Since such findings imply that the organism
is normally hypoxic (in air) they are difficult to accept.

B. Effects of Pure Oxygen

The harmful effects of pure oxygen on a variety of organisms, chiefly
vertebrates, seem to be well established. It is frequently assumed or stated
that insects are also adversely affected by molecular oxygen in high concentrations. However, critical tests are rare, and although paralysis, decreased
respiration, and decreased viability have been reported as symptoms of
oxygen poisoning, the experiments have usually involved the use of several
atmospheres total pressure (e.g, Williams and Beecher, 1944), high temperature (Keister and Buck, 1961), or other treatments which in themselves
may be injurious.

Evaluation of the respective roles of high oxygen concentration and high
pressure per se is difficult, since in most instances neither 100% oxygen at
1 atm nor inert gases at several atmospheres produces the injuries caused
by a combination of the two conditions. Hence it is meaningless to refer
to the toxicity of oxygen without the qualification "at high pressure." The
best work in this area, particularly with respect to respiratory rate, is that
of Clark and his co-workers (Clark, 1959; Clark and Herr, 1954; Clark
and Papa, 1958; Clark et al., 1958; Clark and Cristofalo, 1960, 1961). They

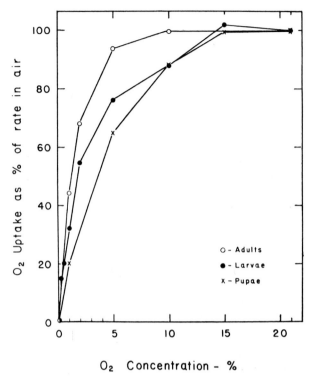

Fig. 8. Effect of oxygen tension on respiratory rates of three stages of one species, *Phormia regina*. Note that rates of larvae and pupae are limited at higher concentrations than the rate of adults. (Larval and adult curves from Keister and Buck, 1961; pupal curve from Park and Buck, 1960.)

found that insect orders differ in their susceptibility, as do developmental stages of the same species. A decrease in respiratory rate followed treatment with pure oxygen at 2–5 atm in pupae of *Grapholitha, Prodenia, Anagasta, Galleria, Tenebrio,* and *Bracon,* but not in *Musca, Drosophila, Phormia,* and *Apis,* or the larvae of *Bracon, Tenebrio,* and *Anagasta.* In some cases, 2 atm of oxygen were more harmful than even 8 atm of nitrogen. However, pupae of *Tenebrio* and *Galleria* were injured more severely by 5 atm of air than by 1 atm of oxygen. These findings seem to indicate that absolute oxygen tension is not solely responsible for the harmful effects but that the high pressure contributes substantially to the damage.

C. Anoxia—Oxygen Debt

Many insects survive hours or days in total absence of oxygen. Some of them apparently accumulate an oxygen debt during the anoxic period, as evidenced by excess oxygen uptake following anoxia, production of carbon dioxide during the anoxic period, and accumulation of lactic acid. The degree of apparent repayment of the debt is extremely variable and its significance is often uncertain because of lack of information about types of anaerobic product formed, amount excreted or sequestered, and other facets of intermediary metabolism.

Excess postanaerobic oxygen uptake has been demonstrated in grasshoppers after submersion (100% repayment; Bodine, 1928); *Blatta* (300–400% repayment; Slater, 1927; Davis and Slater, 1926, 1928); *Cryptocercus* (300% repayment; Gilmour, 1940b); termites (50% repayment; Gilmour, 1940a; see, however, Cook, 1932); and mealworms (150% repayment; Gilmour, 1941). Park and Buck (1960) found 26–33% repayment of oxygen debt in *Phormia* pupae in 7 hours following 4 hours in nitrogen. In a preliminary paper, Kurland *et al.* (1958) reported that pupae of *Cecropia, Polyphemus,* and *Cynthia* survive 3–5 days of anaerobiosis at 25°C while diapausing *Mormoniella* pupae may survive 18 days at 25°C and more than half a year at 5°C. Over a long period, 90–125% of computed debt was repaid by *Cecropia,* lactic acid rising in 2 days to 25 times its original titer.

In pupae of *Galleria* kept in nitrogen up to 16 hours, we found (unpublished) no appreciable increase in oxygen uptake following the anoxic period, but eclosion of the adults was delayed by a period roughly equal to that of anoxia. In this instance development seems simply to be suspended until sufficient oxygen is available to permit completion of development and there appears to be either no mechanism for increased uptake to compensate for the anoxic period or no need for such an accelerated repayment.

V. Respiratory Rate in Relation to Barometric Pressure

The effects of total pressure on insects, especially on their respiration, have been little studied. The literature to 1945 has been reviewed by Wellington (1946a,b). Interest has been mainly in survival at extremes of pressure or the action of fumigants at low or high pressure.

Concerning high pressure there is little to add to what has been said in the section on oxygen at high pressure except that it has been found that 250 psi (17 atm) is required to stop ventilation in grasshoppers (Roller, 1906). The only insects that might normally experience high pressures are certain aquatic species.

The effects of low pressure per se are very difficult to separate from the effects of accompanying desiccation, irritation, low oxygen concentration, and, in connection with reports of natural high-altitude tolerance, low temperature. It is known that insects can withstand very low pressures and sudden decompression without permanent injury, but few measurements of respiration have been attempted at low pressures. Woodworth (1932) found that honey bee carbon dioxide production at 25°C is somewhat higher at 500 mm Hg total pressure than at 600 mm (owing probably to irritation activity) (Cotton, 1932; Woodworth, 1936), then falls progressively with further decrease in pressure. Monro *et al.* (1962) reported analogous observations for *Tenebrioides* larvae in that respiratory rate is lower in pure oxygen at 35 mm total pressure than in air at 150 mm (31 mm pO_2). However, neither degree of muscular activity nor rate of desiccation at the various pressures was measured, the latter being a particularly relevant factor since hypoxia is known to induce spiracular gaping.

Day (1951) subjected *Periplaneta* adults to pressure drops equivalent to an altitude of 50,000 feet (75 mm total pressure). Although he did not measure respiratory rates he studied the tracheae of the tegumen and concluded that natural low pressures could have no mechanical effect on respiration.

The most extensive and best-controlled work on low pressure is that of Galun and Fraenkel (1961) on decapitated adults of *Aedes aegypti* and pupae of *Musca vicina*. These workers devised a method for measuring oxygen uptake at low total pressure and also took into account the deleterious effects of desiccation and hypoxia associated with low pressure. They found that at oxygen tensions above the critical or rate-limiting values, respiratory rate is the same whether a given oxygen tension is attained by decompression in air or by exposure to nitrogen–oxygen mixtures at 1 atm total pressure. Below the critical oxygen tensions for these species (15–30 mm for *Aedes*; 45–60 mm for *Musca*), oxygen uptake is conspicuously lower in decompressed animals than in those subjected to corresponding nitrogen–oxygen mixtures at one atmosphere. A very interesting, though unexplained, observation was that relieving the pressure deficit with nitrogen—i.e., restoring the decompressed insects to atmospheric pressure without increasing the ambient pO_2—induced increases in rates of oxygen uptake of 2–8-fold in the mosquitoes and 2–3-fold in the fly pupae. This seems conclusive evidence that, although desiccation and hypoxia contribute to the effects of low total pressure, low pressure in itself depresses respiratory rate.

VI. Effects of Pharmacological Agents: Metabolic Stimulators and Inhibitors; Insecticides

There is an enormous literature concerning effects of various chemical compounds on respiratory rate in insects, involving much variety in species and developmental stages investigated, techniques of application, concentrations utilized, and details of respirometry. In most instances gas exchange has been measured on the whole insect and only as an index of toxication. Hence, except in the few studies that were supplemented by *in vitro* preparations, effects remain at a primarily descriptive level with respect to both the mechanism of action of the agent being tested and the mechanism of respiratory stimulation or depression. In general the only chemical actions sufficiently studied to be worth discussing are the direct effects produced by inhibitors or uncouplers of oxidative metabolism, and the indirect effects due to stimulation or inhibition of some system, such as muscular contraction, that is maintained by oxidative metabolism.

A. Direct Effects on Oxidative Metabolism

The best-defined direct chemical stimulation of insect respiration is that produced by dinitrophenol and dinitrocresol, uncouplers of phosphorylation. These compounds are reported to stimulate respiration in bees by about 50%, DNP being more effective than DNC (Goble and Patton, 1946); in *Blattella* by about 400%, DNC being the more effective (Harvey and Brown, 1951); and in *Tribolium* up to 8-fold (DNC) (Lord, 1950). DNP stimulates respiration 340% in diapausing grasshopper embryos and 276% in developing embryos, the stimulated respiration being sensitive to cyanide, carbon monoxide, azide, and urethane. Homogenates of embryos, however, show no stimulation (Bodine and Boell, 1936; Bodine, 1950a,b; Bodine and Lu, 1950).

The standard inhibitors of the cytochrome system, cyanide and carbon monoxide, have been studied extensively in insects and have been shown to have conventional effects in general. Aside from work on fumigation the most extensive studies have been those on grasshopper eggs by Bodine and collaborators and those on lepidopterous pupae by Williams and his co-workers (see below). Since development and diapause are treated (in Volume I, Chapter 4), only the barest summaries will be given here.

In the grasshopper egg, cytochrome oxidase is lacking up to the peak of prediapause respiration at about 15 days, rises to a low plateau concentration during diapause while respiratory rate falls concurrently to about a quarter of its prediapause level, then, after break of diapause, increases rapidly to about 10 times its diapause level accompanied by a similar increase in respiratory rate (Allen, 1940). The respiration of whole diapausing eggs is quite insensitive to cyanide (19% inhibition) while that of whole developing eggs is quite sensitive (74% inhibition) (Bodine, 1934a; Bodine and Boell, 1936, 1937). Rather surprisingly, there is no difference in sensitivity between diapausing and developing embryos removed from the egg, both being inhibited about 90%. It would thus seem that the normal oxygen uptake of the whole diapausing egg represents cyanide-insensitive respiration—i.e., a fraction independent of the cytochrome system. The fact that the sensitivities of whole diapausing egg and isolated embryo differ was construed to mean that the yolk plays a larger part in the total respiration than expected. Alternatively, however, Robbie's (1941) work with whole eggs, isolated embryos, and eggs in which the embryos had been destroyed by X rays, suggests that the isolation procedure may activate the embryo. In any case, the need for concentrations of cyanide much higher than is usual in *in vitro* work on enzyme inhibition suggests that penetration barriers are present in the intact egg.

The findings obtained with carbon monoxide support those reported for cyanide. A mixture of 75% CO and 25% O_2 actually stimulated the respiration of diapausing grasshopper embryos, whereas the same gas mixture inhibits the respiration of developing embryos (Bodine and Boell, 1934). The respiration of brainless *Bombyx* pupae (artificial diapause) in $CO/O_2 = 80/20$ is three times as high as in air (Kobayashi and Nakasone, 1960). Respiration of diapausing silkworm eggs is not affected by high concentrations of CO, although that of developing eggs, larvae, and pupae is inhibited, the inhibition being light reversible (Nittono and Nakasone, 1956a,b,c). Artificially bivoltinized (activated) diapause eggs, however, with their higher rate of oxygen consumption, are sensitive to, although not totally inhibited by CO (Wolsky, 1949).

Turning now from eggs to lepidopterous pupae, we find the same sort of insensitivity to cyanide and carbon monoxide during diapause. This was originally attributed to deficiency or lack of cytochrome oxidase (Schneiderman and Williams, 1954; Schneiderman, 1957), but later work has shown that the insensitivity is due to the presence of a large excess of cytochrome oxidase in relation to cytochrome c (Harvey and Williams, 1958; Kurland and Schneiderman, 1959; Shappirio, 1960; Gilbert and Schneiderman, 1961; Harvey, 1962). It appears, therefore, that the low respiratory rate seen in diapausing eggs and pupae does not represent a qualitatively different cyanide- and monoxide-insensitive system, but simply residual cytochrome oxidase-mediated respiration which persists because the inhibitor is tied up by the large excess of enzyme. A low respiratory rate, in other words, escapes inhibition rather than being low because of inhibition.

B. INDIRECT EFFECTS ON RESPIRATION: INSECTICIDES

Many of the bewildering array of contemporary insecticides affect respiratory rate. It is difficult to systematize these effects because of the number of chemical groups involved, the greatly different rates of penetration into different insects, and the problem of achieving comparable concentrations at the only place where it is meaningful—the site of action. Also, as with

many other pharmacological agents, insecticides may either depress or stimulate respiration depending on concentration. We shall accordingly content
ourselves with summarizing the most important effects reported (Table IV
and Fig. 9) and with discussing briefly a few agents having fairly well-
defined effects on respiration, principally neurotoxins. Leads to references
not given, as well as to the chemical properties of the insecticides and the
practical details of dosage, mode of application and effectiveness in control,
can be found in the comprehensive texts of Brown (1951), Shepard (1951,
1958), Metcalf (1955), and O'Brien (1960).

Respiratory depression brought on by general cell toxicants such as cyanide, strychnine, heavy metals, hydrogen sulfide, and the thiocyanates
(Lethanes) present little of physiological interest, since, if the respiring cells

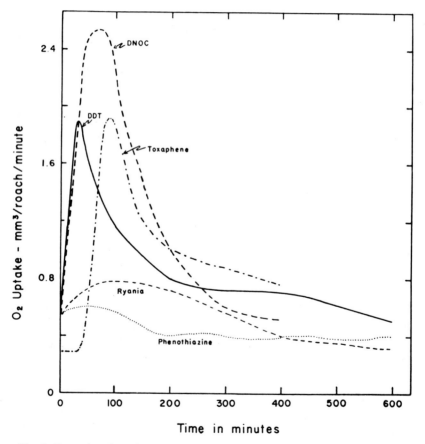

Fig. 9. Examples of respiratory responses of *Blattella* to injection of various insecticides. (Redrawn from Harvey and Brown, 1951.)

are killed, respiration naturally ceases. Most of the common insecticides, however, operate more selectively on the neuromuscular system and produce a wide spectrum of respiratory effects. These have been very difficult to unravel, partly because of uncertainty as to whether acetylcholine or related substances actually function as neuromediators in insects, and because of the noteworthy impermeability of the connective tissue sheath of insect nerve to most substances.

The chlorinated hydrocarbons are probably the best-known neurotoxins in common use, and of these DDT is certainly the most investigated. In low concentrations it acts on sensory nerves, inducing supernumerary impulses which excite motor nerve via the reflex arc. The motor nerves in turn induce muscular hyperactivity. The symptomatology, involving incoordination, high frequency "jitters," prostration and spasm, is well known and seems ample explanation for the quickly apparent and often greatly enhanced concurrent rate of oxygen uptake (Table IV). In higher dosage the motor nerves may be affected directly, and, at still higher levels, the muscle. Although DDT, methoxychlor, and lindane (γ-isomer of benzene hexachloride) were once suspected of being inhibitors of cholinesterase, it appears that they produce their effects rather by increasing the concentration of acetylcholine by promoting its release from reserves (Tobias and Savit, 1946).

A second class of neurotoxins, including sodium fluoride, the phenothiazine derivatives, and the organic phosphates (TEPP, HETP, parathion, DFP) also produce their effects by promoting accumulation of acetylcholine, but by the mechanism of cholinesterase inhibition.

The physiological action of some very effective insecticides is less clear, partly because of variations in time of effect and partly because of differing symptoms in different species. For example, chlordane, although a powerful respiratory stimulant, appears to act on *Periplaneta* like a depressant, decreasing muscle tonus, whereas its action on flies somewhat resembles that of DDT. The chlorinated terpenes (chlordane, aldrin, dieldrin, toxaphene) and parathion have delayed effects in comparison with DDT.

Ryanodine, an extract of the wood of the tropical plant *Ryania*, is a particularly interesting compound because it has been reported to increase respiratory rate in *Periplaneta* up to 9 or 10 times in the presence of a reversible flaccid paralysis (Edwards *et al.*, 1948; Hassett, 1948). Harvey and Brown (1951), however, were unable to obtain stimulation greater than 40% above normal in *Blattella*. In unpublished work we observed flaccid paralysis in *Periplaneta*, several species of grasshopper, adult *Phormia*, and *Galleria* larvae, but contracture in *Phormia* larvae. However, only in the orthopterans did we observe any respiratory stimulation, and this was very variable. Edwards *et al.* (1948) reported that ryanodine does not act on the nervous

TABLE IV
EFFECT OF INSECTICIDES ON RESPIRATORY RATES

Substance	Species	Stage	Effect on respiration	Comments	Reference
DDT	Blattella germanica	Adult	Increased 3–5×	Injected; peak within 30 min	Harvey and Brown (1951)
DDT	Blattella germanica	Adult	Increased about 6×	Sprayed; peak after about 1 hr	Harvey and Brown (1951)
DDT	Popillia japonica	Larva	Increased 2×	Peak after 24 hr	Ludwig (1946)
DDT		Adult	Increased 3–4×	Peak about 3.5 hr	Ludwig (1946)
DDT	Phormia regina	Larva	Increased 2–3×	Injected in kerosene into tracheae; peak in 1 hr	Buck et al. (1952)
		Adult	Increased 5×	Contact; peak at 3.5 hr	Buck and Keister (1949)
DDT	Tribolium castaneum	Adult	Increased about 2×	As dust; peak about 30 min	Lord (1950)
DDT	Musca domestica	Adult	Increased about 5×	Peak about 1 hr	Fullmer and Hoskins (1951)
DDT	Hypothenemus hampei	Adult	33% depression in 24 hours	As dust and in peanut oil solution	Edwards and González (1953)
Methoxychlor	Blattella germanica	Adult	Increased about 4×	Injected; peak in 30 min	Harvey and Brown (1951)
Lindane	Blattella germanica	Adult	Increased about 5×	Injected, peak about 60; sprayed, about 30 min	Harvey and Brown (1951)
γ-BHC	Tribolium	Adult	Increased 3×	Peak in less than 1 hr	Lord (1950)
Benzene hexachloride	Hypothenemus	Adult	Increased 4–5×	Peak in less than 1 hr	Edwards and González (1953)
Toxaphene	Blattella	Adult	Increased about 3×	Peak in 1 hr	Harvey and Brown (1951)

Compound	Insect	Stage	Effect	Timing	Reference
Toxaphene	*Tribolium*	Adult	Increased about 3×	Peak about 40 hr	Lord (1950)
Chlordane	*Blattella*	Adult	Increased 2–6×	Peak 100–400 min	Harvey and Brown (1951)
Chlordane	*Tribolium*	Adult	Increased about 3×	Peak 16–17 hr	Lord (1950)
Heptachlor	*Blattella*	Adult	Increased about 5×	Peak in 2.5 hr	Harvey and Brown (1951)
Aldrin	*Blattella*	Adult	Increased about 5×	Peak in 3–5 hr	Harvey and Brown (1951)
Dieldrin	*Blattella*	Adult	Increased about 5×	Peak in 100 min	Harvey and Brown (1951)
Paradichlorobenzene	*Blattella*	Adult	Increased 3–4×	Peak about 100 min	Harvey and Brown (1951)
Paradichlorobenzene vapor	*Periplaneta, Carabus, Calliphora, Triatoma*	Adults	Increased up to 5–6×	Peak 30–60 min to several hr	Punt (1950)
Dichloroethyl ether	*Blattella*	Adult	Increased 3–4×	Peak ca. 100 min	Harvey and Brown (1951)
Parathion	*Blattella*	Adult	Increased 3×	Peak in 1 hr	Harvey and Brown (1951)
Tetraethylpyrophosphate (TEPP)	*Blattella*	Adult	Increased 3×	Immediate	Harvey and Brown (1951)
Hexaethyltetraphosphate (HETP)	*Tribolium*	Adult	Increased about 4×	Peak in about 1 hr	Lord (1950)
p-Nitrophenyldiethylthion-phosphate	*Tribolium*	Adult	Increased 4–5×	Peak about 100 min	Lord (1950)
Lethane 384 (β-butoxy-β'-thiocyanoethyl ether)	*Blattella*	Adult	Immediate depression	—	Harvey and Brown (1951)
Lethane 60 (Thiocyano-ethyl laurate)	*Blattella*	Adult	Immediate depression	—	Harvey and Brown (1951)
Phenothiazine	*Blattella*	Adult	None	—	Harvey and Brown (1951)
Phenothiazine	*Dixippus, Tachycines, Periplaneta*	Adults	None	—	Janda (1955)
Lauryl thiocyanate	*Tribolium*	Adult	None	—	Lord (1950)
Pyocyanine	*Melanoplus differentialis*	Embryo	Increased 60–130%	Diapause embryos stimulated more than developing ones	Carlson and Bodine (1939)
E605 (*p*-nitrodiethylthion-phosphate)	*Tribolium*	Adult	Increased 4×	—	Lord (1950)

TABLE IV (*Continued*)

Substance	Species	Stage	Effect on respiration	Comments	Reference
Malathion (o, o-dimethyl-dithiophosphate of di-ethylmercaptosuccinate)	Pseudococcus maritimus	Adult	Slight depression	—	Ezzat and Highland (1957)
Systox [o-(2-(ethyl mercapto)ethyl o, o-diethyl thiophosphate]	Pseudococcus	Adult	Slight increase	—	Ezzat and Highland (1957)
Nicotine	Blattella	Adult	Gradual increase to 2×	Peak about 100 min then gradual fall	Harvey and Brown (1951)
Rotenone	Blattella	Adult	Slight transient increase	—	Harvey and Brown (1951)
Rotenone	Periplaneta	Adult	Inhibition	—	Fukami (1955)
Pyrethrins	Blattella	Adult	Transient rise to 5×	Peak within 1 hr	Harvey and Brown (1951)
Pyrethrins	Tribolium	Adult	Increased 3–4×	Peak about 1 hr	Lord (1950)
Pyrethrins	Periplaneta	Adult	Increased 2–3×	Peak about 1 hr	Mahmoud et al. (1953)
Ryanodine	Periplaneta	Adult	Increased up to 9×	—	Edwards et al. (1948)
Ryanodine	Periplaneta	Adult	Increased up to 6.5×	—	Hassett (1948)
Ryania extract	Blattella	Adult	40% increase	—	Harvey and Brown (1951)

system or on neuromuscular transmission, but directly on the contractile process of muscle. Usually it blocks contraction, but may sometimes induce contracture, rigor, or inhibition of relaxation. Type of response may be correlated with muscle phosphagen level (Haslett and Jenden, 1961). In insects, "fast" muscles react more rapidly than "slow" muscles (Becht and Dresden, 1956; Becht, 1959). In frog muscle, degree of shortening apparently controls rate of oxygen uptake, shortening muscles using less oxygen than those that do not shorten (Edwards and Flinker, 1950).

References

Abercrombie, W. F. (1936). *J. Morphol.* **59**, 91.
Agrell, I. (1947). *Ark. Zool.* **39**, 1.
Allen, M. D. (1959a). *J. Econ. Entomol.* **52**, 399.
Allen, M. D. (1959b). *J. Exp. Biol.* **36**, 92.
Allen, T. H. (1940). *J. Cell. Comp. Physiol.* **16**, 149.
Argo, V. N. (1939). *Ann. Entomol. Soc. Amer.* **32**, 147.
Balke, E. (1957). *Z. Vergl. Physiol.* **40**, 415.
Becht, G. (1959). *Bijdr. Dierk.* **29**, 1.
Becht, G., and Dresden, D. (1956). *Nature (London)* **177**, 836.
Beck, S. D. (1960). *Trans. Wis. Acad. Sci., Arts Lett.* **49**, 137.
Bell, J. (1940). *Physiol. Zool.* **13**, 73.
Bialascewicz, K. (1937). *Acta Biol. Exp. (Warsaw)* **11**, 229 (in Polish).
Birch, L. C. (1947). *Ecology* **28**, 17.
Bodenheimer, F. S. (1929). *Z. Angew. Entomol.* **15**, 435.
Bodine, J. H. (1922). *J. Exp. Zool.* **35**, 47.
Bodine, J. H. (1928). *Biol. Bull.* **55**, 395.
Bodine, J. H. (1934a). *J. Cell. Comp. Physiol.* **4**, 397.
Bodine, J. H. (1934b). *Physiol. Zool.* **7**, 459.
Bodine, J. H. (1950a). *Physiol. Zool.* **23**, 63.
Bodine, J. H. (1950b). *Physiol. Comp. Oecol.* **2**, 140.
Bodine, J. H., and Boell, E. J. (1934). *J. Cell. Comp. Physiol.* **4**, 475.
Bodine, J. H., and Boell, E. J. (1936). *Proc. Soc. Exp. Biol. Med.* **35**, 504.
Bodine, J. H., and Boell, E. J. (1937). *Physiol. Zool.* **10**, 245.
Bodine, J. H., and Lu, K.-H. (1950). *Proc. Soc. Exp. Biol. Med.* **74**, 448.
Bodine, J. H., and Orr, P. R. (1925). *Biol. Bull.* **48**, 1.
Brown, A. W. A. (1951). "Insect Control by Chemicals." Wiley, New York.
Buck, J. B., and Keister, M. L. (1949). *Biol. Bull.* **97**, 64.
Buck, J. B., and Keister, M. L. (1955). *Biol. Bull.* **109**, 144.
Buck, J. B., Keister, M. L., and Posner, I. (1952). *Ann. Entomol. Soc. Amer.* **45**, 369.
Bullock, T. H. (1954). *Biol. Rev. Cambridge Phil. Soc.* **30**, 311.
Burkett, B. N. (1962). *Entomol. Exp. Appl.* **5**, 305.
Butler, C. G., and Innes, J. M. (1936). *Proc. Roy. Soc., Ser. B* **119**, 296.
Carlson, L. D., and Bodine, J. H. (1939). *J. Cell Comp. Physiol.* **141**, 159.
Chadwick, L. E. (1947). *Biol. Bull.* **93**, 229.
Chadwick, L. E., and Gilmour, D. (1940). *Physiol. Zool.* **13**, 398.

Chadwick, L. E., and Williams, C. M. (1949). *Biol. Bull.* **97**, 115.

Chino, H. (1958). *J. Insect Physiol.* **2**, 1.

Clare, M. R. (1925). *Biol. Bull.* **49**, 440.

Clark, A. M. (1959). *Ann. Entomol. Soc. Amer.* **52**, 637.

Clark, A. M., and Cristofalo, V. J. (1960). *Amer. J. Physiol.* **198**, 441.

Clark, A. M., and Cristofalo, V. J. (1961). *Physiol. Zool.* **34**, 55.

Clark, A. M., and Herr, E. B. (1954). *Biol. Bull.* **107**, 329.

Clark, A. M., and Papa, M. J. (1958). *Biol. Bull.* **114**, 180.

Clark, A. M., Harwitz, G. A., and Rubin, M. A. (1958). *Science* **127**, 1289.

Clarke, K. U. (1957). *J. Exp. Biol.* **34**, 29.

Cook, S. F. (1932). *Biol. Bull.* **63**, 246.

Cotton, R. T. (1932). *J. Econ. Entomol.* **25**, 1088.

Crescitelli, F. (1935). *J. Cell. Comp. Physiol.* **6**, 351.

Davis, J. G., and Slater, W. K. (1926). *Biochem. J.* **20**, 1167.

Davis, J. G., and Slater, W. K. (1928). *Biochem. J.* **22**, 331.

Davis, R. A., and Fraenkel, G. (1940). *J. Exp. Biol.* **17**, 402.

Day, M. F. (1951). *Aust. J. Sci. Res., Ser. B* **4**, 64.

Dehnel, P. A., and Segal, E. (1956). *Biol. Bull.* **111**, 53.

Dirken, M. N. J. (1922). *Arch. Neer. Physiol.* **7**, 126.

Dobzhansky, T., and Poulson, D. F. (1935). *Z. Vergl. Physiol.* **22**, 473.

Dreyer, W. A. (1932). *Physiol. Zool.* **5**, 301.

Edelman, N. M. (1949). *Rev. Entomol. URSS* **30**, 216 (in Russian).

Edelman, N. M. (1950). *Zool. Zh.* **29**, 427.

Edwards, D. K. (1958). *Can. J. Zool.* **36**, 363.

Edwards, G. A. (1946). *J. Cell. Comp. Physiol.* **27**, 53.

Edwards, G. A. (1953). *In* "Insect Physiology" (K. D. Roeder, ed.), pp. 55–146. Wiley, New York.

Edwards, G. A., and González, M. D. P. (1952). *Bol. Fac. Fil., Cienc. Letras, Univ. Sao Paulo, Zool.* **17**, 211.

Edwards, G. A., and González, M. D. P. (1953). *Bol. Fac. Fil., Cienc. Letras, Univ. Sao Paulo, Zool.* **18**, 77.

Edwards, G. A., Weiant, E. A., Slocombe, A. G., and Roeder, K. D. (1948). *Science* **108**, 330.

Edwards, L. E., and Flinker, M.-L. (1950). *Proc. Soc. Exp. Biol. Med.* **75**, 93.

Edwards, R. W. (1958). *J. Exp. Biol.* **35**, 383.

Ellenby, C. (1939). *Proc. Zool. Soc. London, Ser. A* **108**, 525.

Ellenby, C. (1945a). *J. Exp. Biol.* **21**, 39.

Ellenby, C. (1945b). *J. Exp. Zool.* **98**, 23.

Ellenby, C. (1953). *J. Exp. Biol.* **30**, 475.

Ellenby, C. (1956). *Proc. Int. Congr. Zool., 14th, 1953* p. 282.

Ellinger, T. (1915). *Int. Z. Phys.-Chem. Biol.* **2**, 85.

Enger, P. S., and Savalov, P. (1958). *J. Insect Physiol.* **2**, 232.

Ezzat, Y. M., and Highland, H. A. (1957). *Bull. Soc. Entomol. Egypti* **41**, 473.

Fox, H. M., Wingfield, C. A., and Simmonds, B. G. (1937). *J. Exp. Biol.* **14**, 210.

Fraenkel, G., and Herford, G. V. B. (1938). *J. Exp. Biol.* **15**, 266.

Fraenkel, G., and Herford, G. V. B. (1940). *J. Exp. Biol.* **17**, 386.

Fukami, J. (1955). *Jap. J. Appl. Zool.* **19**, 148 (in Japanese).

Fullmer, O. H., and Hoskins, W. M. (1951). *J. Econ. Entomol.* **44**, 858.

Gaarder, T. (1918). *Biochem. Z.* **89**, 48.

Galun, R. (1960). *Nature (London)* **185**, 391.

Galun, R., and Fraenkel, G. (1961). *J. Insect Physiol.* **7**, 161.

Gilbert, L. I., and Schneiderman, H. A. (1961). *Amer. Zool.* **1**, 11.

Gilmour, D. (1940a). *J. Cell. Comp. Physiol.* **15**, 331.

Gilmour, D. (1940b). *Biol. Bull.* **79**, 297.

Gilmour, D. (1941). *J. Cell. Comp. Physiol.* **18**, 93.

Goble, G. J., and Patton, R. L. (1946). *J. Econ. Entomol.* **39**, 177.

Gromadska, M. (1949). *Stud. Soc. Sci. Torun, Sect. E* **2**, 17 (in Polish).

Gunn, D. L., and Cosway, C. A. (1942). *J. Exp. Biol.* **19**, 124.

Harnisch, O. (1959). *Biol. Zentralbl.* **78**, 315.

Harvey, G. T., and Brown, A. W. A. (1951). *Can. J. Zool.* **29**, 42.

Harvey, W. R. (1962). *Annu. Rev. Entomol.* **7**, 57.

Harvey, W. R., and Williams, C. M. (1958). *Biol. Bull.* **114**, 36.

Haslett, W. L., and Jenden, D. J. (1961). *J. Cell. Comp. Physiol.* **57**, 123.

Hassett, C. C. (1948). *Science* **108**, 138.

Heller, J. (1930). *Z. Vergl. Physiol.* **11**, 448.

Heller, J. (1938). *Acta Biol. Exp. (Warsaw)* **12**, 99.

Hocking, B. (1953). *Trans. Roy. Entomol. Soc. London* **104**, 223.

Hsueh, T. Y., and Tang, P. S. (1944). *Physiol. Zool.* **17**, 71.

Itaya, K. (1940). *J. Sericult. Sci. Jap.* **11**, 113. (In Japanese).

Janda, V. (1952). *Acta Soc. Zool. Bohemoslov.* **16**, 237 (in Czech).

Janda, V. (1955). *Acta Soc. Zool. Bohemoslov.* **19**, 197 (in Czech).

Janda, V. (1961). *Acta Soc. Zool. Bohemoslov.* **25**, 115.

Janda, V., and Kocián, V. (1933). *Zool. Jahrb., Abt. Allg. Zool. Physiol. Tiere* **52**, 561.

Janda, V., and Mrciak, M. (1957). *Acta Soc. Zool. Bohemoslov.* **21**, 244 (in Czech).

Jongbloed, J., and Wiersma, C. A. G. (1934). *Z. Vergl. Physiol.* **21**, 519.

Kalmus, H. (1929). *Z. Vergl. Physiol.* **10**, 445.

Kamioka, S. (1950). *Oyo Kontyu* **6**, 144 (in Japanese).

Keister, M. L. (1953). *J. Morphol.* **93**, 573.

Keister, M. L., and Buck, J. (1961). *J. Insect Physiol.* **7**, 51.

Kennington, G. S. (1957). *Physiol. Zool.* **30**, 305.

Kirberger, C. (1953). *Z. Vergl. Physiol.* **35**, 175.

Kittel, A. (1941). *Z. Vergl. Physiol.* **28**, 533.

Kleiber, M. (1961). *Univ. Mo. Res. Bull.* **767**, 3.

Kobayashi, M., and Nakasone, S. (1960). *Bull. Sericult. Exp. Sta.* **16**, 100 (in Japanese).

Kosmin, N. P., Alpatov, W. W., and Resnitschenko, M. S. (1932). *Z. Vergl. Physiol.* **17**, 408.

Kozhantschikov, I., and Maslowa, E. (1935). *Zool. Jahrb., Abt. Allg. Zool. Physiol. Tiere* **55**, 219.

Krogh, A. (1914). *Z. Allg. Physiol.* **16**, 178.

Krogh, A. (1916). "The Respiratory Exchange of Animals and Man." Longmans, Green, New York.

Krogh, A. (1941). "The Comparative Physiology of Respiratory Mechanisms." Univ. of Pennsylvania Press, Philadelphia.

Krogh, A., and Weis-Fogh, T. (1951). *J. Exp. Biol.* **28**, 344.

Krogh, A., and Zeuthen, E. (1941). *J. Exp. Biol.* **18**, 1.

Krüger, F. (1958). *Biol. Zentralbl.* **77**, 581.

Kurland, C. G., and Schneiderman, H. A. (1959). *Biol. Bull.* **116**, 136.

Kurland, C. G., Schneiderman, H. A., and Smith, R. D. (1958). *Anat. Rec.* **132**, 465.

Lord, K. A. (1950). *Ann. Appl. Biol.* **37**, 105.

Lotz,·R., and Thiel, N. (1962). *Naturwissenschaften* **49**, 378.

Ludwig, D. (1946). *Ann. Entomol. Soc. Amer.* **39**, 496.

Ludwig, D., Fiore, C., and Jones, C. R. (1962). *Ann. Entomol. Soc. Amer.* **55**, 439.

Ludwig, W. (1956). *Z. Vergl. Physiol.* **39**, 84.

Luferov, V. P. (1958). *Dokl. Akad. Nauk SSSR* **119**, 1229 (in Russian).

Lühmann, M., and Drees, O. (1952). *Zool. Anz.* **148**, 13.

Mahmoud, M. I. Z., Dahm, P. A., Hein, R. E., and McFarland, R. H. (1953). *J. Econ. Entomol.* **46**, 324.

Marzusch, K. (1952). *Z. Vergl. Physiol.* **34**, 75.

Melampy, R. M., and Willis, E. R. (1939). *Physiol. Zool.* **12**, 302.

Melvin, R. (1928). *Biol. Bull.* **55**, 135.

Metcalf, R. L. (1955). "Organic Insecticides. Chemistry and Mode of Action." Wiley (Interscience), New York.

Monro, H. A. V., Dumas, T., and Buckland, C. T. (1962). *Enotomol. Exp. Appl.* **5**, 79.

Müller, I. (1943a). *Riv. Biol.* **35**, 48.

Müller, I. (1943b). *Z. Vergl. Physiol.* **30**, 139.

Nittono, Y., and Nakasone, S. (1956a). *J. Sericult. Sci. Jap.* **25**, 317 (in Japanese).

Nittono, Y., and Nakasone, S. (1956b). *J. Sericult. Sci. Jap.* **25**, 1 (in Japanese).

Nittono, Y., and Nakasone, S. (1956c). *J. Sericult. Sci. Jap.* **25**, 5 (in Japanese).

O'Brien, R. D. (1960). "Toxic Phosphorus Esters." Academic Press, New York.

Orr, P. R. (1925). *J. Gen. Physiol.* **7**, 731.

Parhon, M. (1909). *Ann. Sci. Natur., Zool.* [9] **9**, 1.

Park, H. D., and Buck, J. (1960). *J. Insect Physiol.* **4**, 220.

Poláček, I., and Kubišta, V. (1960). *Physiol. Bohemoslov.* **9**, 228.

Poulson, D. F. (1935). *Z. Vergl. Physiol.* **22**, 466.

Punt, A. (1950). *Acta Physiol. Pharmacol. Neer.* **1**, 82.

Raffy, A. (1934). *Ann. Physiol. Physiochim. Biol.* **10**, 437.

Raffy, A., and Portier, P. (1931). *C. R. Soc. Biol.* **108**, 1062.

Richards, A. G. (1957). *In* "Influence of Temperature in Biological Systems" (F. H. Johnson, ed.), pp. 145–162. Amer. Physiol. Soc., Washington, D.C.

Richards, A. G. (1958). *Proc. Int. Congr. Entomol., 10th, 1956* Vol. 2, p. 67.

Richards, A. G. (1959). *Biol. Zentralbl.* **78**, 308.

Richards, A. G. (1963a). *Ann. Entomol. Soc. Amer.* **56**, 355.

Richards, A. G. (1963b). *Entomol. News* **74**, 91.

Richards, A. G., and Suanraksa, S. (1962). *Entomol. Exp. Appl.* **5**, 167.

Robbie, W. A. (1941). *J. Cell. Comp. Physiol.* **17**, 369.

Rogers, E. (1929). *Physiol. Zool.* **2**, 275.

Roller, L. W. (1906). *Univ. Kans. Sci. Bull.* **3**, 211.

Rutschky, C. W., and Joseph, S. R. (1957). *Proc. Pa. Acad. Sci.* **31**, 131.

Sattel, M. (1956). *Z. Vergl. Physiol.* **39**, 89.

Sayle, M. H. (1928). *Biol. Bull.* **54**, 212.

Schmalfusz, H., Buszmann, G., Schmalfusz, H., and Pohl, H. (1939). *Z. Vergl. Physiol.* **27**, 434.

Schneiderman, H. A. (1957). *In* "Physiological Triggers" (T. H. Bullock, ed.), pp. 46–59. Amer. Physiol. Soc., Washington, D.C.

Schneiderman, H. A., and Williams, C. M. (1953). *Biol. Bull.* **105**, 320.

Schneiderman, H. A., and Williams, C. M. (1954). *Biol. Bull.* **106**, 238.

Scholander, P. F., Flagg, W., Hock, R. J., and Irving, L. (1953a). *J. Cell. Comp. Physiol.* **42**, Suppl. 1, 1.

Scholander, P. F., Flagg, W., Walters, V., and Irving, L. (1953b). *Physiol. Zool.* **26**, 67.

Schwan, H. (1940). *Ark. Zool.* [1] **32A**, 1.

Shappirio, D. G. (1960). *Ann. N.Y. Acad. Sci.* **89**, 537.

Shepard, H. H. (1951). "The Chemistry and Action of Insecticides." McGraw-Hill, New York.

Shepard, H. H. (1958). "Methods of Testing Chemicals on Insects." Burgess, Minneapolis, Minnesota.

Simanton, W. A. (1933). *Ann. Entomol. Soc. Amer.* **26**, 247.

Sláma, K. (1959). *Acta Soc. Entomol. Bohem.* **56**, 113 (in Czech.).

Slater, W. K. (1927). *Biochem. J.* **21**, 198.

Slifer, E. H. (1954). *Ann. Entomol. Soc. Amer.* **47**, 265.

Slowtzoff, B. (1909). *Biochem. Z.* **19**, 497.

Szwajsowna, P. (1916). *Spraw. Tow. Nauk. Warz.* **9**, 405 (in Polish).

Taylor, I. R. (1927). *J. Morphol.* **44**, 313.

Taylor, I. R., and Steinbach, H. B. (1931). *Physiol. Zool.* **4**, 604.

Teissier, G. (1931). *Trav. Sta. Biol. Roscoff* **9**, 27.

Tobias, J. M., and Savit, J. (1946). *J. Cell. Comp. Physiol.* **28**, 159.

Trager, W. (1935). *J. Exp. Zool.* **71**, 489.

Vernberg, F. J., and Meriney, D. K. (1957). *J. Elisha Mitchell Sci. Soc.* **73**, 351.

Von Brand, T., McMahon, P., and Nolan, M. O. (1957). *Biol. Bull.* **113**, 89.

von Buddenbrock, W., and von Rohr, G. (1923). *Z. Allg. Physiol.* **20**, 111.

Waku, Y. (1957). *Sci. Rep. Tôhoku Univ., Ser. 4* **23**, 143.

Weis-Fogh, T. (1952). *Phil. Trans. Roy. Soc. London, Ser. B* **237**, 1.

Wellington, W. G. (1946a). *Can. J. Res.* **24**, 51.

Wellington, W. G. (1946b). *Can. J. Res.* **24**, 105.

Will, A. (1952). *Z. Vergl. Physiol.* **34**, 20.

Williams, C. M., and Beecher, H. K. (1944). *Amer. J. Physiol.* **140**, 566.

Wolsky, A. (1949). *Curr. Sci.* **18**, 323.

Woodworth, C. E. (1932). *J. Econ. Entomol.* **25**, 1036.

Woodworth, C. E. (1936). *J. Econ. Entomol.* **29**, 1128.

Yokoyama, T., and Keister, M. L. (1953). *Ann. Entomol. Soc. Amer.* **46**, 218.

Yust, H. R., and Shelden, F. F. (1952). *Ann. Entomol. Soc. Amer.* **45**, 220.

Zebe, E. (1954). *Z. Vergl. Physiol.* **36**, 290.

Zeuthen, E. (1947). *C. R. Trav. Lab. Carlsberg, Ser. Chim.* **26**, 17.

AUTHOR INDEX

Numbers in italics refer to pages on which the complete references are listed.

511

SUBJECT INDEX

A

Abdominal gills, 409–413
Ablabesmyia monilis, 476
Acarapis, 359
Acarus siro, 302
Acentropus, 403
Acentropus niveus, 431, 433
Acethion, 47, 99
Acheta domesticus, 42, 223
Acidia heraclei, 243
Acilius, 431, 490
Aedes, 494
Aedes aegypti, 20, 21, 24, 33, 38, 39,
    43, 56, 58, 59, 72, 79, 82, 84, 89,
    93, 94, 96, 98, 321, 497
Aedes atropalpus, 254
Aedes nigromaculis, 20
Aedes taeniorhynchus, 20
Aeropyle, 360, 435
Aeshna cyanea, 389
Aeshna grandis, 417
Aeshna spp., 220, 349, 364, 386, 416,
    447, 449, 451, 452, 492
Aeshna umbrosa, 417
Agapema, 365, 493
Agapema galbina, 483
Agrianome, 223, 225
Agrianome spinicollis, 218, 224, 229
Agrion pulchellum, 415
Agrion sp., 481
Agriotes, 287
Agrotis, 491
Air sac, 353
Alanine, 225, 255
β-Alanine, 249–250
Aldicarb, 66, 68, 69, 72, 99

Aldrin, 7, 8, 36, 38–40, 41, 43, 86, 99,
    501, 503
Allethrin, 75, 76, 77, 99
Aminocarb, 67, 68, 72, 99
Amphipod, 284, 285
Anabrus, 225, 236
Anabrus simplex, 132, 146, 147, 218,
    223, 250
Anagasta, 495
Anasa tristis, 481
Anax, 416, 447, 448, 451, 452, 455
Anax imperator, 417
Anax junius, 417, 418
Anisopodidae, 422
Anisops, 346, 458, 460
Anisops pellucens, 358, 458, 459
Anopheles albimanus, 24, 72, 89, 90
Anopheles atroparvus, 21, 90
Anopheles claviger, 21
Anopheles gambiae, 24, 27, 84
Anopheles labranchiae, 21
Anopheles maculipenis, 21
Anopheles quadrimaculatus, 24, 33, 38,
    72
Anopheles sacharovi, 21
Anopheles stephensi, 21, 24
Anopheles sundaicus, 21
Anopheles superpictus, 21
Anoxia, 495–496
Ant, 488
Antheraea, 360
Antheraea pernyi, 42, 483, 484
Antherea polyphemus, 41
Anthonomus grandis, 14, 17, 65
Anticholinesterase, 47
Antocha, 433, 439
Antocha bifida, 435, 437

535

A 4
B 5
C 6
D 7
E 8
F 9
G 0
H 1
I 2
J 3